Shock Wave Science and Technology Reference Library

The new Springer collection, Shock Wave Science and Technology Reference Library, conceived in the style of the famous Handbuch der Physik has as its principal motivation to assemble authoritative, state-of-the-art, archival reference articles by leading scientists and engineers in the field of shock wave research and its applications. A numbered and bounded collection, this reference library will consist of specifically commissioned volumes with internationally renowned experts as editors and contributing authors. Each volume consists of a small collection of extensive, topical and independent surveys and reviews. Typical articles start at an elementary level that is accessible to non-specialists and beginners. The main part of the articles deals with the most recent advances in the field with focus on experiment, instrumentation, theory, and modeling. Finally, prospects and opportunities for new developments are examined. Last but not least, the authors offer expert advice and cautions that are valuable for both the novice and the well-seasoned specialist.

Shock Wave Science and Technology Reference Library

Collection Editors

Hans Grönig

Hans Grönig is Professor emeritus at the Shock Wave Laboratory of RWTH Aachen University, Germany. He obtained his Dr. rer. nat. degree in Mechanical Engineering and then worked as postdoctoral fellow at GALCIT, Pasadena, for one year. For more than 50 years he has been engaged in many aspects of mainly experimental shock wave research including hypersonics, gaseous and dust detonations. For about 10 years he was Editor-in-Chief of the journal Shock Waves.

Yasuyuki Horie

Professor Yasuyuki (Yuki) Horie is internationally recognized for his contributions in high-pressure shock compression of solids and energetic materials modeling. He is a co-chief editor of the Springer series on Shock Wave and High Pressure Phenomena and the Shock Wave Science and Technology Reference Library, and a Liaison editor of the journal Shock Waves. He is a Fellow of the American Physical Society, and Secretary of the International Institute of Shock Wave Research. His current interests include fundamental understanding of (a) the impact sensitivity of energetic solids and its relation to microstructure attributes such as particle size distribution and interface morphology, and (b) heterogeneous and nonequilibrium effects in shock compression of solids at the mesoscale.

Kazuyoshi Takayama

Professor Kazuyoshi Takayama obtained his doctoral degree from Tohoku University in 1970 and was then appointed lecturer at the Institute of High Speed Mechanics, Tohoku University, promoted to associate professor in 1975 and to professor in 1986. He was appointed director of the Shock Wave Research Center at the Institute of High Speed Mechanics in 1988. The Institute of High Speed Mechanics was restructured as the Institute of Fluid Science in 1989. He retired in 2004 and became emeritus professor of Tohoku University. In 1990 he launched Shock Waves, an international journal, taking on the role of managing editor and in 2002 became editor-in-chief. He was elected president of the Japan Society for Aeronautical and Space Sciences for one year in 2000 and was chairman of the Japanese Society of Shock Wave Research in 2000. He was appointed president of the International Shock Wave Institute in 2005. His research interests range from fundamental shock wave studies to the interdisciplinary application of shock wave research.

F. Zhang (Ed.)

Shock Wave Science and Technology Reference Library, Vol. 4

Heterogeneous Detonation

With 198 Figures and 42 Tables

Springer

Fan Zhang
Defence Research and Development Canada - Suffield
PO Box 4000, Station Main
Medicine Hat, AB, T1A 8K6
Canada
fan.zhang@drdc-rddc.gc.ca

Fan Zhang

Fan Zhang is a Senior Scientist in the Department of National Defence at Defence Research and Development Canada – Suffield and an adjunct Professor at the University of Waterloo in the Department of Mechanical Engineering. He specializes in shock waves, detonations and explosions, more specifically in multiphase reactive flow, heterogeneous explosives and high energy density systems. He earned his doctoral degree in science in 1989 from the University of Technology Aachen (RWTH), Germany, and received a Borchers Medal, a Friedrich-Wilhelm Prize and several best paper awards. He published more than 150 refereed journal and proceedings papers, book chapters and journal special issues. He has served for a number of international defense panels and academic committees.

ISBN: 978-3-642-10010-9 e-ISBN: 978-3-540-88447-7

Softcover reprint of the hardcover 1st edition 2009

Cover design: eStudio Calamar

Printed on acid-free paper

9 8 7 6 5 4 3 2 1

springer.com

Preface

This book, as a volume in the *Shock Wave Science and Technology Reference Library*, is primarily concerned with detonation waves or compression shock waves in reactive heterogeneous media, including mixtures of solid, liquid, and gas phases.

The topics involve a variety of energy release and control processes through heterogeneous reactive shock waves; a contemporary research field of detonation that has found wide applications in propulsion and power, hazard prevention, as well as military engineering. This volume contains six chapters. The first two chapters describe the detonation behavior of volumetrically dispersed multiphase explosive mixtures, and the subsequent two chapters deal with condensed multiphase composite explosives. Chapter 5 discusses the unique solid-state reactions in microscopic solid particle mixtures under high-pressure shock loading. The final chapter is fundamental in describing shock ignition behavior of solid and liquid particles. Each chapter is self-contained and can be read independently of the others, though they are thematically interrelated. They offer a timely reference for graduate students as well as professional scientists and engineers, by laying out the foundations and discussing the latest developments, including yet unresolved challenging problems.

The first chapter, by S.B. Murray and P.A. Thibault, discusses spray or liquid aerosol detonation. This chapter provides not only an updated review, but also offers a concise heuristic introduction to spray detonation for both high-vapor-pressure and low-vapor-pressure fuel. After an excellent introductory or refresher reading for any class on laboratory phenomena of confined spray detonation, the authors offer a state-of-the-art description on detonation of unconfined fuel spray in air and its future research directions. This is possibly the first article in the open literature that provides such a comprehensive summary of the unconfined spray detonation phenomenon and its various engineering approaches. The chapter, with many materials unpublished before, complements any of the previous review articles in the area of spray detonation.

The second chapter, by F. Zhang, presents an overview of the fundamentals of dust detonation or detonation in gas–particle flow. It begins with a historical introduction and this is followed by an in-depth description of the detonation theory for explosive systems composed of reactive particles dispersed in oxidizing gases and in reactive gases. The chapter further presents a review of profound experiments on the transition to heterogeneous detonation, heterogeneous detonation structure, quasi-detonation, and hybrid detonation. The discussion on theory and experiments has focused on the unique detonation physics and performance behavior inherent to these multiphase explosive mixtures. Most of the descriptions and discussions are valid not only for low-density gas–solid flow, but can also be applied to dense or condensed fluid–solid heterogeneous explosive mixtures. This chapter offers up-to-date information on the fundamentals and a database for the subject area.

The third chapter, by D.L. Frost and F. Zhang, provides a comprehensive topical review of slurry detonation. The term "slurry explosive" is used in a general sense to include water–gel, emulsion, and metalized slurry formulations and blasting agents. This constitutes an attempt to cover not only available commercial slurry explosives, but also to review the current state of the art and fundamentals for possible future candidates. The chapter first describes the composition of and manufacturing procedure for various slurry explosives, and this is followed by an in-depth review of slurry explosive performance with emphasis on the characterization and properties of the nonideal and heterogeneous detonation wave itself. It further offers a review of models and their unique requirements for slurry detonation propagation, which is dominated by mesoscale (grain-scale) shock wave dynamics coherent with local mechanical and thermal response, as well as chemical reaction of heterogeneous material. Many unique detonation behaviors of fuel-rich metalized slurry explosives are discussed on the basis of the authors' own experiences.

The fourth chapter, by M.F. Gogulya and M.A. Brazhnikov, deals with detonation in metalized composite explosives with emphasis placed on micrometric and particularly nanometric aluminum additives. This chapter offers a selective but profoundly informative review of recent advances in this complex area and many of the materials are taken from the authors' own acknowledged studies. It covers both positive and negative oxygen-balance explosives in binary and ternary formulations with aluminum additives of various particle shapes and in a wide size range of 0.04–100 μm. A two-heat release process (explosive detonation and aluminum particle afterburning) has been demonstrated to be a general energy release principle for the detonation of such metalized explosives. It is often characterized by a shock wave followed by a pressure wave. The chapter, with 33 figures and 24 tables, provides one of the most comprehensive summaries of explosive behavior and detonation properties for various aluminized explosives, including mechanical and shock sensitivity, brisance, heat of reaction or explosion, detonation velocity, failure diameter, pressure, temperature and particle velocity history, as well as plate and cylinder acceleration capabilities.

The fifth chapter, by Yu.A. Gordopolov, S.S. Batsanov, and V.S. Trofimov, is a status report on shock-induced solid–solid reactions and possible detonation (a self-sustained shock wave). Unlike the classic thermal detonation where the expansion of high-pressure detonation gas-phase products provides the work required to sustain the propagation of detonation, a solid–solid detonation describes the concept of a highly energetic explosion where a stable supersonic wave exclusively converts mesoscale or microscale solid particle reactants to high-speed solid-phase products. The intriguing phenomena and possibility of this mode of shock-induced energy release without expanding gases appeared in the 1950s and has grown since the 1980s, resulting in advances in solid-state chemistry under high-pressure shock loading down to microscopic scales. Gasless detonation properties were predicted theoretically using Hugoniot analysis in the 1990s, where the detonation speed appears comparable with that in conventional high explosives; however, negligible heat is produced through shock compression in spite of the high pressures. Experimentally, while supersonic reactive shock waves observed in metal–metallic oxide mixtures were often accompanied by the release of gas products, shock-induced inorganic solid-state reactions have been reported on microsecond timescales in various metal–sulfur mixtures and intermetallics, but observations have been scarce and preliminary to date. Recent advances in experimental efforts have been possible owing to the progress of nanometric powder technology. The mechanisms for solid–solid reactions and wave sustaining are not yet clear, but are hypothesized to be driven by high atomic or molecular mobility resulting from shock and particle interactions with subsequent superfast diffusion or high-speed momentum flux transfer, which is supported by chemical energy augmented by bond energy release if the system is under extremely high pressure. The authors are leading experts in this cutting-edge research field. This chapter is a testament not only to the wide-openness of experimental studies, but also to the necessity of exploring possible fundamental theories in order to describe the phenomena and underlying mechanisms.

The sixth chapter, by S.M. Frolov and A.V. Fedorov, is concerned with the shock ignition of particles. It offers a fundamental aspect for all the first five chapters and the field of heterogeneous detonation. This chapter provides a selective but profound review of recent advances in both solid metal particles and liquid fuel droplet ignition after being subjected to shock loading, with emphasis placed on theoretical fundamentals and mathematical models mostly taken from the authors' own in-depth studies. As the authors indicate, for problems dealing with transient modes of combustion such as ignition or extinction, the effects of finite-rate chemical kinetics must be considered. These processes are subjected to a number of local shocked flow and particle interactions where the influences of neighboring particles are considerable. The phenomena are further complicated by the particle breakup, fragmentation, and subsequent mixing with air accompanied by phase changes, thus creating extreme challenges for the mathematical modeling. The authors offer insightful critiques of current thinking, while applying simplified treatments of various shock and particle interactions to explore the fundamental behaviors of shock ignition. Prototypical samples include magnesium and aluminum particles as well as $n-$alkane liquid droplets. The chapter provides advanced reading on the fundamentals of particle shock ignition and therefore the basis for heterogeneous detonation.

The editor is indebted to all authors for their willingness to prepare and make available their timely and authoritative materials to a wide audience.

Medicine Hat, Canada
March 2009

Fan Zhang

Contents

List of Contributors

Stepan S. Batsanov
Center for High Dynamic Pressures
Mendeleevo, Moscow
141570 Russia
batsanov@gol.ru

Michael A. Brazhnikov
Semenov Institute of Chemical
Physics, Moscow
119991 Russia
bragario@inbox.ru

Alexander V. Fedorov
Institute for Theoretical
and Applied Mechanics
Novosibirsk, 630090 Russia
fedorov@itam.nsc.ru

Sergey M. Frolov
Semenov Institute of Chemical
Physics, Moscow
119991 Russia
smfrol@chph.ras.ru

David L. Frost
McGill University
Department of Mechanical
Engineering, 817 Sherbrooke St. W.
Montreal, QB, Canada H3A 2K6
david.frost@mcgill.ca

Michael F. Gogulya
Semenov Institute of Chemical
Physics, Moscow, 119991 Russia
gogul@polymer.chph.ras.ru

Yury A. Gordopolov
Institute of Structural Macrokinetics
and Materials Science
Chernogolovka, Moscow
142432 Russia
gordop@ism.ac.ru

Stephen B. Murray
Defence R&D Canada – Suffield
P.O. Box 4000, Station Main
Medicine Hat, Alberta
Canada T1A 8K6
Stephen.Murray@drdc-rddc.gc.ca

Paul A. Thibault
TimeScales Scientific Limited
554 Aberdeen Street S.E.
Medicine Hat, Alberta
Canada T1A 0R7
paulth@timescales-sci.com

Vladimir S. Trofimov
Institute of Structural Macrokinetics
and Materials Science
Russian Academy of Sciences
Chernogolovka, Moscow
142432 Russia

Fan Zhang
Defence R&D Canada – Suffield
PO Box 4000. Stn Main
Medicine Hat, AB T1A 8K6 Canada
fan.zhang@drdc-rddc.gc.ca

1

Spray Detonation

S.B. Murray and P.A. Thibault

In memory of our late friend and colleague, Dr. Ingar Olav Moen. Ingar made significant contributions to the understanding of spray detonation.

1.1 Introduction

Spray or aerosol detonation is a topic of keen interest to the hazard prevention and military R&D communities. Spray detonation is also potentially relevant to the study of pulse detonation engines (PDEs) and other hypersonic propulsion systems exploiting detonative combustion.

The early work on spray detonation (i.e., circa 1980 or before) has been well documented in review papers by Dabora and Weinberger [50], Nettleton [108], Borisov and Gelfand [21], Sichel [123], and Dabora [49]. The recent review by Kailasanath [70] puts this early work into perspective and summarizes the key issues for spray detonations in PDE tubes. Almost all of the early work involved spray detonations in confined tubes and chambers. Fuel–oxygen mixtures were typically used, although air was employed as the oxidizer in some studies. A brief summary of highlights is presented in Sect. 1.2.

The main focus of spray detonation studies during the 1980s and into the mid-1990s was on the determination of detonability limits for unconfined fuel sprays in air for a variety of industrial chemicals and motor fuels. However, these studies are few in number and very little follow-on work has been reported during the past decade. A review of this work is provided in Sect. 1.3. Another topic of keen interest during the period from the mid-1970s to the late 1980s was so-called chemical or non-explosive initiation of detonation. The goal of this work was to replace the cloud initiator charges in conventional fuel–air explosive (FAE) systems by a chemical initiation system capable of inducing automatic cloud detonation following a prescribed time delay. While attempts to demonstrate chemical initiation in spray systems were not very successful, the lessons learned are nonetheless informative, and a summary of

F. Zhang (ed.), *Shock Wave Science and Technology Reference Library: Heterogeneous Detonation*, DOI: 10.1007/978-3-540-88447-7_1,

chemical initiation studies has not yet been published in the open literature. A brief synopsis is therefore presented in Sect. 1.4. Most detonation propulsion studies employ fuels in the vapour phase. However, there are a couple of notable exceptions and these are reviewed in Sect. 1.5.

Although spray detonation studies per se have become fairly rare, significant progress has continued in related areas, including a much better understanding of droplet behaviour in high-speed flows and advances in the modelling of spray combustion (versus detonation). Some excellent books have been published recently that summarize these achievements. Therefore, the review of modelling presented in Sect. 1.6 will be restricted to spray detonation modelling and the difficult problem of modelling explosive fuel dispersal which is rather unique to FAE applications.

Some concluding remarks and recommendations are given in Sect. 1.7.

1.2 Laboratory Studies of Confined Spray Detonation

1.2.1 Detonability Studies in Tubes and Chambers

One of the earliest spray detonation studies was conducted by Dabora et al. [51]. These investigators employed a vertical shock tube 3.75 m in length having a 4.13 cm × 4.13 cm cross-section, as shown in Fig. 1.1. Monodisperse sprays were produced using a droplet generator positioned at the top of the tube consisting of a cylindrical chamber fitted with several parallel capillary needles at the bottom end. A thin vibrating brass plate at the top of the chamber was used to break up the fuel jets and thereby control the droplet size. Diethylcyclohexane (DECH) fuel was used in an oxygen-filled tube. Three droplet sizes were studied: 290, 940, and 2,600 μm (droplet diameters are used throughout this review). Initiation was accomplished by detonating a hydrogen–oxygen mixture contained in a chamber positioned near the top of the tube 45° off axis. The instrumentation included pressure transducers, thin-film heat transfer gauges, and high-speed photography. The photographs showed that droplet agglomeration increased as the droplet size decreased. Detonation was observed for all droplet sizes and it was found that the smaller the droplets, the more rapidly the detonation developed into a steady state.

Multiple-droplet experiments were also conducted by Lu et al. [87, 88] using a vertical shock tube 4.13 cm × 4.13 cm in cross-section fitted with a hydrogen–oxygen driver, as shown in Fig. 1.2. Two droplet generators were used. A fog with 0.5–10-μm droplets was produced using an ultrasonic nebulizer. Hypodermic needles vibrating at the Rayleigh instability frequency were employed to produce larger uniform droplets of 700- or 1,400-μm diameter. The desired fuel–air ratio was controlled by the droplet size, number of droplets, and flow rate of secondary air. The detonability was assessed by monitoring the wave speeds. For the 1,400-μm droplets, detonation of

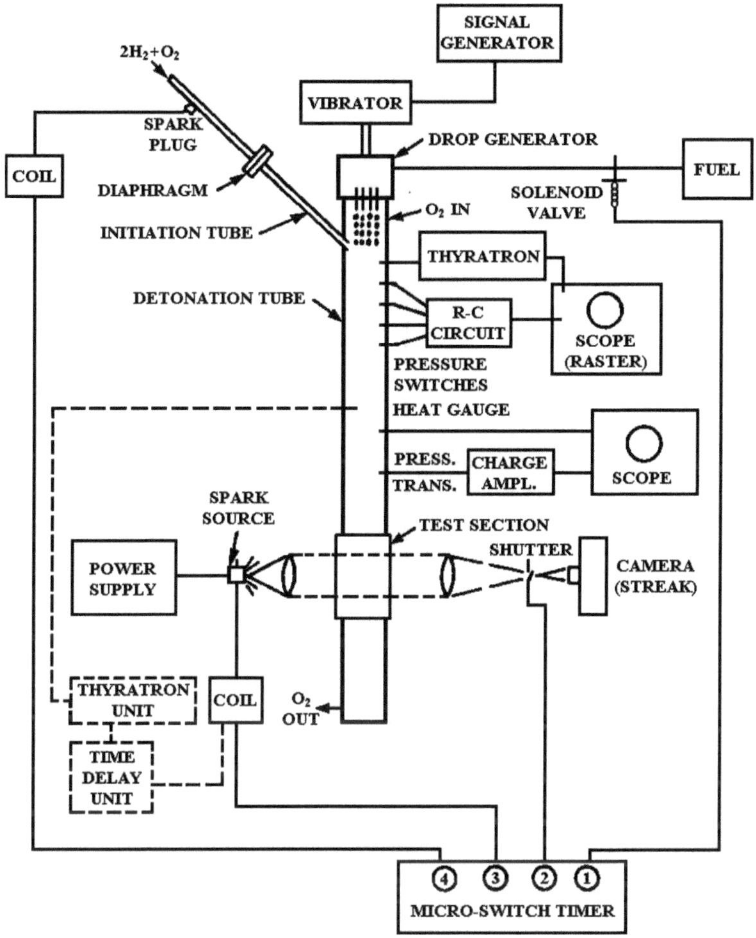

Fig. 1.1. Vertical spray detonation tube and instrumentation set-up of Dabora et al. [51]

heptane–oxygen was successful, but failure was observed for heptane–air when no sensitizer was present. This could have been due to the driver being too weak or the tube cross-sectional area being too small. For a fuel mixture consisting of 10% normal propyl nitrate (NPN) plus 90% heptane, the detonation speed decreased continuously along the tube length and complete failure appeared to occur near the end of the tube. In contrast, self-sustained detonation was apparent for fuel mixtures of 10% butyl nitrite (BN) plus 90% heptane and 25% NPN plus 75% heptane. In both cases, the wave speeds were steady near the end of the tube at about 1,600 m s^{-1}. The improved performance of BN over NPN (for the same amount of additive) was consistent with the results

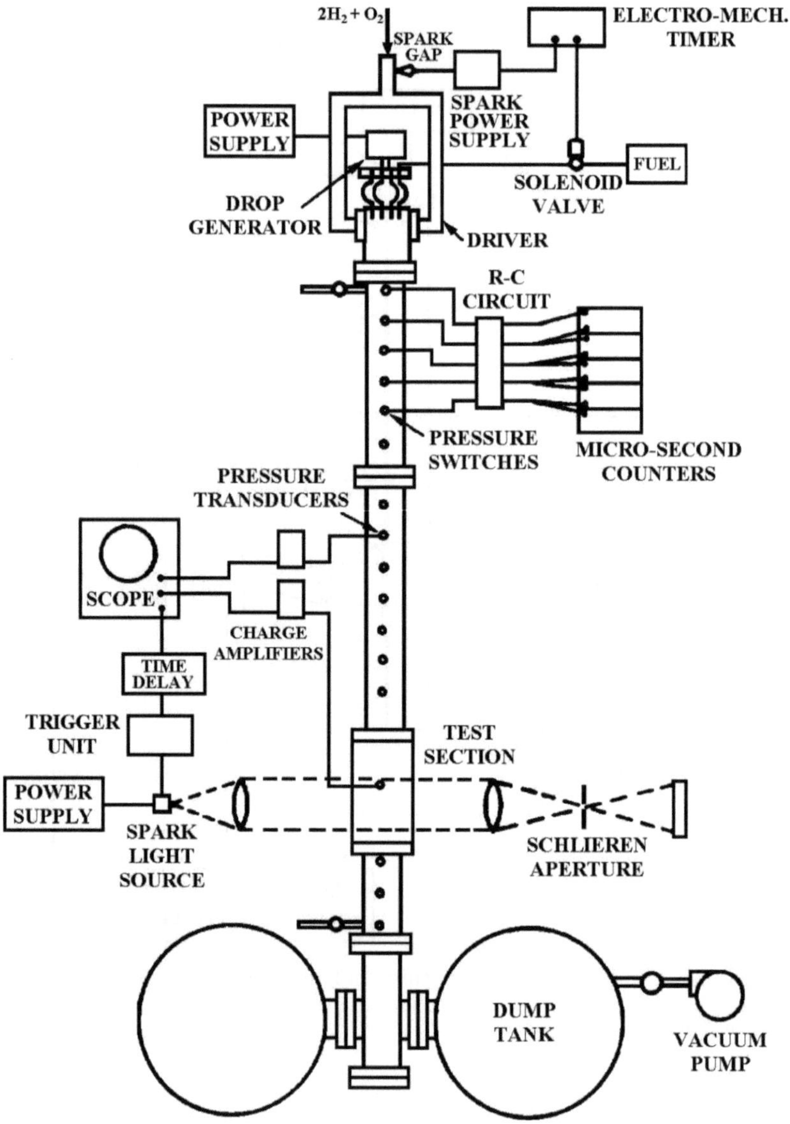

Fig. 1.2. Vertical shock tube used by Lu et al. [88] to study multiple-droplet ignition

from single-droplet experiments. The authors proposed that these chemical sensitizers lead to new chain-branching reactions with higher energy release rates.

The effect of droplet size was investigated by the same authors. In the case of pure heptane fuel in air, detonation was only possible for droplets in the 0.5–10-μm range. Given the high vapour pressure of heptane and its additives, most droplets in this size range probably vaporize prior to detonation.

Detonation was reported for a mixture of 10% NPN plus 90% heptane for 700-μm droplets, but not for 1,400-μm droplets. It was necessary to increase the NPN concentration to 25% to obtain detonation for the larger droplets. These studies showed that the nitrate additive and small droplet size increase the reaction rates and enhance detonability.

Tang et al. [129] later investigated the detonability of low-vapour-pressure decane sprays in oxygen and air. The tube and droplet generator employed were similar to that shown in Fig. 1.1, but the tube was 8.2 m in length and the initiator was a 1.2-m-long slug of a hydrogen–oxygen–helium mixture. The initiator strength was varied by adjusting the initial pressure of the gas. Detonation was observed for both decane–oxygen and decane–air and the critical initiation energies were estimated for a range of equivalence ratios. In some tests, the vapour pressure of decane was increased by heating the tube to 56℃. A decrease in the critical energy and a lowering of the lean detonation limit were both observed. Similar results were obtained when NPN sensitizer was added to decane in room-temperature experiments.

The earliest experiments on spray detonation were conducted in vertical shock tubes. Nicholls et al. [110] later studied cylindrical heterogeneous detonations in a "pie-shaped" or sectored shock tube of 1.4-m radius having a 20° total included angle, as shown in Fig. 1.3. The width of the shock tube was 5.2 cm. Uniform fuel droplets of 400-μm diameter were produced by a series of up to 322 hypodermic needles (0.02-cm inside diameter). The fuels included heptane, kerosene, and a mixture of 25:75 NPN/kerosene. Experiments were conducted in oxygen, nitrogen, and air environments. An explosive charge

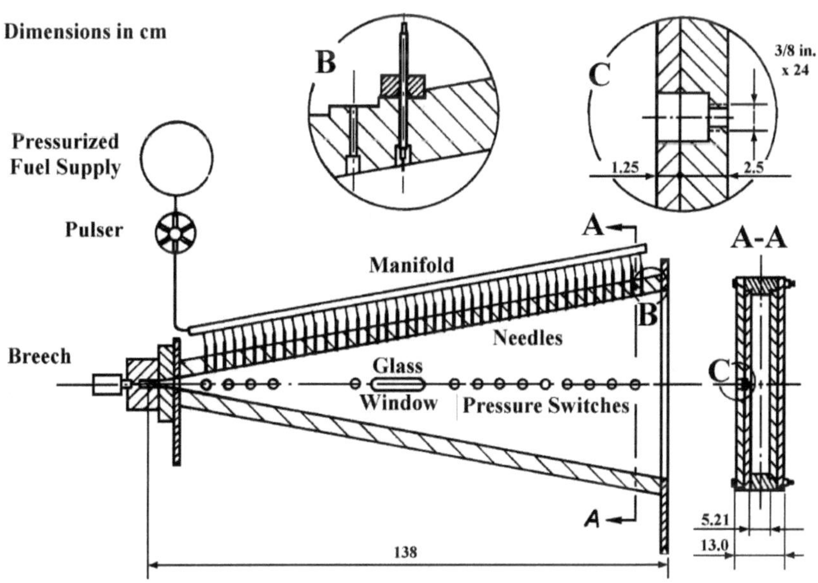

Fig. 1.3. The pie-shaped detonation chamber used by Nicholls and colleagues [17]

consisting of a blasting cap plus up to 5 g of high explosive positioned at the vertex of the shock tube was used as the initiator. A high-speed camera in combination with a parallel-beam xenon source was used for imaging. The wave propagation was monitored by 14 time-of-arrival pressure switches that were used to produce radius–time (R–t) and Mach number–time (M–t) plots.

The initial experiments were conducted using heptane fuel and the highest initiator strength. In one test, the shock tube contained only air (no droplets). Droplets were present in two additional tests in nitrogen and air environments. The strongest blast wave was recorded for the air-only case, while the weakest blast wave was observed for fuel droplets in nitrogen. The difference was attributed to energy and momentum transfer between the phases. The blast wave for fuel droplets in air was slightly stronger than that for fuel droplets in nitrogen, indicating that the wave was partially supported by combustion. However, detonation did not occur.

These experiments showed that detonation was not possible for this droplet size and initiation energy for heptane or kerosene fuels in air. The conditions were reported to be marginal for NPN/kerosene in air on the basis of a low detonation velocity in one test (1,350 $m\,s^{-1}$) and rapid acceleration of the blast wave near the end of the chamber in another test. However, the results did confirm that NPN sensitizes the fuel, as first reported by Lu et al. [88].

The tests also showed that detonation was possible for all three fuels in oxygen. The velocity of heptane–oxygen detonations ($\phi = 0.3$) was determined to be about 2,000 $m\,s^{-1}$. In one test, a heptane vapour–oxygen mixture (no droplets) was detonated at a higher equivalence ratio ($0.5 < \phi < 0.6$) and the observed detonation velocity was found to be in good agreement with the theoretical velocity. The measured detonation velocity for kerosene–oxygen was 1,600 $m\,s^{-1}$ in contrast to the theoretical velocity of 1,876 $m\,s^{-1}$. The authors proposed that losses to the chamber walls were the cause of the velocity deficit. However, Gubin and Sichel [69] have shown that incomplete combustion of the fuel in the reaction zone can also be a significant factor. Cylindrical divergence may have been a contributor as well given the relatively small radius of the chamber. In any event, the deficit was found to increase with increasing reaction-zone length, suggesting that kerosene–oxygen has a longer reaction zone than mixtures of oxygen with heptane or the fuel blend.

A comparison of the R–t plots for the various fuels in an oxygen environment (using a blasting cap initiator) yielded information about the relative reactivity of the fuels. Initiation of heptane–oxygen occurred immediately, while the blast wave in both kerosene–oxygen and NPN/kerosene–oxygen decayed considerably before accelerating to detonation at a larger radius. The acceleration to detonation was more rapid for NPN/kerosene than for unsensitized kerosene.

In an earlier study by Nicholls et al. [109], carried out prior to the sectored shock tube being increased in radius from 0.73 to 1.4 m, these authors reported overdriven detonations in kerosene–air that decayed to the Chapman–Jouguet (C–J) state by the end of the tube. However, after the tube was extended,

Gabrijel and Nicholls [63] carried out similar experiments to assess (1) the degree to which the wave front was cylindrical and (2) the influence of fuel voids on wave propagation. In that study, the authors reported that decane–air detonations had been successfully initiated. Although the wave speeds decreased to $M = 3.2$ by the end of the apparatus (vs. a C–J value of 5.06), the authors attributed these low velocities to boundary-layer effects. The detonations in kerosene–air and decane–air may have been failing. Further tests in a larger facility would be needed to clarify this issue.

1.2.2 Structure of Spray Detonations and Velocity Deficits

One of the earliest studies to reveal insightful information about the phenomena occurring behind the shock front in a spray detonation was conducted by Dabora et al. [51, 116]. These authors employed the detonation tube pictured in Fig. 1.1 and sprays of DECH in oxygen for droplet sizes of 290, 940, and 2,600 μm. Both spark schlieren and combined shadow and direct-light photographs were obtained for a single stream of 2,600-μm droplets. A schlieren photograph from one such experiment is shown in Fig. 1.4.

When the shock wave initially traverses a droplet, it remains stationary because of its inertia, but the droplet deforms instantly. The convective flow behind the front is supersonic and causes a bow shock and wake shock to form in front of and behind the droplet, respectively. The standoff distance of the bow shock increases as the droplet continues to deform. A wake behind the droplet consisting of small particles of fuel that have been stripped from the parent droplet and mixed with the gaseous oxygen ignites and reacts violently. The resulting explosion produces a blast wave which catches up to the main front and constitutes a mechanism for reinforcement of the front [40, 50, 113]. This blast wave also destroys the bow shock of the preceding droplet. Secondary shocks arising from the explosion are also evident. These processes are identical to those reported by Kauffman and Nicholls [71] in their single-droplet studies. However, for a line of droplets, the blast waves

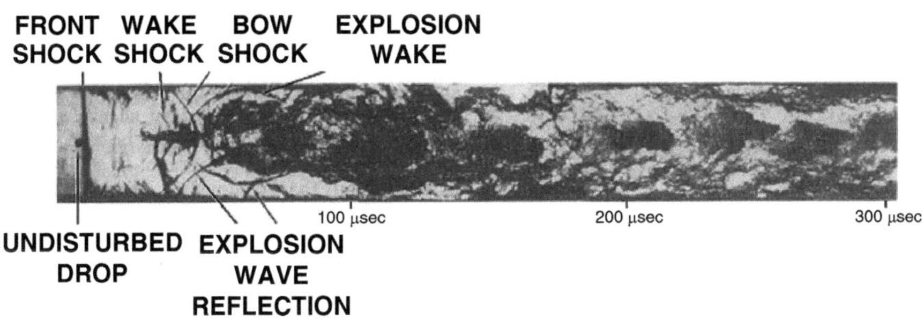

Fig. 1.4. Schlieren photograph of the detonation of a single row of 2,600-μm diethylcyclohexane (DECH) droplets in oxygen [51]

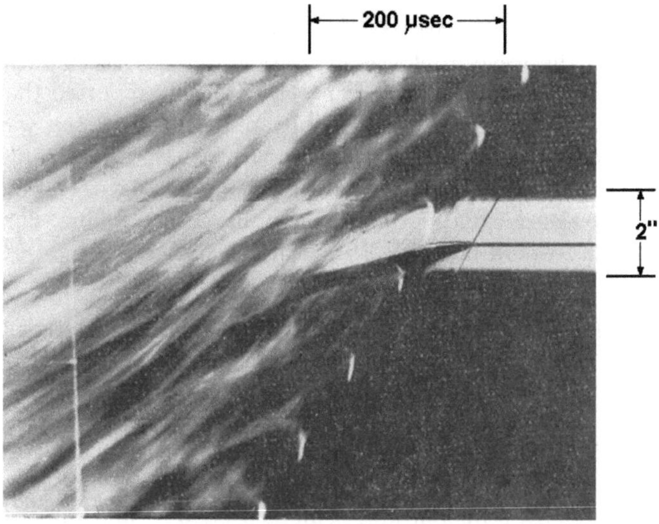

Fig. 1.5. Combined schlieren and direct-luminosity streak photograph for detonation of a single row of 2,600-μm DECH droplets in oxygen [51]

from individual droplets can also interact. Borisov et al. [22] believe that these reinforcing blast waves are necessary for propagation when the droplets are greater than 1 mm in size, but that droplet breakup and vaporization are probably fast enough for 10–100-μm droplets so that local explosions are not necessary. Furthermore, these authors showed that for droplets smaller than 10 μm, vaporization alone is sufficient to ensure homogeneous-like behaviour. Bowen et al. [23] found this to be the case for 2-μm decane fogs in oxygen. The threshold droplet size required for vapour-phase detonation is likely a function of the fuel volatility.

Figure 1.5 shows a combined shadow and direct-luminosity streak photograph of the event shown in Fig. 1.4. The photographic slit is aligned along the droplet's axial position in the tube. Sudden bursts of luminosity indicative of ignition and reaction are apparent. In some cases, the location of ignition appears to coincide with the stagnation point of the droplet.

In the same study, the development of detonation was monitored using pressure switches mounted in the tube walls. The initiating shock strengths were Mach 2.5–3.0. Detonations were observed for all droplet sizes when oxygen was used; however, no detonations developed when air was the oxidizer. It was found that the smaller the droplet size, the faster the detonation developed into a steady state. When the measured propagation velocities for mixtures with equivalence ratios between 0.2 and 1.0 were compared with the C–J velocities for an equivalent gaseous detonation, it was found that the measured velocities were lower by 2–10% for the 290- and 940-μm droplets, and by 30–35% for the 2,600-μm droplets. The large deficits for the largest droplets were proposed to be the result of the increased size of the reaction zone. A model based on heat-transfer measurements, inferred frictional losses to the tube walls in the reaction zone, and a reaction-zone length assumed to be controlled by only the breakup of the droplets was developed to explain the differences. The large deficits were shown to be the result of significant frictional and heat losses in the reaction zone. However, to obtain agreement between the predicted and measured results, it was necessary to use a reaction-zone length approximately twice the observed droplet breakup distance.

Later, Gubin and Sichel [69] shed more light on the matter when they proposed that the large velocity deficits were due to incomplete combustion of the fuel at the C–J plane caused by large unshattered portions of the droplets. These authors employed a Zeldovich–von Neumann–Döring analysis and assumed that the C–J plane was governed by the location of ignition. Any unburned fuel at that point would burn behind the C–J plane where the energy release could not contribute to the propagation velocity. The detonation velocities were calculated for monodisperse kerosene droplets in oxygen for various droplet sizes ranging from 10 μm to 2.7 mm. The calculations showed that the velocity of detonation increased with decreasing initial droplet size and that it approached the thermodynamic velocity for droplets less than 20 μm in size. As shown in Fig. 1.6, the predictions were seen to be in good agreement with experimental velocity deficit data. The data in the figure are for various droplet sizes reported by Stephans and Bowen [127], Dabora et al. [51], and Pierce [112]. When similar calculations were performed for various droplet sizes and fuel loadings ranging from 0.1 to 0.9 kg m^{-3}, it was found that the amount of fuel reacting before the C–J plane was only weakly dependent on the fuel loading, leading to detonation velocities that were relatively independent of fuel–oxidizer ratio for a given droplet size. These trends are shown in Figs. 1.6 and 1.7.

The experiments in the sectored shock tube by Nicholls et al. [110] showed that heptane was more readily detonated than kerosene, but the reasons for this were not entirely clear until further studies on detonation structure were

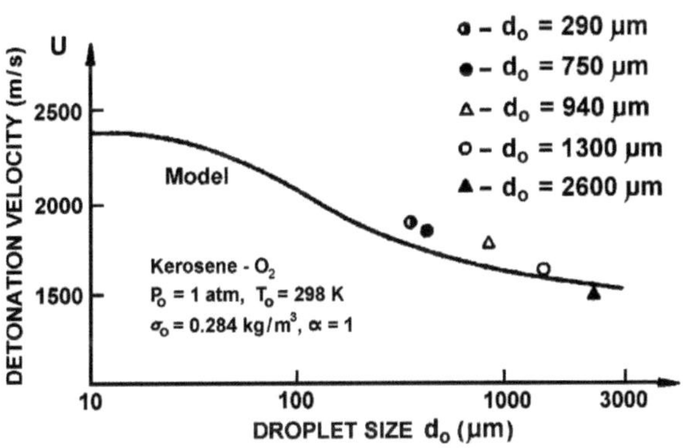

Fig. 1.6. Variation of propagation velocity with liquid droplet diameter [69]

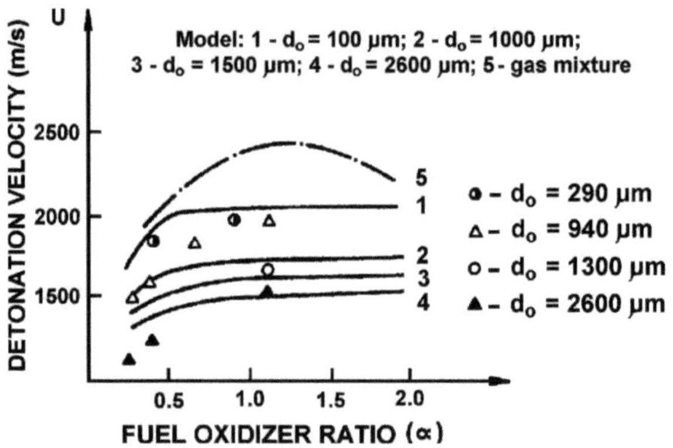

Fig. 1.7. Variation of propagation velocity with fuel–oxidizer ratio [69]

carried out by Bar-Or et al. [16–18] using the same apparatus and experimental procedures. In that work, fuels of low volatility (decane, kerosene, and 25:75 NPN/kerosene) and high volatility (heptane) were sprayed in an oxygen environment to assess the importance of the vapour phase in heterogeneous detonation. The droplet size was 400 μm. Both R–t data and high-speed schlieren streak and framing photographs of the reaction zone were obtained.

Experiments with decane droplets for an equivalence ratio of $\phi = 0.32$ showed that steady propagation velocities were obtained toward the end of the chamber for all initiator masses. However, the observed propagation velocities increased with increasing initiation energy. The measured velocity for a blasting cap alone was 1,430 m s^{-1} (23.5% below the C–J velocity for an equivalent gas-phase detonation), whereas a velocity of 1,661 m s^{-1} (11.2% below the C–J velocity) was measured for an initiator consisting of a blasting cap plus 5 g of high explosive. For initiators less than 1 g, regions of subcritical propagation in which the detonation velocity passed through a minimum were observed. For larger initiators, no minimum was observed and the velocity decayed monotonically to the steady-state value.

Similar experiments were conducted with kerosene ($\phi = 0.354$) and NPN/kerosene mixtures ($\phi = 0.302$) in oxygen to establish the effect of the NPN sensitizer. Both fuels have a low vapour pressure and so are present only as droplets. The results obtained were very similar to those for decane in oxygen. The measured detonation velocities for kerosene were 26.5 and 16% lower than the theoretical velocity for initiation with a blasting cap, and initiation with a blasting cap plus 3 g of explosive, respectively. The corresponding velocities for an NPN/kerosene mixture were 18.6 and 9.8% less than the theoretical velocity, respectively, for the same initiators. Subcritical propagation was observed to be more pronounced for kerosene than for NPN/kerosene mixtures, again confirming that NPN shortens the reaction-zone length.

The schlieren streak photographs verified that the controlling mechanism in the reaction zone is aerodynamically induced droplet shattering followed

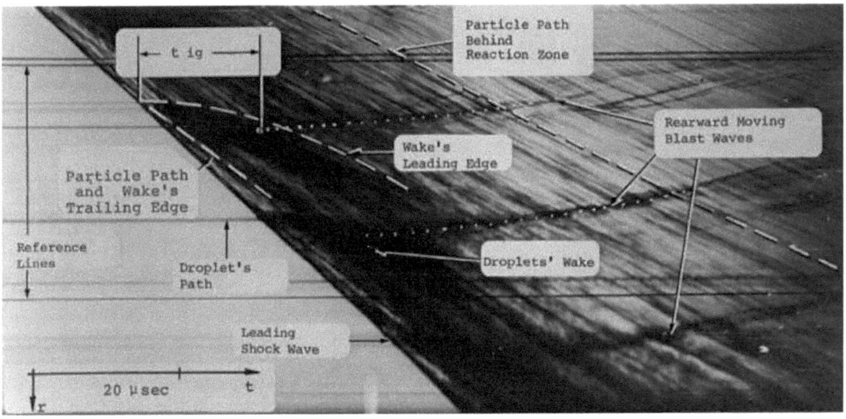

Fig. 1.8. Schlieren streak record of detonating decane spray in oxygen [17]

by explosive ignition in the turbulent wakes behind the parent droplets. One such photograph for a decane–oxygen spray is shown in Fig. 1.8. The leading shock is indicated by a sharp decrease in the intensity of the transmitted light. The breakup of a droplet and the formation of a wake are indicated by the growth of a dark (opaque) region behind the original location of the droplet. Explosive ignition inside the wake is characterized by the emergence of rearward moving blast waves. Two characteristic lengths of significance are the ignition length, l_{ig}, and the length for complete combustion, l_{cc}. The former is defined as the separation between the shock wave and the point of origin of the rearward blast wave, while the latter is defined as the separation between the shock wave and the remainder of the parent droplet when most of the fuel has been consumed. The point of complete combustion is the location where the wake behind the droplet ceases to be completely opaque. Analysis of many streak records showed that l_{ig} is in the range 1.85–2.40 cm, and l_{cc} is in the range 5.0–5.5 cm. These lengths are considerably larger than the reaction zone of gaseous fuel–oxygen mixtures. According to these same authors, the convective flow velocity behind the reaction zone is always subsonic relative to the wave front, indicating the absence of a C–J plane at these small radii.

Similar experiments were performed with high-vapour-pressure heptane fuel. A typical schlieren streak photograph for heptane spray in oxygen is shown in Fig. 1.9. The photograph shows a leading shock wave followed by an immediate reaction of the heptane vapour–oxygen mixture. A transverse wave structure is readily apparent. The shattering and combustion of the droplets was the same as for decane fuel. The convective velocity behind the leading shock was supersonic relative to the shock, indicating the existence of a C–J plane. These results showed that the overall behaviour was dominated by the vapour-phase detonation. The additional heat released from the combustion of the droplets did not appear to contribute to the propagation velocity. This finding is consistent with that of Pierce and Nicholls [114], who found that

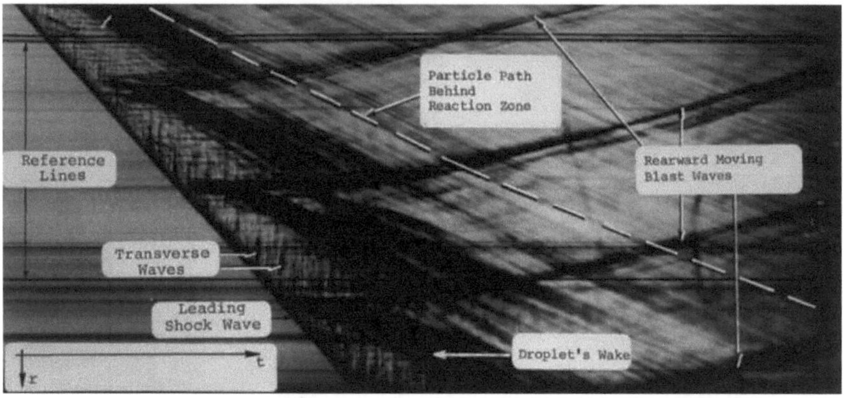

Fig. 1.9. Schlieren streak record of a detonating heptane spray in oxygen [17]

the combustion of non-volatile DECH droplets behind the reaction zone of a hydrogen–oxygen detonation did not affect the propagation velocity.

Schlieren framing records corresponding to the same basic experiments in Figs. 1.8 and 1.9 are shown in Figs. 1.10 and 1.11, respectively. The framing record for decane–oxygen was taken at 9.2-μs intervals and shows a 1,600 m s^{-1} shock wave interacting with three rows of decane droplets. As shown in Fig. 1.10, breakup of the droplets and the formation of misty wakes behind the parent droplets are apparent. The formation and emergence of blast waves from the wakes can also be clearly identified. These blast waves propagate toward the incident shock front and also upstream (backward). The regions originally occupied by the wakes and the fuel mist eventually clear, indicating that only gaseous combustion products remain. The framing record for heptane–oxygen detonation (Fig. 1.11) is drastically different from that for decane. In this case, propagation is completely dominated by the thin vapour-phase detonation preceding breakup and subsequent combustion of the heptane droplets. The transverse wave structure typical of gaseous detonation is again clearly visible.

When these photographs were compared with others for a heptane vapour–oxygen detonation (no droplets), it was clear that a vapour-phase detonation precedes and initiates droplet breakup and burning. The explosion of the droplet wakes occurs almost immediately behind the leading wave, much more rapidly than for low-vapour-pressure decane. The higher temperature behind the vapour detonation was proposed to account for the difference. The experimental R–t trajectories for these experiments were found to be indistinguishable. No significant difference existed between the propagation velocity for pure vapour and vapour–droplet heptane fuels.

The nature of the reaction zones for low- and high-volatility fuels was further clarified in tests where a corresponding schlieren streak record and a pressure trace from a transducer mounted in the side wall of the chamber midway between droplet streams were obtained for a decane–oxygen detonation.

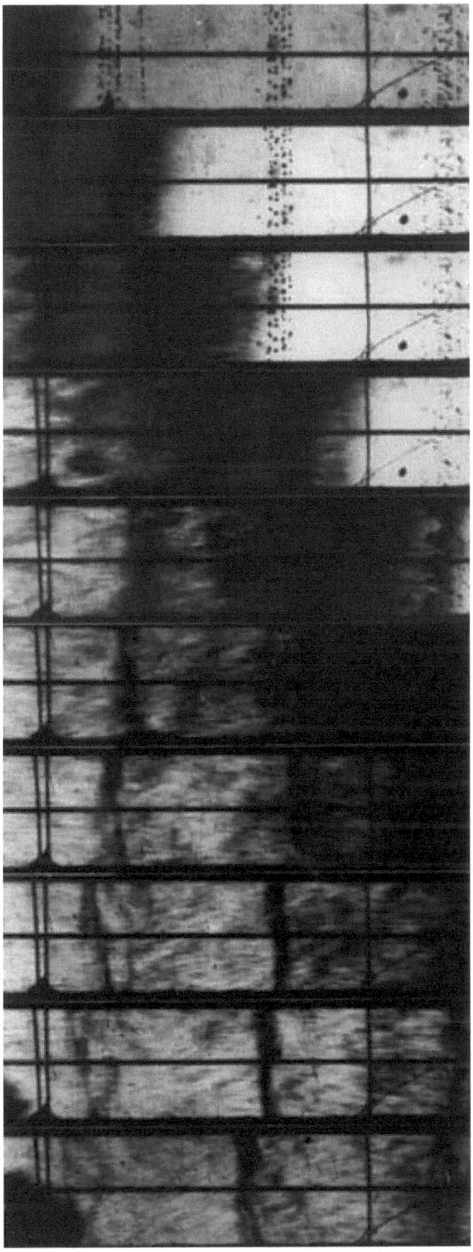

Fig. 1.10. Schlieren framing record for a detonating decane–oxygen spray [18]

The authors, Bar-Or et al. [18], were able to superimpose these records to form a three-dimensional R–t–P plot, making it possible to relate the pressure signals to the structure of the spray detonation. The first pressure spike

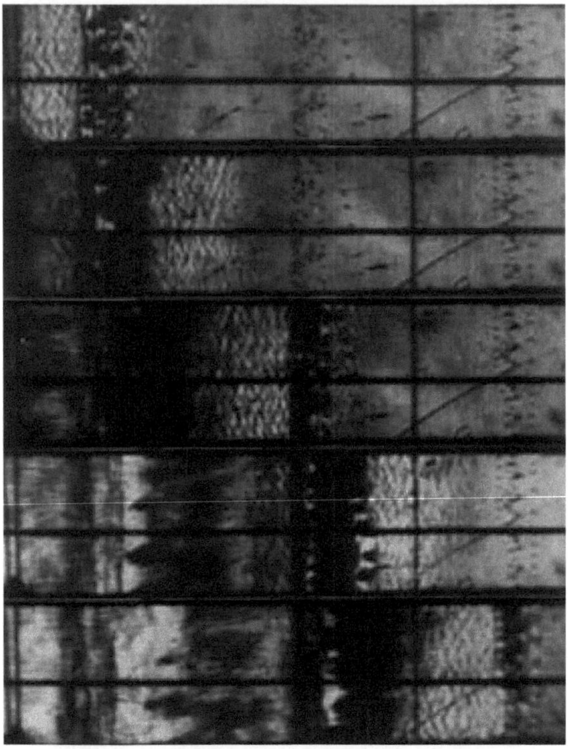

Fig. 1.11. Schlieren framing record for a detonating hexane–oxygen spray [18]

in the P–t signature corresponded to the leading shock wave. Its overshoot above the theoretical C–J pressure (calculated for a gaseous system having the same wave speed) may have been caused by a blast wave from upstream wake explosions overtaking the leading shock just prior to its arrival at the pressure gauge. A second peak in the signature was interpreted as a forward-moving blast wave generated by explosive ignition in the wake of droplets immediately preceding the transducer. The pressure then decreased gradually to about 12 atm at a time that correlates with the moment the back edge of the burning wake passed over the transducer. Oscillations later in the signature were thought to be caused by rearward moving pressure waves, indicating continuing combustion in the wake as it moved downstream.

Schlieren photographs were also obtained for a shock wave interacting with decane droplets, but in a 40:60 O_2/N_2 atmosphere rather than in pure oxygen. The interaction of the droplets with the supersonic flow behind the shock is strikingly illustrated, with the deformation of single droplets and the formation of bow and tail shock waves around the droplets being clearly visible. The interaction of neighbouring droplets and the formation of a micromist behind the original sheet of droplets are again evident. However, there is a clear absence of wake-generated blast waves in this nitrogen-diluted atmosphere. The M–t plot for this experiment suggested that detonation might not have occurred, or that it was in the process of failing. This prompted the same authors to conclude that wake explosions might be an essential element of spray detonations for low-volatility fuels.

1.3 Detonation in Unconfined Fuel Sprays in Air

1.3.1 Sprays Created by Low-Pressure Nozzles

One of the earliest investigations on the detonability of unconfined fuel sprays in air was carried out by Bull et al. [35] during the late 1970s at the Thornton Research Centre in the UK. Strictly speaking, the tests were not "truly unconfined" in that the fuel–air mixture was protected from the effects of ambient winds by spraying the fuel into a polyethylene enclosure or "bag" measuring 1.5 m × 1.5 m × 2.2 m high (125-μm wall thickness). The apparatus is shown in the schematic diagram in Fig. 1.12.

A particular emphasis of the study was on the influence of fuel volatility on the detonability. The fuels employed were high-vapour-pressure n-hexane (C_6H_{14}) and low-vapour-pressure decane ($C_{10}H_{22}$) and dodecane ($C_{12}H_{26}$). An attempt was also made to synthesize fuels having intermediate volatilities by combining hexane and dodecane in various proportions. The sprays were created using four Sonicore^{TM} nozzles positioned at the bottom of the bag pointing vertically upward. These nozzles atomize the fuel through a process in which an air blast is directed through a sonic convergent–divergent nozzle, and they represent a good compromise between the conflicting requirements of high liquid flow rates and narrow droplet size distributions. The global fuel–air concentration in the bag was controlled by spraying for different time

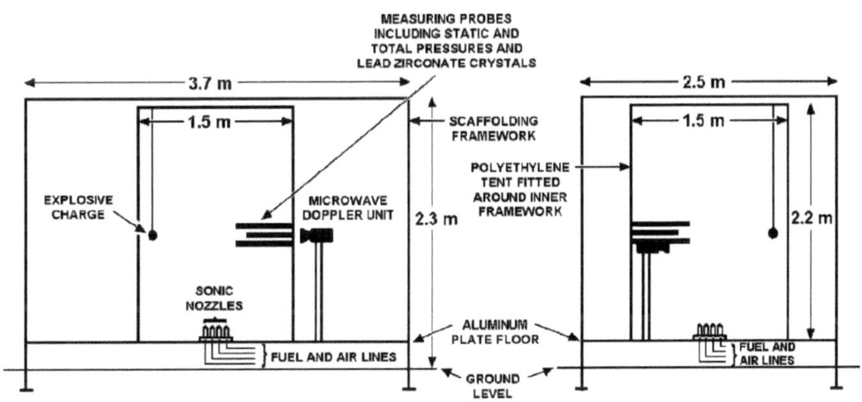

Fig. 1.12. The 5-m^3 apparatus of Bull et al. [35] for the study of unconfined aerosol detonation

intervals, while the droplet size was varied by adjusting the fuel and air flow rates, or by selecting different nozzles. The aerosols were characterized by a Malvern particle size analyser. For the purposes of calibrating the Malvern instrument, commercial-grade hexane and kerosene (the latter being representative of decane and dodecane) were used in place of their more costly, higher-purity counterparts. The calibration experiments showed that the majority of the aerosol mass fell in the 6–17-μm range for hexane and the 9–30-μm range for kerosene.

Initiation of detonation was achieved using a PE4 (88% RDX) plastic explosive charge mounted in one corner of the bag. The blast characteristics of the initiator charges in air were thoroughly quantified in a series of tests beforehand. Piezoelectric pressure transducers and shock pins were used to obtain pressure and time-of-arrival data, while an attempt to track the reaction front using a microwave Doppler sensor was unsuccessful. A high-speed camera was used to capture images in selected tests.

The results for hexane are presented in Fig. 1.13 in terms of the initiator charge mass as a function of the equivalence ratio, ϕ. The solid curve between the "detonation" and "no-detonation" points indicates the approximate detonability limits. The minimum initiator charge mass is about 0.025 kg for an equivalence ratio of $\phi = 1.1$–1.2. These values compare with those of 0.018 kg of Tetryl™ for $\phi = 1.18$ obtained by Bull et al. [34] for gaseous ethane–air mixtures, suggesting that the two systems have similar detonability. The dashed curve in the figure is based on a modification of the Zeldovich criterion, $E \propto \tau^3$, where E is the initiation energy and τ is an induction period

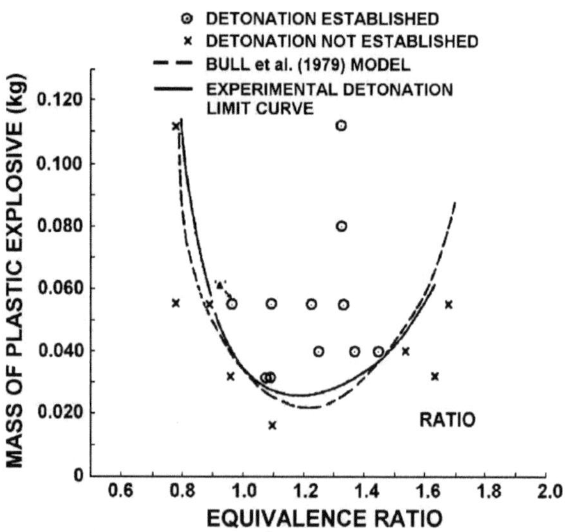

Fig. 1.13. Initiation energy for hexane sprays in air as a function of equivalence ratio for droplet sizes in the 6–17-μm range [35]

given by $\tau = A[O_2]^{-1}\exp(E_A/RT)$, where A is a pre-exponential constant, the activation energy $E_A = 121.4\,\mathrm{kJ\,mol^{-1}}$, T is the temperature in Kelvins, and R is the gas constant. The first-order dependence of τ on oxygen concentration was inferred from the data for other lower paraffin hydrocarbons, and the pre-exponential constant was determined from the data at stoichiometric conditions. The dashed curve describes the trend of the data fairly well.

The detonation velocities for hexane–air detonations were measured in selected tests and found to be 6–13% lower than the C–J values for an equivalent gas-phase system. This observation appears to conflict with earlier investigations that attributed large velocity deficits to momentum and heat transfer losses to the walls. The large detonation wave curvature in the tests of Bull et al. may well have been a factor in these results.

A single experiment was conducted in which the droplets were considerably larger than in the other experiments (i.e., 50–90 μm). Detonation was achieved using a charge mass of 0.055 kg, which was considerably larger than the critical charge mass of about 0.03 kg for the finer aerosol at the same equivalence ratio. The measured detonation velocity of 1,300 $\mathrm{m\,s^{-1}}$ in this experiment was significantly lower than the value of approximately 1,580 $\mathrm{m\,s^{-1}}$ for the finer aerosol. This result emphasizes the influence of the droplet breakup time on the effective reaction-zone length of the detonation.

Detonations could not be initiated for low-vapour-pressure dodecane and decane fuels over a wide range of conditions for charges up to 0.5 kg for dodecane and 0.3 kg for decane. Although the authors, Bull et al. acknowledged that the path length of the experiment (1.83 m from the initiator charge to the bag wall) was too short to ascertain that any detonation wave initiated was truly self-sustained, they concluded that the wave velocity had decayed to the point where no detonation had occurred or could have occurred. For these fuels, insignificant evaporation takes place prior to the arrival of the initiating shock wave, so the droplet sizes are not reduced by vaporization, as is the case for hexane.

In an effort to control the volatility of the fuel, hexane and dodecane were mixed in various proportions in subsequent experiments. The initiator charge mass was 0.12 kg for most experiments, but was reduced to 0.10 kg in some instances. The results are shown in Fig. 1.14. It is evident that as the proportion of dodecane in the mixture increases, the minimum equivalence ratio for detonation increases, and reaches a value of $\phi = 2$ for a 50:50 mixture. The points labelled "delayed detonation" in the figure denote tests in which the blast wave arrived at the transducer locations before the detonation wave. In these cases, the high-speed films showed that detonation had already commenced at another remote site, indicating that the outcome is sensitive to the mixture homogeneity near the concentration limits.

When the results for the hexane–dodecane mixtures are compared with those for pure hexane (Fig. 1.13), it appears that the detonation of these mixtures is effectively a hexane detonation since the quantity of hexane in each of the mixtures is at least that required for the lean detonation limit,

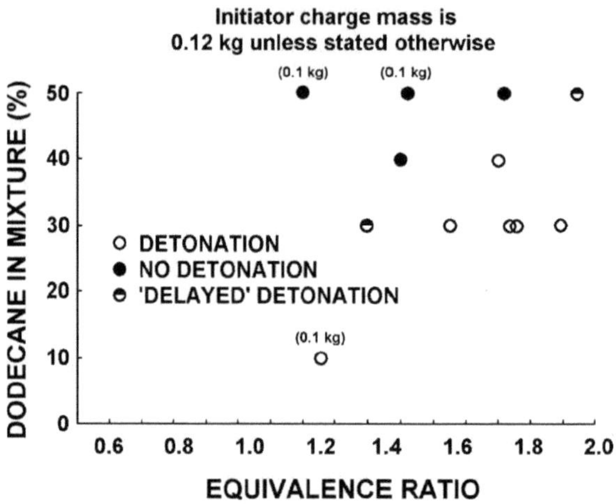

Fig. 1.14. Detonation of mixtures of hexane and dodecane sprayed in air for various mixture proportions [35]

at 0.10–0.11-kg charge mass, in Fig. 1.13. This indicates that dodecane plays little or no part in the detonative combustion and suggests that the hexane emerges from the mixture droplets in vapour form. It also explains why an increasing quantity of fuel must be dispersed as the percentage of dodecane in the mixture increases. Apparently, the idea of synthesizing a fuel with a prescribed volatility by blending two pure fuels of different volatility is not feasible.

A similar spray detonation study was carried out in the mid-1980s by Sandia National Laboratories (S. Tieszen, private communication, 1990). The approach was similar to that taken by Bull et al. [35] in that spray towers consisting of a rectilinear metal frame covered with plastic sheeting were employed to minimize the effects of ambient winds, while still permitting high-speed photography of phenomena occurring inside. One such tower measured 3 m × 3 m × 6 m high (54 m^3), while the other was much smaller, measuring 1.2 m × 1.2 m × 3 m high (4.3 m^3). Each tower was equipped at the top with precision ultrasonic spray nozzles. Early nozzle calibration tests were performed using water. Simulated lean and rich fuel mixtures were produced by varying the water delivery pressure, but water was concluded to be a poor simulant for a combustible liquid fuel and the calibration tests were later repeated with kerosene. Three methods were used to estimate the droplet size distribution and/or mean droplet velocity, and total droplet mass. These included (1) simple attenuation of a laser beam over a known path length, (2) phase Doppler particle size analysis, and (3) use of an absorption technique. In the last method, towels were placed at the bottom of the spray tower. The towels were initially covered by an awning-like device. Once the spray had equilibrated, the fuel flow was abruptly terminated and the awning retracted. The droplet mass that settled on the towels was subsequently weighed. The fuels tested included hexyl nitrate, ethyl hexyl nitrate, methyl cyclopentadiene dimer, dipentene, α-methylstyrene, decalin, 1-decene, 1,2-epoxydecane, 2-(2-butoxyethoxy)ethanol, and JP-10. A number of these fuels were found to be detonable, although many were characterized by critical diameters in excess of 3.7 m. Several differences between the results from trough tests (described later) and tower tests were also identified. These were attributed to differences in confinement, droplet size and concentration, and vapour-phase accumulation time in the two sets of apparatus. Detonability and performance data for these fuels have not been openly published.

Samirant et al. [120] studied heterogeneous fuel–air detonations at the Institute Saint-Louis using a 50-cm diameter, vertical, open-ended steel tube of 4-m length. Methanol, ethanol, propanol, hexanol, and decane fuels were dispersed from the top using a ring of 12 injectors. The droplet number density and size distribution were determined using a stroboscopic photographic technique. The mean droplet size and liquid fuel volume fraction were estimated to be $350\,\mu\text{m}$ and $1.1\times10^{-1}\,\text{cm}^3\,\text{cm}^{-3}$, respectively. The vapour was always saturated by intermittently spraying several times prior to the experiment until an equilibrium temperature of the spray was reached. The vapour-phase equivalence ratio was determined from this temperature. Initiation of detonation was achieved using a 1-kg high-explosive charge suspended 35 cm below the top of the tube. Detonations were observed for all fuels. Three experiments of considerable interest involved decane and hexanol fuels. In these tests, the vapour-phase equivalence ratio was as low as 0.06. The measured detonation velocities were $1{,}580$–$1{,}680\,\text{m}\,\text{s}^{-1}$. These results clearly demonstrate that a two-phase C–J detonation can exist in sprays consisting of relatively large droplets of low-volatility fuel.

1.3.2 Sprays Created by Explosively Driven Nozzles

Experiments on unconfined detonation of motor fuel sprays in air were carried out on a much larger scale by the Kurchatov Institute in Russia during the 1980s. The apparatus, first detailed by Alekseev et al. [3], was capable of forming clouds up to $1{,}500\,\text{m}^3$ in volume, about 300 times the size of the polyethylene bags employed by Bull et al. [35]. The apparatus consisted of a 400-l, hermetically sealed, underground vessel filled with liquid fuel connected to an aboveground manifold fitted with a series of slot nozzles. The details are shown schematically in Fig. 1.15. A smokeless powder charge ignited inside the chamber provided the driving pressure for fuel dispersal. The stoichiometry of the cloud was controlled by varying the mass of fuel in the chamber, as well as the powder charge size. The aerosol clouds created were of hemicylindrical cross-section and measured 8–20 m in length and 4–8 m in radius. The cloud formation time was 0.8 s. Gasoline (type A-76), kerosene (type TS-1), and diesel fuels were tested. The mean droplet size was estimated by an unspecified settling technique and was reported to be 50, 100, and $200\,\mu\text{m}$ for the three fuels, respectively. TNT charges were used to initiate

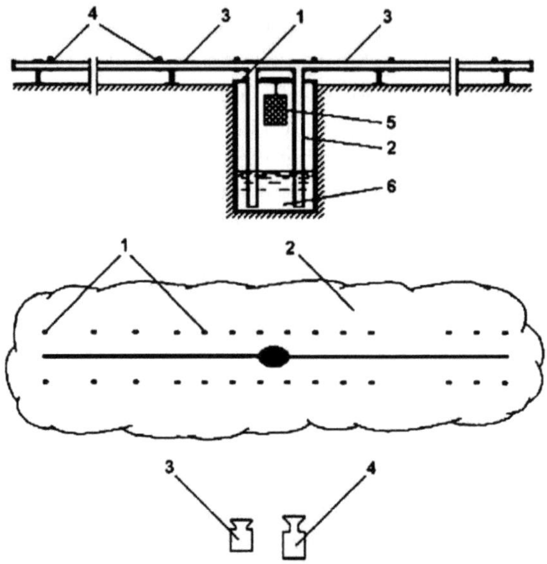

Fig. 1.15. The Kurchatov Institute fuel dissemination apparatus and diagnostic techniques. *Top*: *1* chamber, *2* dip tube, *3* manifold, *4* slot nozzle, *5* powder charge, *6* fuel. *Bottom: 1* pressure transducers, *2* aerosol cloud, *3* high-speed cine camera, *4* video camera [3]

detonation with a delay time of 0.8–4.0 s from the commencement of fuel dispersal. The tests were conducted for ambient temperatures in the 18–22℃ range. The diagnostic techniques included pressure transducers, video, and high-speed cinematography. The original study was later extended to expand the parameter space [4, 6].

The vapour pressure of gasoline is much higher than for kerosene and diesel, so it was expected that the vapour content in the gasoline–air clouds would be considerably higher than for the other fuels. Figure 1.16 shows the critical initiator charge mass versus delay time for the gasoline–air system for fuel concentrations between 0.08 and 0.12 $\mathrm{kg\,m^{-3}}$. It can be seen that the minimum energy for direct initiation of detonation is strongly dependent on the initiator delay time. For sufficiently long delays (2.5–3.7 s), the critical initiator mass is about 0.3 kg of TNT. Detonation was not generally possible for delays less than 1.7–1.8 s for initiator masses up to 6 kg. In cases of successful initiation, the wave was seen to propagate steadily along the full length of the cloud at velocities of 1,500–1,800 $\mathrm{m\,s^{-1}}$. Measured overpressures were in the 1.5–1.8-MPa range. These velocities are slightly lower than the 1,900 $\mathrm{m\,s^{-1}}$ reported earlier by Lobanov et al. [86] for gasoline vapour–air detonations in a 5.2-cm-diameter tube.

The critical initiator mass is plotted as a function of fuel concentration in Fig. 1.17. Here, the delays were 2.4–3.4 s, long enough for significant vapour-phase accumulation. The curve through the data has the typical U-shape

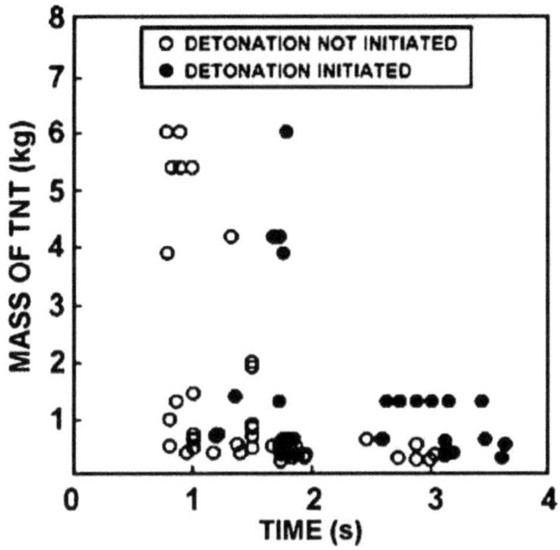

Fig. 1.16. Minimum initiator charge mass dependence on the initiation delay for gasoline sprays in air. The fuel concentration is within the range 0.08–0.12 kg m^{-3}. *Filled circles* detonation, *unfilled circles* no detonation [3]

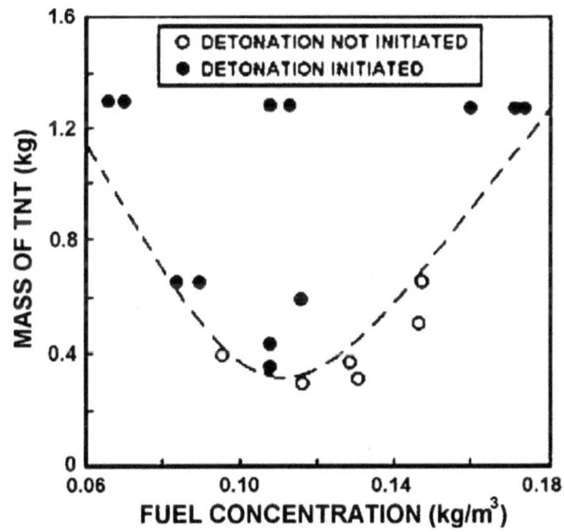

Fig. 1.17. Minimum initiator charge mass dependence on the mean fuel concentration for gasoline sprays in air. The initiation delay is in the range 2.4–3.4 s. *Filled circles* detonation, *unfilled circles* no detonation [3]

with the minimum displaced to fuel-rich mixtures as expected for heterogeneous systems. The minimum initiator mass occurs for a fuel concentration of 0.10–0.11 kg m^{-3}.

Initiation of detonation was not achieved for kerosene and diesel fuels in the original study [3]. However, monitoring the decay of the reactive blast waves for a given initiator mass allowed the conclusion to be drawn that the kerosene–air system was more sensitive to detonation than diesel–air. A similar blast wave decay was observed for the gasoline–air system in tests where the delay times were short.

Successful initiation of kerosene–air was achieved in the follow-on experiments [4, 6] where the critical initiator mass for direct initiation was determined to be 4.5 kg of TNT for a fuel concentration of $0.1\,\mathrm{kg\,m^{-3}}$. This minimum was found to be independent of the delay time, a finding that confirms the results are not dependent on vapour-phase accumulation during fuel dispersal for this low-vapour-pressure fuel. This charge mass is more than an order of magnitude (15 times) larger than the minimum charge mass for the gasoline–air system. An increase in kerosene fuel concentration to $0.14\,\mathrm{kg\,m^{-3}}$ or a decrease to $0.08\,\mathrm{kg\,m^{-3}}$ resulted in the critical initiator mass increasing to 7 kg. These results suggest that, if the scale of the event is sufficiently large, continuous aerodynamic stripping of the droplets will eventually reduce them to a size where detonation becomes possible.

An attempt was made by these investigators to estimate the critical spherical radius and the critical cloud height for detonation propagation. In the case of gasoline–air mixtures, the critical radius was determined by positioning the initiator charge 3 m above the ground and identifying the shock radius for which the wave velocity reached a minimum during the initiation process. The critical radius was estimated to be about 2 m using this procedure. The critical radius could not be measured for kerosene–air mixtures because the cloud height was too small to accommodate the procedure.

The minimum cloud height was determined in a series of experiments in which the height was varied from test to test. The minimum height was found to be 2.5–3.0 m for gasoline–air and 6.5–7.0 m for kerosene–air. Note that the critical energy and minimum cloud height data are consistent with the correlation linking the critical energy to the cube of the critical diameter, as observed in gaseous detonation studies [i.e., $(6.75\,\mathrm{m}/2.75\,\mathrm{m})^3 \cong 14.8 \cong 4.5\,\mathrm{kg}/0.3\,\mathrm{kg}$]. The above results explain the failure to initiate decane and dodecane detonations in the study by Bull et al. [38]; both the cloud size and the initiator charge mass were too small.

Given that the maximum cloud height that can be produced with this apparatus was only 8 m, this means that all of the tests with kerosene resulted in marginal detonations. This explains the longitudinal galloping phenomena reported by the authors where the wave underwent up to four oscillations characterized by velocities in the $700\text{–}900\,\mathrm{m\,s^{-1}}$ range at the low end and in the $1{,}000\text{–}1{,}300\,\mathrm{m\,s^{-1}}$ range at the high end. The existence of marginal detonations also explains the large velocity deficits of about 25% observed for kerosene–air mixtures. Similar velocity oscillations were observed for the gasoline–air system for delay times shorter than 1.7–1.8 s.

An attempt was made in this study to determine the "degree of completeness" of the reaction by measuring air-blast performance at the edge of the cloud and at distances up to 40 m away. For tests involving 90 kg of gasoline, the pressure and impulse TNT equivalencies (at 40 m) were estimated to be

250–300 and 450–500 kg, respectively. For similar tests with 80 kg of kerosene, the corresponding values were 230–270 and 400–450 kg, respectively. Variations were attributed to the effects of ambient winds on the clouds during these relatively long delay times. Nonetheless, these TNT equivalencies confirm that detonation took place and indicate a high degree of reaction completeness. TNT equivalency ratios in the 2.5–5 range are consistent with those reported for FAE canister devices (to be described later).

Estimates of the detonation cell size for stoichiometric gasoline–air and kerosene–air sprays were made on the basis of experimental data for the critical radius, R_c, and critical energy, E_c, using semi-empirical relations from gaseous detonation studies that link these parameters to the cell length, b [58, 83, 85]; notably $R_c = (8–12)b$ and $E_c = 50\rho_o D_o^2 b^3$ where ρ_o and D_o are the mixture density and detonation velocity, respectively. The estimated cell length for gasoline–air was 17–25 mm on the basis of these expressions. The cell length for kerosene–air was estimated at 50 mm on the basis of the expression for E_c (as noted previously, R_c could not be determined experimentally).

The authors concluded that the above estimates for the cell lengths appeared to be reasonable and did not contradict conventional wisdom. However, further work on gasoline–air sprays was later reported by the Kurchatov Institute [8] that called into question the applicability of cell-size-based correlations for heterogeneous systems. The impetus for the study was provided, in part, by the findings of earlier investigations into gaseous detonation. For example, Moen et al. [97] examined the link between the critical tube diameter and the cell size, $d_c = 13\lambda$, and found that the correlation was approximately valid for fuel–air systems having irregular cellular structure, but failed for regular structures typical of mixtures that are highly diluted with argon. Moen et al. also reported higher velocity deficits for propagation in tubes and failure for more sensitive mixtures in systems characterized by a regular cellular structure. More recently, Desbordes [52] reported that the 13λ correlation, as well as the relationship linking the critical initiation energy to the cell size ($E_c \sim \rho_o D_{CJ}^2 \lambda^3$), broke down for oxyacetylene mixtures highly diluted with monatomic inert gases. This author found that the critical energy correlated with the critical shock radius much better.

The same fuel dispersal apparatus shown in Fig. 1.15 was used in this follow-on study to create clouds of 1,100-m^3 volume. Two types of cloud initiator were employed. The first was a "point-source" TNT charge positioned 1 m above the ground, while the second was a "non-ideal" initiator consisting of a spherical polyethylene bag filled with stoichiometric propane–air. The bag was placed on the ground inside the cloud and a centrally positioned 0.14-kg TNT charge was used to initiate the propane–air detonation. As in the earlier tests, a sufficiently long delay time was used (2.5 s) to allow vapour-phase accumulation, and the droplet size was again estimated at 50 μm. The detonation cell sizes were measured directly using smoke foils positioned at ground level 6–8 m away from the initiator charge. Dynamic detonation parameters, including the critical initiator energy and the critical shock radius at the onset of detonation, were determined in the tests.

The minimum initiator mass was found to be 0.64 kg of TNT, corresponding to a critical energy of 2.9 MJ. A charge mass of 0.59 kg failed to initiate detonation. This value is about twice that measured in the earlier studies [3,4,6], with the difference being attributed to a change in the type of gasoline used. The minimum initiating volume of propane–air was determined to be 6 m^3. This corresponds to a critical energy of 19.3 MJ, about 6.7 times the critical energy using a TNT charge. An energy of 16.9 MJ (5.2 m^3 of propane–air) was insufficient to initiate detonation. These results are not entirely surprising, as it is well known that the critical energy for spherical initiation depends on the rate of energy deposition. Bach et al. [15], for example, have shown that the critical energy can vary by 3 orders of magnitude depending on the initiator type.

As in the earlier investigations, the critical shock radius was determined by identifying the location of the minimum shock velocity on x–t diagrams during critical initiation. This procedure yielded $R_c \cong 2.75$ m for high-explosive initiation and $R_c \cong 5.17$ m for initiation by a gaseous charge. Damage to the smoke foils due to blast-propelled sand particles made cell size measurements difficult. However, some records were readable and revealed a cellular structure of poor regularity with a cell size $\lambda \cong 40$–50 mm, as shown in Fig. 1.18. This is consistent with cell sizes of 54–67 mm for common gaseous fuels in the alkane group reported by Knystautas et al. [76], and with a cell size of 42 mm for decane vapour–air detonations reported by Tieszen et al. [132]. The fraction of gasoline present in droplet form may have an influence on the cell size measurements. Papavassiliou et al. [111], for example, have shown that the cell size for 5-μm decane sprays in oxygen is twice as large as that for the equivalent gaseous mixtures.

There are significant differences between the dynamic detonation parameters measured for gasoline aerosol detonations and those reported for gaseous propane–air detonations. For example, even though the cell sizes are similar for these two systems, the critical energy for high-explosive initiation of gasoline–air is about 7 times larger than for propane–air. The critical shock radius is about 20λ for propane–air and 55–69λ for gasoline–air under conditions of high-explosive initiation. Interestingly, the ratio of the critical energy to the cube of the critical radius, E_c/R_c^3, was found to be identical for the two types of initiator used to detonate gasoline–air, despite the values of E_c and R_c being substantially different. The value of this ratio for propane–air, however, is nearly 2 times larger. These results show that the classical relationships linking the dynamic detonation parameters to the cell size do not appear to be applicable to this insensitive heterogeneous system. They also suggest that E_c/R_c^3 may be constant for a given fuel–air system, in accordance with the findings of Desbordes.

Initiation of detonation for diesel fuel sprays in air was not achieved in any of the studies employing the apparatus pictured in Fig. 1.15. As previously noted, the blast wave decay in diesel–air was more rapid than in kerosene–air for subcritical experiments. When combined with the conclusion that the

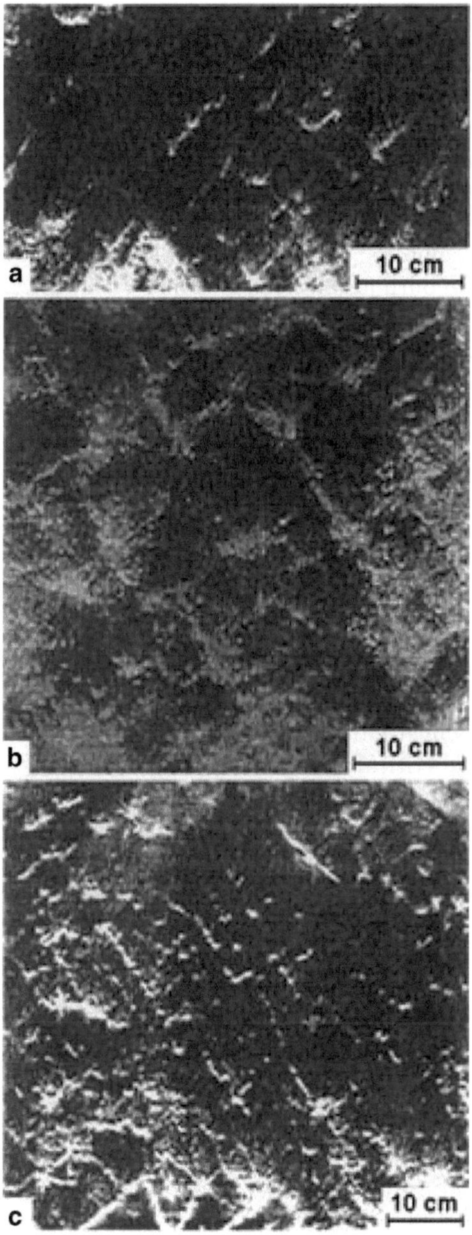

Fig. 1.18. Smoke foils for gasoline–air sprays obtained by the Kurchatov Institute [8]

kerosene–air detonations observed were marginal, this would indicate that the dispersal apparatus was too small to support a diesel spray detonation. However, detonation of diesel aerosols was achieved by Dorofeev et al. [54,55] in much larger experiments to be detailed later.

The effects of preburning the aerosol clouds on the resulting blast field were studied at the Kurchatov Institute by Kuznetsov et al. [79, 80] using the same apparatus and fuels described earlier. In these experiments, the fuels were dispersed to form aerosol clouds, but the clouds were deliberately ignited by a distributed ignition source consisting of a 16-m length of detonating cord suspended 1.5 m above the ground. A blast wave, produced by the detonation of between 0.35 and 10 kg of high explosive outside the cloud, was subsequently monitored as it propagated through the burning cloud. The cloud was allowed to burn for between 0.02 and 1 s. This procedure was found to have a significant effect on the blast propagation. Blast amplification was observed inside the cloud, and the attenuation of the transmitted air blast was reduced under some conditions.

In another study, Alekseev et al. [5, 7] dispersed kerosene fuel in the same facility and studied initiation of detonation under critical conditions. In these experiments, the reaction zone was seen to separate from the leading shock wave 3–4 m from the initiator charge. This was followed approximately 10 ms later by an explosive recoupling of the reaction zone with the leading shock. A peculiarity of the process was the fact that recoupling occurred in a mixture that was only slightly precompressed; i.e. the leading shock wave had decayed to a strength of only 1–2 atm by that time. This was attributed to the rapid amplification of pressure waves in the induction-time gradient present in the decoupled region. These observations provided the impetus for a second series of experiments in which induction-time gradients were intentionally created by detonating a subcritical charge inside the cloud (i.e. less than 40% of the critical charge mass). As the shock wave and reaction zone separated, a second relatively weak shock wave (2–3 atm) was propagated through the decoupled region. This shock was produced by the detonation earlier in time of a high-explosive charge positioned outside the cloud. Rapid amplification to detonation was observed in cases where the second shock arrived after a suitable induction-time gradient had been formed. The reinitiation process was highly three-dimensional and typically occurred in the region of the triple point associated with the reflection of the initial shock wave from the ground.

1.3.3 Sprays Created Explosively in "Trough Tests"

Benedick et al. [19] at Sandia National Laboratories utilized a high-explosive technique for liquid fuel dispersal. In these so-called trough tests, a piece of pentaerythritol tetranitrate (PETN) detonating cord was laid along the bottom vertex of a horizontally positioned V-shaped steel trough (a piece of angle iron) and the trough was subsequently filled with liquid fuel. Propylene oxide, hexane, decane, and an unspecified low-vapour-pressure nitrated hydrocarbon were investigated; however, the main emphasis was on hexane and decane. When the detonating cord was initiated, the resulting explosion propelled the liquid fuel vertically upward to form an aerosol cloud. Clouds of globally stoichiometric composition were obtained when the ratio of fuel mass to explosive mass was about 200, on the basis of the known quantity of fuel dispersed and the observed cloud volume from high-speed imagery. Under these conditions, the cloud was typically 7 m high by about 1.5-m average width (the maximum width was 2.1 m). The length of the trough was 9 m, giving a total cloud volume of about 100 m^3. No direct measurement of droplet size was made by these investigators. However, on the basis of the earlier work of Mayer [92] and Anderson and Wolfe [10], the droplet size was estimated to reside in the 20–50-μm range.

Initiation of detonation was brought about by a high-explosive charge consisting of Detasheet™ mounted on a vertical sheet of plywood at one end of the cloud. The charge mass was typically 0.85 kg. For the less sensitive fuels, the first 1.5-m length of the trough was filled with propylene oxide to serve as a driver. Once initiated, the propylene oxide–air detonation transmitted to that portion of the cloud consisting of the test fuel. This approach minimized the degree of initiator overdrive. For insensitive fuels requiring a larger cloud volume, two parallel troughs spaced 1.2 m apart were used. The instrumentation included smoke foils and high-speed cinematography. The setup is shown in Fig. 1.19.

When high-vapour-pressure propylene oxide was dispersed, the cloud became transparent quickly, indicating that the aerosol had vaporized. The resulting detonation was practically in the gas phase and the propagation velocity of 1,600 m s^{-1} was typical of fuel–air mixtures. In the case of the low-vapour-pressure nitrated fuel, no change in cloud transparency was noted

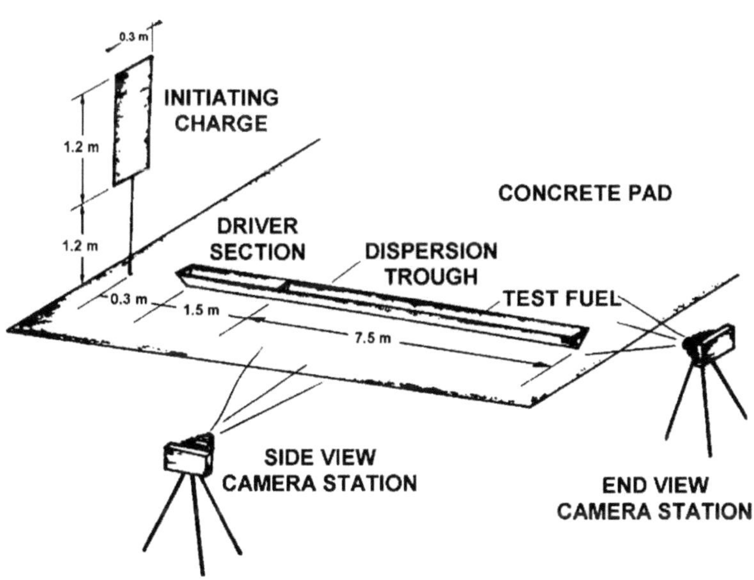

Fig. 1.19. The test setup and camera stations for the Sandia National Laboratories trough tests [19]

following dispersal, suggesting that insignificant fuel vaporization had occurred. The observed detonation was thought to be propagating in a true heterogeneous system.

A completely transparent cloud may be good evidence that complete fuel vaporization has taken place. However, when the cloud is opaque, there is no way of visually assessing the mass fraction of fuel that has actually vaporized. The method of determining vapour content on the basis of temperature measurements described earlier [120] provides more quantitative information.

The main focus of the Sandia National Laboratories experiments was to determine the detonability of hexane and decane aerosols in air with emphasis on clarifying the role of the vapour phase. These fuels were selected because they have very different vapour pressures, yet the detonability of the corresponding gaseous fuel–air mixtures is nearly identical. The vapour pressures for hexane and decane at 20℃ are 16.9 and 0.20 kPa, respectively. Heated detonation tube testing at Sandia National Laboratories [131] had previously shown that both fuel–air mixtures have a detonation cell size of about 50–60 mm at 100℃ under stoichiometric conditions.

Hexane and decane were dispersed in the two-trough configuration and a nitrated hydrocarbon driver was used. The time required for full development of the clouds was approximately 100 ms. At that point in time, the clouds remained predominantly in aerosol form and the driver detonation was unable to transmit to the hexane–air or decane–air mixtures. However, in the case of hexane fuel, when the delay time was increased from 100 to 150 ms, the high-speed film showed that the hexane–air portion of the cloud had become transparent, while the low-vapour-pressure driver portion remained opaque. In this case, successful transmission of the driver detonation to the hexane–air mixture was observed. This was confirmed by smoke foils and cinematography (see Fig. 1.20). The tests were repeated for increasing delay times up to 500 ms, with the same outcome in every test. In the case of decane fuel, the high-speed film showed that the cloud remained in aerosol form for a delay time of 200 ms, and the driver detonation was unable to transmit to the decane–air mixture. No cellular structure was recorded on the smoke foil. When the test was repeated for delay times up to 400 ms, the outcome did not change. These results are consistent with those reported for kerosene by the Kurchatov Institute. Clearly, the cloud size and initiator mass in the Sandia National Laboratories experiments were too small for decane–air detonation to occur.

These tests showed convincingly that the vapour phase plays an important role in determining the detonability of insensitive fuels. While concluding that “detonation can only occur in the vapour phase”, the investigators, Benedick et al. [19], did acknowledge that low-vapour-pressure nitrated fuels can be detonated in aerosol form and added that “the precise mechanisms underlying the detonation of aerosols were not understood.” Presumably, low-vapour-pressure fuels can be detonated if the size of the cloud and the mass of the initiator are sufficiently large that the droplets can be reduced to a size, under

Fig. 1.20. Selected frames from a high-speed film (500 frames per second) showing successful initiation of detonation in a hexane aerosol cloud 150 ms after fuel dispersal. Time is in milliseconds. (From Benedick et al. [19] with special thanks to S. Tieszen)

the influence of aerodynamic stripping, where meaningful vaporization takes place in the postshock region despite the low-vapour-pressure nature of the fuel. In other words, the scale of the experiment is all important.

Benedick (W. Benedick, Sandia National Laboratories, private communication, 1988) has revealed a number of other interesting observations about these trough tests that were not discussed in the paper by Benedick et al. [19]. For example, explosively dispersed low-vapour-pressure fuels produced much higher fuel spray clouds than high-vapour-pressure fuels (as high as 12 m for the same detonating cord size). This was likely the case because the fuel was not evaporating during dispersal and the larger droplets were better able to penetrate the atmosphere. For this reason, the explosive dispersal charge had to be reduced in size to obtain globally stoichiometric clouds. In an unreported test involving hexane fuel, it was possible to initiate a hexane aerosol cloud prior to 100 ms by preshocking the cloud with two arrays of detonating cord positioned on either side of the cloud. The combination of increased air temperature and enhanced droplet shattering/vaporization were likely responsible for this result. In other unreported tests, attempts to initiate dipentene aerosols were unsuccessful.

1.3.4 Sprays Created by Explosive Dispersal from Canisters

The first documented report of a FAE device producing an unconfined detonation in air was by Gey and Mygaard [65] in 1961 at the US Naval Ordnance Test Station, now known as the Naval Air Weapons Center in China Lake, California. Ethylene oxide fuel was employed because it had the widest reported detonability limits in air. A high-explosive charge was used to disperse the fuel into an aerosol cloud, and a high-explosive secondary charge was employed to initiate cloud detonation. The US Navy began weaponizing this technology in 1962 and demonstrated Fuel–Air Explosive Weapon System Number 1 (FWS-1) in 1964. In an independent study carried out at about the same time at Eglin Air Force Base in Florida, Howard (J. Foster, AFATL, Eglin Air Force Base, private communication to L. Josephson, 1978) dispersed propane to form a cloud consisting of vapour and aerosol by exploiting the high vapour pressure of propane for dispersal purposes. Again, a high-explosive secondary charge was used to induce detonation of the cloud. The US Air Force commenced weaponizing this technology in 1964, eventually leading to the BLU-72/B air-dropped munition. Details about these and follow-on studies are not available.

Detonation studies involving explosively dispersed liquid fuels contained in thin-walled cylindrical canisters were first carried out at Defence R&D Canada – Suffield (DRDC Suffield) (formerly, Defence Research Establishment Suffield) during the early 1980s by Moen and Ward [95]. This work was undertaken in support of military engineering applications. Figure 1.21 shows a typical 50-l research device based on a commercially available polyethylene container. The device contains 43 kg of propylene oxide. The optimum standoff (i.e. distance between the ground and the bottom of the canister) for a static device in land mine neutralization studies has been shown to be 1 m. For this standoff, the fully developed cloud is in intimate contact with the ground. In some tests, the canister was scored over its full length in the vertical direction at 45° intervals in azimuth to promote structural weaknesses that result in uniform canister breakup. This is a standard technique used in FAE device design [120]. As shown in the inset in Fig. 1.21, the canister is fitted with an axial fuel dispersal, or "burster", charge consisting of a glass tube filled

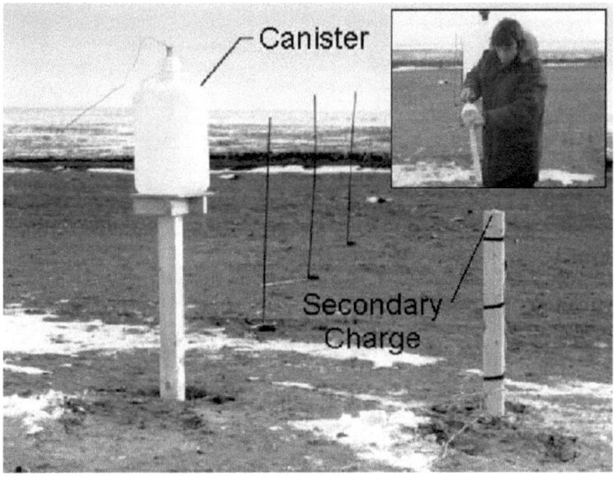

Fig. 1.21. A 50-l fuel–air explosive (FAE) canister device used to assess the utility of FAEs in neutralizing land mines (shown tethered to the vertical stakes). The *inset* shows the burster charge

with DM12, a PETN-based high explosive. A cloud initiator, or "secondary", charge is mounted on a post 1.25 m high at a radial position of 1–2 m. Exploding bridge-wire detonators are used to initiate the charges. The time between fuel dispersal and cloud detonation is typically 100 ms for this size of device. The instrumentation used in these experiments included pressure transducers inside and outside the cloud, as well as high-speed cinematography and video. A typical video sequence showing explosive fuel dispersal and cloud detonation is shown in Fig. 1.22.

Laser fluorescence techniques have been used at the Institute Saint-Louis to understand the canister breakup process and to investigate the various phases of explosive fuel dispersal [120]. A holographic method has also been used to measure droplet sizes during dispersal.

Figure 1.23 shows a plot of the cloud mean radius and height as a function of the elapsed time from detonation of the burster charge for ten canister trials at DRDC Suffield. All measurements were taken from high-speed film records with the aid of a high-resolution film reader. Bearing in mind the errors involved in identifying the "edges" of the clouds, the data are consistent from test to test, exhibiting a scatter of less than $\pm 10\%$ about the mean dimension at any point in time. The plot includes previously unpublished data by one of the authors of the present review (tests 1–8) and by Moen and Ward for a fuel-to-burster (F/B) mass ratio of 100 which are in good agreement. The data for higher (150) and lower (80) values of F/B reported by Moen and Ward [95] emphasize the sensitivity of the final cloud radius to the size of the burster charge. As expected, the final radius increases with decreasing values of F/B. The cloud height, on the other hand, is not particularly sensitive to the burster charge size and appears to be more a function of the canister standoff. Similar data reported by Axelsson and Berglund [14] from two tests involving slightly

Fig. 1.22. Frames from a high-speed video sequence showing dispersal of propylene oxide and subsequent cloud detonation

smaller devices (38 kg) are also included in Fig. 1.23. Assuming that the 0.7 kg of oxygen-balanced ammonium nitrate used in their burster charges is equivalent to 0.7/1.35 kg of DM12 explosive (based on cylinder test data), the comparable F/B is about 70. Thus, their data are consistent with the F/B trends indicated by the DRDC Suffield data. Another point of interest in Fig. 1.23 is that the unscored canister (test 1) yielded the smallest-diameter cloud for $F/B = 100$ in the previously unreported tests by one of the present authors, suggesting that casing effects played a role, even though the canister walls were relatively thin.

The data in Fig. 1.23 reveal final cloud radii of approximately 7.5, 6.25, and 5.5 m, respectively, for F/B values of 80, 100, and 150. Assuming that the clouds are shaped like a right circular cylinder measuring 2 m high and having the above-mentioned radii, the final cloud volumes are 353, 245, and 190 m^3, respectively. A stoichiometric cloud containing 43 kg of propylene oxide has a

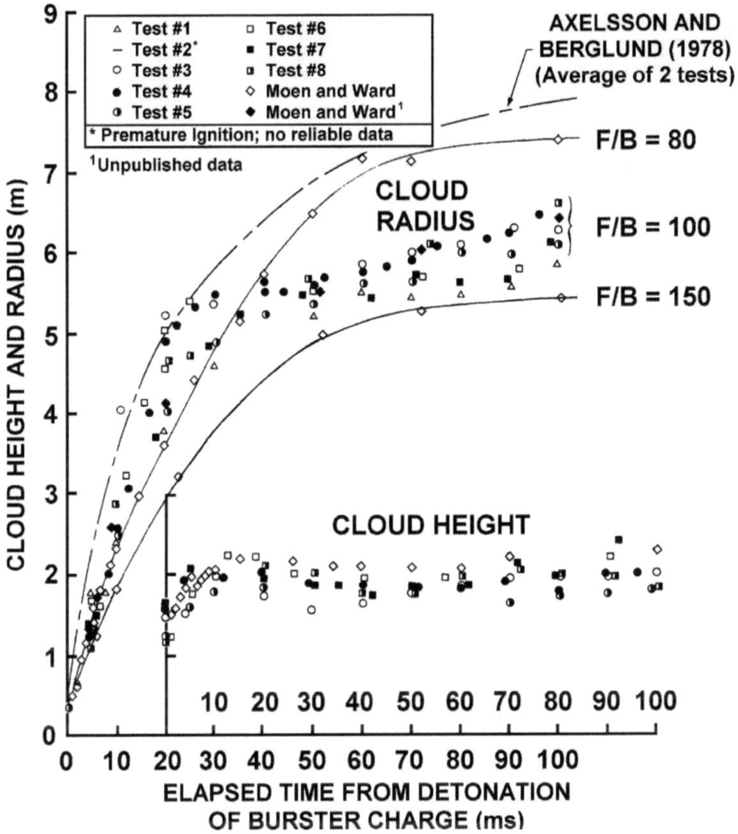

Fig. 1.23. Data showing the development of propylene oxide clouds during explosive fuel dispersal as a function of the burster charge size

volume of $356\,\mathrm{m}^3$ (assuming the fuel remains in droplet form); therefore, the largest fuel dispersal charge produced a cloud near stoichiometric conditions. The clouds were fuel-rich for smaller dispersal charges. The corresponding equivalence ratios are 0.99, 1.45, and 1.87. The ambient pressure and temperature for all tests were 93 kPa and 10–20℃, respectively. All clouds were successfully detonated using a 200-g secondary charge. The best blast performance was obtained for $F/B = 100$. In general, the overpressures were lower near the centre of the cloud, suggesting that the fuel–air mixture was leaner in that region. This conclusion is supported by observations from overhead cinematography that show the clouds are more transparent in the interior owing to more complete fuel vaporization. This gives the cloud a toroidal appearance at late times during fuel dispersal.

In some military applications, reliable cloud-to-cloud transmission of detonation is a requirement. Figure 1.24 consists of frames from a cinematographic record showing simultaneous fuel dispersal from four canisters and subsequent

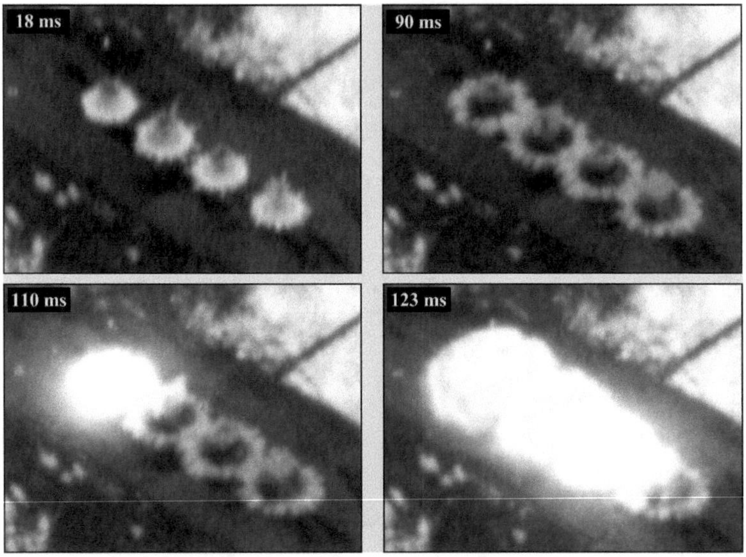

Fig. 1.24. Frames from a high-speed film record showing fuel dispersal from four canisters followed by detonation of four overlapping clouds

detonation initiated by a single charge positioned at one end of the array. The fuel in this test was propylene oxide and the toroidal appearance of the clouds is readily apparent. The collision between the two detonation waves moving circumferentially through the toroidal volume creates a strong Mach reflection that promotes initiation of the adjacent cloud. The ease of transmission depends on the detonability of the fuel–air mixture. The required cloud overlap increases as the mixture sensitivity decreases (a 1-m overlap is sufficient for propylene oxide).

The canister tests summarized above employed large secondary charges to ensure reliable initiation. The detonability limits for propylene oxide sprays in air were investigated in a series of trials employing smaller cylindrical canisters [105]. These were of 10-cm diameter and 38-cm height, for a contained fuel volume of approximately 3 l. Each canister was fitted with an axial burster charge consisting of a glass tube filled with 35 g of DM12 explosive. For these small canisters, the mass of the polyethylene and glass components is significant (approximately 15% of the charge mass). These items are therefore included as part of the "fuel" or propelled mass in the calculation of F/B. The resulting F/B of 85 produces clouds that are near stoichiometric in composition. The canister was placed on a wooden stand and the secondary charge, consisting of a sphere of DM12 explosive, was mounted on top of a post positioned 1 m from the canister axis.

Successful initiation of detonation was observed for secondary charge masses of 10 (on two occasions), 25, 50, and 100 g. Failure of detonation was observed for secondary charge masses of 5, 8, and 9 g. Therefore, the critical charge mass for direct initiation of detonation in a globally stoichiometric, propylene oxide, droplet–air cloud is 10 g of DM12, which contains about 8.5 g of PETN. This is somewhat larger than the 3.5 g of Tetryl^{TM} critical charge mass reported by Vanta et al. [137]. Experiments carried out by Murray et al. [101] using disks of Detasheet^{TM} explosive to initiate ethylene–air detonations have shown that the critical charge mass for a stoichiometric mixture lies between 9 and 18 g. Given that only about two thirds of the Detasheet^{TM} is energetic material (mostly PETN), the critical charge mass for this mixture becomes 9 ± 3 g of PETN. These results indicate that propylene oxide (most of which is assumed to be in the vapour phase) and ethylene have similar detonation sensitivities; therefore, the critical tube diameter for stoichiometric propylene oxide and air is expected to be very close to that for stoichiometric ethylene–air. Moen et al. [96] reported the latter to be 43 cm. According to Murray et al. [99], the critical thickness of a cloud resting on the ground is about one quarter of the critical tube diameter, or about 10–11 cm in this case.

The critical charge mass required for direct initiation is very sensitive to the fuel concentration in the vicinity of the charge. In a recent study at DRDC Suffield [107] to assess the utility of a triple FAE array as an inexpensive blast simulator, as first proposed by Shamshev [122], it was essential that all three clouds be initiated promptly and centrally to guarantee simultaneous spherical detonations and symmetrical blast wave interactions. For this purpose, the secondary charges were positioned on posts directly below and vertically aligned with the fuel canisters. Canisters of 2-, 10-, and 66-l volume were used in the experiments. The respective burster charge masses were 20, 100, and 634 g of C4 explosive (91% RDX), corresponding to F/B ratios of 86, 86, and 89. The time delays for the three canisters were 60, 80, and 110 ms, respectively. Frames from a high-speed video record showing fuel dispersal from 10-l canisters, simultaneous cloud detonations, and the formation of a strong Mach stem along the bisector of the array are shown in Fig. 1.25. When the secondary charge is positioned centrally in this fashion, the fuel–air mixture surrounding the charge is very lean and much more difficult to detonate. In the case of the 10-l canister, for example, it was necessary to increase the size of the secondary charge to 250 g before detonation could be initiated reliably. With this size of charge, explosion centres typically formed at one or two locations inside the toroidal region of the cloud. It was necessary to increase the charge to 500 g before initiation occurred uniformly throughout the toroidal volume.

Although propylene oxide is an acceptable fuel for FAE applications, it does have some disadvantages in terms of toxicity, high vapour pressure, and low boiling point. Brosinsky et al. [33] investigated hexyl nitrate, ethyl hexyl nitrate, nitromethane, and nitropropane as alternative fuels. The first two candidates were selected on the basis of an initial screening of potential fuels performed by Yee (T. Yee, Naval Weapons Center, China Lake, CA, USA, private communication, 1985). The key requirements for alternative fuels were ready

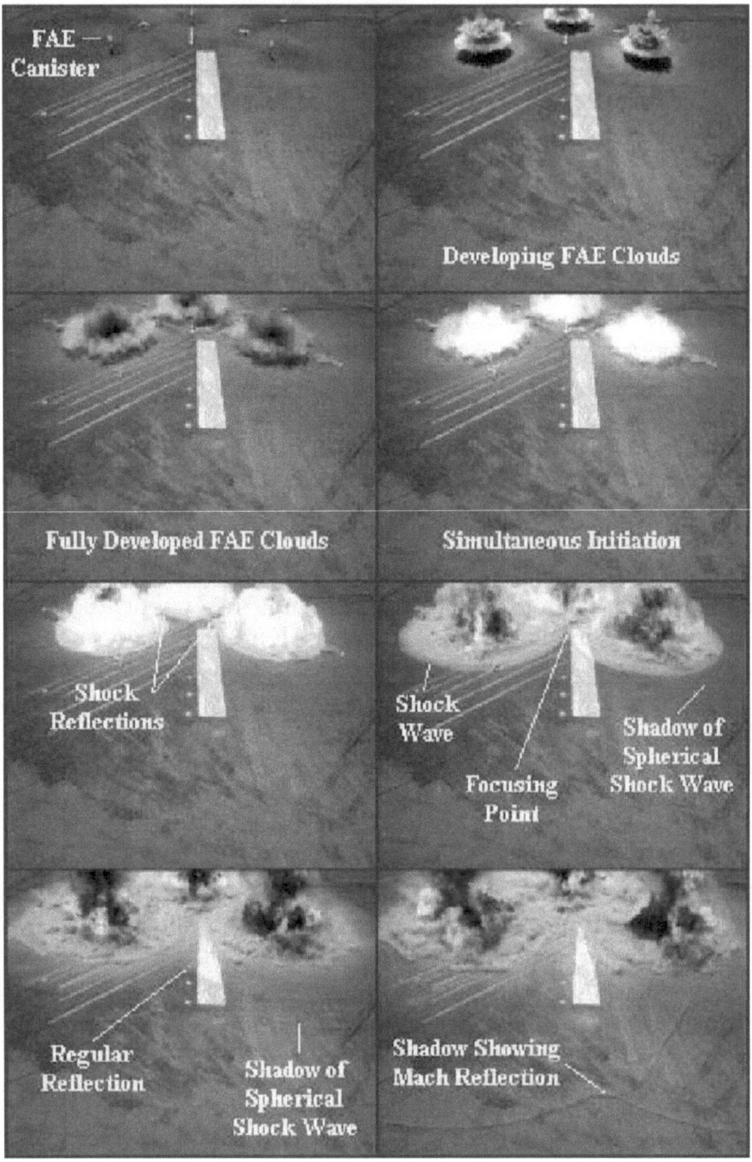

Fig. 1.25. High-speed sequence showing the use of a triple FAE array as a nuclear blast simulator [107]

availability, low cost, safety in handling, and acceptable blast performance, with preference being given to low-vapour-pressure fuels. Hexyl nitrate and ethyl hexyl nitrate are common diesel ignition improvers used in the automotive industry. Both are readily available, less toxic than propylene oxide, and have lower vapour pressures and higher boiling points than propylene oxide.

Liquid nitromethane is very sensitive to detonation. Studies were therefore undertaken that showed the sensitivity of nitromethane can be significantly reduced by diluting it with other nitroparaffins such as nitroethane or nitropropane [104]. For this reason, nitropropane was included in the study of alternative FAE fuels.

The tests were conducted using the same setup as shown in Fig. 1.21, with one exception. For tests involving nitromethane, it was necessary to install an annular buffer around the dispersal charge to prevent high-order detonation of the fuel during dispersal. This buffer took the form of a 10-cm-diameter glass tube positioned coaxially around the burster charge and filled with propylene oxide. The test conditions are summarized in Table 1.1. The ambient pressure and temperature for all tests were 92–93.6 kPa and 5–22℃, respectively. Two tests with propylene oxide were included in the series to confirm consistency with previous results.

The radius and height of the visible clouds were plotted against time into fuel dispersal for the various fuels in the same manner as that shown for propylene oxide in Fig. 1.23. When this was done, it was found that the final cloud

Table 1.1. Test conditions and results for alternative fuels study employing nominal 50-l canisters [33]

Trial reference	Fuel type	Vapour pressure (kPa at 20℃)	Fuel mass (kg)	Fuel-to-burster mass ratio (F/B)	Canister standoff	Secondary charge mass (g)	Time delay (ms)	Result
1/86	PO	58.8	43.0	100	1.0	200	100	Detonation
22/85	PO	58.8	13.7	100	1.0	200	100	Detonation
19.86	HN	0.08	48.5	100	1.0	200	100	Detonation
21/86	HN	0.08	48.5	200	1.0	200	100	Dctonation
22/86	HN	0.08	48.5	250	1.0	200	100	Detonation
15/86	EHN	0.03	48.0	100	1.0	200	100	Detonation
16/86	EHN	0.03	48.0	150	1.0	200	100	Detonation
17/86	EHN	0.03	48.0	200	1.0	200	100	Detonation
18/86	EHN	0.03	48.0	250	1.0	200	100	Detonation
2/86	NM	3.6	57.0	100	1.0	500	100	Marginal
3/86	NM	3.6	57.0	150	1.0	500	100	Detonation
4/86	NM	3.6	57.0	200	1.0	500	100	Marginal
5/86	NM	3.6	57.0	100	0.5	500	100	Marginal
6/86	NM	3.6	57.0	150	0.5	500	100	Marginal
7/86	NM	3.6	57.0	200	0.5	500	100	Marginal
9/86	NM	3.6	57.0	200	0.5	200	100	Marginal
10/86	NM	3.6	57.0	100	0.5	200	50	Detonation
27/86	NP	1.3	49.6	100	1.0	200	100	Failure

PO propylene oxide, *HN* hexyl nitrate, *EHN* ethyl hexyl nitrate, *NM* nitromethane, *NP* nitropropane.

radii were not strongly dependent on the value of F/B over the range from 100 to 250. The final radii and heights fell in the ranges 6.25–7.0 and 1.8–2.2 m, respectively. The clouds were also found to be larger than those for propylene oxide for $F/B = 100$, particularly at late times during dispersal. These results are likely attributable to differences in the vapour pressure of these fuels. The vapour pressure for propylene oxide is 1–3 orders of magnitude higher than for the other fuels, with all other physical properties being quite similar. This hypothesis is supported by the high-speed film records which showed little change in the opacity of the clouds during fuel dispersal. Apparently, evaporation of propylene oxide from the droplets during fuel dispersal has a considerable impact on the droplet dynamics and, hence, the ultimate cloud size. These findings are consistent with those for low-vapour-pressure fuels of Benedick et al. [19] (W. Benedick, Sandia National Laboratories, private communication, 1988). Assuming the cloud shape is that of a right circular cylinder, the above-mentioned dimensions give total cloud volumes between 203 and 338 m^3, corresponding to normalized cloud volumes between 3.9 and 6.5 $m^3\,kg^{-1}$ of fuel. These figures compare with calculated values of 6.2, 7.2, and 1.4 $m^3\,kg^{-1}$ for stoichiometric mixtures of hexyl nitrate, ethyl hexyl nitrate, and nitromethane with air. Accordingly, the clouds were stoichiometric to fuel rich for the diesel ignition improvers and very lean for nitromethane.

The results in Table 1.1 confirm that the initiation requirements for hexyl nitrate and ethyl hexyl nitrate are not excessive. The clouds were all reliably initiated with 200-g high-explosive charges after a 100-ms time delay despite these fuels having a low vapour pressure. Again, these results are consistent with those of Benedick et al. [19] (W. Benedick, Sandia National Laboratories, private communication, 1988) for low-vapour-pressure nitrated fuels. The results for dispersed nitromethane were less impressive. There was considerable scatter in the peak pressure and impulse data. The magnitude of these parameters was also much smaller on average than for the other fuels, but this can be explained by the fact that the mixtures were very lean. Detonation of the nitromethane clouds was non-uniform and characterized by large variations and multiple peaks in the P–t profiles. The high-speed films also showed that detonation occurred in isolated pockets of the cloud at various times. This marginal behaviour was likely the result of the clouds being too small for these lean mixtures. In the single test with nitropropane fuel, the 200-g secondary charge failed to initiate detonation.

A number of other FAE fuels have been evaluated in Russian studies. Shamshev [122] has described several liquefied fuels including Aerozine, a 50:50 mixture of hydrazine and 1,1-dimethylhydrazine. This mixture was described as a liquid analogue for methylacetylene-propadiene (MAPP) gas. Vasil'ev et al. [142] reported aerosol detonation in mixtures of hydrazine, methylhydrazine, 1,1-dimethylhydrazine, 1,2-dimethylhydrazine, trimethylhydrazine, and Aerozine in oxygen and air environments. Some of these mixtures are characterized by detonation cell sizes considerably smaller than those for the most sensitive hydrocarbon fuels. Shamshev [122] has also reported that isopropyl nitrate (IPN) and BN are effective sensitizers for liquid fuels. In the case of IPN, it was stated that the detonability limits for some fuels can be increased by a factor of 3. The effectiveness of IPN as a sensitizer has also been demonstrated by Zhang et al. [156]. In studies of shock ignition and detonation of hydrocarbon–air–IPN mixtures in the vapour phase, it was found that the ignition delay and detonation cell size had maximum values for hydrocarbon–air, minimum values for IPN–air, and varied monotonically between these two extremes. In tests involving the explosive dispersal of IPN by the authors in 1992, the droplet cloud was seen to turn black 50 ms into the event, suggesting that thermal decomposition of the fuel was taking place.

Christensen and Hermansen [43] conducted an extensive study of FAE fuels and sensitizers using canister devices. The canister size (1–4 l), burster charge mass ($40 \leq F/B \leq 120$), time delay (40–250 ms), and secondary charge mass (50–100 g of C4) were varied in the experiments. The results for propylene oxide were similar to those already detailed for the DRDC Suffield programme. The sensitizing effect of *n*-butyl nitrite on gasoline fuel was assessed for a 100-ms time delay in some tests. For neat gasoline, detonation failure occurred for secondary charges of 100 and 300 g, but initiation was successful for 700- and 1,000-g charges. Four tests were conducted with 87.5:12.5 mixtures of gasoline and *n*-butyl nitrite and a 100-ms time delay. Detonation initiation was successful for 200-, 300-, and 1,000-g secondary charges, but failure occurred for a 100-g charge. When the percentage of BN in the blend was doubled, a 100-g secondary charge was sufficient to initiate detonation. The results for neat gasoline are approximately consistent with the findings of the Kurchatov Institute described earlier. In those trials, gasoline–air detonation was initiated with a 300-g secondary charge, but the longer time delay permitted a higher concentration of vapour to accumulate.

Successful initiation of dispersed octyl nitrate fuel was also achieved using a 100-g secondary charge and a 70-ms time delay. However, failure of detonation was observed for ether and xylene fuels for the same test conditions. In three tests employing 1-kg secondary charges, dispersed ethanol and propanol fuels failed to detonate for a 100-ms time delay, while turpentine failed to detonate for a 150-ms time delay. Initiation of dispersed nitromethane was also unsuccessful for a 300-g secondary charge using a 100-ms time delay.

In a final round of experiments by these authors, the detonability of unsensitized and sensitized diesel fuel was assessed for 4-l canisters and a 100-ms time delay. Detonation failure occurred for diesel fuel using a 700-g secondary charge. The earlier work at the Kurchatov Institute suggested that both the cloud size and the secondary charge were too small for diesel–air detonation to occur. Failure was also observed for a 50:50 mixture of diesel fuel and octyl nitrate and a 75:25 mixture of diesel fuel and *n*-butyl nitrite in tests involving 100-g secondary charges.

In a follow-on study by Christensen [42], successful detonation of dispersed propylene oxide was achieved for canister volumes of 50, 100, and 1,000 l.

Initiation of detonation in the above-described canister studies has been typically brought about by the shock wave from a secondary charge. However, other methods of initiation have been reported. For example, initiation by hypervelocity projectiles has been investigated by Vasil'ev for fuel–oxygen mixtures [138–140] and by Chernyi et al. [41] for fuel–air mixtures. Some of the work by Vasil'ev [139] makes reference to experiments involving fuel sprays, but no details are provided. A large number of high-speed projectiles could be used to achieve multipoint initiation or initiation along a locus. This could be important for generating optimal blast effects in certain cloud geometries. See, for example, the multipoint initiation study of toroidal clouds by Koren'kov and Obukhov [77]. The bow shocks produced by multiple projectiles could also be instrumental in generating a multitude of local induction-time gradients that could lead to the onset of detonation in the manner observed by Alekseev et al. [5, 7]. In practice, a fragmentation device could be used to produce a large number of projectiles.

Some very large FAE experiments have been performed in Russia. Shamshev [122] has made reference to tests in the open atmosphere measuring several hundred metres in size. Dorofeev et al. [53–55] have reported the results from some exceptionally large FAE canister tests. In these experiments, gasoline, kerosene, and diesel fuels were contained in steel cylindrical vessels having an aspect ratio near unity. Each cylinder was fitted with an internal TNT charge that was detonated to disperse the fuel into a large heterogeneous cloud. Fuel masses ranging from 0.1 to 100 metric tons were dispersed to study the thermal effects and blast characteristics of fuel-rich deflagrations and detonations. The fuel concentration for the deflagration experiments was 1,000–2,000 $\mathrm{g\,m^{-3}}$, approximately 10–20 times the stoichiometric concentration.

For the detonation experiments, the fuel concentration was 200–300 $\mathrm{g\,m^{-3}}$ or about 2–3 times the stoichiometric concentration. A high-explosive secondary charge was used to initiate detonation. The concentration limits for heterogeneous detonation appeared to be much wider than those for gaseous detonation. This conclusion is in agreement with the observation by Zhang et al. [155] that the rich concentration limit for dust detonations is very high. Gubin and Gelfand [68] have shown that interphase heat and mass transfer in the relaxation zone of a two-phase gas/particle flow can have a significant effect on the detonability of the mixture. The extent depends on the size and concentration of the condensed-phase particles. It is shown, for example, that the presence of large particles (more than 10 μm) which are not able to heat up noticeably behind the shock front can lead to a shortening of the induction period and therefore the reaction zone. This process leads to a widening of the detonability limits.

The detonation of fuel-rich clouds is accompanied by strong heat radiation. Scaling relationships for total radiation energy, heat flux from the fireball surface, duration of the thermal effect, and maximum fireball size have been proposed by Dorofeev et al. [54] on the basis of direct thermal measurements.

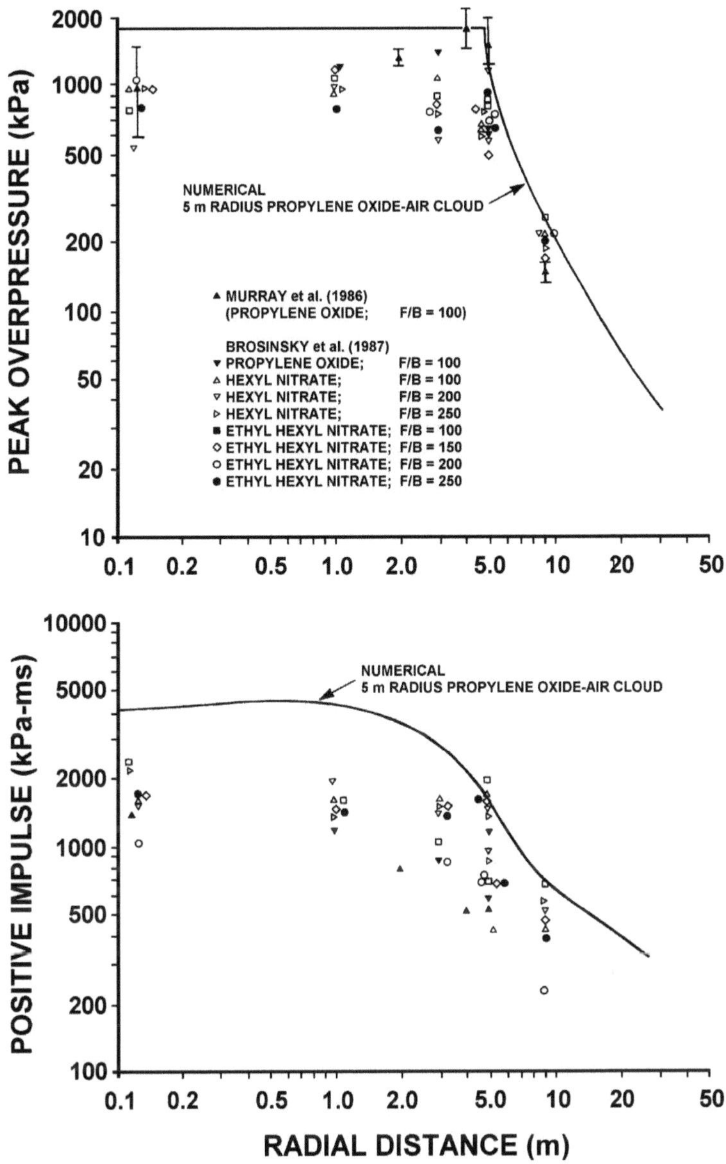

Fig. 1.26. Peak overpressure and positive-phase impulse versus range for FAE clouds produced by 50-l canister devices at Defence R&D Canada – Suffield [33]

A final point to be made in this subsection is the unique and scalable blast field produced by a detonating FAE cloud. Figure 1.26 shows peak overpressure and positive-phase impulse as a function of range for 50-l canisters containing propylene oxide, hexyl nitrate, or ethyl hexyl nitrate fuels [33]. The pressure gauges in these tests were mounted flush with the ground. The solid line in the plots is a numerical prediction for a 5-m-radius hemispherical cloud containing stoichiometric propylene oxide–air. The corresponding volume ($260\,m^3$) is close to that observed from film records for an actual cloud ($F/B = 100$). It can be seen in the figure that the peak overpressure inside the cloud corresponds to the detonation pressure and that the transmitted air shock decays rapidly outside the cloud. One of the attractive features of FAEs is that the impulse profile inside the cloud is relatively flat. This is in contrast to the region of low impulse (known as a "skip zone") that characterizes the blast field produced by high-explosive charges.

Figure 1.27 shows the overpressure and positive-phase impulse data reported by Dorofeev et al. [53, 55] for canisters containing three types of fuel and ranging in mass from 0.1 to 100 metric tons. The parameter C/C_o denotes the fuel mass used in a given test divided by the stoichiometric fuel mass for the same cloud volume. The blast from a fuel-rich cloud is the same as for a stoichiometric cloud because the energy released during the detonation phase of the event is governed by the availability of oxygen in the cloud. The excess fuel produces an intense fireball, but it does not contribute to the blast generated by the initial detonation. The dashed lines in the plots are from calculations for a gaseous detonation in a stoichiometric cloud. The peak overpressure and impulse data from these experiments are seen to scale extremely well over the full range of fuel type, fuel mass, and mixture stoichiometry. Of course, there would be a limit to how rich one could make the cloud and still have this cube-root volume scaling hold true. For exceedingly rich clouds, a significant amount of energy would be consumed through fuel vaporization. The scatter in the data is due to the variable cloud shape from test to test. The near-field blast is dominated by the cloud geometry, a fact duly exploited by weapon designers [77, 122].

1.3.5 Sprays Created by Explosive Dispersal from Hoses

A FAE line charge has been investigated at DRDC Suffield as a candidate technology for minefield breaching operations during combat. The line charge is a hose filled with fuel and fitted with an internal axial fuel dispersal charge along its length. The feasibility of producing a long hemicylindrical FAE cloud that could be detonated was first demonstrated by Ward et al. [148] in 1982. These authors employed propylene oxide fuel contained in light-walled (0.25- or 0.38-mm) polyethylene tubing of 45-mm diameter. The line charges were 7–10 m in length. Moen and Ward [95] later extended the study to include charges up to 25 m in length. This was done because of concern that the differential between the detonation velocities in the dispersal charge (approximately $6{,}000\,m\,s^{-1}$) and the FAE cloud (approximately $1{,}700\,m\,s^{-1}$) could cause an increase in the effective dispersal time that might be detrimental to the blast performance.

In support of the line-charge development, a series of tests were carried out in which water contained in the same 45-mm tubing was explosively dispersed

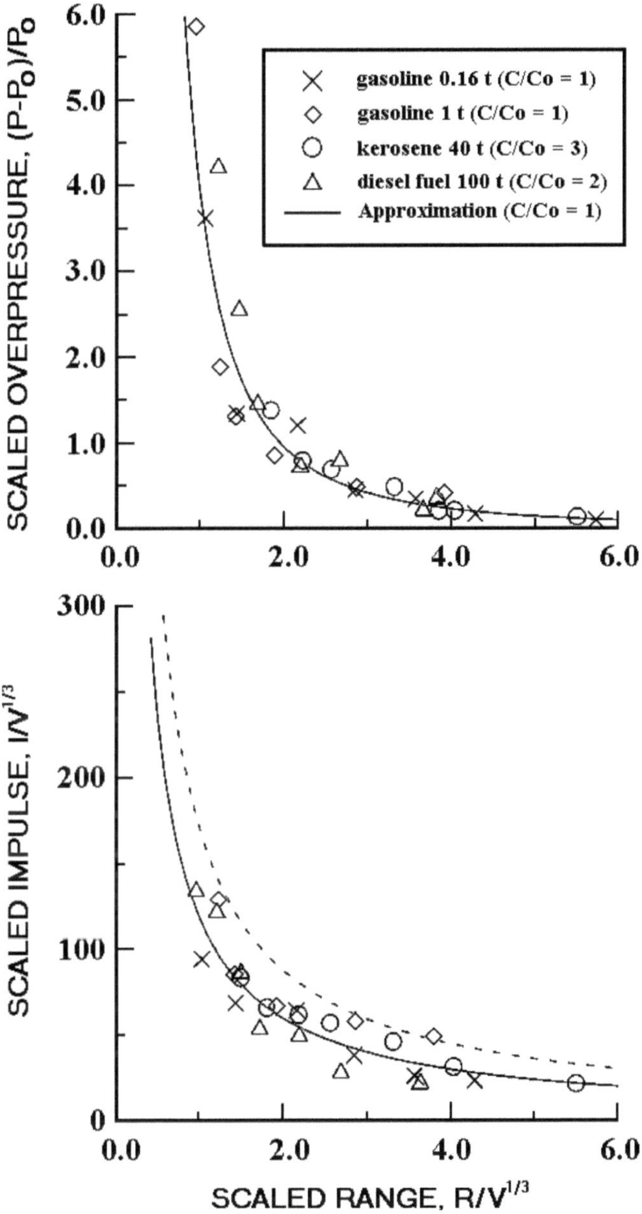

Fig. 1.27. Scaled overpressure and positive-phase impulse versus scaled range for very large FAE canister devices at the Kurchatov Institute [53]

into the atmosphere to study cloud development and the evolution of droplet size distribution. The tests were conducted for F/B values of 50, 75, and 150. A photographic technique was used to measure droplet size at discrete locations inside the clouds. The method involved a tube of 3.18-cm inside diameter equipped with a collimated light source at one end and a high-speed camera at the opposite end. The tube contained a transverse slit. This light-tube–camera apparatus was positioned parallel to the line charge so that droplets could pass through the slit during dispersal. The droplets were found to be spherical in shape toward the end of the dispersal phase (approximately 100 ms) with mean sizes in the 125–200-μm range. Complete details were provided by Cooke et al. [44–46].

An investigation of 65-mm-diameter line charges was subsequently carried out by Murray and Gerrard [102] for charges up to 61 m in length. Successful detonation of 200-m-long clouds produced by 100-mm-diameter line charges was later demonstrated by the same authors.

The most comprehensive of the line-charge studies, for 65-mm-diameter charges, will be described here in some detail. Various commercially available or custom-built industrial or fire-fighting hoses were used for this purpose. Propylene oxide fuel was employed in most experiments, with ethyl hexyl nitrate being used in selected tests. The burster charges were configured from commercially available detonating cord (60:40 TNT/RDX). Both one-strand and two-strand burster charges having a total explosive mass between 32.0 and 85.2 $g\,m^{-1}$ were tested. The detonating cords were bound together when two-strand configurations were employed. Detonation of the clouds was achieved by a pair of DM12 high-explosive secondary charges mounted on posts ranging in height up to 1 m. The secondary charge masses varied between 200 and 500 g. These were detonated after a time delay of between 100 and 200 ms from the commencement of fuel dispersal. Large charges were used to ensure initiation of off-stoichiometric compositions. The instrumentation included high-speed cinematography and a series of ground-level pressure transducers for blast measurements. A total of 40 tests were conducted. Various technical difficulties were encountered in 17 of these developmental tests. The interested reader is referred to [102] for details about and solutions to these problems.

A summary of the successfully executed tests appears in Table 1.2. Sustained detonation, in which the wave propagated the full length of the cloud, was observed in 13 experiments. Partial detonation was observed in six tests. Here, the wave appeared to be initiated successfully, but failed after propagating only a few metres. The conditions for propagation must have been marginal in these tests. Initiation of detonation failed in four experiments.

Selected frames from an overhead cinematographic record, showing the development of a propylene oxide–air cloud for $F/B = 87$ (test 8548), are presented in Fig. 1.28. As was the case for small FAE canisters, the mass used in the calculation of F/B is the total propelled mass (i.e. the hose mass plus the fuel mass). In these tests, the hose mass accounted for between 8 and 31% of the total propelled mass. The cloud in this experiment is approximately 10 m wide by 2 m high just prior to detonation of the secondary charges. It can be seen that the cloud becomes nearly transparent as the fuel evaporates.

Table 1.2. Summary of successfully executed 65-mm line-charge trials [102]

Trial serial	Hose type	Hose length	Fuel type	Fuel-to-burster ratio (F/B)	Initiator delay time (ms)	Ambient temperature (°C)	Mean detonation velocity ($m\,s^{-1}$)
			Sustained detonation				
8524	Linen[a]	30.5	PO	71	100	33	1,900
8528[b]	Chem-Pet[a]	30.5	PO	40	150	21	1,570
8529	Linen[c]	15.2	PO	47	150	21	1,680
8558	GASynCHEM	15.2	PO	38	150	18	1,760
8559	Covered Mill	15.2	PO	49	150	18	1,820
8560	Chem-Pet	15.2	PO	40	150	6	1,510
8561	Chem-Pet	30.5	PO	53	150	11	1,460
8562	Six-Pass	30.5	PO	63	150	12	1,630
8640	Redskin	15.2	PO	51	150	30	1,100
8641	Five-Pass	15.2	PO	53	150	27	1,630
8644	Five-Pass +	30.5	PO	54	150	24	1,660
	Six-Pass +	15.2	PO	63	150	24	1,660
	Burlington Fabric	15.2	PO	56	150	24	1,660
8651[b]	Red Chief	15.2	EHN	46	150	24	1,810
8652[b]	HF-25	15.2	EHN	42	150	29	1,480
			Partial detonation				
8531[b]	Chem-Pet	15.2	PO	40	150	20	1,410
8553	Redskin	15.2	PO	77	100	11	1,340
8555	Five-Pass	15.2	PO	88	100	16	1,470
8556	Covered Mill	15.2	PO	73	100	17	1,900
8557	Red Chief	15.2	PO	84	100	15	1,540
8635	Chem-Pet	15.2	PO	40	100	30	1,500
			Failure to initiate detonation				
8550	Red Chief	15.2	PO	84	100	16	–
8552	Red Chief	15.2	PO	84	100	16	–
8646	Red Chief	15.2	EHN	91	150	26	–
8648	Chem-Pet	15.2	PO	87	150	25	–

PO propylene oxide, *EHN* ethyl hexyl nitrate.
[a]Lined with loose-fitting polyethylene tubing.
[b]Lined with bonded heavy rubber tubing.
[c]Premature ignition occurred but did not prevent successful detonation.

In fact, the lofted soil inside the cloud becomes visible by 103.6 ms. The ambient temperature for this test was 25°C.

Contour plots showing both the side-on and the overhead history of the cloud appear in Fig. 1.29. These data were obtained from high-speed film

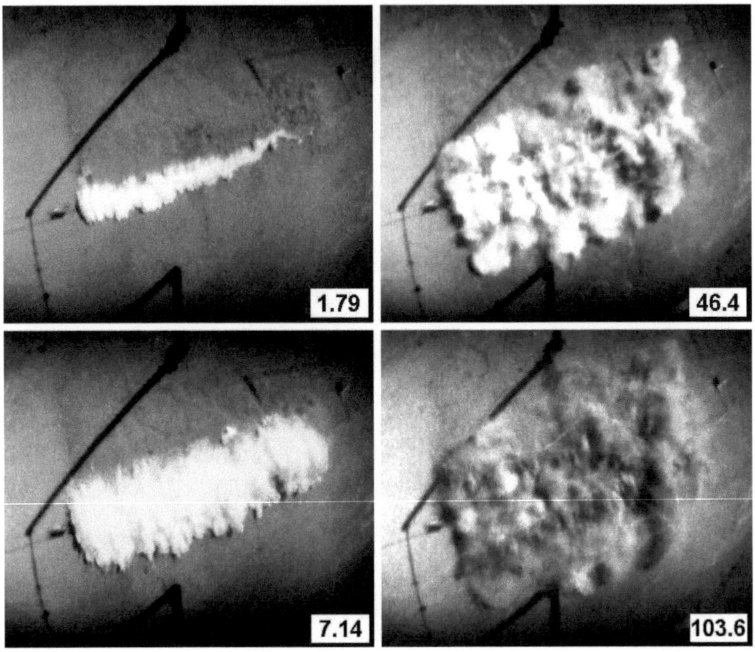

Fig. 1.28. Frames from a high-speed film record showing an overhead view of explosively dispersed propylene oxide for $F/B = 87$. Time is shown in milliseconds [102, 105]

records. The ultimate cloud size is sensitive to both F/B and the dispersal time. This is emphasized in Fig. 1.30 by the cloud height measurements from many of the tests. The two regions highlighted in the graph depict the two general sets of test conditions:

1. Group A: Experiments involving a relatively large burster charge ($40 \leq F/B \leq 53$) and a 150-ms time delay
2. Group B: Experiments involving a relatively small burster charge ($73 \leq F/B \leq 88$) and a 100-ms time delay

A side-on cinematographic sequence showing successful fuel dispersal and cloud detonation is presented in Fig. 1.31 (test 8558). Note that the hose fragments and the soot from the burster charge detonation become visible by 109.8 ms as the fuel evaporates (ambient temperature 18℃). As shown in Table 1.2, this type of sustained propagation corresponded, for the most part, to the group A parameter range.

For the group A tests in which satisfactory side-on and overhead cinematographic records were available, an estimate of the cloud volume was made from the contour plots by dividing the product of the cloud plan and profile areas by the cloud length. The results of this exercise, shown in Fig. 1.32 for clouds

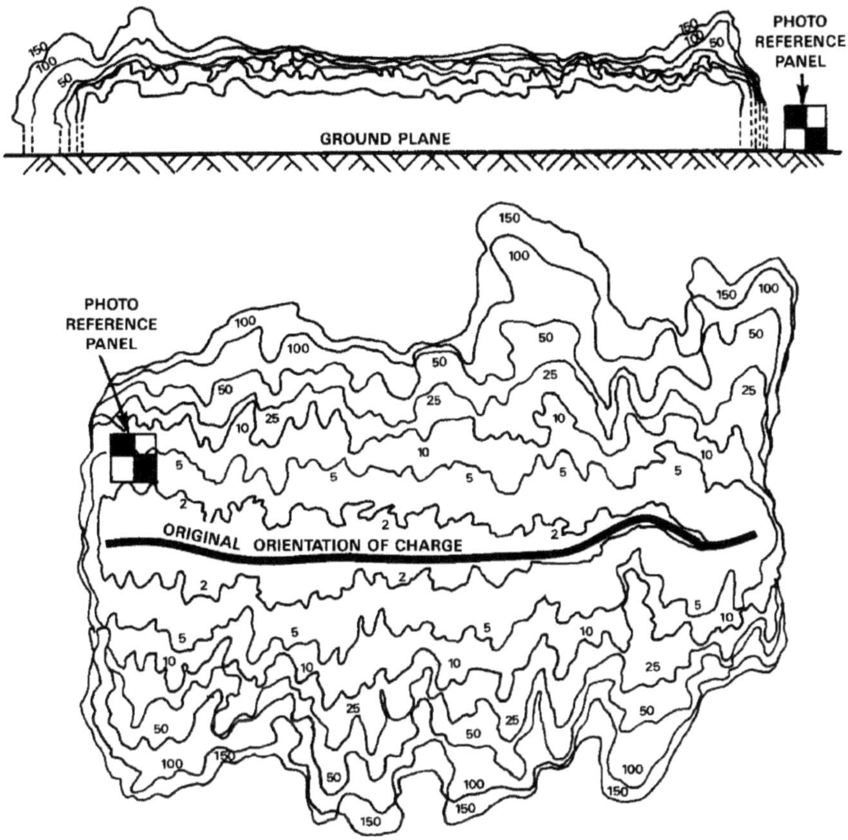

Fig. 1.29. Contour plots showing the development of the cloud as recorded by ground-level (*top*) and overhead (*bottom*) cameras [102]. The photo reference panels measure 1 m × 1 m. Time is shown in milliseconds [102, 105]

in the F/B range from 40 to 53, show the tendency toward globally stoichiometric composition as fuel dispersal progresses. However, it should be borne in mind that the method described above assumes a rectangular cloud cross-section and therefore somewhat overestimates the cloud volume. Results of a similar exercise for the group B tests confirm that the combination of smaller burster charge and shorter time delay leads to clouds that are considerably fuel-rich and, consequently, of reduced detonation sensitivity.

In the six experiments involving partial detonation, initiation of detonation was apparent, but propagation was short-lived in that the wave failed after propagating only a few metres. It can be seen in Table 1.2 that these instances correspond to F/B values near the limits of the range for which sustained detonation was observed; specifically for $F/B \leq 40$ or $F/B \geq 73$. It is likely that the cloud dimensions were too small for detonation to occur at the lean and rich extremes, or that failure occurred in voids of unfavourable mixture.

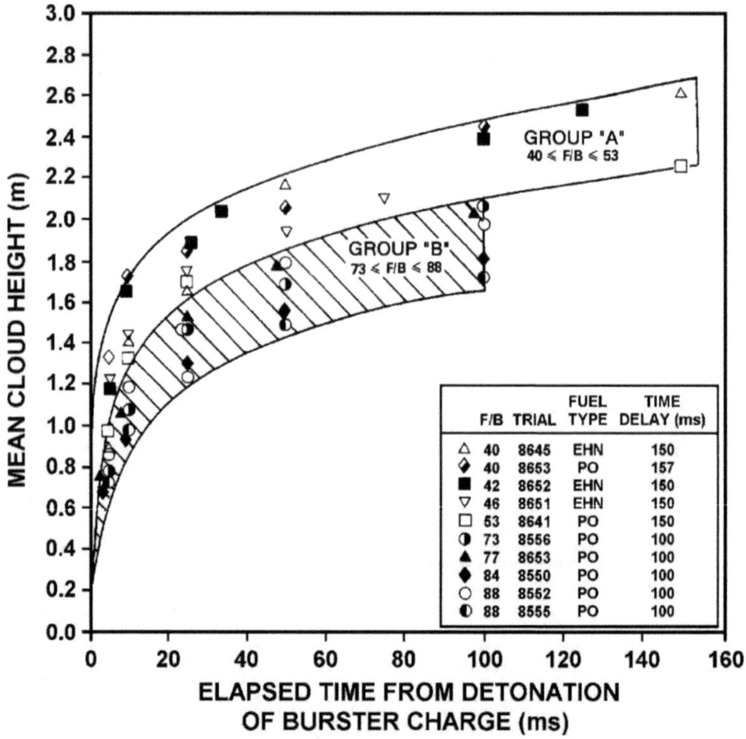

Fig. 1.30. Mean cloud height from high-speed film records for explosive dispersal of fuel from 65-mm-diameter line charges [102, 105]

A graphical summary of the test results is presented in Fig. 1.33. On the basis of these data for 65-mm line charges, the most reliable F/B range for a 150-ms time delay is 45–55.

The peak overpressures generated by a detonating hydrocarbon–air cloud are limited to about 20 atm. As such, FAE canisters and line charges are not capable of defeating hardened (i.e. blast-resistant) or deeply buried land mines. Dispersed monopropellants were considered as a possible alternative. Nitromethane, for example, can be detonated as a neat liquid or as a fuel–air cloud. Therefore, it was believed that detonation might be possible in a dense droplet medium between these two extremes. Under such conditions, it was hypothesized that propagation would be a cyclical process in which the strong shock wave produced by the detonation of a pocket of densely packed droplets would be capable of initiating neighbouring pockets. The droplet density and size of the droplet mass could be controlled through the appropriate selection of F/B and dispersal time. It was thought that atmospheric air would contribute insignificantly to the blast under these conditions.

A study of 100-mm-diameter nitromethane line charges was carried out at DRDC Suffield during the early 1990s [105]. The first practical issue to

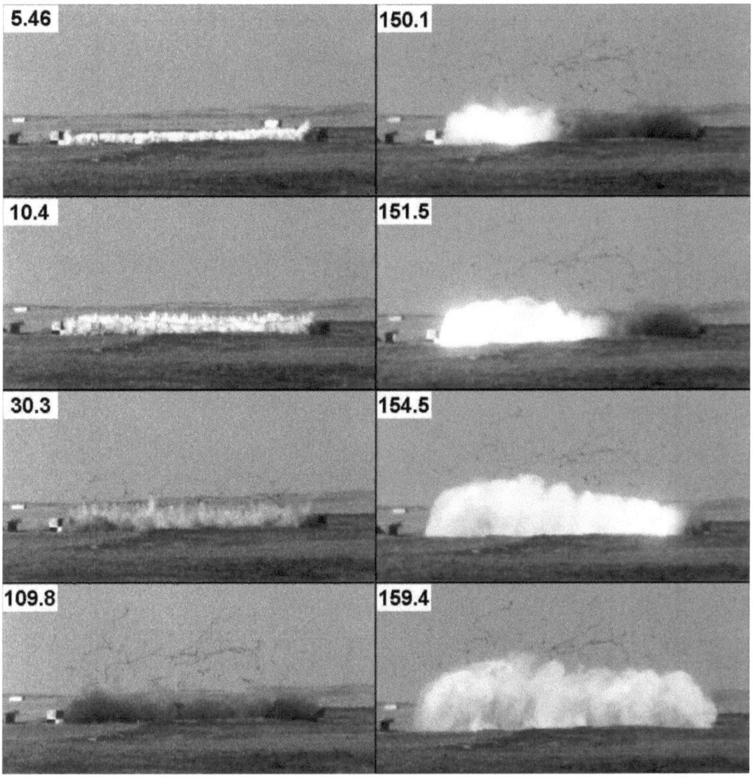

Fig. 1.31. Frames from a high-speed film record showing the side view of explosive dispersal of propylene oxide and subsequent cloud detonation. Time is shown in milliseconds [102, 105]

be addressed was that of explosive fuel dispersal. To disperse nitromethane into a droplet cloud without a bulk detonation occurring, it was necessary to use a relatively small burster charge and/or to encase the charge in a shock-attenuating, or "buffer", material, similar to what had been done to disperse nitromethane from canisters. A number of burster-charge encasement configurations were investigated for this purpose, including a dense rubber jacket of 3-mm thickness, a closed-cell foam-rubber jacket having a wall thickness of 7 mm, and a double-jacket arrangement with the rubber sheath positioned between the burster charge and outer foam jacket. The position of the dispersal charge inside the hose was also found to have a strong influence on the cloud shape for small dispersal times. Various means of tethering the dispersal charge to ensure its vertical position inside the hose were therefore evaluated. The interested reader is referred to [105] for details.

After successful dispersal methods had been identified, a series of 24 experiments were carried out to determine if and how reliably such a medium could be detonated. The charges ranged in length from 5 to 10 m. The secondary

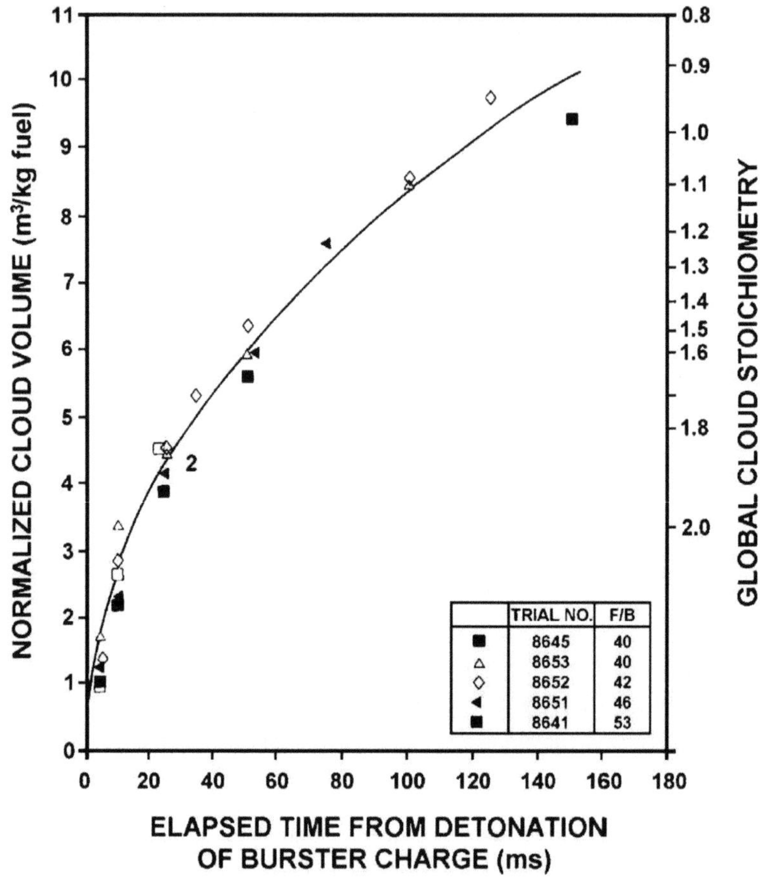

Fig. 1.32. Normalized cloud volume and associated stoichiometry during explosive dispersal of propylene oxide for $40 \leq F/B \leq 53$ [102,105]

charge in all cases was a 1-kg block of DM12 explosive mounted on a post near ground level 30 cm from the line-charge axis. The dispersal times ranged from 5 to 30 ms. The combination of smaller dispersal time and larger F/B resulted in clouds much smaller than those described earlier for 65-mm-diameter charges. The effective cloud radius could be made as small as 1 m.

Detonation was found to be possible for all dispersal times; however, the scatter in the measured detonation velocities was considerable. In the case of a 5-ms dispersal time, the velocities were 3,600 ± 1,200 m s^{-1} on the basis of seven tests. The observed velocities, as well as the scatter, decreased as the fuel dispersal time increased. For dispersal times in the 20–30-ms range, the propagation velocities were 1,500–2,000 m s^{-1}. Detonation pressures were successfully measured in four tests for a 5-ms dispersal time. Again, the data scatter was significant. The detonation pressures were in the 50–200-atm range

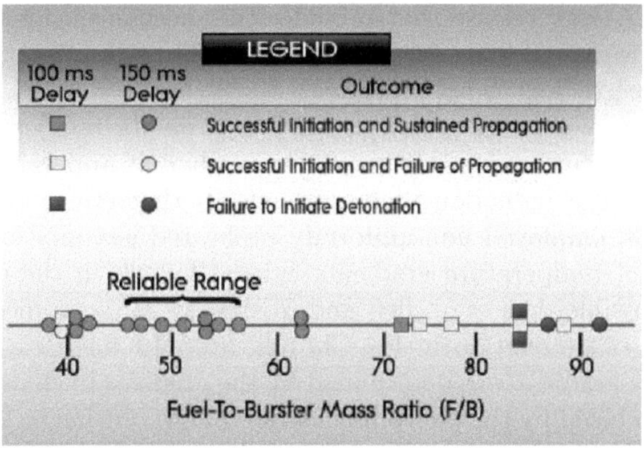

Fig. 1.33. Summary of 65-mm-diameter line-charge trials showing the outcome as a function of F/B and time delay [102, 105]

and in approximate agreement with predictions using the TIGER code [47]. The blast field outside the cloud was characterized by lower overpressures but significantly higher impulses than predicted by numerical simulations. Afterburning of the fuel with air was proposed as a possible explanation for these observations.

1.4 Chemical Initiation of Detonation

"Chemical", or "non-explosive", initiation of detonation (also known as single-event FAE, or FAE III) has been studied in an attempt to eliminate the secondary charge(s) present in conventional two-stage FAE devices. These studies have either been fundamental in nature or focused on single-event device design.

1.4.1 Fundamental Chemical Initiation Studies

Fundamental studies of chemical initiation in heterogeneous systems have been limited and were attempts to build on the earlier results for gaseous systems. A brief summary of the gas-phase studies and the isolated attempts to extend these results to heterogeneous systems is given below.

The first indications that an induction-time gradient is an important factor in this type of initiation process came from induction-time experiments employing the reflected shock technique [37]. Although the induction time is the same for all regions of the mixture behind a shock wave reflected from the end wall of a tube, ignition occurs in a definite time sequence because the gas nearest to the end wall is processed earliest in time. It is this continuous time sequencing of energy release that provides the mechanism for the amplification of shock waves and the eventual onset of detonation in cases where the induction-time gradient is appropriate.

The induction time of a mixture depends on its temperature and the concentration of free radicals present. A number of approaches have been taken to tailor the induction-time gradient. In theoretical work, Zeldovich et al. [152, 153] employed non-uniformly preheated gas mixtures and found that a range of temperature gradients existed for which the onset of detonation was possible. Lee et al. [84] and Yoshikawa [151] studied photochemical initiation of detonation in H_2–Cl_2 mixtures. In these experiments, the induction-time gradient was determined by the gradient in chlorine atom concentration produced by the photodissociation of Cl_2. Owing to the absorption of light by the gas, the Cl_2 concentration decreased in the direction of the light beam. It was found that detonation occurred only for a certain range of flash intensities. The process was described as "shock wave amplification by coherent energy release", or SWACER. Photochemical initiation using an ultraviolet source of a given wavelength was described by Shamshev [122], but details were not provided.

The conditions for the onset of detonation can also be created by the rapid turbulent mixing of combustion products and unburned mixture. In experiments reported by Knystautas et al. for acetylene–oxygen mixtures [75] and acetylene–air mixtures [74], and by Dorofeev et al. [56] for hydrogen–air mixtures, the rapid jetting of hot products from a vessel into a quiescent unburned mixture led to the formation of an induction-time gradient within large turbulent eddies. This gradient was the result of both a temperature gradient and a free-radical concentration gradient caused by the hot products. Amplification occurred with increasing energy release as the shock propagated outward. A similar phenomenon was observed in the large-scale, flame-jet ignition experiments by MacKay et al. [89, 90], Ungut and Shuff [135], and Moen et al. [98], in which a flame jet emerging from a tube was entrained into the vortex ring created by the unburned mixture exiting the tube ahead of the flame.

The required induction-time gradient can also be created in turbulent jets containing reactive chemicals. Knystautas and Lee [74], for example, injected gaseous dimethyl zinc – $Zn(CH_3)_2$ – into low-pressure equimolar oxyacetylene to induce detonation. Lee [82] also demonstrated that the conditions for detonation can be created within the turbulent mixing region created by two opposing jets; one containing propane and the other containing a reactive fluorine–oxygen mixture. The chemistry of the fluorine–propane–oxygen system was studied in detail by von Elbe [143]. Initiation of detonation was also shown to be possible by rapidly venting fluorine–air from a 5-cm-diameter high-pressure (20-atm) chamber into a larger-diameter (31-cm) vessel containing stoichiometric hydrogen–air or ethylene–air [73, 74]. However, the larger vessel in these experiments was on the order of the critical tube diameter for the mixtures tested, so the walls of the vessel may well have played a role in the initiation process. Attempts to initiate butane–air in the same apparatus were unsuccessful.

The above-described fluorine-jet experiments were later repeated using a 15-cm-diameter high-pressure chamber connected to a large plastic bag to eliminate any possible boundary effects. Successful initiation of detonation was observed in stoichiometric hydrogen–air [103] and ethylene–air [100]. In these tests, a large primary vortex was formed in the fuel–air mixture following the rupture of a brass diaphragm separating the mixtures. The induction-time gradient inside the vortex was due mainly to the fluorine concentration gradient. Figure 1.34 consists of selected frames from a high-speed cinematographic record showing the onset of detonation in ethylene–air at a location approximately 12 chamber diameters from the chamber exit. The delay between injection and ignition was found to depend on the initial fluorine concentration in the chamber. Too high a concentration (more than 12%) resulted in the fuel–air mixture igniting before the vortex had formed. Conversely, for too low a concentration (less than 9%), either the fuel–air mixture did not ignite or ignition took place late in time after the induction-time gradient had become too mild for SWACER to occur. There was a narrow range of fluorine concentration for which initiation of detonation could be achieved reliably. Again, butane–air could not be detonated, even when the diameter of the high-pressure chamber was increased to 30 cm. An additional prerequisite

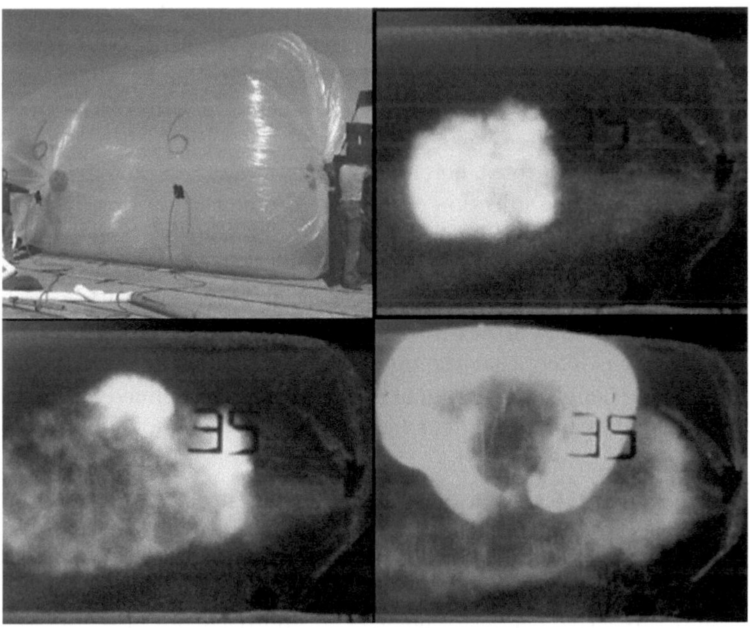

Fig. 1.34. Setup for fluorine-jet initiation of fuel–air mixtures (*top left*) and a high-speed sequence showing initiation of ethylene–air detonation [100]

for the onset of detonation is that the scale of the vortex must be on the order of the critical diameter for the fuel–air mixture being initiated [56]. The failure to initiate butane–air may have been scale-related. Shamshev [122] reported successful initiation of detonation using fluorine in mixtures of acetylene, propane, and butane with either oxygen or air, but no details regarding the scale of the experiments or the degree of confinement were provided.

Urtiew et al. [136] used a chemical inhibitor, rather than a sensitizer, in their chemical initiation studies. Specifically, a *cis*-2-butene inhibitor was used to delay the reaction between tetrafluorohydrazine and silane until a sizeable mixing zone had been created. A suitable gradient in inhibitor concentration led to shock wave amplification and the ensuing onset of detonation.

To the authors' knowledge, there have only been two fundamental studies of chemical initiation involving heterogeneous systems. S.B. Murray, I.O. Moen and C. Brosinsky (Defence Research Establishment Suffield, unpublished results, 1987) attempted to detonate an unconfined propylene oxide droplet–air mixture using the same fluorine-jet apparatus as described above. The cloud was created by explosively disseminating the liquid fuel from a 50-l canister as described in Sect. 1.3.4. The location and direction of venting (either along or normal to the cloud radius), the quantity of fluorine contained in the chamber, and the diluent type (air or helium) were varied in the tests. A rapidly accelerating luminous wave front was observed in some experiments. However, amplification to detonation was not achieved despite the fact that propylene oxide–air has a sensitivity similar to that of the ethylene–air mixtures that had been successfully initiated in the earlier tests. Failure to initiate detonation may have been due to the systems having different chain-branching reactions, or possibly the heterogeneous nature of the propylene oxide–air system.

The second example of a heterogeneous system, reported by Xuezhong et al. [150], involved the injection of liquid bromine trifluoride (BrF_3) from a nozzle into mixtures of MAPP/air or MAPP/oxygen/air contained in a 12-cm-diameter tube. Detonation did not occur for MAPP/air mixtures. However, initiation was apparently successful in oxygen-enriched mixtures as evidenced by pressures in the range from 22 to 28 atm. Given the small diameter of the tube, however, it is likely that the outcome was dominated by wall effects.

In summary, chemical initiation of detonation in unconfined heterogeneous fuel–air systems has yet to be demonstrated in any fundamental study.

1.4.2 Single-Event FAE Device Studies

Most attempts to demonstrate chemical initiation in heterogeneous systems have involved prototype FAE devices designed to explosively codisperse a liquid fuel and a reactive chemical agent. The approach proposed by von Elbe and McHale [144–146] is based on simultaneous fuel dispersal and combustion rather than the delayed onset of detonation. The initial work by these authors involved laboratory tests in which submillilitre quantities of BrF_3 (also referred to as "BTF") or liquid chlorine trifluoride (ClF_3, or CTF) were driven pneumatically into a thimble containing less than 1 g of diesel fuel [144]. Blast gauges and high-speed cinematography confirmed that an explosive reaction took place upon injection of BTF into the thimble. The CTF was found to be even more reactive. However, the high reactivity produced a counterforce at the interface between the liquids that prevented thorough mixing and reaction.

A schematic diagram of a proposed single-event device based on these initial test results is shown in Fig. 1.35 [146]. An annular stainless steel vessel containing liquid CTF surrounded by diesel fuel contains a high-explosive charge at its core. Confinement of the fluids during the period between charge detonation and fluid ejection is provided by massive upper and lower steel blocks. The CTF/fuel reaction, together with the high-explosive detonation, drives the fuel into the surrounding atmosphere, forming a cloud of atomized fuel and generating a strong primary shock wave. The research version of this device employed a 3.3-g RDX pellet to explosively disperse 350 ml of fuel and 32 ml of either BTF or CTF. Tests were performed with diesel and heptane fuels, and both were found to perform similarly. A significant blast was generated when the test was performed in air; no blast was observed when the test was performed in a nitrogen-filled tent; and a very strong blast was

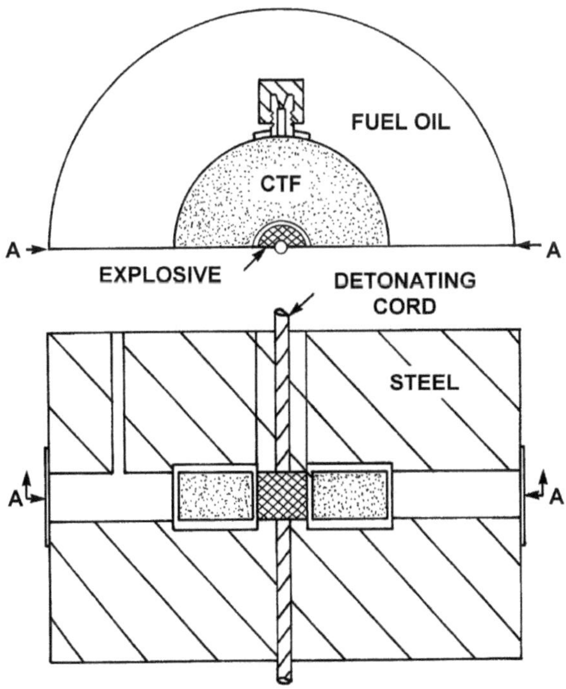

Fig. 1.35. Early FAE III device tested by von Elbe and McHale [146]. *CTF* chlorine trifluoride

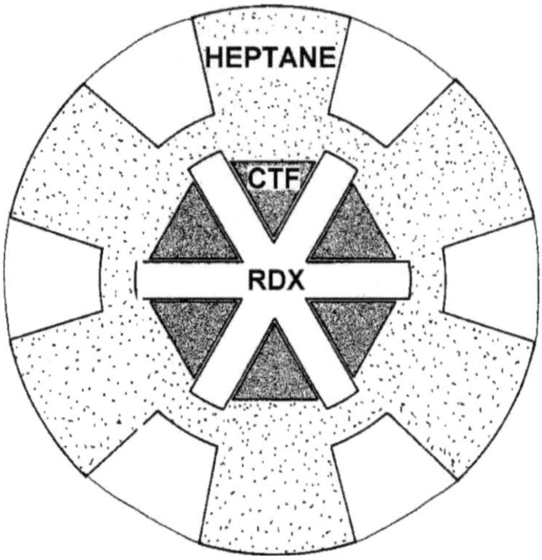

Fig. 1.36. Second-generation FAE III device proposed by von Elbe and McHale [145]

produced when the test was performed in an oxygen-filled tent. These results confirmed that the primary energy release was due to the reaction between the fuel and air. However, in view of the data for the minimum cloud size for detonation in diesel–air mixtures reported by the Kurchatov Institute, it is clear that a device of this scale cannot produce a self-sustained detonation.

Following the above-described tests, an improved device believed to be capable of achieving even more rapid ejection of the fuel and agent into the atmosphere was proposed [145]. A schematic diagram of that device appears in Fig. 1.36. The central charge of high explosive is greatly enlarged and the shaped-charge effect is used to propel jets of fuel and agent into the surrounding air at very high rates. It was anticipated that, in contrast to the earlier design, the CTF/fuel reaction would play no significant role in the process of fuel ejection, but would instead take place in the cloud later, where it could contribute to shock amplification. The authors are unaware if the improved device was ever constructed or tested.

A patent was later issued for a device [121] that specified the use of several possible fuels including volatile liquids of a lower molecular weight alkane or epoxyalkane. The chemical initiators included n-hexylcarborane, carboranylmethyl propionate, ferrocene, and n-butylferrocene. Some of the experiments supporting the patent application employed as little as 1 ml of fuel. Supplementary full-scale tests employed 300 g of fuel and 60 g of ferrocene. The clouds produced by such a minute amount of fuel would be very small (estimated at 2–3 m^3). Although little or nothing is known about the detonability of some of these fuels, there is a high likelihood that the resulting clouds would not exceed the critical dimensions required for detonation. No evidence of successful detonation was presented in the patent.

The approach taken by Tulis [133, 134] is similar to that of von Elbe and McHale in that a hypergolic-oxidizer/fuel reaction produces ignition. However, the delay between dispersal and ignition is achieved by using an implosion mechanism in which dispersal of the oxidizer is delayed by the time required for the compression wave to be relieved. Tulis's model of induced heterogeneous detonation is illustrated in Fig. 1.37. When the explosive is detonated, the fuel is dispersed to form a droplet–air cloud, while the oxidizer is impulsively compressed inward. Once the compression wave reaches the device centre, a shock is formed that propagates outward, convecting the oxidizer particles and atomizing the fuel droplets. It was hypothesized that the intense postshock reaction leads to self-sustained detonation. Although high overpressures were measured a small distance from the device, conclusive evidence of successful detonation was not provided.

On the basis of their fundamental studies with gases, Murray et al. [106] attempted to demonstrate a liquid analogue of the fluorine-jet initiation experiments described in Sect. 1.4.1. The device consisted of a fuel canister fitted with a central explosive dispersal charge and a smaller, internal, coaxially aligned chamber filled with reactive liquid. The components are shown in Fig. 1.38. The fuel was propylene oxide and the reactive liquid was a mixture of triethyl aluminium and heptane. The heptane was used as a diluent to control the reactivity of the triethyl aluminium/heptane mixtures in the same manner that air had been used to control the reactivity of the fluorine/air mixtures in the gas-phase experiments. Explosive dispersal of triethyl aluminium/heptane mixtures in air confirmed that the ignition time could be precisely controlled by varying the ratio of the constituents. The amount of reactive liquid was varied by changing the size of the internal chamber. An attempt was made to produce large vortices during fuel dispersal by placing the assembled device inside a sleeve consisting of heavy steel vertical slats separated by vertical openings. The purpose of this sleeve was to channel the liquids through the openings during dispersal in an effort to induce large vortices during cloud formation (analogous to the vortices produced in the fluorine-jet experiments upon diaphragm rupture). This method of vortex generation was determined to be unsatisfactory. Shamshev [122] briefly described the use of explosive confinement for cloud shaping in past Russian studies. This approach may be more effective than the inertial confinement described here.

An experiment involving a similar device, but without a vortex-inducing sleeve, was reported more recently by Xuezhong et al. [150]. Again, the fuel canister was fitted with a central explosive dispersal charge as well as an internal coaxial chamber to hold the sensitizer. The fuel and sensitizer were propylene oxide and BTF, respectively. The sensitizer volume in the first experiment was 9.4% of the total canister volume. This fraction was increased to 17.9% in a second experiment. The nominal fuel mass was 3 kg. High overpressures (23–32 atm) were measured in close proximity to the devices (approximately 0.5 m). However, at a radius of 3 m, the overpressures had decreased to 3.1 and 1.1 atm for the first and second experiments, respectively. While the dispersal

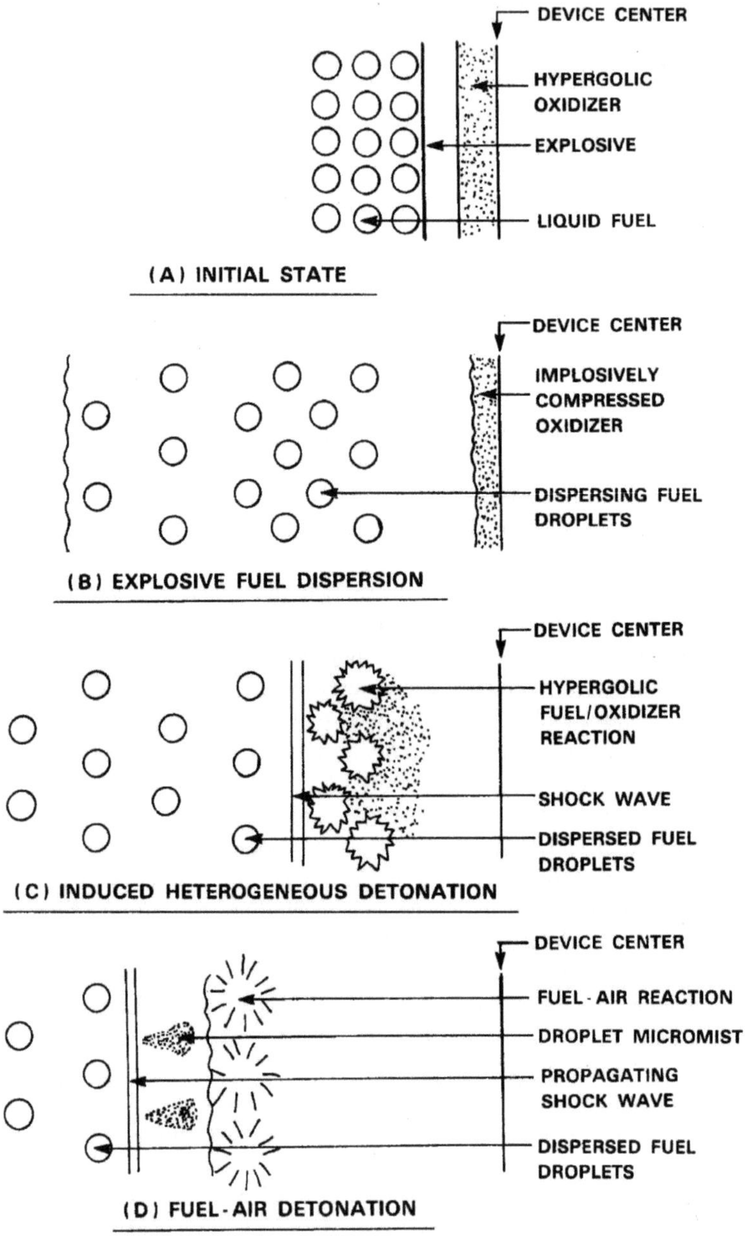

Fig. 1.37. Implosion–explosion FAE III concept proposed by Tulis [133, 134]

charge mass and canister standoff were not given in the paper, the high overpressures measured in the near field could have been due to the burster charge detonation (the size of which was not specified) or the propylene oxide/BTF

Fig. 1.38. FAE III device tested by Murray et al. [106]; vortex-generating cage (*left*) and chamber with initiator fluid surrounding burster charge (*right*)

reaction, or both. It is worth noting that the fuel-sensitizer loading in a sphere of 0.5-m radius for a 3-kg device would be 5.7 $kg\,m^{-3}$, or more than 40 times that for a stoichiometric propylene oxide–air mixture.

In conclusion, early attempts at chemical initiation of detonation based on liquid constituents were founded on the belief that rapid turbulent mixing of the fuel and initiator with air, followed by intense chemical reactions in the resulting fuel–air cloud, were the only prerequisites for the initiation of detonation. The concepts of (1) using an induction-time gradient capable of promoting shock wave amplification and (2) the need for a minimum cloud volume to achieve self-sustained detonation were not appreciated by most researchers. The attempt by Murray et al. [106] to create a liquid analogue of the successful fluorine-jet initiation experiments appeared promising, but this initiative was terminated prematurely owing to changing R&D priorities.

1.5 Detonation Propulsion Studies

Spray detonation is potentially relevant to the study of PDEs and other hypersonic propulsion systems utilizing detonative combustion. However, most current R&D programmes either employ gaseous fuels or preheat a liquid fuel so that it enters the combustor in the vapour phase. There are two notable exceptions. The first is the work being conducted at the US Naval Postgraduate School (NPS) by Brophy et al. [24–32]. These investigators have been optimizing the performance of a driver tube or "predetonator" to initiate JP-10/air aerosol detonations in an experimental PDE.

The initial NPS studies [27] were conducted in a constant-diameter tube with direct JP-10 injection at the head end using commercial atomizers (Fig. 1.39, configuration 1). Air or oxygen was introduced into the combustor through a series of circular holes in the end wall surrounding the atomizer.

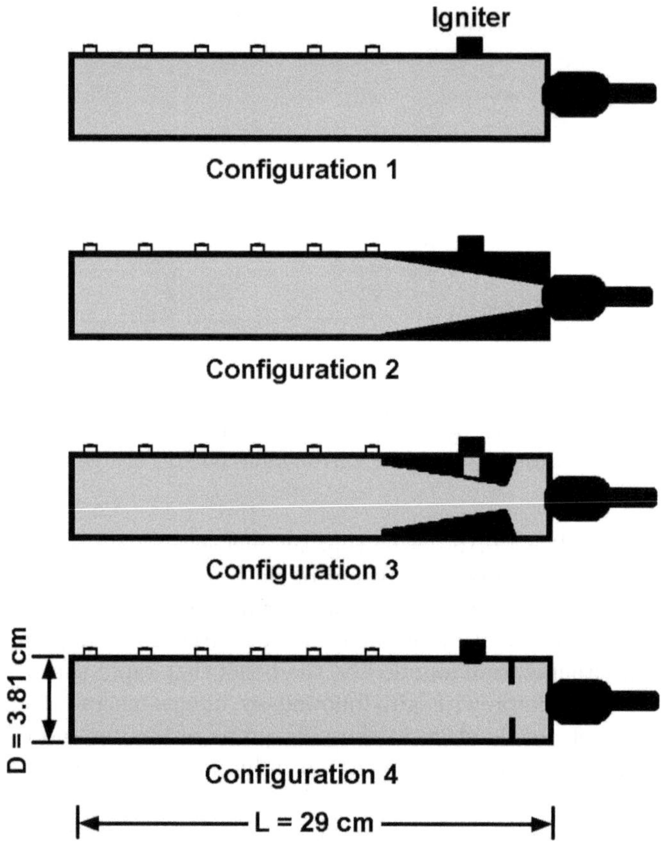

Fig. 1.39. Configurations for 3.81-cm combustor showing stepped ramp and recirculation cavity [27]

Tests with a Malvern instrument indicated that the droplet size in the vicinity of the atomizer was typically 10–20 μm. Spark ignition was possible at one of three locations: 0.25, 1.0, and 1.25 combustor diameters from the head end. A series of eight pressure transducers was used to identify the location of deflagration-to-detonation transition (DDT). Detonation was not observed for JP-10/air, but DDT did occur toward the end of the combustor for JP-10/oxygen. Subsequently, the performance was improved by installing a stepped-ramp diffuser at the head end [26] that prevented any significant recirculation from occurring, but its presence also minimized the degree of tube wall wetting (Fig. 1.39, configuration 2). It was found that the DDT run-up distance decreased as the axial location of ignition increased, with one tube diameter being the optimal position. Moving the igniter further than two tube diameters downstream actually increased the DDT distance. The performance was further improved [25] by introducing a 1.5-cm-long recirculation cavity (the same diameter as the combustor) upstream of the stepped ramp

(Fig. 1.39, configuration 3). However, it was found that no further gains could be realized by increasing the ignition energy above 500 mJ. When the diffuser was removed and the same recirculation cavity was recreated by using an orifice plate instead (Fig. 1.39, configuration 4), the performance was seen to diminish.

Following the initial studies, a new apparatus was constructed to permit the use of a JP-10/oxygen predetonator to initiate JP-10/air mixtures in a larger 12.7-cm-diameter combustor [29,30]. The inlet air was heated to 152℃ and the fuel was introduced into the air stream 45 cm upstream of a plenum chamber containing a perforated cone to induce turbulence. This arrangement allowed plenty of time for mixing and an opportunity for some fuel vaporization to take place. The mixture was then directed into the combustor through openings surrounding a centrally mounted fuel–oxygen predetonator. The predetonator was essentially a 25-cm-long version of the JP-10/oxygen apparatus described in the previous paragraph. Preliminary tests using ethylene–air in the combustor showed that the predetonator could reliably initiate these mixtures. Tests with JP-10/air indicated that detonation could be initiated in a steady-flow configuration with an air inlet temperature of 152℃, providing that the Sauter mean droplet diameter was below 3 μm and 70% of the fuel was in the vapour phase. Detonations were not possible when the inlet air temperature was reduced below 102℃. These results show the importance of fuel vapour on the mixture detonability.

In subsequent studies [28, 31], the combustor was modified to consist of individual segments, one of which contained an optical window that could be located at multiple combustor locations. This allowed the imaging of the predetonator detonation as it diffracted into the main combustor. Images of the fuel distribution, natural emission, CH chemiluminescence, and shock structure were also obtained. On the basis of the diffraction studies, a new predetonator was developed that permitted a more favourable flow path for the reactants and a less severe diffraction at the exit [32]. Thermal imaging was also carried out to determine cooling requirements during continuous engine operation [31, 32]. More recently, tunable diode laser and absorption spectroscopy techniques have been applied in tests using ethylene fuel to assess the effects of fuel distribution on fuel-based specific impulse [24]. These tests were carried out in the most current embodiment of the NPS PDE, shown in Fig. 1.40, at a frequency of 40 Hz. The results showed that a stratified fuel distribution, in which the mixture begins near stoichiometric at the head end of the combustor and gradually becomes fuel-lean near the combustor exit, produces a substantially higher specific impulse than a uniform fuel distribution based on the same amount of fuel. Once a detonation is established in the favourable mixture, the wave can easily propagate into the less reactive mixture. The authors reported that the overall detonability of the system actually improves over that for a uniform fuel distribution.

Given the detonation cell sizes for gaseous JP-10/air measured by Akbar et al. [2] at atmospheric pressure and by Zhang et al. [157] for a range of

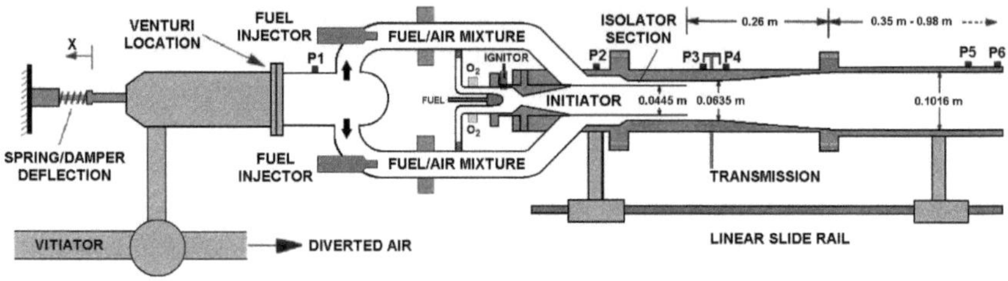

Fig. 1.40. Cross-sectional view of the US Naval Postgraduate School pulse detonation engine (PDE); two of four arms are depicted [24]

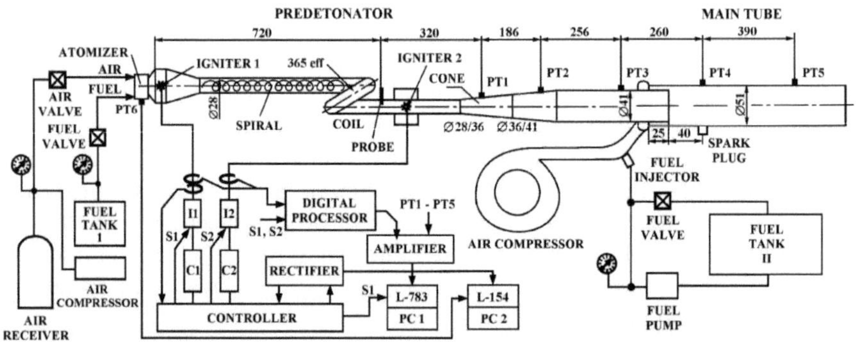

Fig. 1.41. The Semenov Institute PDE demonstrator. Dimensions are in millimetres [62]

combustor stagnation conditions typical of flight at altitude, it is likely that sustained detonation in the NPS prototype would not depend on the combustor walls.

A PDE demonstrator based on aerosol detonation is also being developed at the Semenov Institute of Chemical Physics in Moscow. An excellent review of that work was recently provided by Frolov [62]. A schematic diagram of the system is shown in Fig. 1.41. It consists of a predetonator and a main tube.

The design and operation of the predetonator is quite novel. It consists of a combination of two tubes, 28 and 41 mm in diameter, connected by a tapered transition section. The fuel atomizer at one end of the 28-mm-diameter tube produces very fine droplets (5–6 μm at a distance of 70 mm from the nozzle). A low-energy electrical ignition system is used to ignite the resulting two-phase flow at a position 60 mm downstream from the atomizer. A 400-mm-long Shchelkin spiral inserted in the tube is used to induce flame acceleration. The straight portion of the predetonator is connected to a single coil of 365-mm length. The shock wave produced by the accelerating flame propagates through the coil and subsequently emerges as a detonation wave in a second straight section of 28-mm-diameter tube. Interaction of the shock wave with the coil walls is believed to cause a transverse wave which transitions into a single-head spinning detonation. The detonation propagates through the tapered section and into the 41-mm-diameter portion of the predetonator tube. It subsequently transmits to the 51-mm-diameter main tube which is supplied with a fuel–air mixture through the annular gap between the predetonator and main tubes. The fuel–air mixture in the main tube is supplied by a compressor in conjunction with a standard automotive fuel injector that produces droplets in the 70–80-μm range. The total length of the system is 2.2 m. The fuels tested include n-hexane and n-heptane. Under cold start-up conditions, only a shock wave emerges from the coil (rather than a detonation wave), and a second igniter is fired at the precise moment the shock wave traverses it, resulting in initiation of detonation. The second igniter is not required after seven to eight detonation pulses. Stable operation is observed after 12–15 pulses.

The performance of the PDE was optimized by tuning a number of engine parameters. For example, tests conducted with n-hexane and a predetonator containing multiple coil loops (without a Shchelkin spiral present) showed that the ignition energy could be reduced by a factor of 2 if three loops were used rather than one. Additional tests employing a straight predetonator fitted with a Shchelkin spiral showed that a strong shock wave ($800 \pm 50\,\mathrm{m\,s^{-1}}$) could be generated that was independent of ignition energy over the range 24–130 J. However, DDT was not observed for spiral lengths up to 750 mm.

The Semenov Institute design has the advantage over the NPS system that no supplemental oxygen is required for successful predetonator operation. However, the operating frequency (2.2–3.9 Hz) is very low owing to the time required for flame acceleration and DDT, so the system would not be practical in its current embodiment.

1.6 Modelling of Spray Detonation and Explosive Liquid Dispersal

Spray detonations share many similarities with their gaseous counterparts. In both cases, a minimum initiation energy exists that depends on the combustible mixture, the geometry (e.g. planar, cylindrical, or spherical), and the level of confinement. A minimum volume is also required for detonation propagation in both cases.

Spray combustion involves additional mechanisms and modelling complexities associated with liquid vaporization and droplet breakup. The underlying mechanisms may be divided into fluid dynamic and combustion processes. With respect to the former, spray detonations involve a variety of multiphase fluid dynamic interactions between the condensed and gaseous phases, as is the case for dust detonations. These include mass transfer through combustion, momentum transfer through drag, energy transfer due to heat transfer and radiation, and particle/droplet collision and agglomeration. In the case of spray detonations, the liquid fuel introduces additional complexities associated with mass transfer owing to droplet vaporization, droplet deformation and breakup owing to hydrodynamic forces, droplet–droplet collisions, and droplet interaction with a wall. The combustion of oxygen-free droplets requires liquid vaporization before the fuel can burn with the surrounding oxidizer. The time required to heat and burn a droplet depends on separate time scales associated with the droplet heating and vaporization processes, the time scales associated with mixing the vapour with the surrounding oxidizer, and the chemical kinetics of the fuel–oxidizer mixture.

Excellent reviews of spray flows and combustion are available in the books by Crowe et al. [48], Sirignano [124], and Stiesch [128], so the details will not be reiterated here. The current review will be restricted to spray detonation modelling per se and to the difficult problem of explosive liquid dispersal.

1.6.1 Spray Detonation Modelling Studies

Many of the general concepts used in spray combustion modelling can be applied to spray detonation propagation. From a modelling point of view, detonations differ from flames in that they involve highly compressible flows with shock-induced processes, which play an important role in the detonation propagation and initiation processes. Most contributions to spray detonation modelling emerged during the 1960s, 1970s, and early 1980s, and have been summarized in the excellent review by Sichel [123]. The following section provides a brief summary of the early work and highlights more recent work in spray detonation modelling.

1.6.1.1 Steady Detonation Analyses

Much of the early work on spray detonations addressed the steady propagation of the detonation wave. For the case of very small droplets of volatile fuel, the combustion may be assumed to be completed before the C–J point. This allows the detonation properties to be computed using one of many available chemical equilibrium codes. For larger droplets or low-vapour-pressure fuels, it has been observed that the detonation velocity can be significantly below that computed assuming complete combustion. For gaseous mixtures, detonation velocity deficits can occur owing to boundary effects, including momentum and heat transfer to the wall. The actual detonation velocity depends on the ratio of the tube diameter to the effective reaction-zone length, which, for gaseous detonations, can be related to the detonation cell size [99]. For low-vapour-pressure spray detonations with large droplets, the effective reaction-zone length is usually controlled by the droplet breakup time. As noted by Gubin and Sichel [69], calculations and experiments performed by Ragland et al. [116] indicate that the velocity deficits observed in some experiments cannot be fully attributed to losses to the walls. Cherepanov [40] attributed the velocity deficit to the explosion event that occurs after droplet breakup. This produces shock waves propagating toward and away from the leading shock. Assuming that only half the explosion energy contributes to the detonation front, Cherepanov concluded that the detonation should have a velocity that is $1/\sqrt{2}$ times that of an ideal detonation. Experimental data, however, indicate that the velocity deficit is not a constant, but depends on numerous parameters, including the droplet diameter and the fuel volatility. Gubin and Sichel [69] developed a more comprehensive model that determines the amount of energy contributing to the detonation front on the basis of the rate of droplet shattering and subsequent combustion. The premise behind this analysis is that only the energy from the portion of the droplet that has been shattered into a micromist is released soon enough to contribute to the detonation front. Detonation velocities computed over a range of droplet diameters for kerosene–oxygen, with and without nitrogen dilution, predict velocity deficits as high as 38% for a droplet diameter of 3,000 μm. The calculated results are in good agreement with experimental data available for droplet diameters between 290 and 2,600 μm.

Steady-state detonation profiles were computed by Mitrofanov et al. [93] using three models that differed mainly in their description of droplet deformation and breakup. The purpose of the calculations was to estimate reaction-zone lengths and to compare them with those determined by experiments. Since detonation velocities were computed assuming complete combustion, no attempt was made to account for possible velocity deficits. The first model used the droplet deformation model of Borisov et al. [22] with droplet breakup into secondary droplets triggered at a critical droplet deformation. The model also included droplet boundary-layer stripping using a formula developed by Ranger and Nicholls [117]. Evaporation rates were computed using a simple Reynolds number correlation. The second model extended the first model by allowing secondary and subsequent droplet generations to break up. Finally, the third model modified the drag law to account for the effect of droplet deformation. Whereas the first model predicted very long reaction-zone tails, the more advanced models predicted reaction-zone lengths that were smaller than the experimental values. The authors attributed the differences to uncertainties in the extent of secondary droplet breakup and in the coefficients used in the boundary-layer stripping model.

1.6.1.2 Transient Detonation Initiation and Propagation Analyses

Mitrofanov et al. [93] applied their simplest model described in the previous subsection to the transient problem of point-blast initiation of a cylindrical spray detonation. The governing Euler equations, with phase-interaction terms, were solved using a method of characteristics with a point-blast solution as initial conditions [59]. A simple ignition delay law was added to account for the finite-rate kinetics during the shock velocity undershoot before detonation initiation. The activation energy for this model was calibrated to obtain a minimum initiation energy similar to that obtained by Nicholls et al. [109] for a lean kerosene–air mixture with 384-μm-diameter droplets in a cylindrical geometry. The final velocity deficits at the end of the calculations for near-critical detonation initiation were approximately 4–6%.

Burcat and Eidelman [36, 60] published two papers on the initiation of spherical detonation for a heptane spray in oxygen. As in the case of the Mitrofanov analysis, the calculations solved the Euler equations with phase-interaction terms using a point-blast similarity solution [59]. The governing equations were solved in a Eulerian frame of reference using the flux-corrected transport algorithm. The overall model was somewhat simpler than that used by Mitrofanov et al. The model used the Engel droplet shattering model [61], but neglected the droplet deformation before breakup. Owing to the fast kinetics assumption, a gas-phase combustion model was not included. As expected for a spherical geometry, the shock decay profiles display a velocity undershoot followed by a gradual increase toward the detonation velocity. The latter was somewhat below the experimental and chemical equilibrium values for the stoichiometric heptane–oxygen mixture considered. The detailed pressure profiles for a spray with 200-μm-diameter droplets displayed a secondary wave behind the leading shock due to delayed combustion. Although the exact origin of this front was not fully analysed, it seemed consistent with secondary explosions that have been observed experimentally. The authors also performed calculations for cylindrical detonation initiation in a kerosene–air spray with 380-μm-diameter droplets. They compared their shock front trajectory with the experiments of Gabrijel and Nicholls [63] for a given initiation energy and obtained relatively good agreement between computed and experimental results.

In the second paper by Burcat and Eidelman, the effect of the fuel droplet size on the detonation initiation process was investigated. Computations performed for heptane–oxygen mixtures with droplet diameters between 100 and 1,000 μm indicate that the detonation velocity decreases with increasing droplet radius and with a corresponding longer reaction time. For a droplet diameter of 100 μm, the detonation velocity is close to that for a gaseous system. The reported final leading front velocity deficits for the larger droplets are quite large, reaching 48 and 57% for 600- and 1,000-μm-diameter droplets. For these droplet sizes, it is likely that detonation initiation failed or that the calculation run times were too short to accommodate the full detonation initiation process. Higher-resolution computations with 300- and 600-μm-diameter droplets indicate that the leading front is repeatedly supported by secondary waves emanating from reaction zones, which results in unstable detonation propagation.

Other spray detonation calculations in the 1980s and 1990s include those of Zhdan [159], Voromin and Zhdan [147], and Vasil'ev et al. [141]. The calculations by Zhdan were performed for a kerosene–oxygen mixture in a spherical geometry with droplet diameters ranging between 100 and 700 μm. Over this range, the critical initiation energy for spherical initiation was calculated to increase from 9.2 to 589 kJ. For 500-μm-diameter droplets, the detonation velocity deficit at the end of the calculation was approximately 14% and still decreasing, indicating that detonation initiation had indeed been achieved. Analysis of the flow profiles indicated that 12% of the droplet burning was occurring behind the C–J plane. Subsequent calculations by Voromin and Zhdan were performed for heptane–oxygen and heptane–air mixtures with 1,400-μm-diameter droplets in tubes initiated with a hydrogen–oxygen driver section. The authors investigated the effect of initial pressure on the detonation initiation process. While heptane–air detonations could not be initiated for the highest initial pressure considered (6.8 atm), heptane–oxygen mixtures were more easily detonated with increasing initial pressure. The final velocity deficit was approximately 17%, which is attributed to wall friction and heat transfer effects that were included in the model.

A more recent publication by Chang and Kailasanath [38] presents calculations for 25-μm-diameter droplets, with and without combustion. The calculations for inert droplets use a simple droplet breakup model with a prescribed time delay for breakup. After breakup, a selected fraction of the mass is converted to vapour, while the remaining mass is converted to ten smaller droplets. For combustion calculations, a simple d^2 law is used to model the rate of evaporation. From a numerical simulation point of view, this study differs from previous investigations in that it applies a Lagrangian description for the liquid phase using 100,000 real droplets and a number of "virtual" droplets to decrease the overall computational time. These calculations highlight the competition between the droplet-induced blast decay, due to momentum transfer to the droplets/particles, and the rate of exothermic energy release. For the same particle loading density and particle diameters between 12 and 100 μm, the blast decay rate in an inert system increases with decreasing initial droplet size. It also increases when droplet breakup and vaporization models are included in the calculations. When a combustion model is included, the coupling of the energy release and the leading shock depends on the effective burning time of the droplet, which must be sufficiently short to compete with the droplet-induced blast decay.

A subsequent paper by Cheatham and Kailasanath [39] presented an extensive study of the effects of droplet size and prevaporization on the detonation initiation of JP-10 sprays in oxygen. Since this work was mainly aimed at PDE applications, the authors considered relatively small droplets where the complex problem of droplet breakup could be ignored. The authors also investigated the sensitivity of the computed results to the drag and heat transfer models and grid resolution. The calculations were performed using a 120-cm test section and a detonation initiator consisting of a 1-cm high-pressure region and a 29-cm section of fully vaporized fuel. Leading front velocity profiles revealed a stable vapour detonation in the initiator section, followed by a rapid deceleration due to momentum transfer to the droplets, and a subsequent combustion-driven transition to detonation. With the exception of the overly simple Stokes drag law, similar front velocity profiles were observed for 10-μm-diameter droplets using various drag models that differ in their account of Reynolds number, Mach number, and particle acceleration effects. Similarly, the results did not seem very sensitive to the particular convective heat transfer model used in the calculations. On the other hand, the results were sensitive to the energy release duration after completion of the induction time process. For droplet diameters in the range 2.5–25 μm, the transition to detonation distance decreases with decreasing droplet size. Results for droplet diameters of 2.5–15 μm, where a final stable detonation is observed, indicate a very slight (2% or less) detonation velocity deficit. The detonation initiation process for 20-μm-diameter droplets temporarily displays a double detonation pattern until the sonic point of the first wave recedes behind the second front, allowing eventual coupling of the released energy with the leading front. Such transient double-front patterns were also observed in the calculations of Eidelman and Burcat [60]. For droplet diameters 30 μm or larger, a detonation cannot be initiated even when the tube length is increased to 8 m. This may be partly due to the fact that droplet breakup was not included in the model. For calculations involving hybrid mixtures of droplets and a prevaporized fuel, the detonation transition distance decreases as the equivalence ratio of the prevaporized fuel is increased from 0.015 to 0.9. A higher concentration of prevaporized fuel also makes the detonation transition process more gradual.

The calculations described above solved the one-dimensional inviscid flow equations. Smirnov et al. [126] recently performed two-dimensional computations for a hydrocarbon–air spray detonation initiation including turbulent combustion effects. The model is an extension of that previously used by the same authors to model combustion in turbulent dust–air mixtures [125]. The calculations use an Eulerian frame of reference for the gas and 15,000 Lagrangian droplets of diesel fuel. The droplets are polydispersed in a 0.35-m-long, 20-mm-diameter test section with a triangular distribution having a mean diameter of 50 μm and a range of 10–100 μm. Droplet breakup is modelled using correlations developed by the authors for the postbreakup droplet diameter as a function of the Weber number. Calculations performed with different planar shock initiation energies indicate that the initial decay is much faster for droplets than for solid particles of the same size. This observation, which is consistent with the results of Chang and Kailasanath [38], is due to the shattering of droplets into smaller droplets. The shock propagation into the droplet cloud generates a high-density "ρ-layer" first observed theoretically by Korobeinikov [78] and confirmed experimentally in tests with heterogeneous explosives. This phenomenon was also observed in the calculations of Chang and Kailasanath. Ignition occurs downstream of the shock, resulting in a local explosion which produces a second front that catches up with the leading shock when detonation initiation is successful. Droplet combustion occurs faster along the axis of the tube than at the walls, where unburned droplets remain.

None of the above-mentioned calculations investigated the possibility of a cellular structure for spray detonations. Zhdan and Prokhorov [160] computed two-dimensional cellular structures for a cryogenic hydrogen–oxygen mixture with liquid oxygen droplets. Detonation propagation in this mixture behaves quite differently from that observed in hydrocarbon–oxygen systems. One-dimensional calculations display an oscillatory detonation propagation with an average detonation velocity that is approximately 9% larger than the theoretical C–J value. Two-dimensional calculations performed with 50- and 100-μm-diameter droplets display a cellular structure with respective cell widths of 21 and 44 mm. These cell sizes for a low-temperature cryogenic heterogeneous system are of course much larger than the 1-mm cell size observed for a gaseous hydrogen–oxygen system at room temperature and 1-atm pressure [91]. The cellular structure for the cryogenic system differs from its gaseous counterpart in that the induction-zone length for the heterogeneous system does not decrease as dramatically near the triple point. The detonation cell is also more elongated in the axial direction for the heterogeneous system. The average velocity of the two-dimensional front remains approximately 3% higher than the theoretical C–J velocity, indicating that no velocity deficit is observed for this mixture.

1.6.2 Modelling of Explosive Liquid Dispersal

Detonation initiation in a liquid spray is usually preceded by a liquid dispersal phase. For PDEs, this is typically accomplished through spray nozzles. Military FAE devices, on the other hand, disperse the fuel explosively using a small high-explosive charge.

Previous work on explosive liquid dispersal divided the controlling process into three phases [1, 81]. The first is an ejection phase, which occurs in the very near field when the liquid is accelerated by the expansion of the explosion products. This phase includes the detonation of the burster charge, the rupture of the casing, and the initial acceleration and breakup of the liquid. The second phase is a transition regime where the aerodynamic drag on the fluid starts to become important. The final phase is an expansion phase where the fluid droplets decelerate owing to drag. This phase accounts for seven eighths of the total volume expansion.

Some of these phases introduce new mechanisms that require specialized modelling techniques, as described in the following.

1.6.2.1 Fluid Dynamic Mechanisms

The dispersal of a liquid, using a central explosive charge, triggers a series of events that include:

1. Shock propagation in the liquid
2. Cavitation of the liquid when the shock emerges from the liquid and sends a cavitation-inducing expansion wave propagating back into the liquid
3. Rayleigh–Taylor (R–T) and/or Richtmyer–Meshkov (R–M) instabilities due to the sudden acceleration and expansion of the liquid sheet

4. Formation of primary droplets triggered by the instability processes
5. Kelvin–Helmholtz instabilities resulting from the shear created by the R–T and R–M instabilities
6. Flow-induced breakup of the primary droplets into secondary droplets
7. Droplet collision and coalescence
8. Evaporation of the smaller droplets

The processes described above are extremely complex and require very fine-grid resolution to resolve the various instabilities using available computational fluid dynamics (CFD) methods. For this reason, work performed on this difficult problem has usually combined analytical and CFD methods to construct an overall solution.

1.6.2.2 Modelling Approaches

Shock propagation and cavitation. The initial shock propagation phase in the liquid can be computed using specialized CFD or "hydrocodes" that can model the detonation propagation in the explosive, the wave transmission into the liquid, and the shock transmission into the air. This requires the use of suitable equations of state for the explosive, a real-gas equation for the air, and an equation of state for the liquid that is not only suitable for high pressures, but can also accommodate negative overpressures that are responsible for cavitation and vapour formation. Diffusive heat transfer between the explosion products and the liquid may also act to enhance vaporization. The overall process is further complicated by the fragmentation of the casing used to contain the liquid.

Early instability event. Since the liquid contains two interfaces, R–T instabilities can be triggered during the initial acceleration and subsequent deceleration. Although R–T and R–M instabilities can be observed using hydrocodes, the complex three-dimensional process leading to primary droplet formation would require prohibitively high resolution. The accuracy in modelling of the instabilities is also restricted by the lack of a surface tension model in most hydrocodes. For these reasons, this phase of the dispersal process is usually modelled using analytical theories with input from the hydrocode calculation just prior to the formation of the instability. Available instability theories for liquid interfaces include the linear theories of Taylor [130], Richtmyer [118], and Gardner [64]. Richtmyer's model was improved by Zhang and Sohn [158] to include non-linear effects. More recently, Wu [149] developed a critical Weber number theory for an expanding liquid torus. Although available linear and non-linear theories will provide information on critical conditions and the rate of instability growth, additional information is required to predict the diameter of the eventual droplet generated by the breakup. This usually involves some knowledge of the perturbation wavelength. Many theories assume that the droplet will be of the order of the liquid film thickness. For liquid dispersal problems involving casings, it is reasonable to assume that the wavelength will be associated with the casing fragment sizes. Suitable fragmentation models can be found in the work of Grady [67]. Borg [20] combined these models with the Zhang and Sohn nonlinear instability model to determine the breakup diameter of an impulsively driven smooth cylinder.

Spray development and evaporation. Once the initial droplet size, number density, and velocity have been determined, the subsequent spray evolution can be modelled using a suitable combination of relevant multiphase models.

1.6.2.3 Liquid Dispersal Modelling Studies

Explosive dispersal calculations have been performed using both simplified models and detailed CFD analysis. With respect to the former, Anderson et al. [9–13] developed simple correlations to compute the mean droplet diameter, the initial cloud expansion velocity, V_o, and the maximum cloud radius, R_{max}, on the basis of the fuel properties, such as density, viscosity, and surface tension, and on the burster charge and casing properties. Labbé [81] compared results from these correlations with his experimental studies and found relatively good agreement for V_o and R_{max}. Droplet diameters were not measured in the experiments.

The first detailed multidimensional explosive dispersal calculations were performed in 1976 by Rosenblatt et al. [119] at the Air Force Armament Laboratory using the DICE-FAE code. This model simulated the far-field expansion regime and included submodels for evaporation and droplet breakup. This work was followed in 1990 by calculations performed by Gardner [64] and Glass [66] at Sandia National Laboratories using the CTH and KIVA codes.

Gardner addressed the early phase of the dispersal process by performing CTH calculations that provided flow-field information as well as the liquid sheet thickness profile. The calculations indicated that the liquid volume expands into a very thin shell separating the high-explosive combustion products from the surrounding air. This information was used as input into a linear film instability model developed as part of the study. Since casing fragmentation was not considered, the droplet diameter was estimated on the basis of the sheet thickness. Gardner compared his results with the ethylene oxide experiments of Samirant [120]. The average droplet diameter of 4.4 mm predicted by the instability model was slightly less than the experimental value of 5 mm. Owing to the linear instability analysis, the computed droplet breakup time of 0.11 ms was much shorter than the experimental value of 1.3 ms.

Glass performed two-dimensional CFD calculations using the KIVA-II code which combined a κ–ϵ turbulence model with submodels for droplet evaporation, breakup, and collision (including coalescence). The probability distributions were computed using a stochastic parcel method. The report presented a table comparing computation times with different models turned on and off. Computationally intensive calculations included those related to aerodynamic droplet breakup, which doubled the overall calculation time, while the inclusion of droplet coalescence further extended the calculation by a factor of 4. Initial conditions for the flow field and the droplet distribution were estimated using simplified profiles based on the kinetic and internal energies proportions used by Rosenblatt. Calculations were performed using low- and high-vapour-pressure fuels (decane and propylene oxide) and for solid flaked aluminium. Sensitivity studies were performed to investigate the role of the initial droplet size on the overall dispersion profile. Calculations included monodispersed decane droplets with Sauter mean radii of 3, 10, and 30 mm and polydispersed χ^2 distributions with Sauter mean radii of 10, 20, and 30 mm. The final droplet size distribution of 5–20 μm was relatively insensitive to the initial distribution assumed. On the other hand, computations with smaller initial droplet sizes produced flatter clouds and richer regions than those produced with larger initial droplets. Although the far-field KIVA calculations seem to display late-time jet structures, it is not clear whether the jets are due to an intrinsic physical instability and are of a similar nature to those observed experimentally by Labbé [81].

More recently, Doustdar et al. [57] modelled the explosive liquid dispersal problem by representing the effect of the burster charge as a distribution of injector nozzles. The calculations were performed with a modified version of KIVA and a discrete-droplet model that accounts for droplet collision, coalescence, breakup, and evaporation. Calculations were performed with different initial injection velocities for a 15-kg cylinder of gasoline or kerosene at a 1-m height above the ground. The calculated flow fields reveal two main vortices generated at the top and bottom edges of the charge. These vortices control the fuel concentration distribution and the "effective volume" available for detonation. This volume increases with time and reaches a maximum as the upper vortex eventually penetrates the cloud to create a lean section at the centre of the cloud, thereby decreasing the effective volume. The time at which the maximum effective volume is reached represents the optimum time to detonate the cloud with a secondary charge. For the 15-kg charge considered, this time increases from approximately 65–120 ms for injection velocities between 150 and 350 m s^{-1}. The authors also compared computed cloud radius–time histories with experimental data for a 120-kg gasoline container dispersed with a C4 charge. This comparison, showing good agreement between computed and experimental results, was performed by assuming that 70% of the burster energy is converted to fuel kinetic energy, and by defining an expression between the injection velocity, the specific enthalpy of the explosive and the fuel-to-burster mass ratio.

It is clear from the above discussions that typical FAE devices are extremely complex to model. To the authors' knowledge, no one has attempted to model the complete FAE event from the initial cloud dispersal to the subsequent detonation initiation and propagation. This would require not only a very advanced computer code, but would also require extremely high resolution to resolve all the physical phenomena related to casing rupture, fuel jetting and breakup, and the highly complex spray combustion process.

1.7 Concluding Remarks

The studies on spray detonation to date point to the existence of two propagation regimes. In the case of high-vapour-pressure fuels (e.g. propylene oxide, heptane, hexane, and gasoline), significant fuel vapour accumulates during the fuel dispersal phase (i.e. 0.1–3.0-s duration). The resulting detonation takes place in the vapour–air mixture and a significant portion of the droplet breakup and combustion occurs in the wake of the detonation. For a high-volatility fuel, an appropriately sized burster charge, and a sufficiently long dispersal time, virtually all of the fuel evaporates and the optimal cloud size corresponds to the stoichiometric volume for the amount of fuel dispersed.

In the case of low-vapour-pressure fuels (e.g. decane, dodecane, kerosene, and diesel), the quantity of vapour that forms during the dispersal phase is low to insignificant. Under these conditions, initiation of detonation requires large initiator charges in large clouds. If the scale of the event is sufficiently large, aerodynamic stripping and atomization of the droplets persists until they become small enough for meaningful vaporization to take place in the shock-heated air. The reaction is controlled by turbulent mixing. Optimal detonation performance is observed for fuel-rich mixtures because not all of the fuel takes part in the detonation reaction.

The fact that low-vapour-pressure nitrated fuels (e.g. hexyl nitrate, ethyl hexyl nitrate, and IPN) detonate quite readily suggests that chemical kinetic effects inside the droplets themselves may be playing a role in increasing the detonation sensitivity of the fuel–air clouds. There is some experimental evidence indicating that nitrated fuels begin thermal decomposition immediately after being shocked by the fuel dispersal charge. The resulting heat release in the droplets may promote more rapid vaporization of the fuel. The free radicals formed during decomposition may also be contributing to the increased detonability. Future investigations should aim to clarify these issues.

In the case of gaseous detonations, there is a reasonably strong correlation between the critical tube diameter for detonation transmission and the detonation cell size for a broad range of fuel–oxidizer [94] and fuel–air [96] mixtures. Likewise, the explosion length characteristic of a powerful energy source has been shown to correlate well with the detonation cell size under conditions of critical initiation for selected fuel–air mixtures [115]. With the exception of the studies by Alekseev et al. [8] and Papavassiliou et al. [111], detonation cell size data for spray detonations are virtually non-existent. One study by the Kurchatov Institute concluded that the dynamic detonation parameters for gasoline–air sprays correlated with the detonation cell size in accordance with the relationships developed for gaseous fuel–air detonations.

However, a later study by the same group called into question the validity of cell-size-based correlations. Table 1.3 summarizes the available data for cell sizes and dynamic detonation parameters for fuel sprays in air. Similar data for fuel blends are compiled in Table 1.4. There is a clear need to measure the detonation cell sizes for the various spray systems and to determine definitively whether or not cell-size correlations are applicable. In the event these are not applicable, alternative scaling relationships will have to be developed.

Future experimental studies must strive to accurately measure the amount of vapour present in the mixture. Monitoring the transparency/opacity of the cloud is subjective and non-quantitative. The cloud temperature was used to estimate the amount of fuel that had vaporized in tests at the Institute Saint-Louis. This method should be further explored and others devised.

Two-front detonations have been observed for particle suspensions in combustible gas mixtures. The existence of such detonations is due to two combustion time scales; one for the gas and one for the particles. Although transient double-front profiles have been observed in spray detonation initiation computations and experiments, there is no experimental evidence to suggest that stable two-front detonations exist for sprays. Nonetheless, there are two combustion time scales associated with spray detonations as well; one for the vapour phase and one for the droplets that burn in the wake of the vapour-phase detonation. This begs the question: "Can stable two-front spray detonations exist?"

A number of issues concerning spray detonation have yet to be resolved. A hypothetical experimental study that might help improve our understanding of spray detonation and resolve some of the outstanding issues would involve a suspension of low-vapour-pressure fuel droplets in a combustible gaseous mixture (say, diesel droplets in ethylene–air). The variables in such a study would be the diesel droplet size and number density, the ethylene–air ratio, and the overall system stoichiometry. In essence, adjusting the ethylene concentration is a way of mimicking the fuel evaporated from the droplets. In the limit of a thin gas-phase reaction zone, the scenario of a high-volatility fuel would be recovered; i.e. non-reacting droplets in a gaseous detonation characterized by transverse waves. In the limit of a thick gas-phase reaction zone, the scenario of a low-vapour-pressure droplet suspension in air with almost no fuel vapour present would be recovered. Somewhere between these two extremes, new propagation mechanisms (e.g. double-front detonations) may become evident. While the fundamental aspects of heterogenous detonations were not their main interest, Knappe and Edwards conducted experiments of the type being suggested using gaseous propane and liquid decane [72]. They were able to vary the fuel present in the liquid phase from 0 to 100%. Although recent experimental and theoretical studies have been performed on such "hybrid" systems, it would be useful to conduct experiments on a large enough scale to eliminate boundary effects. Such experiments could be supported by numerical calculations investigating the cellular structure, and its regularity, as a function of the prevaporized fuel content.

Table 1.3. Summary of fuel properties and dynamic detonation parameters for fuel sprays in air

Fuel	Boiling point at 1 atm (°C)	Vapour pressure (kPa at °C)	Fuel dispersal time (s)	Critical initiator mass (kg)	Critical cloud height (m)	Detonation cell size (mm)	References
Decane	174	0.13 at 16.5	∞[a]	>0.3 (PE4)	>4	~41[b]	[19,35,132]
			∞[a]	<1[c]			[120]
Diesel	170–390	<0.09 at 20	<4	>6 (TNT)			[3,4,6]
			0.1	>0.7 (C4)			[43]
Dodecane	216	0.13 at 47.8	∞[a]	>0.5 (PE4)			[35]
Ethanol	78	5.3 at 19	0.1	>1 (C4)			[43]
			∞[a]	<1[c]			[120]
Ether	34.6	59 at 20	0.07	>0.1 (C4)			[43]
Ethyl hexyl nitrate	100	0.03 at 20	0.1	<0.2 (DM12)	<2		[33]
Gasoline	32–225	54–103 at 32	>1.8	0.3 (TNT)	2.5–3	40–50	[3,4,6]
			~2.5	0.59–0.64 (TNT)			[8]
			0.1	0.3–0.7 (C4)			[43]
			0.1	>2.5 (DM12)			[95]
Hexane	69	17.6 at 20	∞[a]	0.025 (PE4)			[35]
			0.15	FAE driver[d]		~55[b]	[19]
			∞[a]	FOE driver[e]		~40[f]	[156]
Hexanol	157	0.13 at 24.4	∞[a]	<1[c]			[120]
Hexyl nitrate	154	0.08 at 20	0.1	<0.2 (DM12)	<2		[33]
Isopropyl nitrate	101		0.06	0.25 (DM12)	<1	~10[f]	Labbé et al.[g]
			∞[a]	FOE driver[e]		~8[f]	[156]
Kerosene	171–325	0.13 at 20	<4	4.5 (TNT)	6.5–7		[3,4,6]
Methanol	65	13.3 at 21.2	∞[a]	<1[c]			[120]
Nitromethane	101	3.7 at 20	0.05	<0.2 (DM12)	<2		[33]
			0.1	>0.3 (C4)			[43]
Nitropropane	132	1.0 at 20	0.1	>0.2 (DM12)			[33]
Octyl nitrate	110		0.07	<0.1 (C4)			[43]
Propanol	97	1.3 at 14.7	0.1	>1 (C4)			[43]
			∞[a]	<1[c]			[120]
Propylene oxide	33.9	53.3 at 17.8	0.06	0.01 (DM12)	<1		[33,95,105]
			0.04	<0.05 (C4)	<0.5		[43]
			–	0.0035 (Tetryl)			[137]
Turpentine	154–170	<0.4 at 20	0.15	>1 (C4)			[43]
Xylene	138–144	1.3 at 27–32	0.07	>0.1 (C4)			[43]

FAE fuel-air explosive, *FOE* fuel–oxygen explosive.
[a]Continuous dispersal of fuel from nozzles.
[b]Obtained for a vapour-phase detonation in a heated tube at 100°C [19].
[c]Obtained from tests in a 0.5-m-diameter open-ended tube. Initiator explosive not specified [120].
[d]Initiator was an established detonation in a nitrated fuel droplet–air cloud.
[e]Initiator was an established detonation in a hydrogen–oxygen slug.
[f]Obtained for a vapour-phase detonation in a 30-cm heated tube at 115°C [156].
[g]J. Labbé, S. Murray, and K. Gerrard, Defence Research Establishment Suffield, unpublished results, 1992.

Table 1.4. Summary of dynamic detonation parameters for fuel blends sprayed in air

Fuel	Fuel dispersal time (s)	Critical initiator mass (kg)	Critical cloud height (m)	Detonation cell size (mm)	References
50% hexane + 50% dodecane	∞[a]	<0.12 (PE4)	<1.5		[35]
87.5% gasoline + 12.5% *n*-butyl nitrite	0.1	0.1–0.2 (C4)	<1		[43]
75% gasoline + 25% *n*-butyl nitrite	0.1	<0.1 (C4)	<1		[43]
50% diesel + 50% octyl nitrate	0.1	>0.1 (C4)			[43]
75% diesel + 25% *n*-butyl nitrite	0.1	>0.1 (C4)			[43]
90% hexane +10% isopropyl nitrate	∞[b]	FOE driver[c]		~33	[156]
50% hexane + 50% isopropyl nitrate	∞[b]	FOE driver[c]		~23	[156]

[a]Continuous dispersal of fuel from nozzles.
[b]Obtained for a vapour-phase detonation in a 30-cm heated tube at 115℃.
[c]Initiator was an established detoation in a high-pressure hydrogen–oxygen mixture.

There are three competing theories for why the deficits are large for spray detonations: (1) momentum and heat losses to the walls in the lengthy reaction zones characterizing these detonations, (2) incomplete burning of the fuel in the reaction zone, and (3) energy losses associated with the rearward propagating blast waves from droplet wake explosions. Further work is necessary to determine the relative contributions of the above-mentioned phenomena to these deficits. It is also interesting to note that the velocity deficits reported in this review paper vary widely depending on the fuel volatility and initial droplet diameters. Experimental values as high as 50% have been reported in the literature. A theoretical analysis by Zhang [154] suggests that detonation velocity deficits in heterogeneous dust–air systems can reach a maximum value of 48%, which corresponds to the so-called quasi-detonation limit where constant-volume explosion conditions occur at the sonic plane. Further studies on this problem for spray detonations are required to be able to distinguish stable detonations with large velocity deficits from failing waves in experimental and computational studies.

References

1. Abrahamson, G., Gates, R., Muller, G., Schreiner, K.: Explosive dissemination. Technical report 18. Stanford Research Institute, Menlo Park (1967)
2. Akbar, R., Thibault, P., Harris, P., Lussier, L.S., Zhang, F., Gerrard, K., Murray, S.: Detonation properties of unsensitized and sensitized JP-10 and jet A fuels in air for pulse detonation engines. In: Proceedings of the Joint Propulsion Conference and Exhibit, Huntsville, 15–19 Jul 2000
3. Alekseev, V., Dorofeev, S., Sidorov, V., Chaivanov, B.: Experimental study of detonation initiation in motor fuels sprayed in air. Report IAE-4872/13. Kurchatov Institute, Moscow (1989)
4. Alekseev, V., Dorofeev, S., Sidorov, V., Chaivanov, B.: Experimental study of large scale unconfined fuel spray detonations. Report IAE-5227/13. Kurchatov Institute, Moscow (1990)
5. Alekseev, V., Dorofeev, S., Sidorov, V., Chaivanov, B.: Investigation on blast waves transformation to detonation in two-phase unconfined clouds. Report IAE-5228/13. Kurchatov Institute, Moscow (1990)
6. Alekseev, V., Dorofeev, S., Sidorov, V., Chaivanov, B.: Experimental study of large-scale unconfined fuel spray detonations. In: Progress in Astronautics and Aeronautics, vol. 154, pp. 95–104. American Institute of Aeronautics and Astronautics, Reston (1993)
7. Alekseev, V., Dorofeev, S., Sidorov, V., Chaivanov, B.: Investigation on blast waves transformation to detonation in two-phase unconfined clouds. In: Progress in Astronautics and Aeronautics, vol. 154, pp. 105–116. American Institute of Aeronautics and Astronautics, Reston (1993)
8. Alekseev, V., Dorofeev, S., Sidorov, V.: Direct initiation of detonations in unconfined gasoline sprays. Shock Waves **6**, 67–71 (1996)
9. Andersen, W.: Model for the explosive dissemination of Newtonian and viscoelastic liquids. Contract report ARCSL-CR-81009. U.S. Army Armament Research and Development Command, Aberdeen Proving Ground, Maryland (1980)
10. Anderson, W., Wolfe, H.: Aerodynamic breakup of liquid drops. ARS J. **31**, 1783–1785 (1965)
11. Anderson, W., Louie, N., Ialongo, G.: Investigation of the aerodynamic breakup of viscoelastic liquids, phase I: Subsonic dissemination. Technical report, U.S. Army Armament Research and Development Command (1977)
12. Anderson, W., Lawton, E.: Literature survey on explosive dissemination. Contract Report ARCSL-CR-79064. U.S. Army Armament Research and Development Command, Aberdeen Proving Ground, Maryland (1979)
13. Andersen, W., Zernow, L., Gillespie, F., Twayne, M., Wong, J.: Explosive dissemination of thickened liquids. Contract report ARCSL-CR-79063. U.S. Army Armament Research and Development Command, Aberdeen Proving Ground, Maryland (May 1978–Jun 1979)
14. Axelsson, H., Berglund, S.: Cloud development and blast wave measurements from detonating fuel air explosive charges. FOA report C 20225-D4. Forsvarets Forskningsanstalt Huvudavdelning 2, Stockholm (1978)
15. Bach, G., Knystautas, R., Lee, J.: Initiation criteria for diverging gaseous detonations. In: Proceedings of the 13th Symposium (International) on Combustion, pp. 1097–1110. The Combustion Institute, Pittsburgh (1971)

16. Bar-Or, R.: Experimental study of cylindrical two-phase detonations in monodisperse sprays. Ph.D. thesis, University of Michigan, Department of Aerospace Engineering (1979)
17. Bar-Or, R., Sichel, M., Nicholls, J.: The propagation of cylindrical detonations in monodispersed sprays. In: Proceedings of the 18th Symposium (International) on Combustion, pp. 1599–1606. The Combustion Institute, Pittsburgh (1981)
18. Bar-Or, R., Sichel, M., Nicholls, J.: The reaction zone structure of cylindrical detonations of monodisperse sprays. In: Proceedings of the 19th Symposium (International) on Combustion, pp. 665–673. The Combustion Institute, Pittsburgh (1982)
19. Benedick, W., Tieszen, S., Knystautas, R., Lee, J.: Detonation of unconfined large-scale fuel spray-air clouds. In: Progress in Astronautics and Aeronautics, vol. 133, pp. 297–310. American Institute of Aeronautics and Astronautics, Reston (1991)
20. Borg, J.: Estimating the break-up diameter of an impulsively driven initially smooth cylinder. In: Proceedings of the Conference of the American Physical Society Topical Group on Shock Compression of Condensed Matter. AIP Conference Proceedings, vol. 706, pp. 715–718. American Institute of Physics, Melville (2004)
21. Borisov, A., Gel'fand, B.: Review of papers on detonation of two-phase systems. Arch. Thermodyn. Spal. **7**, 273–287 (1976)
22. Borisov, A., Gel'fand, B., Gubin, S., Kogarko, S., Podgrebenkov, A.: The reaction zone of two-phase detonations. Astronaut. Acta **15**, 411–419 (1970)
23. Bowen, J., Ragland, K., Steffes, F., Loflin, T.: Heterogeneous detonation supported by fuel fogs or films. In: Proceedings of the 13th Symposium (International) on Combustion, pp. 1131–1139. The Combustion Institute, Pittsburgh (1971)
24. Brophy, C., Hanson, R.: Fuel distribution effects on pulse detonation engine operation and performance. J. Propuls. Power **22**, 1155–1161 (2006)
25. Brophy, C., Netzer, D.: Effect of ignition characteristics and geometry on the performance of a JP-10/O_2 fueled pulse detonation engine. AIAA 99-2635 (1999)
26. Brophy, C., Netzer, D.: Effects of head-end geometry and ignition location on the operation of a JP-10/O_2 fueled pulse detonation engine. In: Proceedings of the 17th ICDERS, Heidelberg, 25–30 Jul 1999
27. Brophy, C., Netzer, D., Forester, D.: Detonation studies of JP-10 with oxygen and air for pulse detonation engine development. AIAA 98-4003 (1998)
28. Brophy, C., Netzer, D., Sinibaldi, J., Johnson, R.: Detonation of a JP-10/air aerosol for pulse detonation applications. In: High-Speed Deflagration and Detonation: Fundamentals and Control. Selected papers from the colloquium on the control of detonation processes, Moscow, 4–7 July 2000. ELEX-KM (2000)
29. Brophy, C., Sinibaldi, J., Netzer, D., Johnson, R.: Operation of a JP-10/air pulse detonation engine. AIAA 2000-3591 (2000)
30. Brophy, C., Netzer, D., Sinibaldi, J., Johnson, R.: Detonation of a JP-10/aerosol for pulse detonation applications. In: Proceedings of the 13th ONR Propulsion Meeting, Minneapolis, 10–11 Aug 2000, pp. 190–197
31. Brophy, C., Netzer, D., Sinibaldi, J.: Detonation studies of JP-10 sprays and initiator development for pulse detonation engines. In: Proceedings of the 14th ONR Propulsion Meeting, Chicago, 8–10 Aug 2001, pp. 171–176

32. Brophy, C., Sinibaldi, J., Damphousse, P.: Initiator performance for liquid-fueled pulse detonation engines. AIAA 2002-0472 (2002)
33. Brosinsky, C., Moen, I., Murray, S.: Dispersed fuel–air explosives: Evaluation of alternate fuels and fuel combinations. Memorandum 1189. Defence Research Establishment Suffield, Ralston (1987)
34. Bull, D., Elseworth, J., Hooper, G.: Concentration limits to unconfined detonation of ethane–air. Combust. Flame **35**, 27 (1979)
35. Bull, D., McLeod, M., Mizne, G.: Detonation of unconfined fuel aerosols. In: Progress in Astronautics and Aeronautics, vol. 75, pp. 48–60. American Institute of Aeronautics and Astronautics, Reston (1981)
36. Burcat, A., Eidelman, S.: Evolution of a detonation wave in a cloud of fuel droplets: part II. Influence of fuel droplets. AIAA J. **18**, 1233–1236 (1980)
37. Burcat, A., Scheller, K., Lifshitz, A.: Shock-tube investigation of comparative ignition delay times for C_1–C_5 alkanes. Combust. Flame **16**, 29–33 (1971)
38. Chang, E., Kailasanath, K.: Shock wave interactions with particles and liquid fuel droplets. Shock Waves **12**, 333–341 (2003)
39. Cheatham, S., Kailasanath, K.: Numerical modelling of liquid-fuelled detonation in tubes. Combust. Theory Modell. **9**, 23–48 (2005)
40. Cherepanov, G.: The theory of detonations in heterogeneous systems. Prik. Mech. Tek. Fiz. **4**, 163 (1965). The preview of the English translation of this paper is also available at: http://www.springerlink.com/content/j0g8018066p05l6p/
41. Chernyi, G., Chernyavskii, S., Baulin, N.: The motion of high-speed bodies in a hydrogen-air mixture. Dokl. Akad. Nauk SSSR **212**, 316–319 (1973)
42. Christensen, S.: Detonasjonsforsok med propylenoksid. Fortifikatorisk notat NR 279/99. Forsvarets Bygningstjeneste Teknisk Avdeling, Oslo (1999)
43. Christensen, S., Hermansen, T.: Detonasjonsforsok, raufoss 1997. Fortifikatorisk notat NR 249/97. Forsvarets Bygningstjeneste Teknisk Avdeling, Oslo (1997)
44. Cooke, N., Khandhadia, P.: Unconfined vapour clouds, I: kinetics of dispersed clouds of liquid. In: Proceedings of the International Conference on Vapor Cloud Modeling, Boston, pp. 597–624 (1987)
45. Cooke, N., Khandhadia, P.: Unconfined vapour clouds, II: kinematics of explosively dispersed clouds of liquid. In: Proceedings of the International Conference on Vapor Cloud Modeling, Boston, pp. 625–666 (1987)
46. Cooke, N., Khandhadia, P., Brown, D., Eyre, F., Mansour, M.: Experimental facility for fuel–air explosive research. Contract report. SNC Incorporated, Montreal (1985)
47. Cowperthwaite, M., Zwisler, W.: Tiger computer program documentation. Report Z106. Stanford Research Institute, Menlo Park (1973)
48. Crowe, C., Sommerfeld, M., Yutaka, T.: Multiphase flow with droplets and particles. CRC, Boca Raton (1997)
49. Dabora, E.: Fundamental mechanisms of liquid spray detonations. In: Proceedings of the International Conference on Fuel–Air Explosions, McGill University, Montreal, study no. 16, pp. 245–264. University of Waterloo Press, Waterloo, Ontario (1982)
50. Dabora, E., Weinberger, L.: Present status of detonations in two-phase systems. Acta Astronaut. **1**, 361–372 (1974)

51. Dabora, E., Ragland, K., Nicholls, J.: Drop-size effects in spray detonations. In: Proceedings of the 12th Symposium (International) on Combustion, pp. 19–26. The Combustion Institute, Pittsburgh (1969)
52. Desbordes, D.: Failure of the classical dynamic parameters relationships in highly regular cellular detonation systems. In: Progress in Astronautics and Aeronautics, vol. 153, pp. 347–362. American Institute of Aeronautics and Astronautics, Reston (1993)
53. Dorofeev, S.: Blast effects of confined and unconfined explosions. In: Proceedings of the 20th Symposium (International) on Shock Waves, Pasadena, pp. 77–86 (1995)
54. Dorofeev, S., Sidorov, V., Efimenko, A., Kochurko, A., Kutznetsov, M., Chaivanov, B., Matsukov, D., Pereversev, A., Avenyan, V.: Fireballs from deflagration and detonation of heterogeneous fuel-rich clouds. Fire Saf. J. **25**, 323–336 (1995)
55. Dorofeev, S., Sidorov, V., Kuznetsov, M., Dvoinishnikov, A., Alekseev, V., Efimenko, A.: Airblast and heat radiation from fuel-rich mixture detonations. Shock Waves **6**, 21–28 (1996)
56. Dorofeev, S., Bezmelnitsin, A., Sidorov, V., Yankin, J., Matsukov, I.: Turbulent jet initiation of detonation in hydrogen–air mixtures. Shock Waves **6**, 73–78 (1996)
57. Doustdar, M., Mazaheri, K., Hosseinalipour, M.: Computational analysis of injection-velocity effects on dynamic parameters of unconfined fuel–vapor clouds. Combust. Explos. Shock Waves **41**, 510–520 (2005)
58. Edwards, D., Hooper, G., Morgan, J.: An experimental investigation of spherical detonations. Acta Astronaut. **3**, 117 (1976)
59. Eidelman, S., Timnat, Y., Burcat, A.: The problem of a strong point explosion in a combustible medium. In: Proceedings of the Sixth Symposium on Detonation, Coronado, California, 24–27 Aug (1976)
60. Eidelman, S., Burcat, A.: Evolution of a detonation wave in a cloud of fuel droplets: part I. Influence of ignition explosion. AIAA J. **18**, 1103–1109 (1980)
61. Engel, O.: Fragmentation of water drops in zone behind air shock. J. Res. Natl. Bur. Stand. **60**, 245–280 (1958)
62. Frolov, S.: Liquid-fueled, air-breathing pulse detonation engine demonstrator: Operation principles and performance. J. Propuls. Power **22**, 1162–1169 (2006)
63. Gabrijel, Z., Nicholls, J.: Cylindrical heterogeneous detonation waves. Acta Astronaut. **5**, 1051 (1978)
64. Gardner, D.: Near-field dispersal modeling for liquid fuel–air explosives. Report SAND90-0686. Sandia National Laboratories, Albuquerque (1990)
65. Gey, W., Mygaard, M.: Feasibility study of FAX explosive. Technical report NAVWEPS-8065. U.S. Naval Ordnance Test Station, China Lake (1963)
66. Glass, M.: Far-field dispersal modeling for fuel–air-explosive devices. Report SAND90-0528. Sandia National Laboratories, Albuquerque (1990)
67. Grady, D., Kipp, M.: Geometric statistics and dynamic fragmentation. J. Appl. Phys. **58**, 1210–1222 (1985)
68. Gubin, S., Gel'fand, B.: A possible mechanism for reducing the lower limit of detonation in hybrid systems. Khim. Fiz. **4**, 572–576 (1985)
69. Gubin, S., Sichel, M.: Calculation of the detonation velocity of liquid droplets and gaseous oxidizer. Combust. Sci. Technol. **17**, 109–117 (1977)
70. Kailasanath, K.: Liquid-fueled detonations in tubes. J. Propuls. Power **22**, 1261–1268 (2006)

71. Kauffman, C., Nicholls, J.: Shock-wave ignition of liquid fuel drops. AIAA J. **9**, 880–885 (1971)
72. Knappe, B., Edwards, C.: Results of decane/propane split-fuelled spray detonations. In: Proceedings of the 16th ONR Propulsion Meeting, Arlington, pp. 117–122 (2003)
73. Knystautas, R., Lee, J.: Catalytic initiation and transmission of detonation, flame acceleration and transition to detonation in fuel–air explosive clouds (FAE III). Progress report. McGill University, Shock Wave Physics Group, Montreal (1984)
74. Knystautas, R., Lee, J.: Initiation, transition and transmission of detonation in fuel–air vapour clouds. Technical report. McGill University, Shock Wave Physics Group, Montreal (1986)
75. Knystautas, R., Lee, J., Moen, I., Wagner, H.: Direct initiation of spherical detonation by a hot turbulent gas jet. In: Proceedings of the 17th Symposium (International) on Combustion, p. 1235. The Combustion Institute, Pittsburgh (1979)
76. Knystautas, R., Guirao, C., Lee, J., Sulmistras, A.: Measurements of cell size in hydrocarbon–air mixtures and predictions of critical tube diameter, critical initiation energy, and detonability limits. In: Progress in Astronautics and Aeronautics, vol. 94, pp. 23–37. American Institute of Aeronautics and Astronautics, Reston (1985)
77. Koren'kov, V., Obukhov, A.: Influence of the location of detonation initiation in a cloud of fuel–air mixture on the characteristics of the explosion field: numerical experiments. Combust. Explos. Shock Waves **37**(6), 688–697 (2001)
78. Korobeinikov, V.: On arising in dusty gas zones with high concentration of particles. Arch. Combust. **9**, 149–152 (1989)
79. Kuznetsov, M., Dorofeev, S., Alekseev, V., Abyzov, M.: Blast wave interactions with burning cloud of fuel sprays. Arch. Combust. **18**, 7–20 (1998)
80. Kuznetsov, M., Dorofeev, S., Alekseev, V., Abyzov, M.: Blast wave interactions with burning cloud of fuel sprays. In: Proceedings of the Colloquium on Gas, Vapor, Hybrid and Fuel–Air Explosions, International Symposium on Hazards, Prevention and Mitigation of Industrial Explosions, Schaumberg, 21–25 September 1998
81. Labbé, J.: Impulsive dispersal of liquid fuels in air. Master's thesis. Department of Mechanical Engineering, Royal Military College of Canada, Kingston (1991)
82. Lee, J.: Initiation of detonation by opposing reactive jets. In: Report on the AFOSR Workshop on FAE III, Fort Walton Beach, Jan 1979
83. Lee, J.: Dynamic parameters of gaseous detonation. Annu. Rev. Fluid Mech. **16**, 311–336 (1984)
84. Lee, J., Knystautas, R., Yoshikawa, N.: Photochemical initiation of gaseous detonations. Acta Astronaut. **5**, 971–982 (1978)
85. Lee, J., Knystautas, R., Guirao, C.: The link between cell size, critical tube diameter, initiation energy and detonability limits. In: Proceedings of the International Conference on Fuel–Air Explosions, McGill University, Montreal, Canada, study no. 16, pp. 157–188. University of Waterloo Press, Waterloo, Ontario (1982)
86. Lobanov, D., Fonbershtein, E., Ekomasov, S., Tregub, A.: Detonation of gasoline–air mixtures. Combust. Explos. Shock Waves **14**, 119–120 (1978)
87. Lu, P., Slagg, N., Fishburn, B.: The effects of chemical and physical processes on two-phase detonations. Technical report. Energetics Materials Division, U.S. Army Armament Research and Development Command (Jun 1978)

88. Lu, P., Slagg, N., Fishburn, B.: Relation of chemical and physical processes in two-phase detonations. In: Proceedings of the Sixth ICDERS, Stockholm, Aug 1977. Acta Astronaut., **6**, 815–826 (1979)
89. Mackay, D., Murray, S., Moen, I.: Flame jet ignition of large fuel air clouds. Memorandum 1274. Defence Research Establishment Suffield, Ralston (1988)
90. Mackay, D., Murray, S., Moen, I.: Flame jet ignition of large fuel air clouds. In: Proceedings of the 22nd Symposium (International) on Combustion, pp. 1339–1353. The Combustion Institute, Seattle (1988)
91. Manzhalei, V., Mitrofanov, V., Subbotin, V.: Measurement of inhomogeneities of a detonation front in gas mixtures at elevated pressures. Combust. Explos. Shock Waves **10**, 89–95 (1974)
92. Mayer, E.: Theory of liquid atomization in high velocity gas streams. ARS J. **31**, 1783–1785 (1961)
93. Mitrofanov, V., Pinaev, A., Zhdan, S.: Calculations of detonation waves in gas-droplet systems. Acta Astronaut. **6**, 281–296 (1979)
94. Mitrofanov, V., Soloukhin, R.: The diffraction of mutifront detonation waves. Sov. Phys. Dokl. **9**, 1055 (1979)
95. Moen, I., Ward, S.: Minefield breaching with fuel–air explosives: Effectiveness of FAE line charges. Report 406. Defence Research Establishment Suffield, Ralston (1984)
96. Moen, I., Funk, J., Ward, S., Rude, G., Thibault, P.: Detonation length scales for fuel–air explosives. In: Progress in Astronautics and Aeronautics, vol. 94, pp. 55–79. American Institute of Aeronautics and Astronautics, Reston (1985)
97. Moen, I., Sulmistras, A., Thomas, G., Bjerketvedt, D., Thibault, P.: The influence of cellular regularity on the behaviour of gaseous detonations. In: Progress in Astronautics and Aeronautics, vol. 106, pp. 220–243. American Institute of Aeronautics and Astronautics, Reston (1986)
98. Moen, I., Bjerketvedt, D., Engebretsen, T., Jenssen, A., Hjertager, B., Bakke, J.: Transition to detonation in a flame jet. Combust. Flame **75**, 297–308 (1989)
99. Murray, S., Lee, J.: The influence of physical boundaries on gaseous detonation waves. In: Progress in Astronautics and Aeronautics, vol. 106, pp. 329–355. American Institute of Aeronautics and Astronautics, Reston (1986)
100. Murray, S., Thibault, P.: Chemical initiation of detonation in hydrocarbon–air mixtures. Paper presented at Nagoya International Center, Nagoya (1991)
101. Murray, S., Gottlieb, J., Coffey, C., Moen, I., Lee, J., Remboutsikas, D.: Direct initiation of detonation in ethylene-air mixtures – influence of bag size. In: Proceedings of the Seventh MABS, Medicine Hat 1981
102. Murray, S., Gerrard, K.: Field studies of fuel–air explosive line charges: a summary report for the 65-mm charge. Report 534. Defence Research Establishment Suffield, Ralston (1990)
103. Murray, S., Thibault, P., Moen, I., Knystautas, R., Lee, J., Sulmistras, A.: Initiation of hydrogen–air detonations by turbulent fluorine-air jets. In: Progress in Astronautics and Aeronautics, vol. 133, pp. 91–117. American Institute of Aeronautics and Astronautics, Reston (1991)
104. Murray, S., Briosi, G., Pinco, M., Anderson, C.: Sensitized nitroparaffin blends: a practical approach to the use of liquid explosives in a military context. Report 601. Defence Research Establishment Suffield, Ralston (1994)
105. Murray, S., Gerrard, K.: On the detonability and blast from propylene-oxide and nitromethane droplet-air clouds. In: Proceedings of the Second International Specialist Meeting on Fuel–Air Explosions, Bergen, 26–28 June (1996)

106. Murray, S., Moen, I., Lee, J., Knystautas, R.: Chemical initiation of detonation in fuel air explosive clouds. Canadian Patent 2,021,396 and US Patent 5,168,123, 13 Aug (1996)
107. Murray, S., Thibault, P., Zhang, F., Bergeron, D., Gerrard, K.: Blast simulators based on fuel–air explosives and shock focusing. In: Proceedings of the 18th MABS, Bad Reichenhall, 27 Sep–1 Oct 2004
108. Nettleton, M.: Shock-wave chemistry in dusty gases and fogs: a review. Combust. Flame **30**, 3–16 (1977)
109. Nicholls, J., Sichel, M., Fry, R., Glass, D.: Theoretical and experimental study of cylindrical shock and heterogeneous detonation waves. Acta Astronaut. **1**, 385–404 (1974)
110. Nicholls, J., Bar-Or, R., Gabrijel, Z., Petkus, E.: Recent experiments on heterogeneous detonation waves. In: Proceedings of the 17th Aerospace Sciences Meeting, New Orleans, 15–17 Jan 1979
111. Papavassiliou, J., Makris, A., Knystautas, R., Lee, J.: Measurements of cellular structure in spray detonation. In: Progress in Astronautics and Aeronautics, vol. 154, pp. 148–169. American Institute of Aeronautics and Astronautics, Reston (1993)
112. Pierce, T.: Experimental and theoretical study of the structure of two-phase detonation in sprays. Ph.D. thesis, University of Michigan (1972)
113. Pierce, T., Nicholls, J.: Time variation in the reaction zone structure of two-phase spray detonations. In: Proceedings of the 14th Symposium (International) on Combustion, pp. 1277–1284. The Combustion Institute, Pittsburgh (1973)
114. Pierce, T., Nicholls, J.: Hybrid gas phase two-phase detonations. Combust. Sci. Technol. **9**, 119 (1974)
115. Radulescu, M., Higgins, A., Murray, S., Lee, J.: An experimental investigation of the direct initiation of cylindrical detonations. J. Fluid Mech. **480**, 1–24 (2003)
116. Ragland, K., Dabora, E., Nicholls, J.: Observed structure of spray detonations. Phys. Fluids **11**, 2377–2388 (1968)
117. Ranger, A., Nicholls, J.: Aerodynamic shattering of liquid drops. AIAA J. **7**, 285–290 (1969)
118. Richtmyer, R.: Taylor instability in shock acceleration of compressible fluids. Commun. Pure Appl. Math. **13**, 297–319 (1960)
119. Rosenblatt, M., Eggum, G., Kreyenhagen, K.: DICE-FAE analysis of fuel dispersal and detonation from a fuel–air-explosive device. Contract report AFATLTR-76-33. U.S. Air Force Armament Laboratory, Eglin Air Force Base (25 Oct 1975–25 Feb 1976)
120. Samirant, M., Smeets, G., Baras, H., Royer, H., Oudin, L.: Dynamic measurements in combustible and detonable aerosols. Propellants Explos. Pyrotech. **14**, 47–56 (1989)
121. Sayles, D.: Method of generating single-event, unconfined fuel–air detonation. US Patent 4,463,680 (1984)
122. Shamshev, K.: Munitions based upon low-density explosives. In: Proceedings of the conference on the role of advanced conventional weapons in joint peacekeeping and anti-terrorism operations. United States, Oct 1996
123. Sichel, M.: Modeling of gaseous and heterogeneous detonation phenomena. In: 27th Conference of Army Mathematicians, West Point, 1981

124. Sirignano, W.: Fluid Dynamics and Transport of Droplets and Sprays. Cambridge University Press, Cambridge (1999)
125. Smirnov, N., Nikitin, V., Legros, J.: Ignition and combustion of turbulized dust-air mixtures. Combust. Flame **123**, 46–67 (2000)
126. Smirnov, N., Nikitin, V., Khadem, J., Shevtsova, V., Legros, J.: Detonation initiation in dispersed fuel–air mixtures. In: Proceedings of the 20th ICDERS, Montreal, 2005
127. Stephans, F., Bowen, J.: Decane aerosol detonations. In: Proceedings of the 21st Annual Meeting of the American Physical Society, 1963
128. Stiesch, G.: Modeling engine spray and combustion processes. Springer, Berlin (2003)
129. Tang, M., Nicholls, J., Lin, Z., Sichel, M., Kauffman, C.: Direct initiation of detonation in a decane spray. In: Progress in Astronautics and Aeronautics, vol. 106, pp. 474–489. American Institute of Aeronautics and Astronautics, Reston (1985)
130. Taylor, G.: The instability of liquid surfaces when accelerated in a direction perpendicular to their planes, 1. Proc. R. Soc. Lond. A **201**, 192–196 (1950)
131. Tieszen, S., Stamps, D., Westbrook, C., Pitts, W.: Gaseous hydrocarbon–air detonations. Report SAND88-1925J. Sandia National Laboratories, Albuquerque (1988)
132. Tieszen, S., Stamps, D., Westbrook, C., Pitts, W.: Gaseous hydrocarbon–air detonations. Combust. Flame **84**, 376–390 (1991)
133. Tulis, A.: Induced heterogeneous detonation in hypergolic fuel-oxidizer dispersions. In: Chemical and Physical Processes in Combustion, pp. 26-2–26-4. Proceedings of the Fall Technical Meeting of the Eastern Section of the Combustion Institute, Miami Beach, 29 Nov 29–1 Dec 1978
134. Tulis, A.: A model for induced heterogeneous detonation. In: Report on the AFOSR Workshop on FAE III, Fort Walton Beach, Jan 1979
135. Ungut, A., Shuff, P.: Deflagration to detonation transition from a venting pipe. Combust. Sci. Technol. **63**, 75–87 (1989)
136. Urtiew, P., Lee, E., Walker, F.: Chemical initiation of gaseous detonation in a small spherical volume. Pol. Arch. Thermodyn. Combust. **9**, 261–272 (1978)
137. Vanta, E., Parsons, G., Collins, P.: Detonability of propylene-oxide/air and n-propyl nitrate/air mixtures. Report AFATL-TR-73-3. U.S. Air Force Armament Laboratory (1973)
138. Vasil'ev, A.: Initiation of gaseous detonation by a high speed body. Shock Waves **3**, 321–326 (1994)
139. Vasil'ev, A.: Detonation combustion of gas mixtures using a hypervelocity projectile. Shock Waves **33**, 583–597 (1997)
140. Vasil'ev, A.: Detonation hazards of gaseous mixtures. In: Prevention of Hazardous Fires and Explosions, pp. 93–108. Kluwer, Dordrecht (1999)
141. Vasilev, A., Zhdan, S., Mitrofanov, V.: Detonation initiation in gaseous and heterogeneous systems. Gaseous and Heterogeneous Detonations: Science to Applications, pp. 25–38. ENAS, Moscow (1999)
142. Vasil'ev, A., Valishev, A., Vasil'ev, V., Panfilova, L.: Combustion and detonation characteristics of hydrazine and its methyl derivatives. Combust. Explos. Shock Waves **3**, 358–373 (2000)
143. von Elbe, G.: A study of the effect of free radical generation by fluorine and fluorine producers on hydrocarbon combustion. Technical report. Atlantic Research Corporation, Alexandria (1973)

144. von Elbe, G., McHale, E.: Chemical initiation of FAE clouds. Annual interim report. Atlantic Research Corporation, Arlington (1978)
145. von Elbe, G., McHale, E.: Chemical initiation of FAE clouds. Annual interim report. Atlantic Research Corporation, Arlington (1979)
146. von Elbe, G., McHale, E.: A proposed FAE III device. In: Report on the AFOSR Workshop on FAE III, Fort Walton Beach, Jan 1979
147. Voromin, D., Zhdan, S.: Calculation of heterogeneous detonation initiation for a hydrogen-oxygen mixture in an explosion tube. Combust. Explos. Shock Waves **20**, 461–465 (1985)
148. Ward, S., Funk, J., Moen, I.: Minefield breaching with fuel–air explosives: Preliminary study of propylene oxide line charges. Memorandum 1086. Defence Research Establishment Suffield, Ralston (1984)
149. Wu, Z.N.: Approximate critical Weber number for the breakup of an expanding torus. Acta Mech. **166**, 231–239 (2003)
150. Xuezhong, X., Boatian, H., MingJing, P., Shan, W., Huaquan, H.: Study on chemical initiation of fuel air explosive. Arch. Combust. **21**, 93–100 (2001)
151. Yoshikawa, N.: Coherent shock wave amplification in photo-chemical initiation of gaseous detonations. Ph.D. thesis, Department of Mechanical Engineering, McGill University, Montreal (1980)
152. Zel'dovich, Y.: Regime classification of an exothermic reaction with non-uniform initial conditions. Combust. Flame **39**, 211–214 (1990)
153. Zel'dovich, Y., Librovich, V., Makhviladze, G., Sivashinsky, G.: On the development of detonation in non-uniformly preheated gas. Acta Astronaut. **15**, 313–321 (1970)
154. Zhang, F.: Detonation waves in dust media: A review. In: Proceedings of the 40th Aerospace Sciences Meeting and Exhibition, AIAA-2002-0772, Reno, 14–17 Jan 2002
155. Zhang, F., Greilich, P., Gronig, H.: Propagation mechanism of dust detonations. Shock Waves **2**, 81 (1992)
156. Zhang, F., Akbar, R., Thibault, P., Murray, S.: Effects of nitrates on hydrocarbon–air flames and detonations. Shock Waves **10**, 457–466 (2001)
157. Zhang, F., Murray, S., Gerrard, K.: JP-10 vapour detonations at elevated pressures and temperatures. In: Proceedings of the 18th ICDERS, Seattle, 29 Jul 29–3 Aug 2001
158. Zhang, Q., Sohn, S.: An analytical nonlinear theory of Richmyer–Meshkov instability. Phys. Lett. **212**, 149–155 (1996)
159. Zhdan, S.: Calculation of the initiation of a heterogeneous detonation with a charge of condensed explosive. Combust. Explos. Shock Waves **17**, 674–679 (1981)
160. Zhdan, S., Prokhorov, E.: Calculation of the cellular structure of detonation of sprays in an H_2–O_2 system. Combust. Explos. Shock Waves **36**, 777–784 (2000)

2

Detonation of Gas–Particle Flow

F. Zhang

2.1 Introduction

Fine organic or metallic particles suspended in an oxidizing or combustible gas form a reactive particle–gas mixture. Explosion pressures in such mixtures are remarkably higher than those of gaseous fuel–air mixtures because of the high energy content of particles and the initial particle mass that transitions to explosion product gases. According to the component reactivity, detonation in particle–gas mixtures may be classified as:

1. "Heterogeneous detonation" in a reactive particle–oxidizing gas mixture
2. "Hybrid detonation" in a reactive particle–reactive gas mixture
3. "Dusty detonation" in an inert particle–reactive gas mixture

Reactive particles can be fuel particles or monopropellant particles that contain both fuel and oxidizer.

While dust explosions have been recognized from the beginning of coal mine exploitation, fundamental studies of heterogeneous detonation in gas–particle flow may trace their origin to the experimental work of Strauss [1] in 1968 for aluminum particle–oxygen mixtures, Nettleton and Stirling [2] in 1973 for coal dust–oxygen mixtures, Cybulski [3] in 1971 for coal dust–air mixtures, and Bartknecht [4] for other organic dust–air mixtures. Since then the fundamental heterogeneous detonation studies might be divided into two periods: global phenomenon studies between the 1970s and 1980s and transverse wave detonation studies since the late 1980s. Representative works in the first period include Wolanski and his coworkers [5,6] for coal dusts, Kauffman et al. [7] for grain dusts, and Peraldi and Veyssiere [8] for cornstarch. Most of the experiments were conducted in small tubes of a few centimeters in cross-sectional dimension by a few meters long, where solid particles were dispersed in pure oxygen to increase mixture sensitivity. The deflagration-to-detonation transition (DDT) observed in these experiments mainly exhibited a progressive nature where the flame was gradually accelerated to a quasi steady state without an abrupt onset of overdriven detonation accompanied

F. Zhang (ed.), *Shock Wave Science and Technology Reference Library: Heterogeneous Detonation*, DOI: 10.1007/978-3-540-88447-7_2,

by a retonation wave. The detonation velocities determined were up to 20–40% less than those predicted by the equilibrium Chapman–Jouguet (CJ) theory. While this phenomenon was termed "quasi-detonation" [9], its propagation mechanism remains a subject of current research. Some global phenomena of marginal spinning detonation were also observed in certain oxygen experiments [1]. In the same period, Wolanski et al. [10], Lee and Sichel [11], and Fan and Sichel [12] extended the one-dimensional Zeldovich–von Neumann–Döring (ZND) detonation model to the heterogeneous detonation where a generalized CJ condition was assumed at the phase-frozen or gaseous sound speed. Perhaps the most fundamental lesson learned in this period was that even when the explosion pressure is invaluably high, micrometric organic or metallic particles are not sensitive to detonation owing to the additional time scales inherent in the mass, momentum, and heat transfer between the finite-sized particles and the gas. Large tubes would be required to observe the DDT and propagation of a self-sustained heterogeneous detonation wave with a strong initiation and a sufficient tube length.

In 1982 Tulis and Selman [13] reported aluminum–air experiments in a 152-mm-diameter tube with a short length of 5.5 m and a 3-g condensed explosive for initiation. It was found that detonation was only achieved for flaked aluminum with a surface area to mass ratio of $3-4\,m^2\,g^{-1}$, equivalent to spherical particles of diameter less than 1 μm. The observed detonation velocity fluctuated between 1,350 and $1{,}640\,m\,s^{-1}$ with a deficit of 10–30% relative to the CJ values. Later, in 1991, Borisov et al. [14] reported more consistent aluminum–air detonation velocities between 1,700 and $1{,}800\,m\,s^{-1}$ in a 122-mm-diameter tube for flakes and 1-μm atomized particles. A strong direct initiation of the detonation technique was employed owing to the short tube length of 4.2 m; thus, it was unclear whether the detonation wave observed was still influenced by the initiation. In 1986, Gardner et al. [15] for the first time clearly recorded the transition with a violent onset to detonation for a coal dust–air mixture in a tube 600 mm in diameter by 42 m in length. At the end of the tube, the transient velocity of the combustion wave reached $2{,}850\,m\,s^{-1}$ and a peak pressure of 80 atm was measured. However, the tube, with a length–diameter ratio of 70, was still too short to record a self-sustained detonation wave. Since 1987, Zhang, Grönig, and their coworkers [16–20] have reported a number of conclusive results for the existence of a self-sustained detonation wave for particles suspended in oxygen and air through observation of the detailed transverse wave structure (spinning and cellular detonation) using a 140-mm- and a 300-mm-diameter tube with a length–diameter ratio larger than 120. The detonation waves observed for aluminum, anthraquinone, and cornstarch particles in air had a transverse wave spacing of 0.4–1 m with a velocity deficit less than 10% relative to the CJ values; the DDT was mostly characterized by an abrupt onset of overdriven detonation accompanied by a retonation wave. The large transverse wave spacing suggests that observation of unconfined detonation in reactive particle–air mixtures would require a much larger scale than that of confined tubes and much stronger initiation. Evidence of the transverse wave structure has recently arisen for unconfined heterogeneous detonation [21, 22].

While a large transverse wave spacing is inherent to detonation in reactive particle–air flow, reactive particles added to a detonable gas mixture can cause a variety of detonation modes as a result of the interaction between the gas reaction and the additional physical processes involved in the mass, momentum, and heat transfer between the two phases. Fast deflagration of particles in gas detonation flow may enhance gas detonation or the DDT and has been referred to as "hybrid detonation" or "hybrid DDT" [23–25]. In 1982, Veyssiere [23, 24] reported for the first time the observation of a detonation wave composed of a double shock structure when aluminum particles were suspended in a lean reactive gas mixture in a 69-mm-diameter tube. In the same time period, Afanasieva et al. [26] theoretically postulated the existence of "double-shock" detonation in multiphase media due to two successive energy releases. Khasainov and Veyssiere [27] applied a two-phase ZND model to show that a "steady" double-shock detonation structure can exist, in which the two fronts are stabilized by a generalized CJ condition for the particle–gas mixture at two subsequent phase-frozen sound speed locations. Their further analysis explored the multiplicity of steady solutions for given initial conditions and nonmonotonic behavior of the heat release process [28]. Wolinski et al. [29] reported that the addition of oat particles into methane–air mixtures may promote methane–air detonation and that a secondary compression wave appeared owing to the late particle combustion. Recent experiments have provided more conclusive evidence on the self-sustained propagation of double-shock detonation for aluminum particles suspended in various detonable gas mixtures [30–33]. In an 80-mm-diameter, 10-m-long tube, Zhang et al. [32, 33] found that the double-shock detonation can quasi-steadily propagate in two modes: either the second shock has the same velocity as the leading shock, or the second shock velocity is less than the leading shock velocity. It was explained as the weak detonation solutions supported by the particle reaction in different time delays and energy release rates. The variety and complexity of hybrid detonation waves and their propagation mechanisms have yet to be fully understood and remain an active area of current detonation research.

It is noticeable that monopropellant or explosive particles can be added into particle–gas mixtures to increase the detonation sensitivity and readers can find relevant results, for example, in the work of Tulis et al. [34]. Detonation in layered dust is less sensitive than in the suspension and the DDT in layered grain dust was investigated by Li et al. [35] in a 300-mm-diameter, 70-m-long air-filled tube.

This chapter focuses on the physical phenomena of heterogeneous detonation and hybrid detonation for particles suspended in gas, while the features of dusty detonation are described briefly. The gas–particle detonation theory is introduced in Sect. 2.2, where particular characteristics of the equilibrium CJ model, the two-phase ZND model, and the unsteady two-phase fluid dynamics model are discussed according to detonation types in gas–particle flow. Sections 2.3 and 2.4 concentrate on the heterogeneous detonation. The types of DDT in tubes and initiation of unconfined detonation are described, followed by a summary of the heterogeneous detonation velocity and pressure. The transverse wave structure that characterizes the heterogeneous detonation is then reviewed and the detonation dynamic parameters are discussed. In Sect. 2.5, the characteristics and limit of the quasi-detonation in reactive particle–oxidizing gas mixtures are examined. The important types of hybrid detonation and their influencing parameters are described in Sect. 2.6. This is followed by a brief concluding remark on the current state of the art and possible developments in the future. Finally, the governing equations of the two-phase fluid dynamics, the equations of state, and the transfer functions between two phases can be found in the Appendix for specific descriptions of modeling dilute and dense particle–gas flow topology.

2.2 Detonation Theory of Gas–Particle Flow

2.2.1 Equilibrium CJ Detonation Model

The equilibrium CJ detonation theory assumes a detonation wave to be a strong discontinuity within which the chemical reaction has ended and produces a final equilibrium detonation products state at an equilibrium sonic locus with respect to the detonation wave. A unique steady solution to the one-dimensional conservation equations across the discontinuity can be found to correspond to the minimum detonation velocity solution where the Rayleigh line is a tangent to the final equilibrium Hugoniot curve. The CJ steady solution has been remarkably successful in predicting the detonation velocity in uniform gas, liquid, and solid matter for conditions well within the detonation limits, given a reasonable equation of state for the detonation products. The detonation pressure, temperature, and flow velocity obtained from the CJ theory can be considered as the mean values at a mean equilibrium sonic locus averaged over the cross section perpendicular to the direction of propagation. It has been experimentally difficult, however, to determine the sonic locus where the averaging can be taken behind the shock front [36–38], since in reality detonation waves have a limited thickness and a three-dimensional structure. The success of the CJ theory lies in the simplicity of assuming a strong discontinuity without the need to consider the details of chemical nonequilibrium processes and the detonation wave structure. This assumption, however, prevents one from gaining insights into the detonation initiation and propagation mechanisms responsible for predicting detonation velocity deficit, failure limits, and other dynamic parameters.

Apart from the chemical nonequilibrium process, detonation of a solid particulate two-phase mixture comprises other nonequilibrium processes of mass, momentum, and energy transfer between the two phases owing to the finite sizes of solid particles. A full or final equilibrium state includes all chemical, mechanical (pressure and particle velocity), and thermal (temperature) equilibria between the phases. In general, the nonequilibrium momentum and heat transfer depend on the physical properties of the particles and do not have the same relaxation length scales as that of the mass transfer or chemical nonequilibrium processes. Since the CJ theory assumes a unique final equilibrium detonation products state at the equilibrium sonic locus, it cannot predict the detonation velocity precisely for finite-sized particulate mixtures with large interphase momentum and heat transfer length scales.

Figure 2.1 illustrates the predictability of the CJ theory by comparing the theoretical predictions from the Cheetah code with the experimental results obtained in various aluminum particle–RDX ($C_3H_6N_6O_6$) mixtures at a common initial mixture density $\rho = 1.66\,\mathrm{g\,cm^{-3}}$ [39]. While dense matter is used here owing to available experimental data, the conclusion is generally valid for dilute hybrid reactive mixtures as well. RDX was chosen owing to its oxygen deficiency. For sufficiently small particles (e.g., 0.1 µm), an increase in aluminum mass fraction results in a decrease in detonation velocity. The experimental detonation velocities are in agreement with the theoretical prediction, regardless of the reactive or chemically frozen nature of the particles. This fact clearly indicates the significance of the momentum and heat transferred to the particles during the process towards the mixture equilibrium state as the flow approaches the sonic locus. The momentum and heat

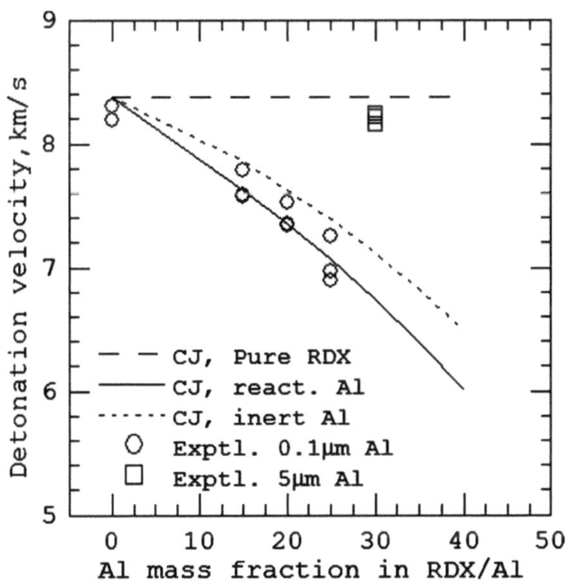

Fig. 2.1. Comparison of experimental detonation velocities with the equilibrium Chapman–Jouguet (CJ) theory for aluminum particle–RDX mixtures at $1.66\,\mathrm{g\,cm^{-3}}$ initial density

transferred are responsible for the velocity deficit with respect to pure RDX detonation. In contrast, for sufficiently large particles (e.g., 5 μm), the experimental velocity is much higher than the final equilibrium prediction and is close to that of pure RDX, thus suggesting a nearly frozen transfer of momentum and heat between the two phases within the detonation zone. The experimental detonation velocity is therefore a strong function of particle size and ranges between the final equilibrium value and the phase-frozen limit.

Detonation of reactive solid particulate mixtures can also result in condensed-phase products that do not contribute to gas pressure, thus reducing expanding work in sustaining the propagation of the detonation wave. For example, the CJ detonation can exist in an aluminum–oxygen mixture owing to the gas-phase detonation products of AlO and Al_2O at high detonation temperature. As the aluminum mass fraction increases, the condensed-phase products such as liquid Al_2O_3 and aluminum increase and the minimum detonation velocity as a unique steady solution is not attainable at the sonic point and instead is subsonic. Hence, there exists a limit for the amount of condensed-phase detonation products above which the CJ detonation solution does not exist.

2.2.2 Two-Phase ZND Detonation Model

The ZND detonation model in uniform matter developed by Zeldovich, von Neumann, and Döring assumes a detonation wave structure that consists of a leading shock front followed by a continuous reaction zone. The ZND model provides a mechanism for detonation propagation. The leading shock front adiabatically compresses a material to an autoignition temperature to initiate chemical reaction, while the expansion of high-pressure reacting gases in turn provides work to sustain the propagation of the shock front. A unique steady solution is obtained by integrating the one-dimensional ordinary differential conservation equations along the reaction progress path to the CJ sonic locus, where the Rayleigh line is tangential to the final equilibrium Hugoniot curve. To calculate detonation velocity deficits and detonation limits, Zeldovich and others [40, 41] further proposed a quasi-one-dimensional model in which the source terms are introduced in the conservation equations to consider lateral boundary effects such as friction and heat loss to the tube wall or expansion into the surroundings. Owing to the presence of loss source terms competing with exothermic reactions, the flow may become sonic prior to the final chemical equilibrium such that the ideal CJ equilibrium sonic condition is no longer valid. An alternative, referred to as the "generalized CJ condition," was introduced as a mathematical saddle point on which the exothermic heat release rate equals the energy loss rate at the gaseous sonic locus with respect to the shock front. For a reactive system with nonmonotonic heat release behavior, Kuznetov [42] demonstrated that the steady ZND solution may not be unique and multiple detonation solutions are possible for given initial conditions. Theoretically, the one-dimensional ZND structure can be unstable [43–49], while observed detonation waves have an unsteady three-dimensional structure. Hence, the one-dimensional detonation wave structure (the profiles of pressure, temperature, and flow velocity along the reaction path) obtained from the steady ZND model may be regarded as a mean structure averaged over the cross section perpendicular to the direction of propagation, and over an unstable period in the propagation direction.

For solid particle–fluid mixtures, strictly speaking, a steady solution cannot be achieved a priori without integration along the reaction path to determine the mechanical and thermal partial equilibrium between the two phases. Hence, a two-phase ZND model has been introduced with a generalized CJ condition as a rear boundary condition, where the net heat release rate resulting from the chemical reactions and interphase nonequilibrium mass, momentum, and heat transfer approaches zero at the phase-frozen or gaseous sonic locus [10–12,27,28,50,51]. The two-phase ZND model can be derived from the one-dimensional, two-phase fluid dynamics governing equations based on the control volume analysis of the continuum theory. In this theory, the fluid and the solid particles are treated as two separated continua with mass, momentum, and energy conservation equations for each phase, continuity equations for species, and conservation equations for solid particle numbers. The interactions between the two continua are described using the source terms for the rate of mass, momentum, and energy transfer. Particle agglomeration or breakup is controlled through a source term for the rate of particle number change. When the solid particle flow is granular or extremely dense, a dynamic compaction equation can be employed where a source term is introduced to describe the rate of solid volume compaction [51–54]. The latter is caused by mechanical nonequilibrium between the internal stresses and the forces exerted by neighboring particles and the interpore fluid. If the reaction zone is large, the loss due to the tube wall or other lateral boundary conditions must also be included and modeled by the source terms for the rate of the momentum and heat exchange with the lateral boundaries. Various forms of source term functions can be found in [10–12,27,28,50–56]. Whereas in a rigorous multiphase continuum theory the source terms must follow constraints imposed by the conservation laws and the entropy inequality of the mixture, they are modeled on the basis of first principles physical rules, often in the form of empirical correlations. Therefore, appropriate choice of the source term functions for a particular flow topology is crucial for the reliability of the solution. The detailed description of the governing equations and some of the source term functions can be found in the Appendix.

From the governing equations in the coordinate frame with respect to the leading shock front propagating at velocity D, a system of ordinary differential equations can be derived in which the change of the fluid velocity u_{g} along the propagation distance x is given by

$$\frac{\mathrm{d}u_{\mathrm{g}}}{\mathrm{d}x} = \Phi/\eta, \tag{2.1}$$

where

$$\eta = 1 - u_g^2/a_g^2 \tag{2.2}$$

is a sonic parameter of the flow with respect to the phase-frozen or gaseous sound speed a_g. The quantity Φ represents the "thermicity," a measure of the rate of net energy release from all nonequilibrium processes to molecular and bulk translational energy. A "generalized multiphase CJ condition" serves as the rear boundary condition at the phase-frozen or gaseous sonic point imbedded in the reaction zone by finding the common zeros of the thermicity Φ and the sonic parameter η:

$$\Phi = 0 \text{ at } \eta = 0. \tag{2.3}$$

The detailed expression of the thermicity depends on the nonequilibrium processes and equations of state. To elucidate the physical meaning of the net energy release rate, an analytical expression is given below for a simple system comprising a perfect gas with single exothermic (heat $q_g > 0$ in joules per kilogram), irreversible gaseous reaction (rate $w_g > 0$) and a negligible volume fraction of incompressible solid particles with exothermic particle combustion (heat $q_p > 0$, rate of mass transfer $J_p < 0$) and conservation of particle number:

$$\Phi = \frac{\gamma - 1}{\rho_g a_g^2} \left(\begin{array}{l} q_g w_g - (q_p + cT_p) J_p \\ \qquad - \left[\frac{(u_g - u_p)}{2} \left(\frac{\gamma+1}{\gamma-1} u_g - u_p \right) J_p - \left(\frac{\gamma}{\gamma-1} u_g - u_p \right) f_p + Q_p \right] \\ \qquad + \left(\frac{\gamma}{\gamma-1} u_g - D \right) f_W + Q_W \end{array} \right). \tag{2.4}$$

Variables ρ, u, and T here are the material density, flow velocity, and temperature, respectively; γ stands for the ratio of specific heats for the fluid and c is the specific heat for the solid phase. The subscripts p and g refer to the solid particle and fluid phase, respectively.

The first term on the right-hand side of (2.4) describes the energy release rate of the gas-phase reaction. The second term (with source term J_p) represents the rate of energy release into the gas due to particle reaction. Evaporation and combustion of solid particles are included in the rate of mass transfer J_p, which adopts a negative value when particle depletion occurs. The third term (in square brackets) corresponds to the rate of gas energy change caused by the nonequilibrium flow velocity and temperature between the two phases. The rate of momentum transfer f_p and the rate of energy transfer Q_p have the same sign as the phase velocity difference $u_g - u_p$ and phase temperature difference $T_g - T_p$, respectively. The last two terms (with f_W and Q_W) represent the rate of gas energy change due to the momentum and heat transfer to the tube wall or other lateral boundaries, and they become important in describing detonation limits. The value of f_W or Q_W is negative if the exchange results in a loss to the lateral boundaries and hence further competes with the exothermic terms for the flow to reach the sonic locus earlier. For finite-sized particle–gas flow, the nonequilibrium momentum and heat transfer between the two phases described in the third term on the right-hand side of (2.4) can result in an energy loss rate competing with exothermic reaction rates of the first and second terms at the phase-frozen sonic locus, thus providing a mechanism that is possible to satisfy the generalized CJ condition shown in (2.3). To elucidate the intrinsic mechanism for the multiphase detonation, all discussions in this chapter, except when specifically noted, do not include lateral boundary source terms.

The generalized CJ condition (2.2)–(2.4) at the frozen sonic locus is derived from the steady gas-phase conservation equations of two-phase fluid dynamics equations and contains no information from the conservation equations for the solid particle phase. This limits the application of the model, particularly to detonation with very small particles whose velocity and temperature are rapidly equilibrated with gas velocity and temperature behind the shock front. In this case, the third term on the right-hand side of (2.4) would disappear. Equations (2.1)–(2.4), without lateral boundary source terms, indicate that the flow would approach the phase-frozen or gaseous sound speed as the chemical reactions [the first and second terms on the right hand side of (2.4)] reach an equilibrium state. This solution, however, conflicts with the CJ equilibrium solution where the flow approaches the phase-equilibrium sound speed as the mechanical, thermal, and chemical equilibrium is achieved. In fact, following a control volume analysis in a steady detonation frame one can find that the steady solution with respect to the phase-frozen sonic locus does not exist in this case as demonstrated in Fig. 2.2 (the minimum detonation Mach number as the unique solution is attainable at the control volume exit local Mach number $M_1 = 1$ with respect to the full equilibrium sound speed, while it is subsonic relative to the phase-frozen sound speed).

The phase-equilibrium sound speed a_e can be expressed by [57]

$$\left(\frac{a_e}{a_g}\right)^2 = \frac{(1-\alpha)}{(1-\phi_p)^2}\,\Gamma/\gamma, \quad \Gamma = \gamma\frac{1-\alpha+\alpha c/c_{pg}}{1-\alpha+\gamma\alpha c/c_{pg}}, \tag{2.5}$$

where $\alpha = \sigma_p/(\sigma_g + \sigma_p)$ is the mass fraction of the solid phase (σ denotes the partial density or mass concentration), ϕ_p refers to the solid volume fraction, c_{pg} stands for the specific heat of the gas phase at constant pressure, and Γ is defined as the ratio of the specific heats for the two-phase mixture. The solid particle material was assumed to be incompressible. From (2.5), the equilibrium sound speed is less than the frozen sound speed and their ratio decreases with increasing mass fraction of the solid phase. Hence, the fully equilibrated two-phase flow is still subsonic with respect to the frozen sonic speed as the CJ equilibrium detonation solution is achieved.

For the two-phase flow approaching $u_g = u_p = u$ and $T_g = T_p = T$ before the equilibrium sonic locus, the thermicity and the sonic parameter in (2.1) and (2.3) must therefore be rederived from the conservation equations of both the gas and the solid phase. They are

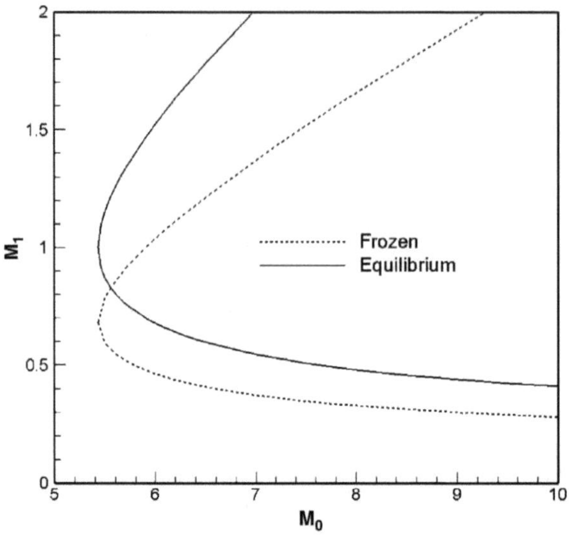

Fig. 2.2. Steady control volume solution for the local Mach number at the exit, M_1, as a function of detonation Mach number, M_0, with respect to the gaseous sound speed. The control volume has the initial 500 g m^{-3} aluminum and air at the entry and the detonation products including nitrogen and condensed aluminum and Al_2O_3 at the exit. The velocity and temperature equilibrium between the two phases is assumed at the exit [22]

$$\Phi = \frac{(\Gamma-1)}{\rho a_e^2}\left[q_g w_g - (q_p + cT)J_p - \frac{1-\Gamma/\gamma}{\Gamma-1}c_{pg}TJ_p + \left(\frac{\Gamma}{\Gamma-1}u - D\right)f_W + Q_W\right] \tag{2.6}$$

and

$$\eta = 1 - u^2/a_e^2 \tag{2.7}$$

for the simple system used in deriving (2.4). Here, $\rho = \sigma_g + \sigma_p$ is the mixture density. The third term on the right-hand side of (2.6) is attributed to the change of the rate of the ratio of the specific heats for the two-phase mixture.

The ordinary differential equation system deduced from the governing equations, the equations of state for the particles and gas phase, together with the generalized CJ condition (2.3) form the closure of the mathematical description of the two-phase ZND model, given the source terms for the exchange between the two phases (and to the lateral boundaries if included). Here, the generalized CJ condition (2.3) contains Φ and η from (2.2) and (2.4) if $u_p \neq u_g$ or $T_p \neq T_g$ at the phase-frozen sonic locus, or Φ and η from (2.6), (2.7) if $u_p = u_g$ and $T_p = T_g$ at the full equilibrium sonic locus. Thus, under the initial conditions of the postshock state, a steady ZND solution can be obtained for the propagation velocity and reaction zone structure of the detonation wave in an inert particle–reactive fluid system, in a reactive particle–oxidizing fluid system, or in a reactive particle–reactive fluid system.

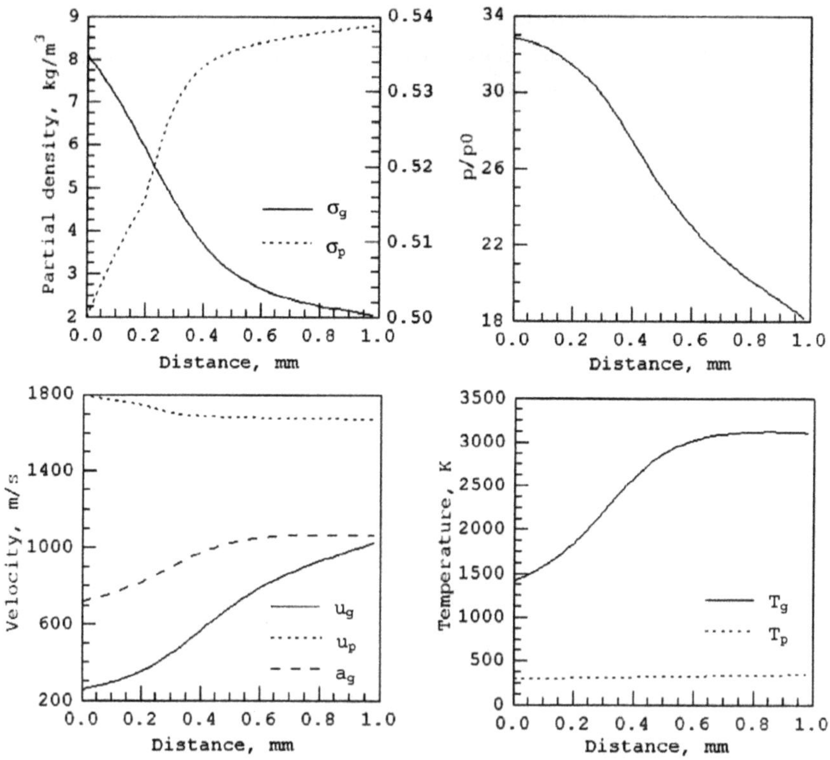

Fig. 2.3. Zeldovich–von Neumann–Döring (ZND) detonation structure in a mixture of stoichiometric C_2H_2–air and $\sigma_p = 500\,\mathrm{g\,m^{-3}}$, $d_p = 10\,\mu\mathrm{m}$ inert aluminum particles [55]

The steady ZND detonation structure in a dilute, inert particle–reactive gas system is analogous to a frictional detonation [58, 59], in which the frictional force is replaced by a drag force determining the momentum transfer between the two phases. Figure 2.3 shows an example in a mixture of stoichiometric acetylene–air (modeled by a single-step Arrhenius rate law) and 10-μm inert aluminum particles at $0.5\,\mathrm{kg\,m^{-3}}$ concentration, where the detonation structure is terminated at the frozen sonic locus with the generalized CJ condition (2.2)–(2.4) [55]. The drag force influences the wave structure in two aspects. While it causes a shock velocity deficit with respect to the gas CJ detonation and therefore a drop in shock gas pressure and temperature, the drag compression in the gas reaction zone behind the shock front gradually amplifies the gas pressure and temperature. For the current particle size and concentration, the velocity and temperature relaxation length scales of the particle flow are 2 orders of magnitude larger than the gas reaction zone length. Thus, the drag compression is more than compensated for by the gaseous combustion expansion, so that the pressure monotonically decreases from the shock front to the frozen sonic locus. While the combustion expansion causes the gas density to decrease behind the shock front, the particle concentration is gradually increased from the initial value owing to the velocity relaxation time lag in which the drag force drives the particles. The competition of the gas-phase chemical energy release with the momentum and heat transfer to the solid particles results in a detonation velocity $D = 1{,}800\,\mathrm{m\,s^{-1}}$ that has a mild deficit of 3.2% relative to the CJ velocity of the gas detonation ($D_{\mathrm{g-CJ}} = 1{,}860\,\mathrm{m\,s^{-1}}$), but an increase of 10% with respect to the full or final equilibrium CJ velocity ($D_{\mathrm{CJ}} = 1{,}632\,\mathrm{m\,s^{-1}}$).

A steady ZND detonation structure in a dilute, reactive particle–oxidizing gas system is illustrated in Fig. 2.4 for a $\sigma_{\mathrm{p}} = 0.5\,\mathrm{kg\,m^{-3}}$, $d_{\mathrm{p}} = 1\,\mu\mathrm{m}$ aluminum–air mixture modeled by the diffusion reaction model (2.45) and (2.46). The detonation structure consists of a shock front followed by an induction and reaction zone successively and is terminated at the equilibrium sonic locus where the generalized CJ condition (2.3) in combination with (2.6) and (2.7) is satisfied. In the induction zone, while the particles are heated through convective heat transfer from the shocked gas, the drag compression continuously increases the gas pressure, temperature, and particle concentration. Significant heat release of the particles into the gas takes place after the induction, and the resultant gas expansion causes the pressure to decrease

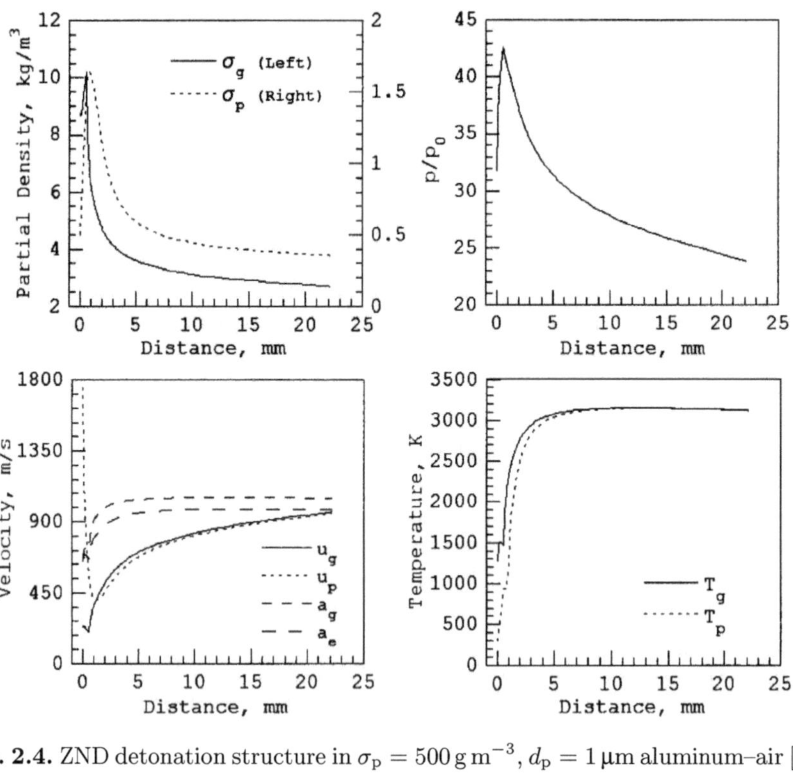

Fig. 2.4. ZND detonation structure in $\sigma_{\mathrm{p}} = 500\,\mathrm{g\,m^{-3}}$, $d_{\mathrm{p}} = 1\,\mu\mathrm{m}$ aluminum–air [55]

when the flow moves towards the sonic locus. Note that the two-phase velocity and temperature equilibrium are reached much before the equilibrium sonic locus in this example. Hence, the generalized CJ condition (2.2)–(2.4) at the frozen sound speed is not valid, as discussed before. In fact, the phase-equilibrated flow is still subsonic with respect to the frozen sonic speed a_g as the CJ equilibrium detonation solution [compatible to (2.3), (2.6), (2.7) with respect to the equilibrium sound speed a_e] is achieved. The expansion of the high-pressure gases provides the work to sustain the propagation of the shock front, which, in turn, supports the momentum and heat transfer between the two phases behind the shock front and the ignition of particles.

In general, this two-phase ZND structure has several significant differences from a ZND structure for homogeneous gas detonation. First, the shock front pressure (i.e., the von Neumann spike) in a gaseous detonation corresponds to the maximum pressure, while the maximum pressure for a reactive particle–oxidizing gas detonation wave may be found behind the shock front at a point at which the combustion expansion balances the drag compression. Second, the equivalence ratio of the particle–gas mixture behind the shock front does not remain the same as the initial value ahead of the shock front. The velocity relaxation time lag, in which the particle is accelerated towards the gas velocity before burning, results in a nonuniform equivalence ratio field behind the shock with a shift from the initial equivalence ratio of fuel particles to oxidizing gas towards a leaner value. This is clearly indicated in the partial density histories behind the shock front in Fig. 2.4 (the preshock initial values are $\sigma_g = 1.17\,\mathrm{kg\,m^{-3}}$ and $\sigma_p = 0.5\,\mathrm{kg\,m^{-3}}$). Third, the ZND structure can also be characterized by a melting phase change and an evaporation phase change (not included in this example) depending on the phase change models. Furthermore, combustion of metal particles such as aluminum may result in a large quantity of condensed metal oxide and therefore a mole decrement, thus leading to a pathological detonation when the products Hugoniot shifts to the left of the reactants Hugoniot. Finally, a high particle concentration layer can be formed downstream in the shocked flow under the appropriate conditions of a velocity relaxation time lag. Such a high particle concentration layer behind the shock was called a "ρ-layer" by Korobeinikov [60]. The late combustion of the dense ρ-layer and its influence on the detonation flow remain a subject of current research.

The generalized CJ locus determined by (2.2)–(2.4) is a mathematical saddle point, after which the subsonic flow relative to the shock front can become supersonic as it reaches the weak detonation branch of the full or final equilibrium Hugoniot curve. Two important conditions must be met for a steady weak detonation solution as follows:

1. The necessary conditions are:
   - Within the reaction zone, there is at least one phase-frozen sonic point imbedded at which the generalized CJ condition is satisfied.
   - The final equilibrium Hugoniot is not the upper bound of all partial equilibrium Hugoniot curves.

2. The uniqueness of a steady weak detonation solution depends on the flow or boundary conditions behind the generalized CJ point.

A simple illustrative example is the well-known detonation wave in a perfect gas with an irreversible exothermic reaction followed by a secondary irreversible endothermic reaction (heat releases $q_a > 0$, $q_b < 0$, $|q_a| > |q_b|$) [61]. It has a saddle point featured with the generalized CJ condition (2.3) imbedded in the reaction zone due to the endothermic reaction rate competing with the exothermic reaction rate. A steady solution can be obtained by integration of the ZND model from the postshock state downstream to satisfy the generalized CJ condition at which the Rayleigh line is a tangent to a partial equilibrium Hugoniot curve. This partial equilibrium Hugoniot corresponds to the highest attainable heat release of the system, $Q_{max} = q_a + q_b[1 - \exp(q_a/q_b)]$, which is larger than the heat release $Q_f = q_a + q_b$ in the final equilibrium Hugoniot. The detonation velocity D_m corresponding to the Rayleigh line tangential to the Q_{max}-Hugoniot is greater than the final equilibrium detonation velocity D_{CJ}. Therefore, the solution satisfies the necessary conditions for a weak detonation. Depending on the rear flow boundary conditions, the solution can return either to a strong detonation point along the D_m-Rayleigh line or continue the integration from the saddle point downstream, as the flow smoothly transits from subsonic to supersonic until it meets the final equilibrium Hugoniot curve. Hence, the solution is incomplete without taking into account the second condition stated above. A variety of weak detonation solutions can be obtained when the detonation wave is followed by a piston of specified constant velocity by adjusting the piston velocity [61].

For a gas–particle flow, a partial equilibrium state also includes that of mass, momentum, and energy transfer processes between the two phases. In fact, detonation in an inert particle–reactive gas system as shown in Fig. 2.3 satisfies the necessary conditions for a weak detonation. It has a saddle point imbedded in the reaction zone as depicted by the generalized CJ criterion (2.2)–(2.4). Secondly, the final equilibrium Hugoniot lies below some partial equilibrium Hugoniot curves because the final equilibrium detonation velocity ($D_{CJ} = 1{,}632\,\mathrm{m\,s^{-1}}$) is less than that shown in Fig. 2.3 ($D = 1{,}800\,\mathrm{m\,s^{-1}}$). After the gaseous sonic locus, the solid particle velocity and temperature will further equilibrate with that of the gas phase towards the final equilibrium Hugoniot as the flow becomes supersonic with respect to the leading shock front.

Detonation in a reactive particle–reactive gas system is analogous to the above example of two-irreversible-reaction gaseous detonation with the second reaction endothermic, followed by a piston of specified constant velocity. The momentum and heat loss from the gas to the particles provides a mechanism to satisfy the necessary conditions for a weak detonation, while various weak detonation solutions can be realized by the late particle energy release, in analogy to the piston, behind the saddle point. The energy release rate of particles represented by the second term in the thermicity (2.4), denoted now as $q_p{}'J_p$ in joules per cubic meter per second, is a characteristic parameter to specify a possible solution. The delay time and the magnitude of $q_p'J_p$ can be adjusted through the particle material, size, or concentration as well as gaseous detonation parameters or product compositions.

2.2.3 Unsteady Two-Phase Fluid Dynamics Model

A steady solution can also be obtained by the long-time asymptotic solution of the one-dimensional unsteady two-phase fluid dynamics equations described in the Appendix, as the induction-to-reaction length ratio is below the value for the stability limit. A set of solutions for the detonation in a reactive particle–reactive gas system (lean acetylene–air with aluminum particles) are obtained as displayed in Fig. 2.5, serving for the quantitative description of physical

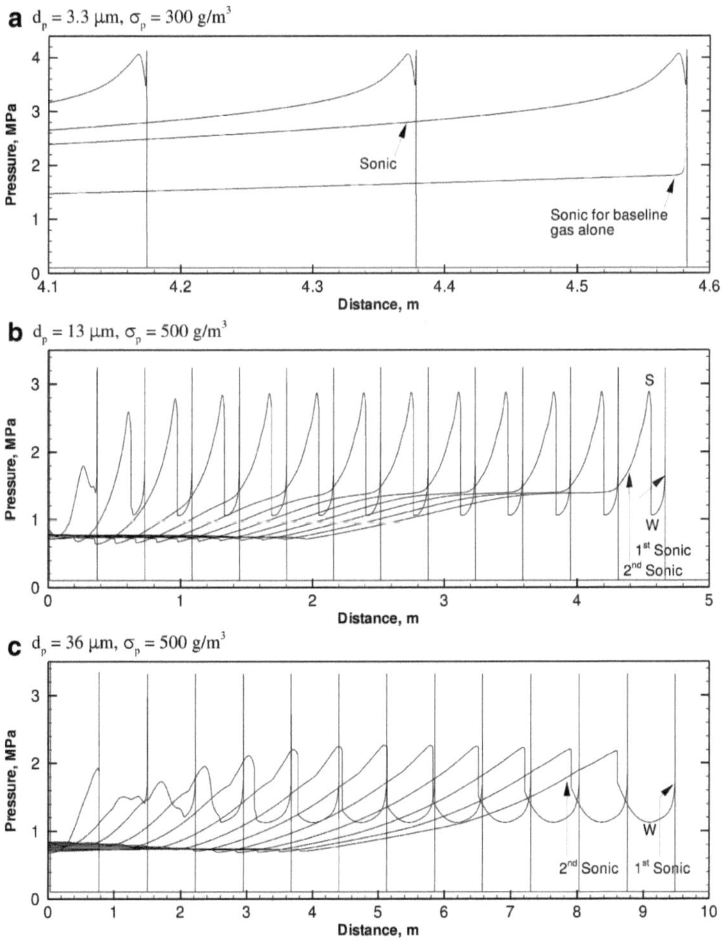

Fig. 2.5. Hybrid detonation solutions in a mixture of lean C_2H_2–air ($\varphi = 0.8$) and aluminum particles: **a** strong detonation, **b** type-I weak detonation, **c** type-II weak detonation

phenomena [32, 33]. The acetylene–air reaction is modeled by a single-step Arrhenius rate law and the aluminum reaction is modeled by the hybrid reaction model (2.49) to (2.52). The supersonic or subsonic flow terminologies used in the following discussions are with respect to the leading shock front:

1. If $q_p{}'J_p$ rises early and significantly, particle reaction can produce a compression wave in the gas reaction zone to increase the detonation velocity and pressure (Fig. 2.5a: 3.3 μm Al at $0.3\,\mathrm{kg\,m^{-3}}$). The entire subsonic reaction zone is substantially extended owing to particle combustion and a steady solution is reached when the generalized CJ condition (2.2)–(2.4) is satisfied at the gaseous frozen sonic locus. There exists a minimum in the pressure profile within the reaction zone when the net heat release reaches a local maximum at thermicity $\Phi = 0$ before the sonic locus. This solution was referred to as "single-front detonation" by Veyssiere and Khasainov [28], but is termed a "strong hybrid solution" in [32, 33] in the sense that particle combustion within the reaction zone overdrives the gas detonation. This terminology comes from the analogy with detonation in an exothermic–endothermic two-reaction gas followed by a piston moving faster than the flow velocity of the strong detonation point. However, unlike the usual overdriven detonation where the entire flow is subsonic with respect to the leading shock front, a strong hybrid detonation will not be disturbed by the supersonic rear flow behind the sonic locus. A strong hybrid detonation wave usually occurs for reactive particles suspended in a lean reactive gas mixture, where the rate of particle energy release into the gas overcomes the loss rate due to the momentum and heat transferred from gas to particles within the gas reaction zone. In a system of dense reactive particles suspended in a very rich reactive gas mixture, the strong hybrid detonation is unlikely to take place owing to the prevailing rate of momentum and heat loss that leads to a detonation velocity deficit and instability.
2. When $q_p{}'J_p$ is delayed and reduced to enable particle reaction behind the gas reaction zone, particles behave as though they are inert within the gas reaction zone and the necessary conditions for a steady weak detonation can be satisfied, where a generalized CJ point appears for the first time (Fig. 2.5b: 13-μm Al at $0.5\,\mathrm{kg\,m^{-3}}$). In the supersonic gas flow behind the first sonic point, heat release from the particles would cause a continuous decrease in gas flow velocity to subsonic levels, and an increase in gas pressure. This, however, will not match the downstream unsteady supersonic flow required by the rear boundary condition, and instead will result in thermal choking. Consequently, a second shock wave is necessary to adjust the gas flow behind the gas reaction zone from supersonic to subsonic. The flow, with the heat release from the particles, is then able to expand towards the second sonic locus, where the generalized CJ condition is satisfied a second time to match the downstream unsteady supersonic flow. Thus, a double-shock solution can be achieved that consists of the steady gas reaction zone followed by a secondary shock. The $q_p{}'J_p$-induced

second shock wave corresponds to a postshock subsonic state S and a preshock supersonic end point W of the steady gas reaction zone. If the Rayleigh line SW coincides with that for the leading front, the second shock moves with the same velocity as the leading front. This solution is referred to as the "type-I double-shock weak solution," analogous to the double-shock solution in the two-reaction gas followed by a piston velocity equal to the flow velocity of the strong detonation point.

3. When $q_{\mathrm{p}}'J_{\mathrm{p}}$ decreases, the velocity of the $q_{\mathrm{p}}'J_{\mathrm{p}}$-induced second shock wave is reduced. The shock therefore recedes from the supersonic end state W of the steady gas reaction zone to produce an ever-widening region of supersonic flow between state W and itself (Fig. 2.5c: 36-μm Al at 0.5 kg m^{-3}). As $q_{\mathrm{p}}'J_{\mathrm{p}}$ further decreases, the strength of the secondary shock decreases and recedes more rapidly. This solution is called the "type-II double-shock weak solution," in analogy to the solution in the two-reaction gas followed by a piston with velocity between that of the strong detonation point and the weak detonation point. Unlike the weak detonation in the two-reaction gas followed by a piston, the ever-widening region of the supersonic flow is unsteady. The initial particle combustion increases the pressure and decreases the flow velocity upstream of the secondary shock. Furthermore, the particle reaction zone length between the second shock and the second sonic locus increases continuously as the shock recedes. Rigorously speaking, a steady solution does not exist after the end point W of the steady gas reaction zone.
4. As $q_{\mathrm{p}}'J_{\mathrm{p}}$ is further delayed and reduced, the supersonic end point W of the steady gas reaction zone is connected to the supersonic rear flow imbedded with a weak compression wave caused by the particle combustion. While the detonation front propagates steadily and satisfies the generalized CJ condition (2.2)–(2.4) at the frozen sonic locus, the particle-reacting rear flow of this type of weak solution is unsteady and subject to the rear boundary condition. The particles become chemically inert as $q_{\mathrm{p}}'J_{\mathrm{p}}$ is reduced to approach a null value.

The one-dimensional multiphase ZND model contains some intrinsic features in detonation instability. While reaction of particles within the fluid reaction zone stabilizes the detonation (e.g., Fig. 2.5a), the momentum and heat transferred from the fluid to the particles within the fluid reaction zone destabilize the detonation for any solid particle–reactive fluid systems. This momentum and heat transfer causes a velocity deficit with respect to the CJ detonation velocity of the pure fluid. For detonation in an inert particle–reactive fluid system, the magnitude of the velocity deficit increases with a decrease in a ratio of the velocity relaxation length scale of solid particles, L_{p}, to the fluid ZND detonation zone length, L_{gr}, that is,

$$\frac{L_{\mathrm{p}}}{L_{\mathrm{gr}}} \sim \frac{d_{\mathrm{p}}^{n} \rho_{\mathrm{s}}^{m}}{\sigma_{\mathrm{p}}^{t} L_{\mathrm{gr}}}, \tag{2.8}$$

where $n > 0$, $m > 0$, and $t > 0$ [55]. Thus, increasing particle concentration σ_p, or decreasing particle diameter d_p and material density ρ_s, will increase the detonation velocity deficit and instability. One can use numerical solutions of the unsteady two-phase fluid dynamics model to examine the validity of a steady ZND solution. For instance, for detonation in the inert particle–reactive gas system displayed in Fig. 2.3, L_p was 2 orders of magnitude larger than L_{gr}. Consequently, a small velocity deficit of 3.2% resulted and the long-time asymptotic unsteady solution appeared in accordance with the steady ZND solution. When $L_p/L_{gr} < 1$, rapid momentum and heat transfer within the gas reaction zone can result in a large velocity deficit, leading to failure of the detonation wave. For a range of intermediate values of L_p/L_{gr}, the detonation executes an unsteady oscillatory behavior and the oscillation irregularity increases as L_p/L_{gr} decreases, indicating that the ZND multiphase model is unstable for a range of intermediate velocity deficits. The generalized CJ condition fails in unstable detonation waves and the detonation limits predicted by the unsteady solution appear to be more restricted than that obtained from the steady solution [55,58]. Caution must therefore be taken when using the steady solution to predict the detonation limits.

For detonation in a reactive particle–reactive gas system, Fig. 2.6 illustrates a numerical simulation for a large concentration of 10-μm aluminum particles suspended in a lean acetylene–air system [32]. In comparison with the steady double-shock detonation wave shown in Fig. 2.5b, an increase in particle concentration results in an increase in velocity deficit to 8%, thus causing the detonation wave to propagate in an unstable oscillatory mode. While the particles still burn behind the gas reaction zone, the energy release from the particle combustion is coupled with the unsteady rear flow of the

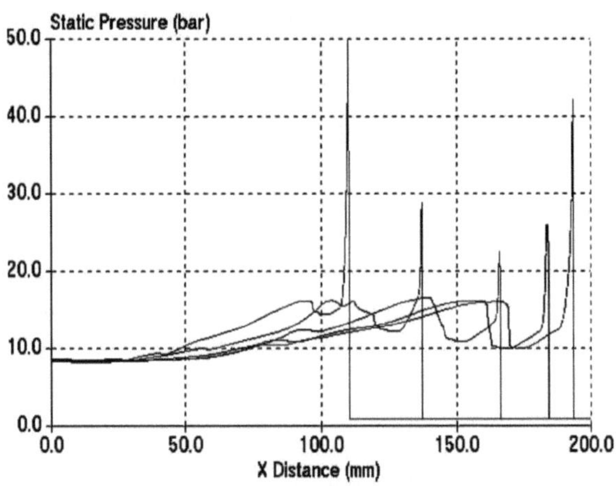

Fig. 2.6. Numerical simulation of unsteady weak hybrid detonation with a transient secondary pressure wave in a mixture of lean C_2H_2–air and 1,000 g m^{-3}, 10-μm aluminum particles [32]

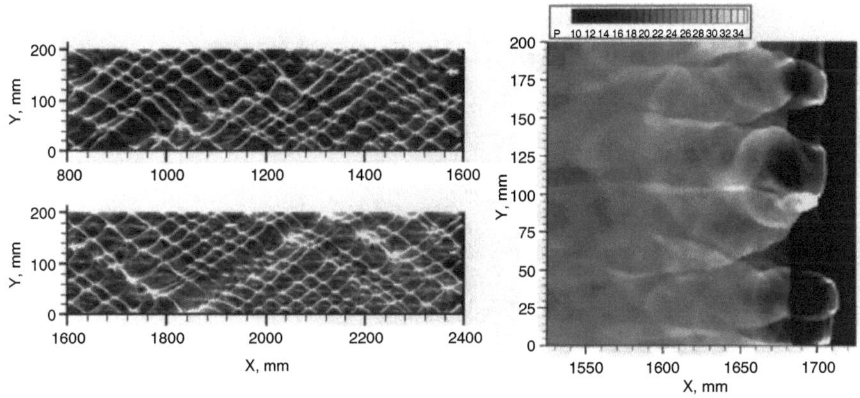

Fig. 2.7. 2D numerical cellular detonation in a mixture of lean H_2–air and $\sigma_p = 0.3\,\mathrm{kg\,m^{-3}}$, 13 µm aluminum. *Left*: Maximum pressure tracks. *Right*: Pressure distribution at 1 ms [62]

gas detonation. This leads to a transient pressure wave in the detonation flow in an oscillatory cycle with an acceleration phase followed by a deceleration phase. The acceleration phase leads to the formation of a shock wave before the deceleration phase commences. Figure 2.7 shows a cellular detonation solution in a mixture of 13-µm aluminum suspended in lean hydrogen–air using a two-dimensional two-phase fluid dynamics computation [a single-step Arrhenius rate law for gas and the diffusion reaction model (2.45), (2.46) for aluminum] [62]. The frontal transverse wave structure from the gas detonation is followed in a distance of about 20–30 mm by a nonplanar secondary shock due to aluminum combustion subjected to the transverse wave flow conditions. The detonation has a velocity deficit of 7.8% relative to the baseline gas detonation, thus leading to an unstable transverse wave structure that results in a more irregular cellular detonation than that of the baseline gas detonation.

A cellular detonation wave in a reactive particle–oxidizing gas system can also be simulated in a multidimensional instability analysis of the two-phase fluid dynamics model. Numerical studies have been conducted in an attempt to capture the nature of the cellular detonation wave in such a system using the Arrhenius reaction models [63–65] or the diffusion reaction model (2.45) and (2.46) [66]. For micrometric and nanometric aluminum–air mixtures, as to be reviewed in the next sections, experimental evidence showed strong dependence of detonation sensitivity on initial pressure and highly nonlinear behavior of detonation initiation and an abrupt DDT. This indicates the dependence of the aluminum detonation mechanism on chemical kinetics. On the other hand, the observed aluminum–air detonation manifested itself in a weak transverse wave structure as revealed by the small amplitude oscillation, which rapidly degenerates behind the shock front, and weak cell traces in the smoke foil records. This could suggest a functional dependence weaker

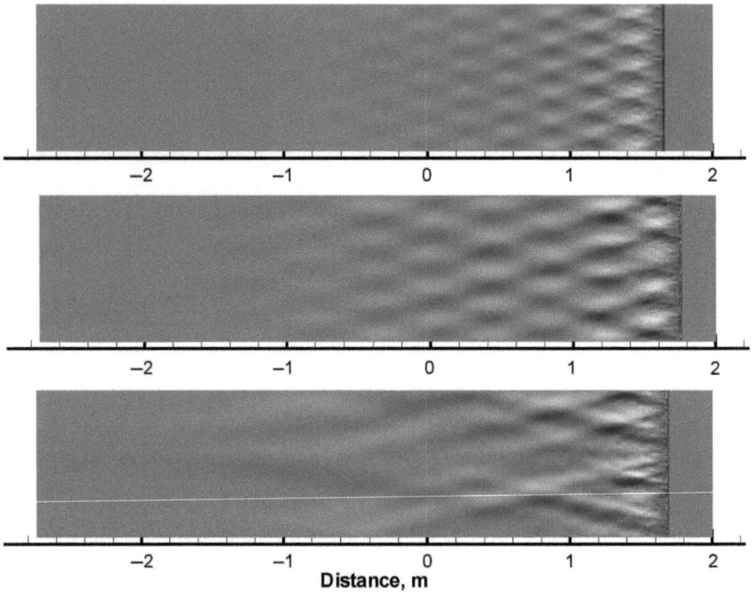

Fig. 2.8. Numerical shadow graph from the pressure data of 2D detonation simulation in a $\sigma_p = 1{,}250\,\mathrm{g\,m^{-3}}$, 2-μm aluminum–air mixture at 2.5-atm initial pressure. *Top*: Activation energy $E = 71.1\,\mathrm{kJ\,mol^{-1}}$. *Middle*: $E = 95.5\,\mathrm{kJ\,mol^{-1}}$. *Bottom*: $E = 120\,\mathrm{kJ\,mol^{-1}}$ [67]

than the highly nonlinear Arrhenius kinetics for the later aluminum combustion. Hence, a surface kinetic oxidation and diffusion hybrid reaction model was suggested as described in (2.49) to (2.52) in the Appendix [22]. The hybrid aluminum reaction model that provides a kinetics-controlled induction and a diffusion-dominant combustion stage is successful in capturing both the kinetics-limited transient processes of detonation initiation, abrupt DDT, and detonation instability, and the diffusion-limited combustion of aluminum in the long reaction zone supporting the weak transverse wave structure [67]. Figure 2.8 presents two-dimensional cellular solutions using the hybrid reaction model in a rich 2-μm aluminum–air mixture at $\sigma_p = 1{,}250\,\mathrm{g\,m^{-3}}$ and an initial pressure of $p_0 = 2.5$ atm. As the activation energy in the aluminum induction stage increases, detonation instability indicated by the cell irregularity increases with an increase in detonation cell width from 0.18 to 0.22 to 0.28 m, respectively. For all activation energies used, the transverse waves are generally weak and rapidly degenerate behind the shock front. This is attributed to the slower diffusion-dominated combustion of the majority of aluminum mass after the kinetic induction and a considerable amount of condensed aluminum oxide without direct contribution to the gas pressure.

Finally, noting that the source term functions are modeled according to first-principles physical rules and empirical co-relations, we see that the reliability and the predictability of the two-phase continuum theory are strongly determined by the choice of source term functions for a particular flow topology. For instance, in handling the momentum transfer in detonation of solid particles suspended in low-density gas flow, the shock interaction time in which the shock front crosses a particle is several orders of magnitude smaller than the velocity relaxation time related to the drag. Thus, a solid particle is assumed to remain stationary as the shock front crosses it [57]. In contrast, for detonation in high-density gas flow or condensed matter containing light metal particles, the shock interaction time can be comparable to the drag-induced velocity relaxation time owing to a significant increase in the initial material density ratio of fluid to particles. Mesoscale modeling showed that the post-shock velocity for aluminum particles achieved 70–80% of the shocked flow velocity of a liquid (Fig. 2.9) and the momentum transferred during the shock interaction time was a strong function of the initial fluid-to-particle material density ratio and the volume fraction of solid particles [39,68]. Hence, caution must be taken in employing appropriate source term functions with respect to the two-phase flow topology involved. Figure 2.9 also indicates that the hot spots are formed in the front of particles owing to the shock reflection and focusing effect as the shock front passes the particles. For condensed explosive mixtures containing metal particle additives, the critical charge diameter for detonation failure may decrease or increase, depending on the competing effects of the sensitization due to the formation of hot spots and desensitization from the momentum and heat transfer to the added mass.

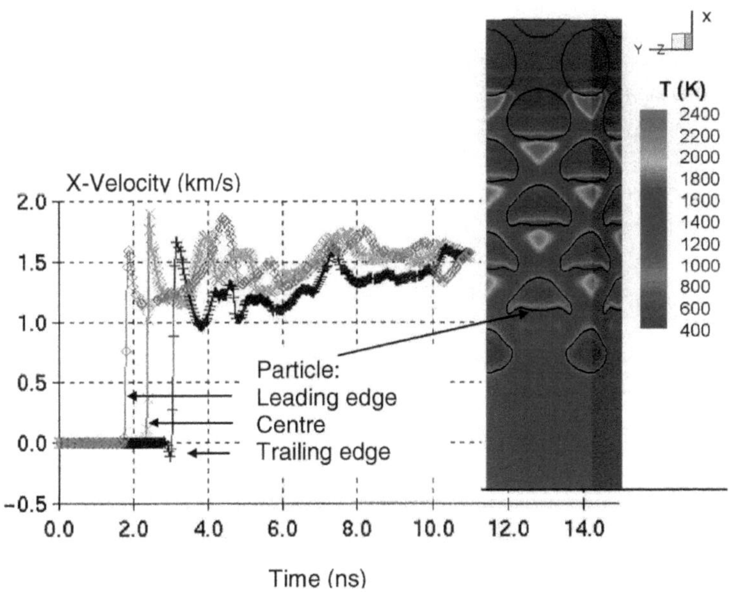

Fig. 2.9. Numerical velocity histories for the leading particle in a 1 g cm^{-3} liquid and aluminum particle system subjected to a 101.3-kbar shock [39,68]

2.3 Transition to Detonation

2.3.1 Progressive DDT

In reactive particle–oxidizing gas mixtures, two types of DDT can be observed: progressive DDT and abrupt DDT via an explosion center. The progressive DDT typically occurs in small tubes, but also for high mass concentrations or large sizes of particles in large tubes.

Figure 2.10 shows a streak photograph registering a progressive DDT process obtained by Strauss [1] in 1968 for a nearly stoichiometric flaked aluminum–oxygen mixture in a 26.4-mm-diameter, 2.7-m-long vertical glass tube. After initiation using a detonator, the flame gradually accelerates towards a spinning detonation without abrupt explosion and a backwards-propagating retonation wave. The final detonation velocity is 1,436 m s^{-1} with a 15% deficit relative to the equilibrium CJ value. Figure 2.11 records a progressive DDT observed by Fangrat et al. [5] for highly volatile brown coal dust–oxygen mixtures in a 50 mm × 50 mm cross section, 3.2-m-long vertical steel tube with a full-length glass window, using 350 J copper wire discharge initiation. While an increase in particle concentration decreases the transition distance, the flame acceleration gradually proceeds towards a detonation without abrupt explosion and retonation. The detonation velocity was recorded to be 1,550, 1,700, and 1,900 m s^{-1} for particle concentrations of 0.28, 0.54, and 1.4 kg m^{-3} respectively. Figure 2.12 shows the evolution of pressure histories in a DDT process reported by Peraldi and Veyssiere [8] for a starch–oxygen mixture in a 53 mm × 53 mm cross section, 4.5-m-long vertical tube using a $2H_2$–O_2 detonation driver as the initiation source. The shock wave is progressively enhanced towards a "quasi-steady" detonation wave with a relatively smooth pressure history behind the shock front without a distinct backwards-

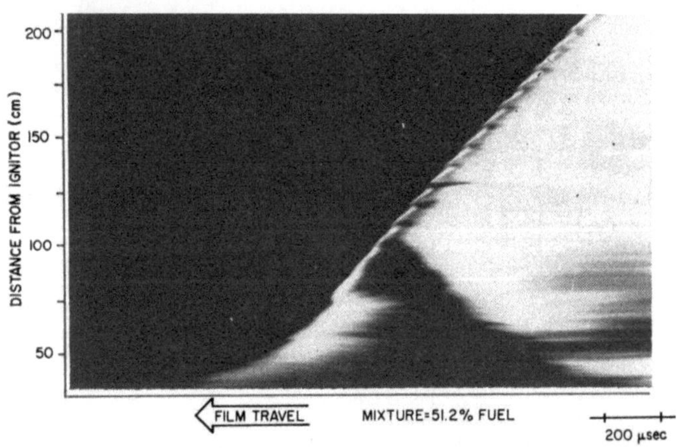

Fig. 2.10. A streak photograph of a progressive deflagration-to-detonation transition (DDT) for a flaked aluminum–oxygen mixture in a 26.4-mm-diameter tube [1]

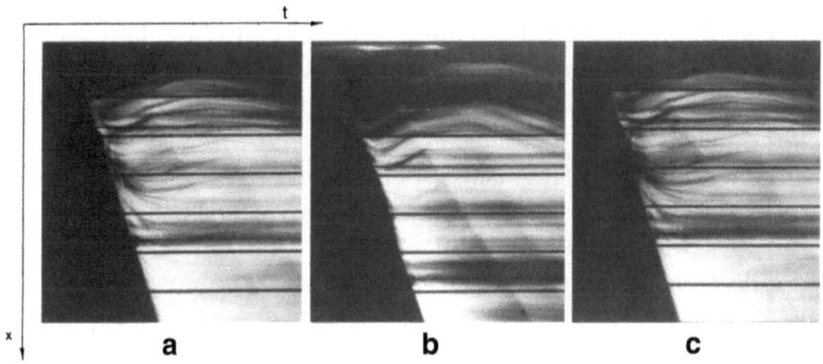

Fig. 2.11. Streak photographs of a progressive DDT for less than 71-μm Egyptian brown coal–oxygen mixtures in a 50 mm × 50 mm cross section tube. **a** $\sigma_p = 0.28\,\mathrm{kg\,m^{-3}}$, **b** $\sigma_p = 0.54\,\mathrm{kg\,m^{-3}}$, and **c** $\sigma_p = 1.4\,\mathrm{kg\,m^{-3}}$ [5]

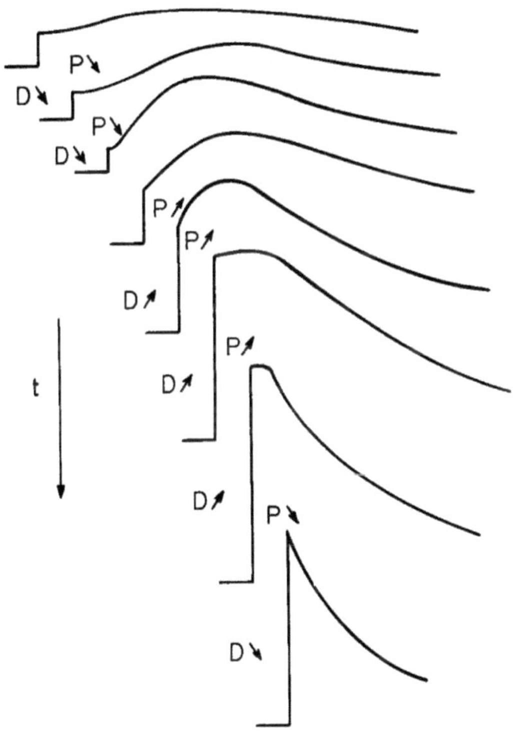

Fig. 2.12. Evolution of pressure history versus propagation distance of an accelerating shock in a $\sigma_p = 1.124\,\mathrm{kg\,m^{-3}}$, 20-μm starch–oxygen mixture [8]

propagating retonation shock wave. In a concentration range of 1–3 $\mathrm{kg\,m^{-3}}$, the propagation velocity reaches 1,300–1,480 $\mathrm{m\,s^{-1}}$ with a deficit relative to the CJ value as large as 30–40%.

Thus, the progressive DDT is characterized by gradual flame acceleration without abrupt onset from an autoexplosion center that forms an overdriven detonation and retonation. The detonation wave developed is often featured with a large deficit of the detonation velocity and a relatively smooth pressure history without a distinguished, periodically oscillating transverse wave structure behind the shock front. The progressive DDT has only been observed in tubes where the tube wall provides confinement of the flame and repeated reflections of the transverse shock waves to progressively amplify the reaction.

2.3.2 Abrupt DDT

For most reactive gases, a detonation wave manifests itself by a transverse wave structure with a detonation cell size of a few millimeters in fuel–oxygen mixtures and a few centimeters in fuel–air mixtures. For reactive solid particles suspended in pure oxygen, however, experiments in tubes of a few centimeters in cross-sectional dimension described earlier found a progressive DDT or very marginal detonation without a distinct transverse wave structure. An abrupt DDT via explosion centers leading to a detonation wave with a transverse wave structure was observed in cornstarch–oxygen mixtures when using a 140-mm diameter, 17.4-m-long tube [70]. Tests using large and sufficiently long tubes are necessary to achieve a DDT in reactive particle–air mixtures.

Using a tube 600 mm in diameter by 42 m in length, Gardner et al. [15] recorded a transition with a violent onset to detonation in a coal dust–air mixture using a 30 $\mathrm{m\,s^{-1}}$ flame jet initiation. As shown in Fig. 2.13, an abrupt DDT takes place near the end of the tube where a peak pressure of 81 atm was measured. The tube, with a length–diameter ratio of 70, was still too short to observe a self-sustained detonation wave.

Complete DDT processes in air via an autoexplosion center were observed for cornstarch and flaked aluminum particles in a 300-mm-diameter, 37-m-long tube (with a tube length–diameter ratio of 123) [20]. The observed DDT process can be divided into a slow-reacting compression stage and a fast-reacting shock stage using relatively weak initiation, as shown in Figs. 2.14–2.16 for mixtures at 1-bar initial pressure. Early in the reacting compression stage, the compression wave is slowly amplified. The compression wave amplification coupled with the chemical energy released by the flame can be clearly recognized through the flame front trajectory crossing the backside of the compression waves. For the lean aluminum–air mixture shown in Fig. 2.15, the reacting compression stage becomes a multiple-compression one, in which the second compression wave is amplified at $50 < x/d < 105$ behind the precursor shock front that was developed from the first compression wave. The multiple-compression stage was also observed in lean cornstarch–oxygen mixtures [70]. As a result of the reacting compression stage, a critical shock wave forms with a Mach number between 3.1 and 3.5 (at $x/d \sim 95$ and 105 in

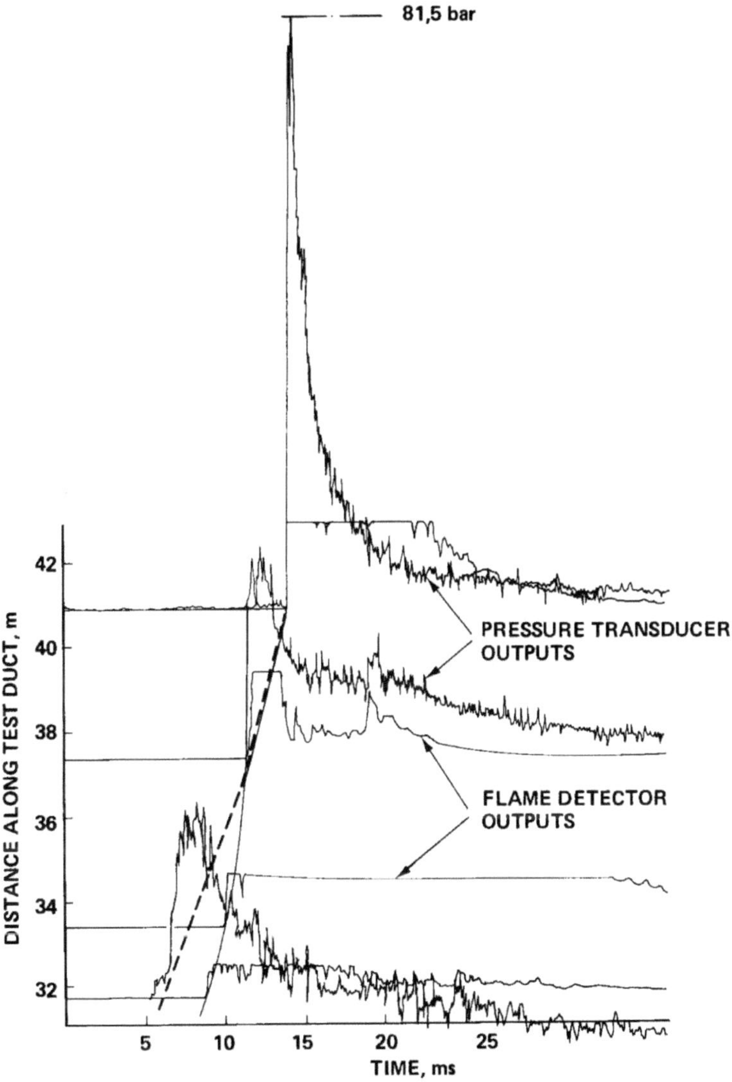

Fig. 2.13. DDT for a $\sigma_p = 0.27\,\mathrm{kg\,m^{-3}}$, less than $75\,\mu\mathrm{m}$ (grade 250) of the US western subbituminous coal–air mixture in a 0.6-m-diameter tube [15]

Figs. 2.14, 2.15 for less sensitive mixtures and at $x/d \sim 35$ in Fig. 2.16 for more sensitive mixtures). Near the formation of the critical shock wave, the flame velocity rapidly increases, indicating that the critical shock wave would be defined not only by a Mach number but also by the temperature gradients behind the shock front. An understanding of the underlying mechanism of detailed turbulent combustion of particles in the reacting compression stage would require further research with advanced diagnostics.

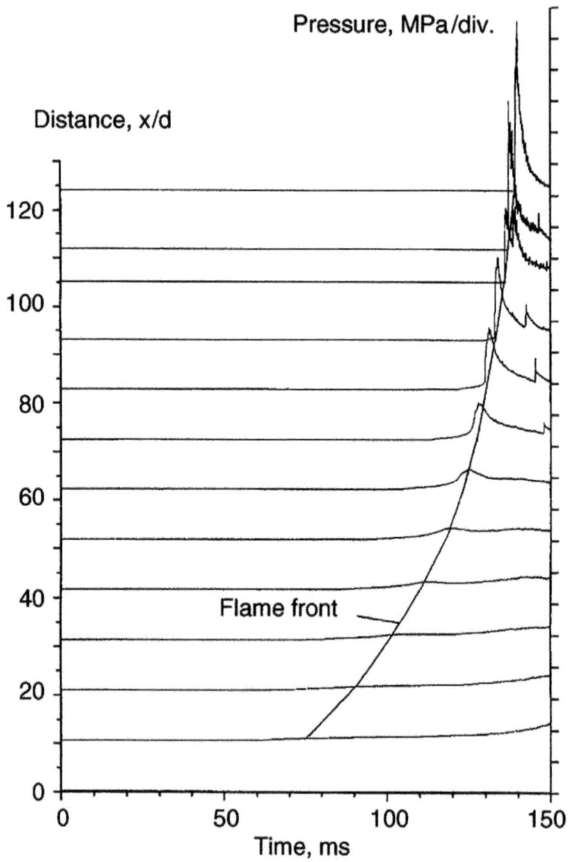

Fig. 2.14. DDT in a $400\,\mathrm{g\,m^{-3}}$, 10-μm cornstarch–air mixture ($\phi = 1.72$) at 1-bar initial pressure using a 0.3 m inner diameter (ID) tube and four 300-J detonators [20]

The formation of the critical shock marks the beginning of the reacting shock stage in which the flame accelerates rapidly owing to close coupling with the shock amplification, as observed in a homogeneous gas DDT [71,72]. Within a propagation distance of the reacting shock of about 20 tube diameters in Figs. 2.14 and 2.15 and less than ten tube diameters in Fig. 2.16, an abrupt onset of overdriven detonation takes place and brings the flame velocity to its maximum. Afterwards, the overdriven detonation wave begins to relax towards a self-sustained transverse wave detonation mode. If it is normalized with the characteristic detonation cell size (to be reviewed in Sect. 2.3.3), the propagation distance of the reacting shock, which begins from the formation of the critical shock and ends at the onset of the maximum overdriven detonation, amounts to about six detonation cell sizes for all three mixtures. While the onset of the overdriven detonation for less sensitive mixtures such as cornstarch–air and lean aluminum–air is clearly accompanied by a retonation

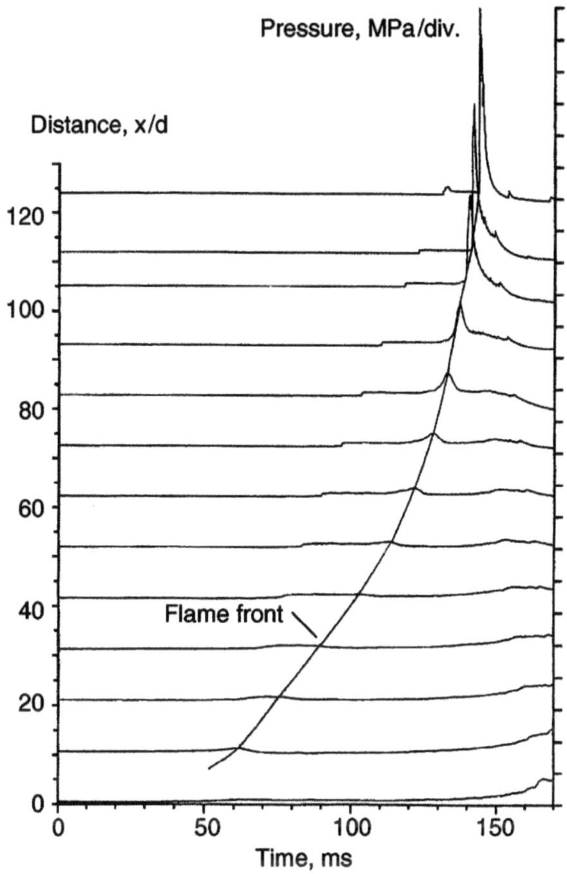

Fig. 2.15. DDT in a 200 g m^{-3}, 36 μm × 36 μm × 1 μm flaked aluminum–air mixture ($\phi = 0.645$) at 1-bar initial pressure using a 0.3-m-ID tube and a 300-J detonator [20]

wave propagating backwards (Figs. 2.14, 2.15), the retonation wave in rich aluminum–air is rather weak as shown in Fig. 2.16.

The self-sustained detonation structure is characterized in the shock wave front and the oscillation frequency recorded in multiple pressure profiles distributed on a tube circumference. Figure 2.17 displays a single-head spinning wave in the 300-mm-diameter tube for the rich cornstarch–air mixture that was used in the DDT experiment shown in Fig. 2.14. The single transverse wave head can be recognized between the profiles numbered 1 and 8. The single-head spinning mode was observed over a wide range of fuel equivalence ratios between 0.7 and 3 at 1-bar initial pressure.

Figure 2.18 shows a detonation wave with multiple transverse wave heads in the 300-mm-diameter tube for the rich flaked aluminum–air mixture with a DDT displayed in Fig. 2.16. Four transverse wave heads can be recognized on one tube circumference at traces 1, 3, 5, and 6. This, together with a

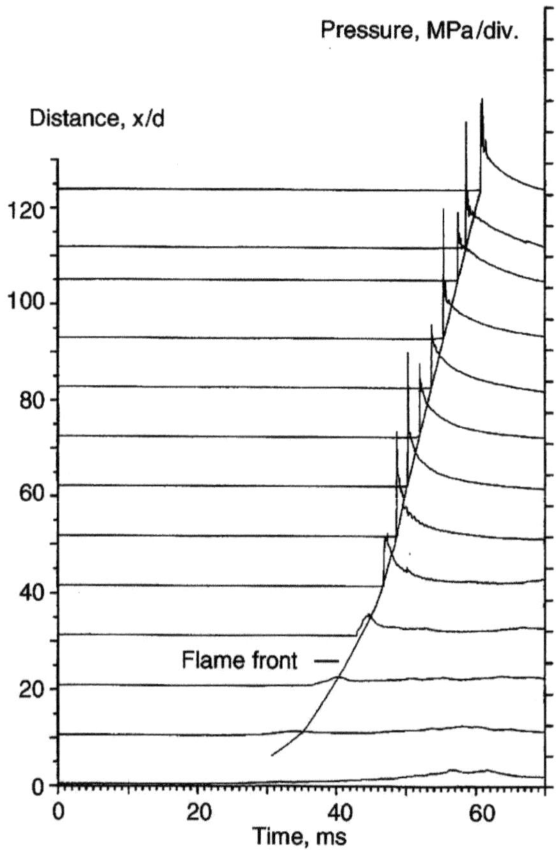

Fig. 2.16. DDT in a 500 g m^{-3}, 36 µm × 36 µm × 1 µm flaked aluminum–air mixture ($\phi = 1.61$) at 1-bar initial pressure using a 0.3-m-ID tube and a 300-J detonator [20]

pressure oscillation period of about 200 µs, indicates at least two detonation cells around the tube circumference with a mean cell size of approximately 0.47 m. The pressure oscillation behind the aluminum detonation front appears weaker than that of the single-head spinning cornstarch detonation displayed in Fig. 2.17, likely caused by the high volatile content of cornstarch particles.

The insensitivity of aluminum–air detonation was further shown through the DDT in 0.1-µm atomized aluminum particles, known as "Alex" made by an exploding wire process, suspended in air using an 80-mm diameter, 10-m-long tube (with a tube length–diameter ratio of 125) [73]. At 1-atm initial pressure using a 6-kJ detonator, a multiple-compression DDT process was observed and abrupt transition to a spinning detonation occurred near the end of the tube, where the propagation distance of the reacting shock was again confirmed to be six characteristic detonation cell sizes. As the initial pressure

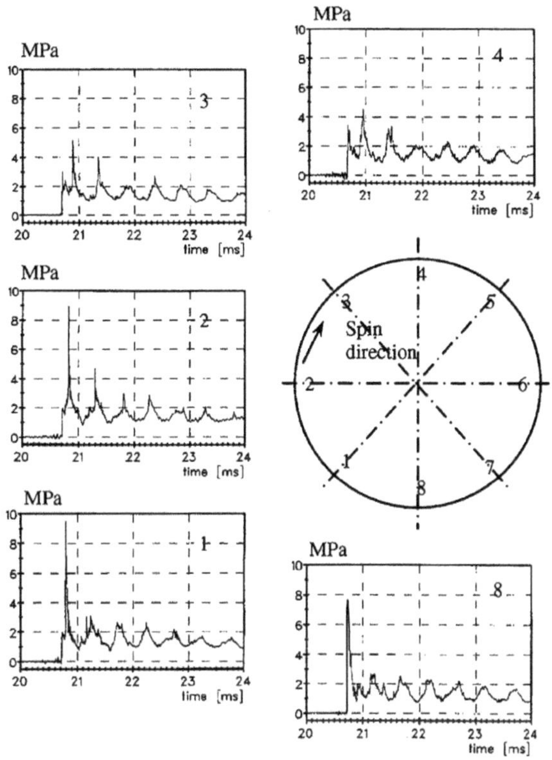

Fig. 2.17. Pressure profiles on a 0.3-m-ID tube circumference in a 400 g m^{-3}, 10-μm cornstarch–air mixture ($\phi = 1.72$) at 1-bar initial pressure [20]

increased, the DDT distance was reduced to the half of the tube length and the weak primary pressure oscillation of the established spinning detonation became more distinguished with an increase in amplitude, indicating a stronger transverse wave (Fig. 2.19). A primary oscillation period of 140 μs can be determined in the pressure history at $p_0 = 2.5$ atm. When multiplied by the propagation velocity at that location, this oscillation period results in a pitch of 230 mm that is consistent with the smoke foil record. In summary, the strong dependence of detonation sensitivity on initial pressure and the highly nonlinear abrupt DDT nature in the micrometric and nanometric aluminum–air mixtures suggest that the aluminum reaction mechanism of the detonation waves depends on chemical kinetics. However, the weak transverse wave structure of the aluminum–air detonation at 1 atm indicates some weaker functional dependence than the Arrhenius kinetics and is likely associated with some slower diffusional combustion of the part of the aluminum mass at a later time. The insensitivity of aluminum–air detonation might be attributed not only to heterogeneous transport processes but also to a high-melting-point oxide layer that passivates the surface of each particle.

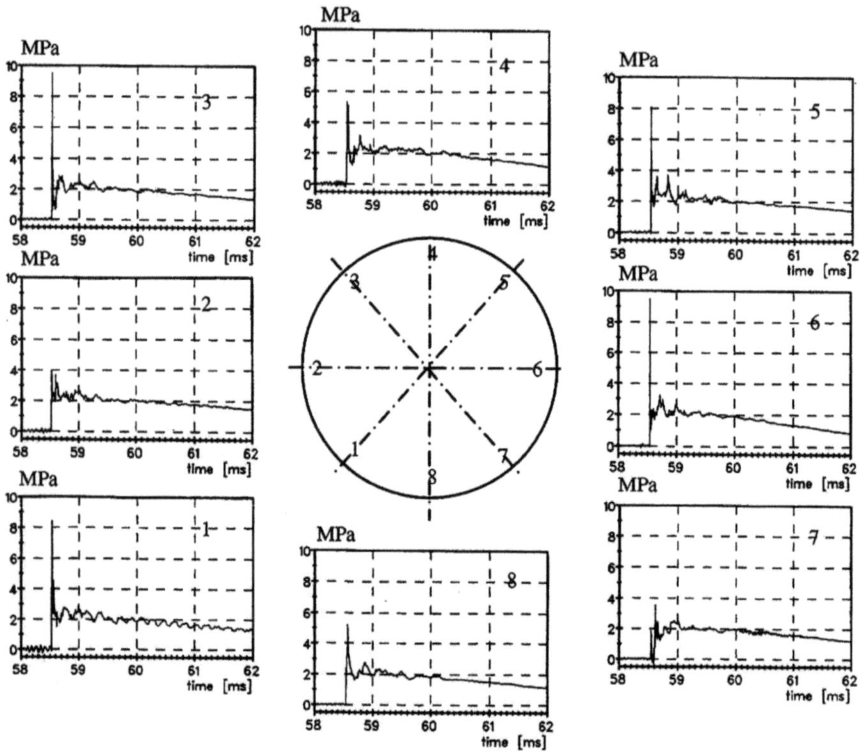

Fig. 2.18. Pressure profiles on a 0.3-m-ID tube circumference in a $500\,\mathrm{g\,m^{-3}}$, $36\,\mu\mathrm{m} \times 36\,\mu\mathrm{m} \times 1\,\mu\mathrm{m}$ flaked aluminum–air mixture ($\phi = 1.61$) at 1-bar initial pressure [20]

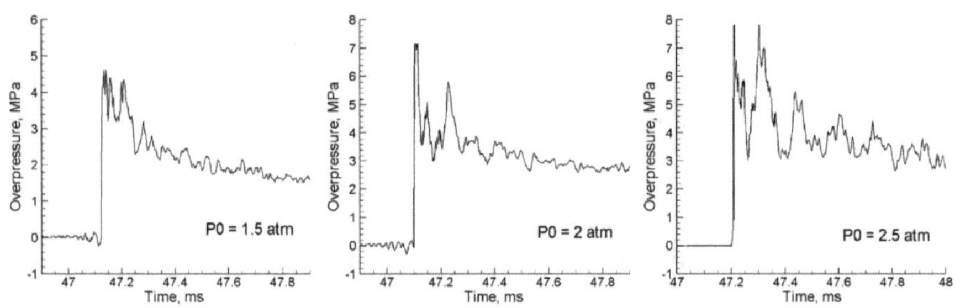

Fig. 2.19. Detonation pressure history in an 80-mm-ID tube at 8.72 m in a 0.1-μm aluminum–air mixture with particle concentrations of 600, 800, and 1,000 $\mathrm{g\,m^{-3}}$ at 1.5, 2, and 2.5 atm, respectively (normalized to be $400\,\mathrm{g\,m^{-3}}$ per atmosphere)

The single-head spinning detonation has been accepted as the lowest stable, self-sustained detonation mode in a tube for homogenous gas mixtures. The corresponding tube diameter is therefore referred to as the "minimum tube diameter" d_{min}, a necessary boundary condition for the transition to and propagation of a stable detonation [77]. The minimum tube diameter for detonation in solid particle–gas flow is at least 1 order of magnitude greater than the minimum tube diameter observed for most detonable gas mixtures.

2.3.3 Transition to Detonation Near the End Wall

Considering a deflagration wave propagating in a tube, the flame Mach number does not reach its critical value when the precursor shock impacts the end wall. In this case, Craven and Greig [74] suggested a DDT scenario with the onset of detonation on the reflected shock region between the flame front and the end wall. The resultant detonation then propagates into the reflected shock region and reflects on the end wall, thus resulting in a very high peak pressure. Such a DDT may interpret an earlier experimental observation of Kogako [75], in which the high pressure in a 6.8% acetylene–air mixture caused destruction of the end section of a steel tube (305 mm in diameter and 10 mm in wall thickness), with heavy fragments flying out in all directions. A Hugoniot analysis indicates that for a flame Mach number of 2.5 observed before the rupture, the peak pressure generated by the reflected detonation on the reflected shock state achieves a value 10 times that of the reflected CJ detonation in the initial mixture where $p_0 = 1\,\text{atm}$ [25].

The pressure generated in the Graven–Greig DDT can be further increased if a dense reactive particle suspension is introduced in a reactive gas mixture in the front of the end wall. The reflected peak pressure behind a dense particle suspension layer on the end wall is considerably higher than that for gas normal reflection on a rigid wall as predicted in Fig. 2.20, owing to wave reflections and focusing acting on the voids in the particle system. As the

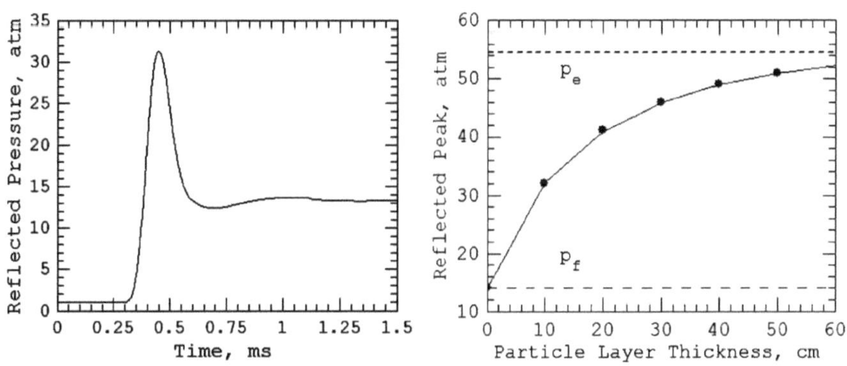

Fig. 2.20. Computed wall-reflected pressure for an inert shock ($M = 1.9$, $\gamma = 1.29$, and $a_g = 345\,\text{m}\,\text{s}^{-1}$) running in a $10\,\text{kg}\,\text{m}^{-3}$, 5-μm aluminum particle suspension layer in front of the wall. *Left*: Pressure history for the 100-mm layer thickness. *Right*: Peak pressure versus layer thickness. p_e phase-equilibrium Hugoniot pressure, p_f gaseous Hugoniot pressure [25]

particle layer thickness increases, the wall-reflected peak pressure increases towards the phase equilibrium Hugoniot value for the given particle concentration. One can expect that the DDT on the reflected shock region in a dense reactive particle–reactive gas suspension near a wall can generate a very significant pressure. Such a hybrid DDT was observed in an 80-mm-diameter tube filled with a lean acetylene–air mixture and a dense aluminum particle suspension layer ($\sigma_p = 10\,\mathrm{kg\,m^{-3}}$, 100 mm thick) located in the front of the tube end wall [25]. A DDT occurred after the wall reflection of the precursor shock and the reflected detonation produced a wall peak pressure of 700 atm (Fig. 2.21). In comparison, the DDT after the wall reflection for the same gas system without particles achieved a wall peak pressure between 260 and 300 atm. Thus, addition of a dense aluminum particle layer provides a pressure enhancement factor of more than 2. The enhancement effect results from both the higher reflected gas pressures due to multiple shock interactions with dense particles and the particle combustion.

2.3.4 Initiation of Unconfined Detonation

Very few experimental studies have been published for initiation of unconfined particle-gas detonation without the influence of tube confinement. Tulis [76] attempted to detonate 4.54 kg flaked aluminum explosively dispersed in air in a 1-m-high, 3-m-radius cylindrical cloud. The cloud was initiated near the center using 2.27 kg solid explosive; detonation was not observed and the ground pressure decayed from 1.8 to 0.9 MPa in a radius from 1.3 to 2.7 m. Veyssiere et al. [21] detonated a rich flaked aluminum–oxygen cloud (with an equivalence ratio larger than 1.6) in a vertical cylindrical polyethylene bag, 1 m high and 0.7 m in diameter, initiated at the top using 150 g TNT. The detonation observed near the bottom of the bag reached a velocity of 1,650 $\mathrm{m\,s^{-1}}$ and a peak pressure of 3.6–5.2 MPa. For the same arrangement, only a decaying blast wave was observed, followed by particle combustion for 20-μm starch and atomized aluminum particles.

To observe the propagation of detonation, a large-scale cloud was generated through a charge configuration arranged in an 18-m-long, 90° V-shaped steel trough line, in which a pentaerythritol tetranitrate cord (21.3 $\mathrm{g\,m^{-1}}$, 6.1 mm in diameter) was located at the bottom vertex covered by a layer of aluminum powder [22]. Detonation of the pentaerythritol tetranitrate cord dispersed the aluminum powder in air to a 3-m-radius cross section and an 18-m-long suspension at a given dispersal time. The aluminum–air cloud was then initiated at one end using 8 kg C4 explosive located 1.5 m from the end as well as 1 m from the ground and the steel line. Figure 2.22 shows an example of high-speed photographs for a DDT in a 31.5 kg flaked aluminum–air cloud with an average particle concentration of 250 $\mathrm{g\,m^{-3}}$ (or an equivalence ratio of 0.9 based on the local air pressure of 91 kPa). From the ground pressure histories along the propagation distance shown in Fig. 2.23, the shock front velocity is 1,600 $\mathrm{m\,s^{-1}}$ at 4 m from the C4 location and decays to 1,380 $\mathrm{m\,s^{-1}}$

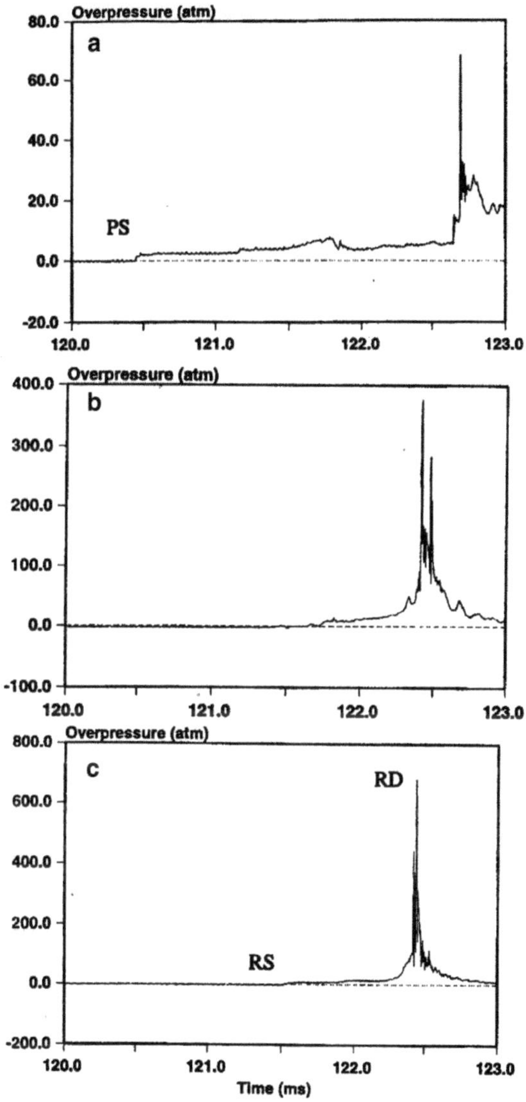

Fig. 2.21. Pressure histories for the DDT in a 6.75% C_2H_2–air mixture at 1-atm initial pressure with a 100-mm layer of 10 kg m^{-3}, 5-μm aluminum particles in front of the end wall of an 80-mm-ID tube: **a** 580 mm from the end wall; **b** 35 mm from the end wall; **c** at the end wall. *PS* precursor shock, *RS* reflected shock, *RD* reflected detonation [25]

at 7 m. At 11 m, the abrupt onset of detonation occurs with a peak pressure of 8.4 MPa. The 2-m-averaged detonation velocity reaches a maximum value of 1,533 m s^{-1} and remains at 1,460–1500 m s^{-1} in the further propagation. In the oscillating pressure history after 11 m, a transverse wave structure is distinguishable with a primary period of about 350–400 μs for the signal at

Fig. 2.22. DDT in an unconfined 3 m × 3 m × 18 m flaked aluminum–air cloud. *Top left*: Aluminum suspension. *Top right*: 8 kg C4 explosive initiation. *Bottom left* and *bottom right*: Detonation propagation [22]

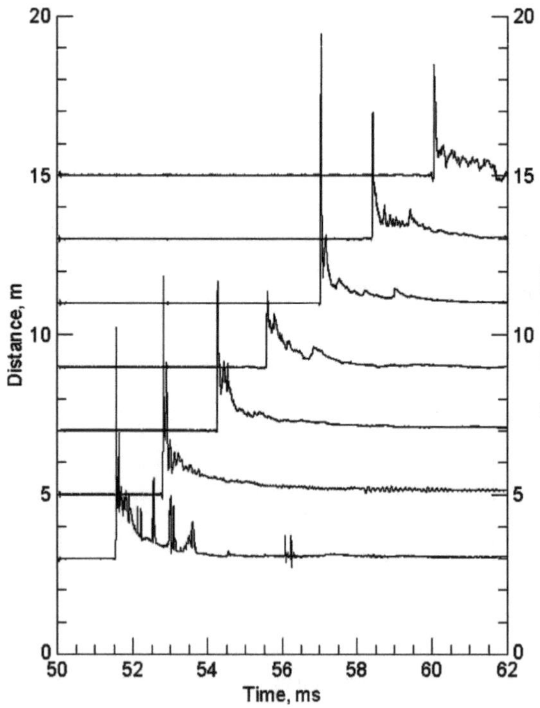

Fig. 2.23. DDT pressures in an unconfined 3 m × 3 m × 18 m flaked aluminum–air cloud using 8 kg C4 explosive initiation [22]

15 m. The smoke foil located at the ground between 13 and 15 m registers a detonation cell width of 0.52–0.6 m, consistent with the pressure oscillation period. This experiment indicates that the 8 kg C4 is near the critical charge for direct initiation of flaked aluminum–air detonation. Replacing the flaked aluminum by 47.6 kg atomized aluminum with a mean diameter of 1.6 μm by number and a mean diameter of 3.3 μm by weight (known as H-2 by Valimet),

abrupt DDT phenomena took place at 13 m with a 5.5-MPa peak pressure. However, in this case the cloud length is too short to observe a self-sustained detonation wave. A DDT was not observed when using 5 kg C4 initiation charge for H-2 aluminum particles. Note that without aluminum particles, the air blast overpressure from the 8 kg C4 explosion decays rapidly to 0.186, 0.046, and 0.023 MPa at 5, 10, and 15 m, respectively.

2.3.5 Detonation Velocity and Pressure

For the solid particle–air detonation waves with spinning structures or transverse wave structures, some experimental detonation velocities are summarized in Figs. 2.24–2.27. These experimental velocities are in agreement with the values computed from the equilibrium CJ theory within about 10% deviation (the low velocity values in Figs. 2.26, 2.27 will be discussed in Sect. 2.5). The 0.1-μm Alex aluminum particles were passivated with an oxide coating to a mass fraction of about 10%, and so its detonation performance was expected to be less energetic than that of the pure aluminum used in the calculation. Agreement with equilibrium CJ theory indicates that the transverse-wave-structured detonation velocity is mainly determined by the energetics. The CJ detonation velocities display a shift towards the lean side and the reasons are twofold: (1) a shift in postshock equivalence ratio from the initial towards a leaner value owing to a particle velocity relaxation time lag and (2) a shift in real particle concentration owing to sedimentation and adhesion of particles to the tube wall during experimental dispersal. As the mixtures become richer in cornstarch and aluminum, the calculated CJ velocity drops; however, the experimental results show a slight decrease or a

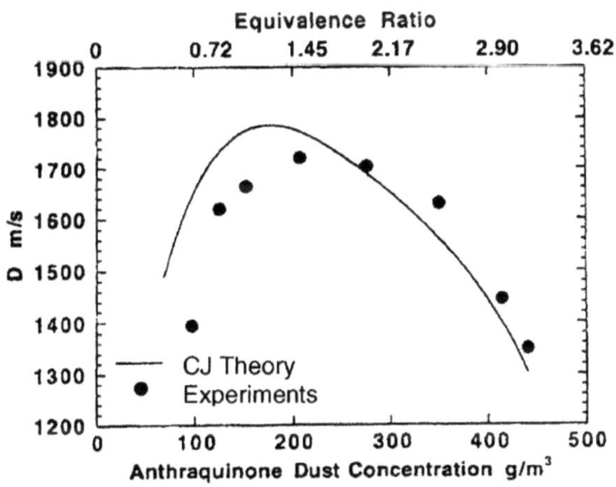

Fig. 2.24. Experimental detonation velocities in 22 μm × 6 μm × 6 μm anthraquinone particle–air mixtures at 1.15-bar initial pressure in a 0.14-m-ID tube and the equilibrium CJ calculations [18]

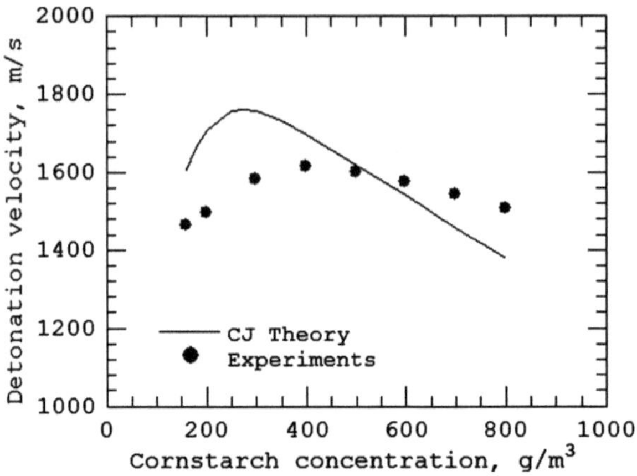

Fig. 2.25. Experimental detonation velocities in 10-µm cornstarch–air mixtures at 1-bar initial pressure in a 0.3-m-ID tube and the equilibrium CJ calculations. The stoichiometric concentration is $233\,\mathrm{g\,m^{-3}}$ [20]

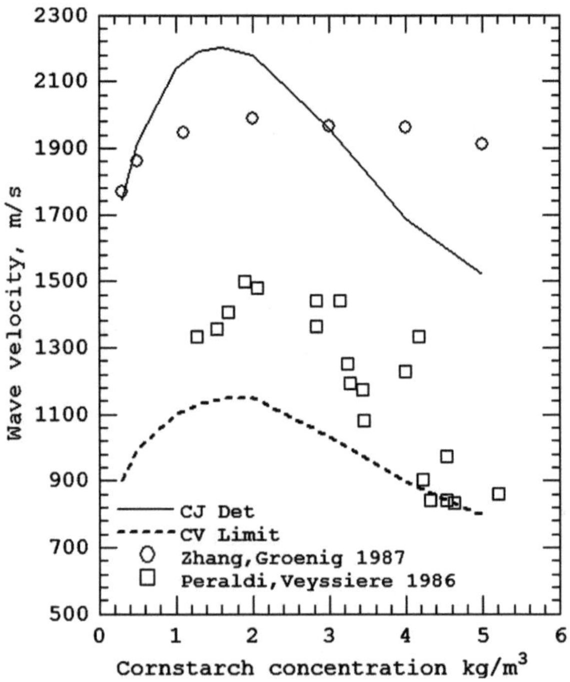

Fig. 2.26. Detonation and quasi-detonation velocities in cornstarch–oxygen mixtures at 1 bar initial pressure. (Data from Zhang and Grönig [16] used a 0.14-m-ID tube and data from Peraldi and Veyssiere [8] used a 53 mm × 53 mm tube)

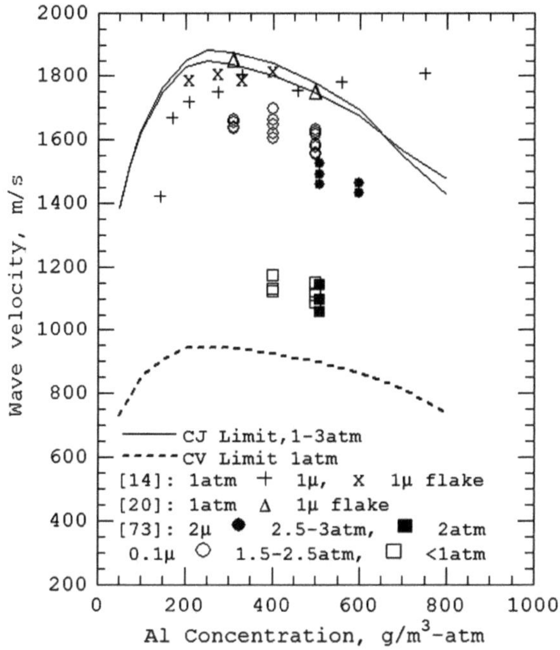

Fig. 2.27. Detonation and quasi-detonation velocities in aluminum–air mixtures. The stoichiometric particle concentration is $310\,\mathrm{g\,m^{-3}}$. (Data from Borisov et al. [14] used a 0.12-m-ID tube, data from Zhang et al. [20] used a 0.3-m-ID tube, and data from Zhang et al. [73] used an 80-mm-ID tube)

"plateau" extending from the peak values. It is probably due to the time lag of momentum and heat transfer from gaseous products to unburned particles and their agglomerates in the hot products. Hence, the equilibrium CJ theory may not be applicable to very rich concentrations of finite-sized particles.

Apart from particle sedimentation and adhesion, which cause experimental difficulties in achieving a uniformly dispersed two-phase mixture as assumed in the equilibrium CJ theory, there are two inherent reasons responsible for detonation velocity deviations, as discussed in Sect. 2.2. First, contrary to the fundamental postulate of the equilibrium CJ theory, detonations in reactive particle–gas mixtures are essentially nonideal, where momentum, heat, and chemical equilibrium between the two phases may not be achieved at the phase-frozen or gaseous sonic locus owing to the finite sizes of particles. Second, momentum and heat loss induced by lateral boundary layer effects behind the shock front increases with larger reaction zones corresponding to an increase in transverse wave spacing.

Experimental determination of the CJ detonation pressure is more difficult than that of the detonation velocity in relatively insensitive particle–gas mixtures, since the large transverse wave spacing of the detonation front provides various pressures along the three-dimensional shock front and a number of oscillations in the pressure profiles behind the shock front (Figs. 2.17, 2.18). The detonation peak pressures are invariably higher than the equilibrium CJ values by a factor of 2 or more. Rather than attempting to interpret the CJ pressure from an oscillatory pressure record, one can average the experimental "peaks" and "valleys" over a period behind the front, and the resultant median was comparable to the CJ pressures [18].

2.4 Detonation Structure

2.4.1 Spinning and Cellular Detonation

For homogeneous gases, the single-head spinning and cellular detonation waves have been considered as the stable transverse wave detonation modes with a triple-point configuration as the basic feature [77–89]. Owing to detonation insensitivity and high detonation pressures as well as in situ particle dispersal, it has been difficult to register a detonation structure in reactive particles suspended in oxygen or air. There are, however, a few experimental studies available for single-head spinning and cellular detonation waves [16–20].

Figure 2.28 displays a single-head spinning structure at the detonation front around the 140-mm-diameter tube on the circumference in a 0.5-bar stoichiometric cornstarch–oxygen mixture. This structure was observed through the use of a large number of pressure-ion double front gauges distributed on the tube circumference at several axial cross sections within a pitch distance [16]. The corresponding pressure profiles distributed on one circumference are similar to those in Fig. 2.17. In Fig. 2.28, the frontal structure manifests itself primarily in a transverse shock wave that propagates into the induction zone behind the incident shock and produces a traveling triple-point configuration at the front. The transverse wave velocity observed on the periphery is approximately equal to the axial propagation velocity, thus resulting in a spinning track angle, $\alpha = 45°$, at the front around the tube wall.

A two-head spinning or a single-cell detonation wave was also recorded in the 140-mm-diameter tube with the same diagnostic method for a rich cornstarch–oxygen mixture (Fig. 2.29) [17]. The collision of two triple points leads to an overdriven wave followed by a transient decoupling between the shock and chemical reaction. Since the two triple-point configurations move in opposite directions on the circumference, successive collisions make the continual reignition to sustain the detonation wave possible. The triple-point trajectories form a single cell with a cell width of $\lambda \sim \pi d$. The experiments produced a transverse wave velocity of $C \approx 0.62D$ and a mean track angle of $\alpha = \tan^{-1}(C/D) \approx 32°$, where D is the mean axial propagation velocity. The single-head and the two-head detonation wave cannot exist without boundary confinement.

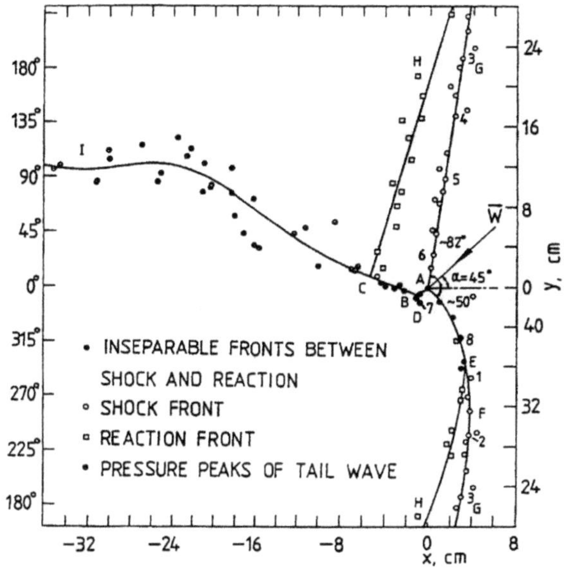

Fig. 2.28. Spinning detonation front structure on a 0.14-m-ID tube circumference in a $554\,\mathrm{g\,m^{-3}}$, 10-μm cornstarch–oxygen mixture ($\phi = 1$) at 0.5-bar initial pressure. x axial coordinate, y circumferential coordinate [16]

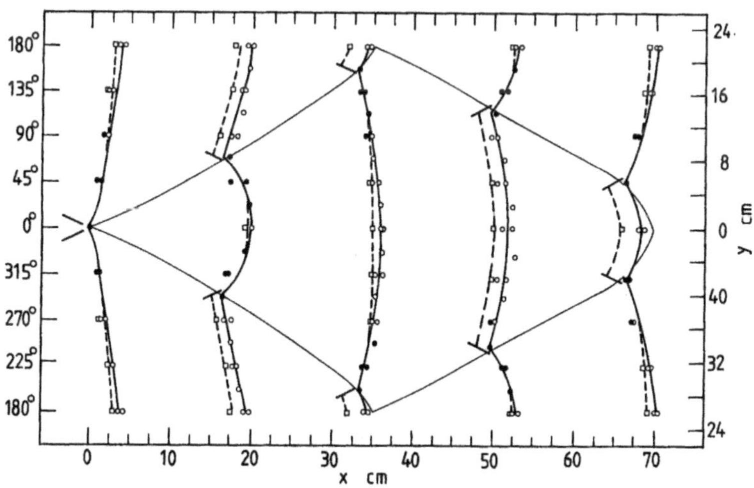

Fig. 2.29. Two-head detonation front structure on a 0.14-m-ID tube circumference in a 10-μm cornstarch–oxygen mixture ($\phi = 3.1$) at 1.15-bar initial pressure. x axial coordinate, y circumferential coordinate. *Open circles* shock front, *squares* flame front, *filled circles* irresolvable between shock and flame front [17]

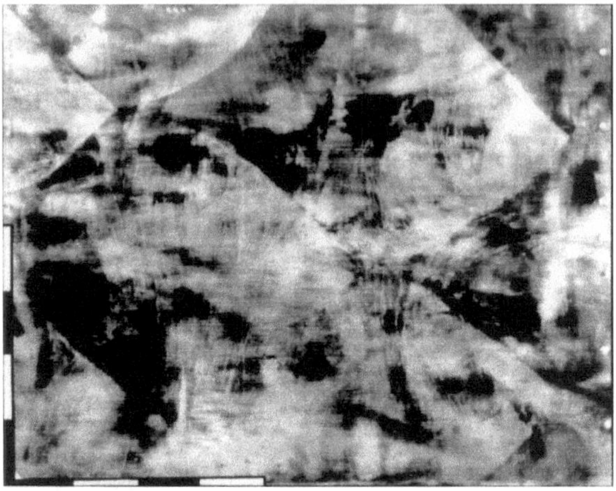

Fig. 2.30. Cellular detonation structure from a smoke foil record on a 0.3 m-ID tube circumference in a 440 g m^{-3}, 10-μm cornstarch–oxygen mixture ($\phi = 0.8$) at 0.5-bar initial pressure. The *scale bar* is 0.1 m and detonation propagates from left to right [20]

Figure 2.30 shows smoke foil records of a cellular detonation structure observed from two experiments in the 300-mm-diameter tube for a $\phi = 0.8$ cornstarch–oxygen mixture at 0.5-bar initial pressure [20]. The soot photographs display more than 1.5 cells in the 0.8-m-wide foil. The average cell width and length measured from the soot photographs are $\lambda = 0.50$ m and $L = 0.77$ m. This results in a mean track angle of the triple-point trajectory of $\alpha = \tan^{-1}(\lambda/L) \approx 33°$ and an average transverse wave velocity of $C = (\tan\alpha)D \approx 0.65D$.

2.4.2 Detonation Dynamic Parameters

In the theory of gas detonation, the correlation by Zeldovich et al. [90] links the minimum tube diameter, d_{min}, for propagation of a stable, self-sustained detonation wave to the cell width of the cellular detonation by

$$d_{\mathrm{min}} \approx \lambda/\pi. \tag{2.9}$$

This defines a characteristic cell size λ for the single-head spinning detonation since it has been considered to be the lowest stable, self-sustained detonation mode [77]. Numerous gas detonation experiments have proven that relation (2.9) is appropriate for fuel–air detonation, but it may underpredict the minimum tube diameter for fuel–oxygen mixtures [123, 124]. For the cornstarch–oxygen mixtures as reviewed in Sect. 2.4.1, the measured cell size, $\lambda = 0.50$ m at the equivalence ratio $\phi = 0.8$ (Fig. 2.30), is consistent with the characteristic cell size of the single-head spinning mode, $\lambda \sim \pi d = 0.44$ m at $\phi = 1$

Table 2.1. Dynamic parameters for detonation of solid particles in oxygen or air

Material	Equivalence ratio	One-head spin tube diameter $d_{\min}$ (m)	Detonation cell size λ (m)	Direct initiation energy E_{cr} (MJ)	Direct initiation charge M_{cr} (kg TNT)	References
10-μm cornstarch–air	1.7	0.3	0.94	293	63	[20]
10-μm cornstarch–O_2	1	0.14	0.44	60	12.8	[16]
Anthraquinone–air (6-μm strips)	1.5	0.14 at 1.16 atm	0.5	40	8.5	[95]
0.1-μm Al–air	1.6	0.08	0.25	4.8	1.0	[73]
Flaked 1-μm Al–air	1.3–1.6	0.12–0.14	0.38–0.44	24–36	5.1–7.7	[14, 20]
Flaked Al–air	0.9	Unconfined	0.55	44	8	[22]
2-μm Al–air	1.3	Unconfined	0.6	62	13	[22]
2 μm Al–air	1.6	0.08 at 2.5 atm	0.62	74	15.8	[73]
Flaked Al–O_2	1–1.6	0.0264	0.08–0.1	0.27–0.8	0.06–0.17	[1, 21]
US W. subbituminous <75μm coal–air	≥1	0.6	1.2–1.8	470–1,500	100–300	[15]
Egyptian brown <75-μm coal–O_2	≥1	0.08	0.25	7.1	1.5	[94]

The initial pressure is 1 atm if not specified. The direct initiation charge is converted on the basis of a TNT detonation energy of 4.67 kJ g^{-1} from the Cheetah equilibrium calculation [130].

(Fig. 2.28). While more experiments are required to draw a firm conclusion, this consistency suggests that relation (2.9) is applicable to the heterogeneous detonation of reactive particles in air and oxygen. This could be attributed to the fact that detonation in reactive particle–gas flow possesses a large transverse wave spacing imbedded with numerous distributed hot or flame spots induced by the particles. Therefore, the detonation is relatively insensitive to small disturbances in the boundary layer on the wall of large tubes. Relation (2.9) has been applied to the single-head spinning detonation waves to evaluate characteristic cell sizes for various mixtures. The resulting characteristic cell sizes for several types of reactive particles suspended in air or oxygen at various equivalence ratios are listed in Table 2.1. The evaluated cell sizes for the particle–air detonation waves range typically between 0.25 and 1 m at 1-atm initial pressure, thus being at least 1 order of magnitude larger than that for the detonation in most gas–air mixtures.

The long DDT distances shown in Figs. 2.14–2.15 were obtained using pyrotechnical igniters with an initiation energy of 1.2 kJ for cornstarch–air and 0.3 kJ for flaked aluminum–air. Hence, an initiation energy of 10^2–10^3 J can be considered a "weak" initiation for a reactive particle–air DDT that starts from slow deflagration during the initial stage. This fact indicates that the initiation energy for the reactive particle–air DDT is at least 3 orders of magnitude greater than that for the gaseous DDT using weak initiation.

Alternatively, as indicated in Sect. 2.3.4, direct initiation of an unconfined flaked aluminum–air detonation requires a critical charge near 8 kg, 3 orders of magnitude greater than the critical charge for direct initiation of a gaseous fuel–air detonation that can be found in [91]. The requirement of a 3 orders of magnitude higher initiation energy was also found when comparing dust explosions to gas explosions in closed vessels [4]. Considering the 1 order of magnitude larger cell size in the particle–gas detonations compared with gas detonations, one can reasonably scale the initiation energy to the cube of the characteristic detonation cell size for reactive particle–oxidizing gas mixtures, a power scaling rule well demonstrated for gaseous detonations [90–93]. Because of the lack of direct experimental data, the correlation of the critical energy, E_{cr}, for direct initiation of gaseous detonation [92]

$$E_{\mathrm{cr}} = A\rho_0 D^2 \lambda^3 \tag{2.10}$$

has been assumed to estimate the critical energy and critical charge mass for direct initiation of unconfined reactive particle–oxidizing gas detonation. Here, the values of ρ_0, D, and λ are taken from the initial mixture density, the experimental detonation velocity and the characteristic cell size, respectively. The coefficient $A = 82$ is a fitting to the initiation charge mass (8 kg C4 with a TNT mass equivalency of 1.19) for the unconfined flaked aluminum–air detonation experiment shown in Fig. 2.23. The estimates of the critical energy and charge mass obtained are listed in Table 2.1. As for a range of aluminum particle sizes, the critical initiation energy and charge mass obtained typically range between 20 and 200 MJ and 5 and 50 kg, respectively, for unconfined aluminum particle–air detonation at 1-atm initial pressure. This estimate only serves as a reference; the accurate values must be obtained from direct measurements performed in large-scale experiments, which are current subjects of research. For a number of experiments in tubes, a hydrogen–oxygen or acetylene–oxygen detonation driver was also often used to initiate the reactive particle–gas detonation directly. A method to evaluate the initiation energy for a detonation driver can be found in [70].

For organic particles with a high volatile content, detonation sensitivity is increased with increasing initial pressure p_0. In the same cornstarch–air mixtures, while the single-head spinning detonation was found in the 300-mm-diameter tube, it was observed in the 140-mm-diameter tube only when the initial pressure was raised to between 2 and 2.5 bar [70]. These results suggest that the high volatile organic particle–gas detonation approximately follows the scaling rule of gaseous detonation, in which the detonation cell size is inversely proportional to the initial pressure [96–98]:

$$\lambda \sim p_0^{-m}, \quad m = O(1). \tag{2.11}$$

Correlation (2.11), together with relation (2.9), links the minimum tube diameter d_{min} with the detonation cell size λ at an initial pressure p_0. The approximately inverse dependency of λ on p_0 arises from the approximate linear dependency of p_0 on the oxygen concentration above a certain lower initial pressure limit. Hence, $m = O(1)$ indicates a reaction mechanism strongly dependent on the gas-phase kinetics for these organic particle–gas detonation waves.

Aluminum particles possess a high-melting-point oxide coating that must be melted or cracked open before aluminum can react. Because sound knowledge of the aluminum reaction mechanism under detonation conditions is lacking, the droplet diffusion combustion theory has been applied to aluminum particles–gas detonation problems in the literature [14, 27, 66]. The diffusion theory states that the particle burning time is proportional to a power of its initial diameter, $t_b \sim d_{p0}^n$ $(n = 2)$, as the temperature of a particle exceeds the oxide melting point [99]. The theory assumes infinite kinetics and is essentially independent of temperature and pressure. It applies according to the classic experimental observations for combustion of large aluminum particles in quiescent atmospheres [100]. Under conditions of 1–100 atm and 1–50-μm particles, however, it has been found that the burning rate of aluminum particles is greater and increases with the initial pressure, thus resulting in a smaller power n between 1 and 2 and a dependence on pressure [101–108]. The power $n < 2$ implies the contribution of finite gas-phase kinetics and possibly convective flow effects [99]. Since the burning aluminum mass flux in the particle radial direction is inversely proportional to the particle radius (yielding $t_b \sim d_{p0}^2$) in the diffusional transport and is independent of the particle radius (yielding $t_b \sim d_{p0}$) in the kinetic process, the diffusional transport rates approach infinity as the particle diameter approaches zero, while the kinetic process rates do not increase with decreasing particle size. Therefore, at sufficiently small particle diameters, the use of the d^2 law becomes incorrect and the particle combustion must become kinetics-dependent.

The high-momentum flow immediately behind the shock, moreover, also changes the physical properties of particles and consequently influences the ignition and reaction of particles [109–114]. In quiescent or low-speed flow, a threshold ignition temperature was observed to be above 2,100 K for 2–30-μm atomized aluminum particles at 6–11 bar, with a measured ignition delay time proportional to a power of the particle diameter, $t_I \sim d_{p0}^2$, following the convective heat law [115]. In shock tube experiments, however, Borisov et al. [109] reported a drop of the ignition temperature to about 1,400 K for 15–20-μm aluminum particles after a reflected shock. The shock ignition temperature would be expected to be even lower for 1–2-μm particles. A temperature of 1,000–1,400 K results in a vapor pressure of 10^{-5}–10^{-1} Pa only, thus further indicating the improbability of diffusion-limited evaporation reaction in shocked conditions. Experiments were also conducted to ignite aluminum particles by a shock wave immediately followed with a detonation products expansion flow, using a hydrogen–oxygen detonation driver connected with an air-filled driven section in which particles were initially dispersed at the end of the driver section [114]. The observed ignition delay time tended towards $t_I \sim d_{p0}$ for a wide aluminum particle diameter range of 2–110 μm, further suggesting possible influences of the kinetics-limited ignition.

Since aluminum–air detonation only occurs for a particle characteristic size less than a few micrometers and the particles ignite and burn under the shocked (above $40p_0$ and 1,500–2,000 K) and subsequent detonation (above $20p_0$ and 3,000–4,000 K) conditions, one would expect that finite kinetics could affect the aluminum ignition and reaction. Chemical kinetics dependence was evidenced by the abrupt aluminum–air DDT in both tubes and an unconfined atmosphere as described in Sects. 2.3.2 and 2.3.4. Experiments also showed that an increase in initial pressure significantly reduces the DDT distance for 0.1-μm aluminum particle–air mixtures in an 80-mm-diameter tube [73]. Through an increase of the aluminum particle size to about 2 μm (known as H-2 by Valimet), a transition to single-head spinning detonation was observed only when the initial pressure was increased to 2.5 atm, while at 2 atm and below no transition to detonation occurred within the tube length, as displayed in Fig. 2.31.

If the above discussions are taken into account, correlations for the ignition delay and burning time of fine aluminum particles can be proposed as $t_{\mathrm{I}} \sim d_{p0}^{n}/p_0^{m}$ and $t_{\mathrm{b}} \sim d_{p0}^{n'}/p_0^{m'}$, where n, $n' \leqslant 2$, m, $m' \leqslant 1$ under detonation conditions. Assuming that the minimum tube diameter for aluminum particle–gas detonation $d_{\min}$ or the detonation cell size λ is proportional to the particle reaction times t_{I} and t_{b}, one obtains

$$\pi\, d_{\min} \sim \lambda \sim d_{p0}^{n} \,/\, p_0^{m} \quad \text{with } n \leqslant 2,\ m \leqslant 1. \tag{2.12}$$

Relation (2.12) is an analogy to that of gaseous detonation where the detonation cell size is scaled to the induction time assuming that the induction time

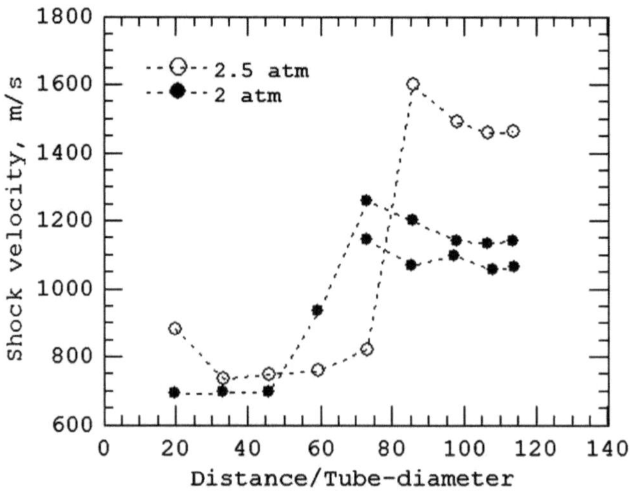

Fig. 2.31. Transition from deflagration to quasi-detonation at 2-atm initial pressure and to detonation at 2.5 atm in 500 g m^{-3} per atmosphere, 2-μm (H-2) aluminum–air mixtures ($\phi = 1.61$) using an 80-mm-ID tube [73]

amounts to most of the detonation time [90–93]. Applying this relation to the available experimental data in the literature results in $m = O(1)$ and $n = O(1)$ for the 1–2-μm aluminum–air detonation [73], thus indicating a dependence on initial pressure and gas-phase kinetics. It is noted that the analysis here is limited to a rather global and qualitative approach to elucidate the importance of finite chemical kinetics on the aluminum reaction mechanism in detonation. A combined surface kinetics-diffusion reaction model (2.49)–(2.52) was further suggested to take aluminum chemical kinetics into account, as shown in Sect. 2.2.3 and in the Appendix. The model has been successful in describing both the kinetics-limited transient processes (detonation initiation, abrupt DDT, and detonation instability) and the diffusion-limited combustion of aluminum in the long reaction zone supporting the weak transverse wave structure. More advanced solution will rely on the development of detailed kinetic and diffusional reaction schemes of aluminum particles under conditions of deflagration and detonation. The influence of high-momentum flow and high pressure immediately behind the shock on the mechanical damage and subsequent reaction of the particles must also be quantitatively determined. Recent flyer plate impact experiments simulating detonation in condensed phase matter showed that the atomized aluminum particles were subjected to severe surface damage and breakup to expose fresh bare aluminum, while aluminum flakes were completely broken into nanometric particles [116]. This helped understand why aluminum reacts much faster under high-pressure condensed detonation conditions. There have been no direct experiments simulating the aluminum particle-gas detonation conditions to recover and analyze the particle morphology subjected to a shock interaction.

2.5 Quasi-Detonation in Tubes

In homogeneous gas mixtures, it is well established that detonation waves propagate at less than the equilibrium CJ detonation velocity as the tube diameter is reduced to around the detonation cell size, owing to boundary layer effects and losses to the tube wall [117–124]. Moen et al. [123, 124] indicated that mixtures with irregular cellular structures are influenced less by the confining tube walls than regular-cell mixtures. While fuel–air detonation waves with irregular structures exhibit velocity deficits within 3% of the theoretical CJ value and fail at $d \leqslant \lambda/\pi$, detonations in fuel–oxygen and argon-diluted mixtures with fairly regular structures show larger velocity deficits above 10% and fail in a larger tube at $d \leqslant (1.3\text{–}2)\lambda/\pi$. In general, detonation failure in gaseous mixtures is associated with the disappearance of the transverse wave structure.

Unlike for homogeneous gaseous detonations, a shock-induced supersonic combustion wave can propagate quasi-steadily in tubes much smaller than the detonation cell size in a reactive particle–oxidizing gas flow [8, 70, 73], owing to distributed particle-induced hot or explosion spots that make the combustion less sensitive to the disturbance originating in the boundary layer on the tube wall. The observed shock-induced combustion waves were characterized by a shock velocity much less than the equilibrium CJ detonation value, and a pressure profile behind the shock front without a fully developed, self-organized transverse wave structure. Such a shock-induced combustion wave may be referred to as "heterogeneous quasi-detonation."

Quasi-detonation can be achieved through progressive acceleration, as shown in Fig. 2.12, in rich cornstarch–oxygen mixtures in a 53 mm × 53 mm cross section tube ($d \sim 0.4\lambda/\pi$) and was also found in rich cornstarch–air mixtures in a 140-mm tube ($d \sim 0.46\lambda/\pi$) [8, 70]. Figure 2.31 shows two experiments of quasi-detonation development in an 80-mm-diameter tube filled with a 2-μm (H-2) atomized aluminum–air mixture at 2-atm initial pressure [73]. After propagating through 70 tube diameters at low velocities, the wave accelerates more progressively to a shock velocity of 1,080–1,140 m s^{-1} and propagates thereafter quasi-steadily to the tube end. The wave has a velocity deficit of nearly 40% with respect to the theoretical CJ value, and displays compression waves behind the shock front but without an inherent, periodically oscillating transverse wave structure (Fig. 2.32). In fact, this shock speed is close to the critical Mach number that could lead to a DDT. Hence, behind

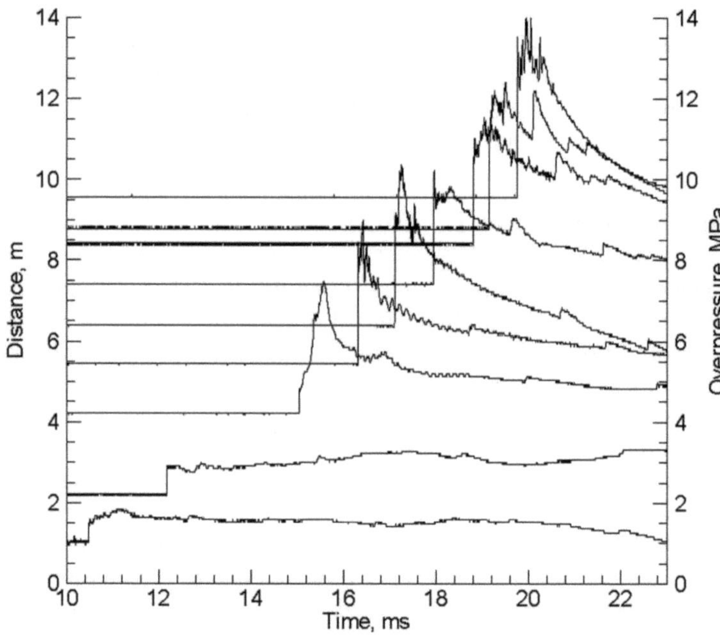

Fig. 2.32. Transition from deflagration to quasi-detonation at 2-atm initial pressure in 1000 g m^{-3}, 2-μm (H-2) aluminum–air mixtures ($\phi = 1.61$) initiated by a 1.1-g explosive detonator in an 80-mm-ID tube. The shock velocity versus distance is given in Fig. 2.31 [73]

the shock front local explosions can take place that are identified by the high compression peaks with subsequent retonation waves propagating backwards. However, the wave is not able to further accelerate to a transverse wave mode detonation because of the small tube confining the development of intrinsic transverse waves. Note that the test section length of 125 diameters may still be insufficient to conclude whether or not the quasi-detonation wave can maintain its quasi-steadiness.

A quasi-detonation wave fails when the distributed particle explosions are suppressed by the momentum and heat loss from gas to particles as well as the expansion and turbulent quenching originating in the boundary layer on the small tube wall. One may predict quasi-detonation limits using the steady two-phase ZND model or, more restrictively, using the unsteady two-phase fluid dynamics model. In the one-dimensional theory, the gas flow velocity with respect to the tube wall, v_{g}, is positive but decreases with the distance behind the shock front, particularly owing to the gas-phase momentum losses to the particles and to the tube wall. Hence, a lower limit can be proposed in which the gas flow velocity with respect to the tube wall is decreased towards zero at the phase-frozen or gaseous sonic plane [125]. Any dust quasi-detonation wave cannot propagate below this absolute limit, because the entire flow would become subsonic with respect to the shock front and the generalized CJ condition (2.2)–(2.4) would be no longer satisfied. Applying $v_{\mathrm{g}} = 0$ at the sonic plane to the same two-phase conservation equations of mass, momentum, and energy used for the ZND model as well as the generalized CJ condition (2.2)–(2.4), one obtains at the phase-frozen sonic plane:

$$\rho_{\mathrm{cr}} = \sigma_{\mathrm{g0}} + \sigma_{\mathrm{p0}}, \tag{2.13}$$

$$e_{\mathrm{cr}} = \frac{(\sigma_{\mathrm{g0}} e_{\mathrm{g0}} + \sigma_{\mathrm{p0}} e_{\mathrm{p0}})}{\sigma_{\mathrm{g0}} + \sigma_{\mathrm{p0}}} + \frac{\int_0^{x_{cr}} Q_{\mathrm{W}}\, \mathrm{d}x}{(\sigma_{\mathrm{g0}} + \sigma_{\mathrm{p0}}) D_{\mathrm{cr}}}, \tag{2.14}$$

$$D_{\mathrm{cr}} = a_{\mathrm{g,cr}}, \tag{2.15}$$

and

$$\left[-\left(q_{\mathrm{p}} + c_{\mathrm{p}} T_{\mathrm{p}}\right) J_{\mathrm{p}} - Q_{\mathrm{p}} + Q_{\mathrm{W}}\right]_{\mathrm{cr}} = 0, \tag{2.16}$$

where the variable e denotes the specific internal energy. The subscript *0* denotes the initial state in front of the shock and the subscript cr represents the critical state at the phase-frozen sonic plane. Equations (2.13)–(2.15) mean that the wave structure begins with a shock front propagating at a critical velocity D_{cr} and ends with a constant-volume combustion boundary at the phase-frozen or gaseous sonic plane. Equations (2.15) and (2.16) satisfy the generalized CJ condition (2.2)–(2.4) and the lower limit solution is therefore a steady solution. Note that for simplicity, the lower limit model (2.13)–(2.16) was obtained by further assuming that at the sonic plane the solid particle velocity with respect to the tube wall approaches zero. Thus, it may not be applicable to mixtures of large particulates.

Under the assumption of an adiabatic process and a full equilibrium critical state, the generalized CJ conditions (2.2)–(2.4) must be replaced by (2.3),

(2.6), and (2.7) and the absolute lower limit model (2.13–2.16) therefore becomes

$$\rho_{\mathrm{cr}} = \sigma_{\mathrm{g0}} + \sigma_{\mathrm{p0}}, \tag{2.13}$$

$$e_{\mathrm{cr}} = \frac{(\sigma_{\mathrm{g0}} e_{\mathrm{g0}} + \sigma_{\mathrm{p0}} e_{\mathrm{p0}})}{\sigma_{\mathrm{g0}} + \sigma_{\mathrm{p0}}}, \tag{2.17}$$

and

$$D_{\mathrm{cr}} = a_{\mathrm{e,cr}}, \tag{2.18}$$

where the subscript cr represents the critical state at the phase-equilibrium sonic plane and the variable $a_{\mathrm{e,cr}}$ is the full equilibrium sound speed. The absolute lower limit based on (2.13), (2.17), and (2.18) can therefore be calculated using an equilibrium constant-volume combustion that results in pressure p_{cr}, temperature T_{cr}, and sound speed $a_{\mathrm{e,cr}}$ equal to the critical shock velocity D_{cr}. The calculated CJ detonation velocities and quasi-detonation critical shock velocities are displayed in Figs. 2.26 and 2.27 for cornstarch–oxygen and aluminum–air mixtures, respectively. As shown in the two figures, the experimental propagation velocities of detonations and quasi-detonations in tubes of various size under different particle dispersion and initiation conditions are bounded between the equilibrium CJ detonation values and the equilibrium constant-volume explosion lower limits. A maximum velocity deficit of about 10% generally holds for the detonation with a transverse wave structure. In the case of the quasi-detonation waves, however, measured shock velocities indicate a deficit as much as beyond 45% with respect to the equilibrium CJ value, yet bounded by the constant-volume lower limit. Between these two limits, the detonation wave undergoes a transition from transverse wave modes to shock-induced quasi-detonation modes. Although the heterogeneous quasi-detonation can be grossly described by the two-phase ZND model, it is essentially unsteady.

2.6 Hybrid Detonation

2.6.1 Hybrid Detonation Modes

While fine solid particles suspended in air are not sensitive to detonation owing to a large transverse wave spacing, their combustion in gaseous detonation products may support so-called hybrid detonation and a hybrid DDT. Theoretically, Afanasieva et al. [26] used the one-dimensional similarity theory to show a spherically or cylindrically diverging "double detonation" wave in a homogenous reactive mixture (Fig. 2.33), in which the first CJ detonation wave is generated from an instantaneous energy release followed by a second shock wave supported by a late energy release. They further indicated the feasibility of such a double-shock detonation solution in a planar wave geometry only with the presence of wall friction and heat losses or when a homogeneous

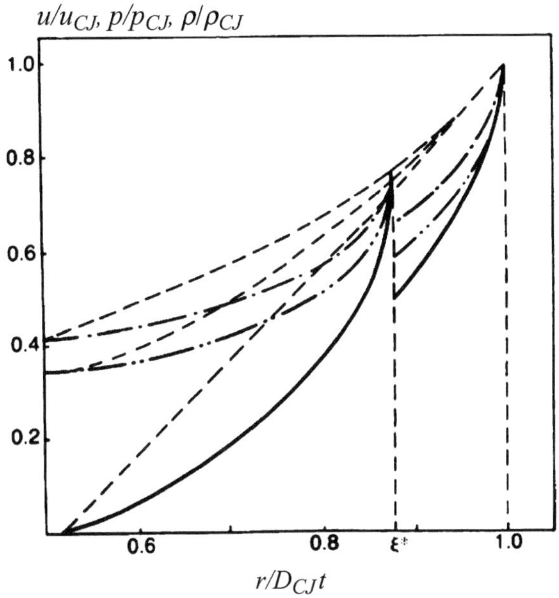

Fig. 2.33. Solution of a spherically diverging double-shock detonation wave using two successive heat releases. u/u_{CJ} *solid line*, p/p_{CJ} *double-dotted dashed line*, ρ/ρ_{CJ} *dotted dashed line*, ξ nondimensional length. [26]

mixture is replaced by a two-phase mixture. Veyssiere [24] first reported experimental evidence of the double-shock detonation in a two-phase mixture of lean hydrogen–air and aluminum particles ($\sigma_{\mathrm{p}} = 55\,\mathrm{g\,m^{-3}}$, $10\,\mu\mathrm{m}$) in a 69-mm-diameter tube. Since then hybrid detonations have been investigated in several reactive particle–reactive gas systems [29–33].

According to the analysis in Sects. 2.2.2 and 2.2.3, the necessary conditions for a weak detonation solution can be satisfied in reactive gas–particle flow under an appropriate choice of physical and chemical properties of solid particles as well as reactive gas. A set of possible solutions can be realized by selecting the late energy release of particles to meet rear flow or boundary conditions behind the gas reaction zone. Experimentally, three most important hybrid detonation modes (as defined in Sect. 2.2.3) that enhance the impulse loading are introduced in this section over an aluminum concentration range of 25–2,000 $\mathrm{g\,m^{-3}}$ [31–33].

Figure 2.34 displays a steady strong hybrid detonation for $\sigma_{\mathrm{p}} = 500\,\mathrm{g\,m^{-3}}$, about $2\,\mu\mathrm{m}$ atomized aluminum particles (known as H-2 by Valimet) suspended in lean acetylene–air using an 80-mm diameter, 10-m-long tube. The strong hybrid detonation is characterized by the leading shock front followed by a compression wave in the gas detonation zone, caused by a sufficiently large heat release rate of the small particles within the gas reaction zone. This increases the pressure and decreases the detonation cell size with respect

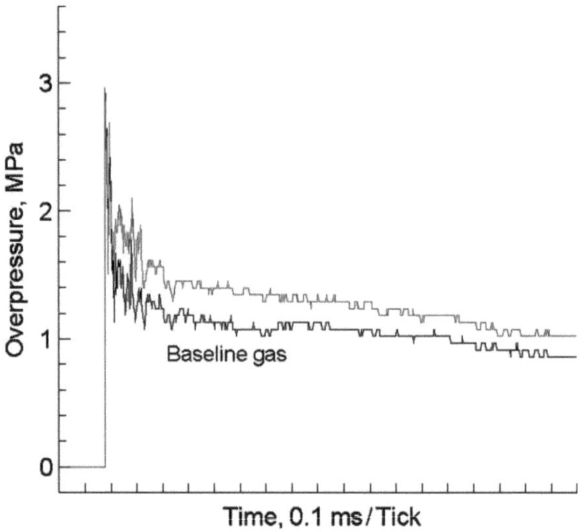

Fig. 2.34. Strong hybrid detonation characterized by the first shock followed by a pressure wave in the gas reaction zone in a mixture of $\phi = 0.8\,C_2H_2$–air and $500\,g\,m^{-3}$, 2-μm (H-2) aluminum particles recorded at $x = 9\,m$ (*upper curve*) compared with detonation in $\phi = 0.8\,C_2H_2$–air (*lower curve*) [31,32]

to the baseline gas detonation (Figs. 2.34, 2.35) and therefore overdrives gas detonation. The detonation velocity is measured to be $1{,}800\,m\,s^{-1}$, increased by 1.5% from the value of the baseline gas detonation. The secondary compression wave can be identified behind the shock and its inclusion in the gas detonation zone can be more clearly resolved through numerical simulations such as shown in Fig. 2.5a, where the compression wave penetrates two thirds within the detonation zone of the baseline gas alone. The numerical prediction in Fig. 2.5a indicates that the rear flow pressure can be enhanced by a factor of 1.6 with respect to the baseline gas detonation when using a particle concentration of $300\,g\,m^{-3}$.

When the aluminum particle size is increased to about 10 μm (known as H-10 with a mean diameter of 6.5 μm by number and a mean diameter of 13 μm by weight), the particles behave inertly within the gas reaction zone and particle heat release takes place after the gas reaction zone. Therefore, a steady hybrid weak detonation wave results and its propagation along the tube length and comparison with the baseline gas detonation are shown in Figs. 2.36 and 2.37, respectively, from two experiments. This is the type-I double-shock weak solution characterized by a two-shock structure, where the second shock behind the gaseous frozen sonic plane has the same velocity as the leading shock (Fig. 2.38). Aluminum combustion is in a fast deflagration mode for this particle size. Owing to insignificant momentum and heat transferred to the larger particles within the gas detonation zone, the detonation velocity is almost the same as that of the baseline gas detonation and the

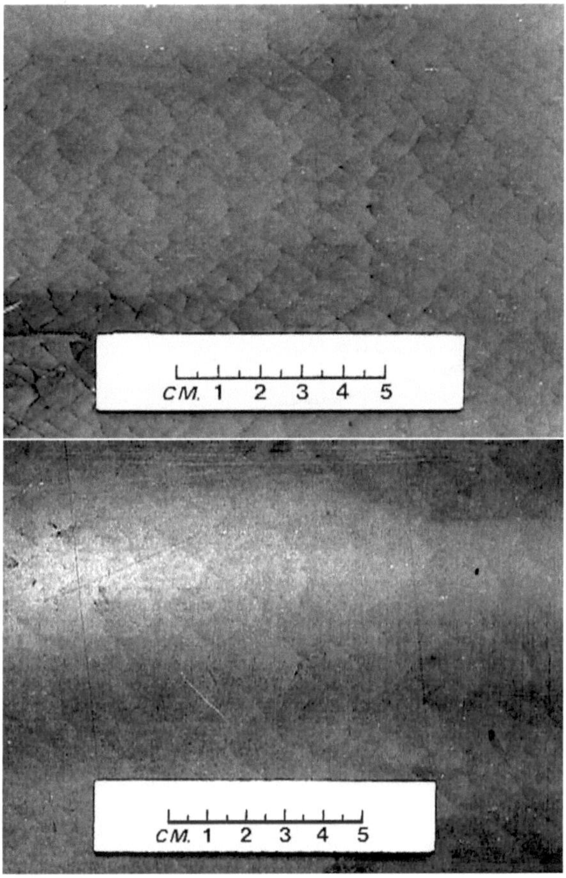

Fig. 2.35. Cellular detonation structure for the strong hybrid detonation in a mixture of $\phi - 0.8\,C_2H_2$–air and $500\,\mathrm{g\,m^{-3}}$, 2-μm (H-2) aluminum particles with a cell size of $\lambda = 7 \pm 2\,\mathrm{mm}$ (*upper*) compared with detonation in $\phi = 0.8\,C_2H_2$–air with $\lambda = 8 \pm 2\,\mathrm{mm}$ (*lower*) [31, 32]

apparent detonation cell size printed on the smoke foil is mainly governed by the gas detonation (Fig. 2.39). This suggests that deflagration of the aluminum particles may only result in a planar secondary shock front subjected to the disturbance of upstream transverse detonation flow conditions.

When the aluminum particle size was increased to about 30 μm (known as H-30, with a mean diameter of 16 μm by number and a mean diameter of 36 μm by weight) to further delay and reduce the heat release rate of the particles, a type-II double-shock weak detonation was observed. This is characterized by two shock fronts, where the second front behind the gaseous sonic plane has a velocity less than the leading front, as demonstrated in Figs. 2.40–2.42. Hence, the second shock recedes from the gas reaction zone to produce an ever-widening region of supersonic flow between the end of the gas reaction

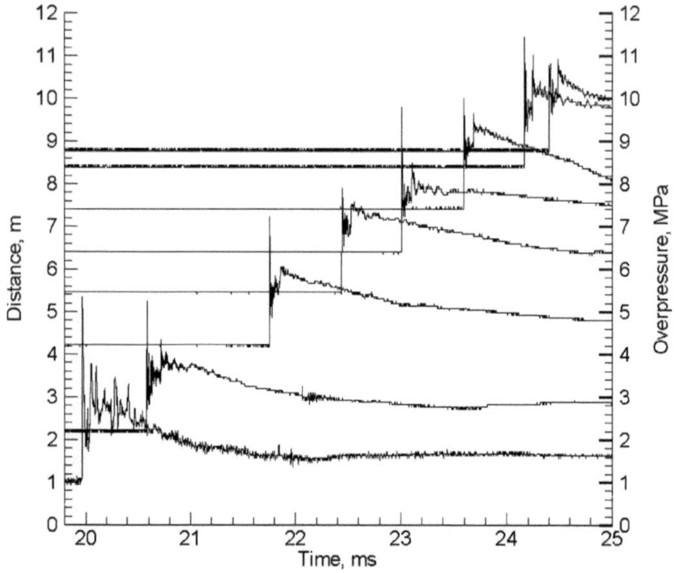

Fig. 2.36. Type-I weak hybrid detonation characterized by two shock fronts where the second shock behind the gas reaction zone has the same velocity as the first in a mixture of $\phi = 0.8\,C_2H_2$–air and $500\,g\,m^{-3}$, 10-μm (H-10) aluminum particles initiated by a 5-g explosive detonator in an 80-mm-ID tube [31, 32]

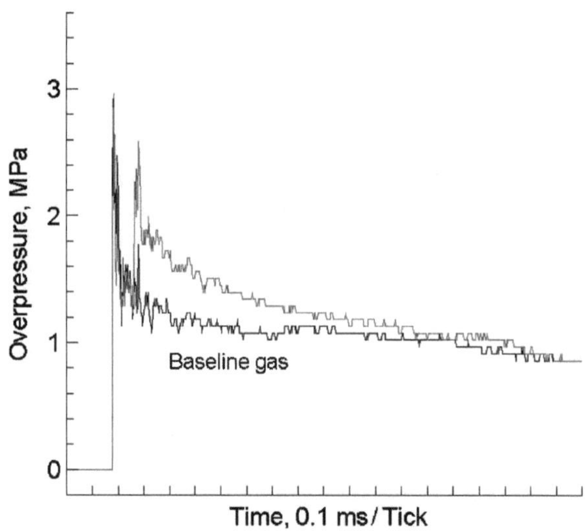

Fig. 2.37. Type-I weak hybrid detonation pressure profile in a mixture of $\phi = 0.8\,C_2H_2$–air and $500\,g\,m^{-3}$, 10-μm (H-10) aluminum particles compared with detonation in $\phi = 0.8\,C_2H_2$–air recorded at x = 8.8 m [31, 32]

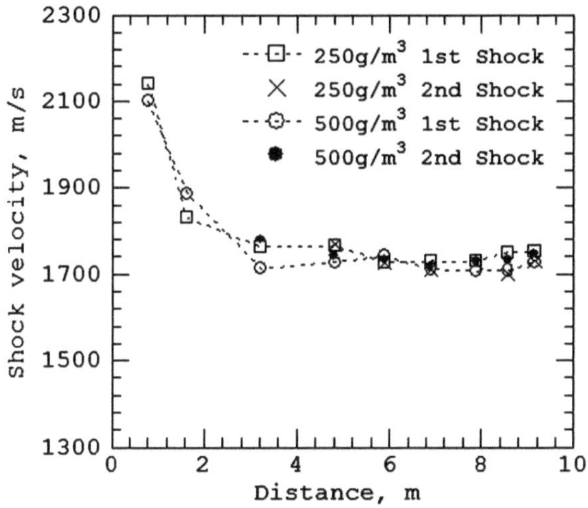

Fig. 2.38. First and second shock velocities versus propagation distance for the type-I weak hybrid detonations in mixtures of $\phi = 0.8\,C_2H_2$–air and 10-μm (H-10) aluminum particles [32]

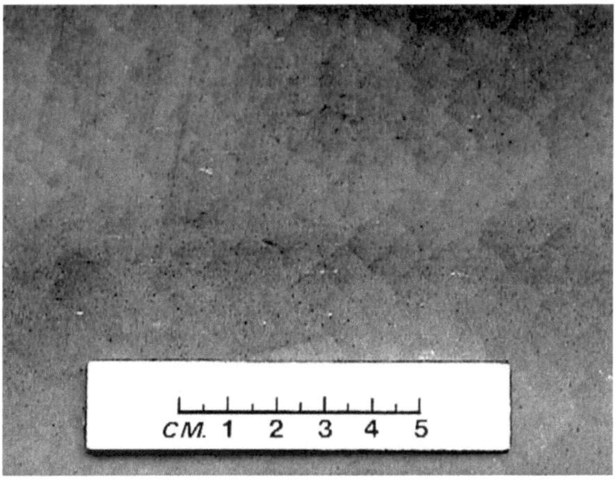

Fig. 2.39. Cellular detonation structure for the type-I weak hybrid detonation in a mixture of $\phi = 0.8\,C_2H_2$–air and $500\,g\,m^{-3}$, 10-μm (H-10) aluminum particles [32]

zone and itself. While rigorously speaking the flow in this widening region in front of the second shock is unsteady, the experimental velocity of the second shock appears quasi-steady as the wave propagates from $x = 4\,m$ to the end of the 10-m-long tube. Again owing to the large particles, the first shock velocity and the detonation cell size recorded on the smoke foil correspond to those of the baseline gas detonation. Figure 2.43 shows a double-shock detonation obtained by Veyssiere and Ingignoli [30] in a 69-mm-diameter tube

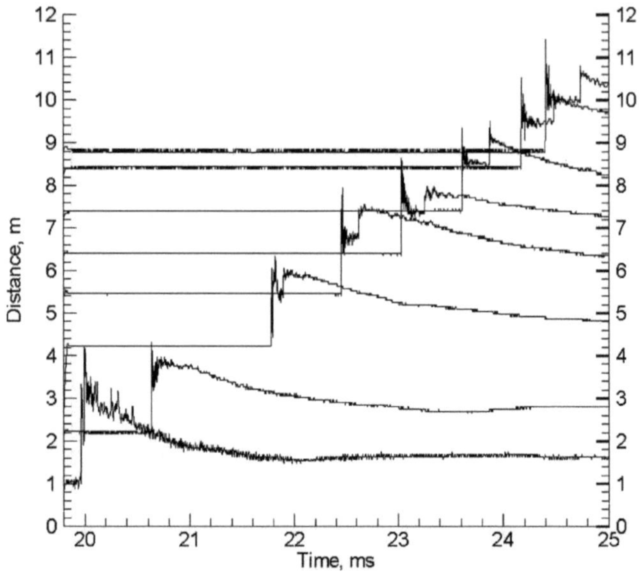

Fig. 2.40. Type-II weak hybrid detonation characterized by two shock fronts where the second shock behind the gas reaction zone has a velocity less than the first in a mixture of $\phi = 0.8\,C_2H_2$–air and 1,000 g m^{-3}, 30-μm (H-30) aluminum particles initiated by a 5-g explosive detonator in an 80-mm-ID tube [32]

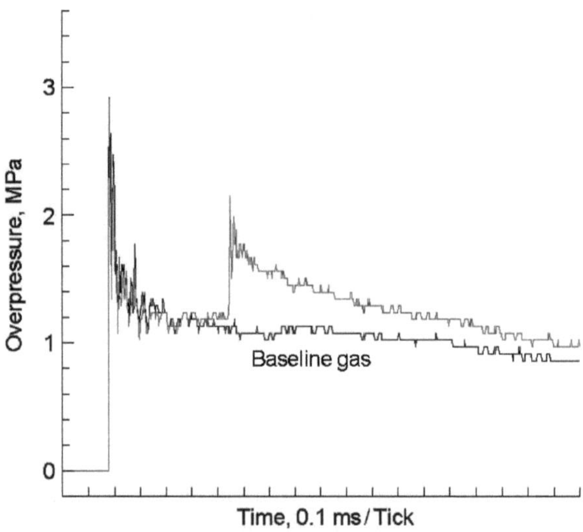

Fig. 2.41. Type-II weak hybrid detonation pressure profile in a mixture of $\phi = 0.8\,C_2H_2$–air and 500 g m^{-3}, 30-μm (H-30) aluminum particles compared with detonation in $\phi = 0.8C_2H_2$–air recorded at x = 8.8 m [32]

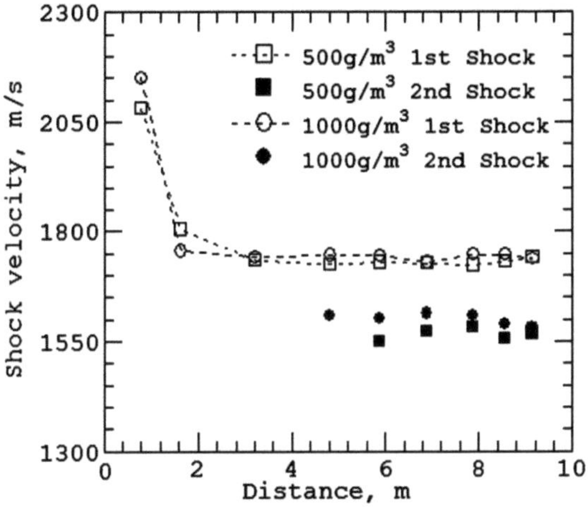

Fig. 2.42. First and second shock velocities versus propagation distance for the type-II weak hybrid detonation in mixtures of $\phi = 0.8\,C_2H_2$–air and 30-μm (H-30) aluminum particles [32]

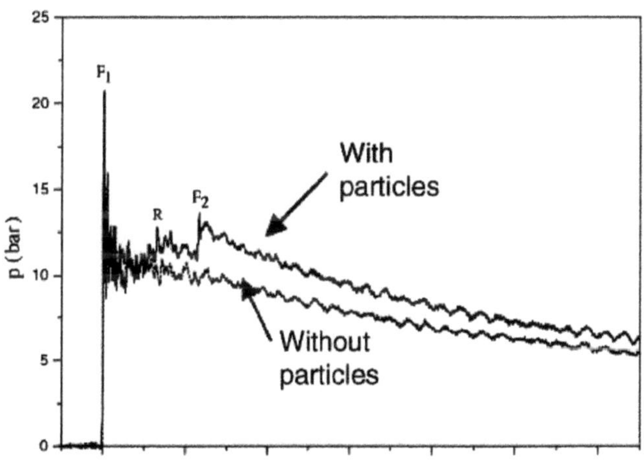

Fig. 2.43. Double-shock detonation in a mixture of $\phi = 0.87\,H_2$–air and $60\,\mathrm{g\,m^{-3}}$, 13-μm aluminum particles, compared with detonation in $\phi = 0.87\,H_2$–air using a 69-mm-ID tube [30]

for $\sigma_p = 60\,\mathrm{g\,m^{-3}}$, 13-μm atomized aluminum suspended in lean hydrogen–air. Between the two shock fronts, a compression wave appears in front of the second shock. The experimental hybrid detonation modes displayed above justify the theory discussed in Sect. 2.2. The double-shock hybrid detonation waves have also been observed recently in condensed-phase explosives with aluminum particles [126].

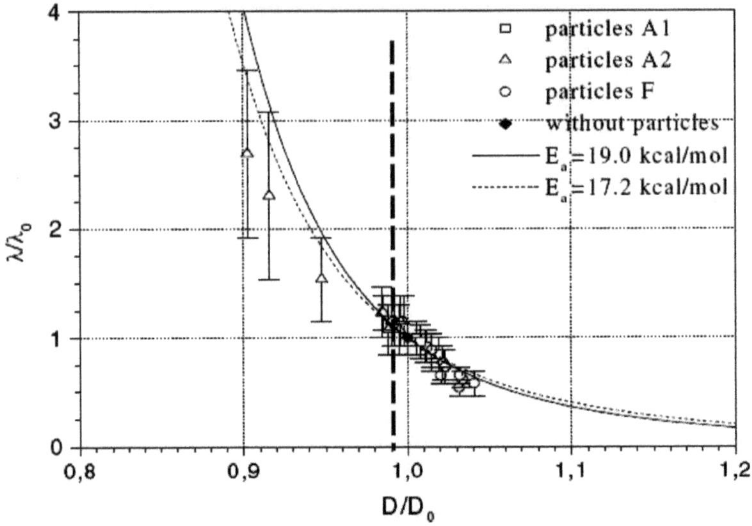

Fig. 2.44. Variation of detonation cell width with detonation velocity in hybrid mixtures of H_2–air and aluminum particles [30]

From the studies of the hybrid detonation for various aluminum particles (flaked, 3.5 and 13 μm atomized in a concentration range of 25–450 g m^{-3}) suspended in hydrogen–air mixtures (with equivalence ratios of ($\phi = 0.87$, 1.06, and 1.32), Veyssiere and Ingignoli [30] correlated the experimental cell size with the detonation velocity according to the Desbordes relation obtained from gaseous overdriven detonation data:

$$\frac{\lambda}{\lambda_{\mathrm{gCJ}}} = \frac{D}{D_{\mathrm{gCJ}}} \mathrm{e}^{E_{\mathrm{g}}[(D_{\mathrm{gCJ}}/D)^2-1]/(RT_{\mathrm{vN}})}, \qquad (2.19)$$

where parameters E and R are the activation energy and universal gas constant. The subscript gCJ refers to the baseline gas CJ detonation, and vN refers to the von Neumann shock state. A comparison of relation (2.19) with the experimental data is shown in Fig. 2.44.

2.6.2 Influencing Factors

Apart from the effects of particle size and concentration discussed already, the influence of the gaseous detonation parameters and the composition of the products on the hybrid detonation was further investigated for aluminum particles suspended in various fuel–air mixtures with 4–38-mm detonation cell sizes for an initial pressure range of 1–2.5 atm [31–33]. Table 2.2 summarizes the gas detonation properties and product compositions used in the experiments. H_2–air ($\phi = 1.5$), 0.97CO–air + 0.03C_2H_2, and C_2H_2–air ($\phi = 2$)

Table 2.2. Gas detonation properties and product compositions used in hybrid detonation tests

Properties	C_2H_2–air ($\phi = 0.8$)	C_2H_4–air ($\phi = 0.9$)	H_2–air ($\phi = 1$)	H_2–air ($\phi = 1.5$)	0.97CO–air $+0.03C_2H_2$	C_2H_2–air ($\phi = 2$)
D_{CJ} (m s^{-1})	1,765	1,792	1,979	2,087	1,762	2,029
P_{CJ}/p_0	17.56	17.58	15.60	15.20	16.91	21.64
T_{CJ} (K)	2,940	2,860	2,970	2,902	3,003	3,240
Cell size (mm)	8	38	15	15	38	6
O_2 (mol%)	4.127	2.414	1.018	2.529×10^{-2}	6.878×10^{-1}	1.243×10^{-2}
H_2O (mol%)	5.479	10.53	29.496	28.384	2.491	4.105
CO (mol%)	2.989	2.604	0	0	18.307	23.961
CO_2 (mol%)	9.729	9.011	0	0	18.911	1.686
O (mol%)	4.409×10^{-1}	2.474×10^{-1}	2.579×10^{-1}	3.232×10^{-1}	2.284×10^{-1}	5.792×10^{-1}
OH (mol%)	1.136	1.133	2.043	6.874×10^{-1}	5.672×10^{-1}	4.043×10^{-1}
HO_2 (mol%)	9.791×10^{-4}	7.660×10^{-4}	8.175×10^{-4}	4.363×10^{-1}	1.924×10^{-4}	1.963×10^{-5}
NO (mol%)	1.982	1.344		1.276×10^{-1}	7.710×10^{-1}	1.382×10^{-1}
H (mol %)	1.507×-10^{-1}	1.556×10^{-1}	6.266×10^{-1}	1.128	2.217×10^{-1}	1.848
H_2 (mol%)	2.357×10^{-1}	4.367×10^{-1}	3.004	14.630	3.319×10^{-1}	7.593
N_2 (mol%)	73.728	72.117	63.554	54.985	67.482	60.192
Notes				H_2O dominant	CO_2 present	CO dominant
Double-shock detonation [31, 33]	Yes	Yes	Yes	Yes	Yes	Fail

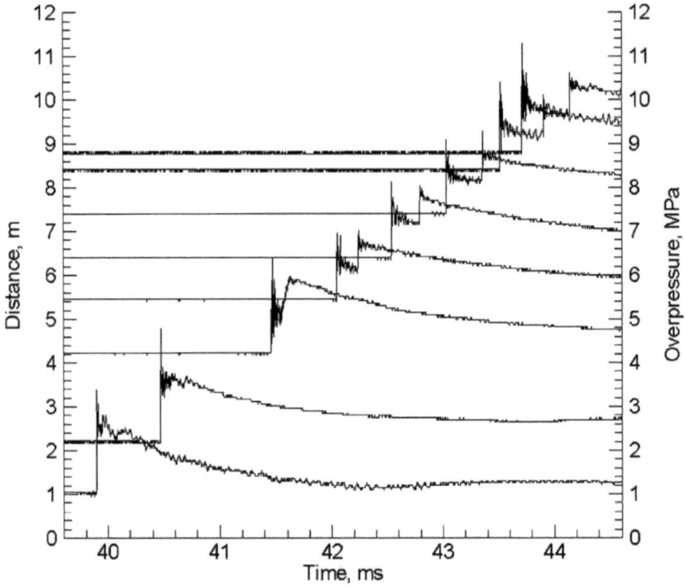

Fig. 2.45. Type-II weak hybrid detonation in a mixture of $\phi = 1.5H_2$–air and $500\,g\,m^{-3}$, 30-μm (H-30) aluminum particles initiated by a 5-g explosive detonator in an 80-mm-ID tube [33]

were specifically chosen to exclude the oxygen to determine the dependence of the hybrid detonation on water vapor, carbon dioxide, or carbon monoxide, respectively. It was found that the double-shock detonation waves can propagate in detonation products with the presence of oxygen (Fig. 2.40), water vapor, or carbon dioxide using the same initiation source (Figs. 2.45, 2.46). However, as displayed in Fig. 2.47, propagation of the secondary shock fails in detonation products dominated by carbon monoxide even initiated by an acetylene–oxygen detonation driver, a stronger initiation source than used in the other mixtures [33]. The observed secondary shock was the weakest for the C_2H_4–air mixture among the double-shock detonations observed in all the fuel–air mixtures reported in Table 2.2, possibly because of its low detonation temperature.

Owing to the detonation insensitivity of micrometric grades of aluminum particles, the gaseous fuel–air detonation alone may not be sufficient to initiate and accelerate combustion of relatively large aluminum particles quickly enough to form a secondary shock in the gas detonation products flow. It was experimentally found that the C_2H_2–air detonation was initiated directly with a 0.2-g explosive charge mass and propagated steadily, but it cannot initiate the about 30 μm aluminum particles (H-30) suspended in the detonation wave to form the secondary shock until 5 g explosive charge or a stoichiometric C_2H_2–O_2 detonation driver is used for initiation [31]. These results clearly indicate that the expanding fuel–air detonation products alone are insufficient

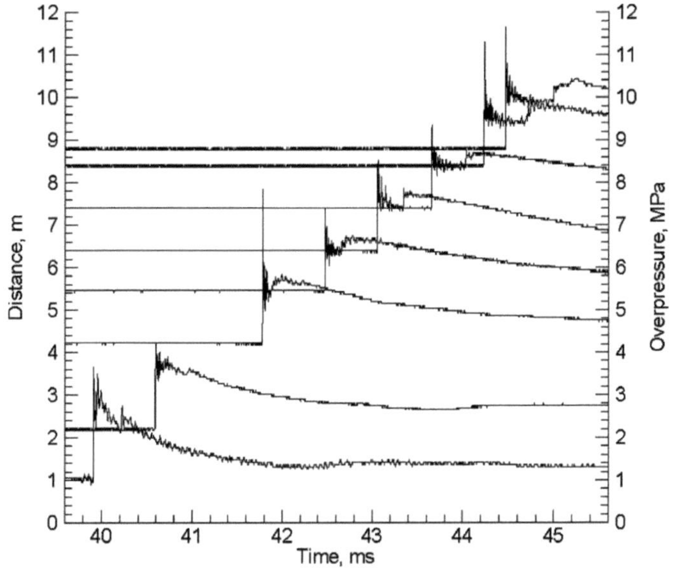

Fig. 2.46. Type-II weak hybrid detonation in a mixture of 0.97CO–air + $0.03C_2H_2$ and $500\,\mathrm{g\,m^{-3}}$, 30-μm (H-30) aluminum particles initiated by a 5-g explosive detonator in an 80-mm-ID tube [33]

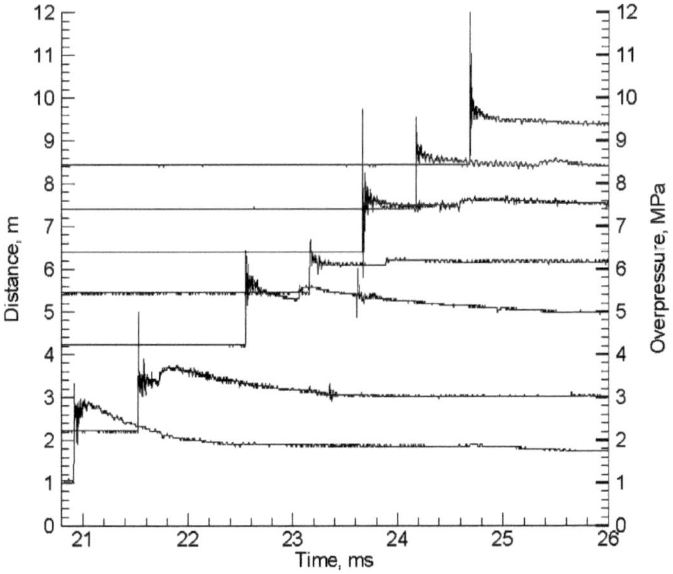

Fig. 2.47. Type-II weak hybrid detonation in a mixture of $\phi = 2C_2H_2$–air and $500\,\mathrm{g\,m^{-3}}$, 30-μm (H-30) aluminum particles initiated by a $C_2H_2 + 2.5O_2$ detonation driver in an 80-mm-ID tube [33]

and additional initiation charges (energy power and high temperature) are required to initiate and sustain fast combustion of large aluminum particles to form the secondary shock leading to a double-shock weak detonation.

An increase in initial pressure results in an increase in the heat release rate of particles as discussed in Sect. 2.4.2 and therefore facilitates a hybrid detonation wave with large particles. Experiments showed a stronger two-shock detonation as the initial pressure increases from 1 to 2.5 atm [32].

2.7 Concluding Remarks

The recent efforts in pursuing experiments have led to progress in the understanding of the fundamental mechanisms of detonation waves in dilute solid particle–gas flows. It appears that detonations in fine organic or metallic particles suspended in an oxidizing gas can be divided into heterogeneous detonation waves and quasi-detonation waves. The macroscopic propagation mechanisms for the heterogeneous detonation waves are similar to those for homogeneous gas mixtures, that is, the transverse waves in the single-head spinning or cellular structure provide the coupling between the shock and the reaction. The difference is that the characteristic scale of the detonation cell size for particle–gas mixtures is at least 1 order of magnitude larger than that typically encountered in gas mixtures, owing to the additional time scales introduced by the physical processes of mass, momentum, and heat transfer between the particles and the gas. The time scale of particle reaction is basic and leads to the main time scale for the mass transfer and therefore for the energy release of particles.

The heterogeneous quasi-detonation waves that propagate in tubes much smaller than the characteristic detonation cell size are essentially shock-induced supersonic combustion waves without a fully developed transverse wave structure. The quasi-detonation waves propagate at a shock velocity below the transverse wave mode detonations but above a lower limit characterized by the constant-volume combustion at the sonic plane. While the quasi-detonation wave is unique for reactive particle–gas mixtures, presumably owing to the distributed hot spots or local explosions induced by particles, more investigations are required to understand its propagation mechanisms.

Hybrid detonation waves occur in reactive particles suspended in a detonable gas. A variety of hybrid detonation modes can exist and the solution is a function of the gas reaction time scales and the additional time scales of the mass, momentum, and heat transfer between the particles and the gas. The rate of particle energy release, including its delay time and magnitude, represents a characteristic parameter to specify a possible solution. Among various modes, a strong hybrid detonation wave and two types of double-shock weak detonation wave are most important in practice owing to their enhancement of the gas detonation impulse. The strong hybrid detonation is characterized by the leading shock front followed by a compression wave resulting from particle combustion in the gas detonation zone, thus overdriving the gas detonation. The two types of double-shock weak detonation wave are featured by a two-shock structure where the second shock front caused by the particle combustion behind the gas reaction zone has a propagation velocity either the same as or less than that of the leading shock front, thus enhancing the impulse loading of the gas detonation. While the strong hybrid detonation reduces the cell size of the baseline gas detonation, the cell sizes in the hybrid weak detonation waves are mainly governed by the gas detonation and provide little information about the sensitivity of aluminum ignition and combustion, which often requires a strong initiation. The variety of hybrid detonation modes and their propagation mechanisms are still the subject of current research efforts.

While the fundamental studies of detonation in solid particle–gas flow have made significant progress, many problems remain to be resolved. The detonation velocity and pressure are experimentally available only for a few solid particle–gas mixtures. More detonation dynamic parameters, including the ordinary cell size and the critical initiation energy, will need to be directly measured. Although the macroscopic mechanisms for the DDT, detonation onset, and propagation seem similar to those for the homogeneous gas detonation, the mechanisms of the heterogeneous detonation and quasi-detonation cannot be fully established without mesoscale studies on the shock and gas flow interactions with the combustion dynamics of distributed particles as well as the influence of the boundary layer. While the heat release rate of particles under detonation conditions is a key parameter in controlling the detonation phenomena in gas–particle flow, there is a lack of direct experimental measurements for a quantitative description of the reaction mechanism and heat release rate even for the popular aluminum particles under shock and detonation conditions. The reliability and predictability of two-phase continuum detonation theory and numerical modeling are currently challenged by the uncertainty of ignition and reaction mechanisms of solid particles under shock and detonation conditions and, for dense solid particle–gas flow, the interphase interaction functions and the equations of state for solid particle flow. Experimental determination and mesoscale modeling would be two critical approaches for resolving these difficult issues.

Appendix

A Two-Phase Fluid Dynamics Equations

A control volume is considered that contains a large number of solid particles immersed in a fluid, but has a small dimension with respect to the macroscopic two-phase flow. The fluid and solid particles are treated as two separate flows which are coupled through their interactions associated with the exchange of mass, momentum, and energy between the two phases. For the supersonic

detonation flow involved, the viscosity and conductivity in the fluid as well as particle gravities are not considered. From the control volume analysis, the one-dimensional conservation equations can be derived as listed below in the laboratory coordinate frame.
Solid phase:

$$\text{Mass}: \frac{\partial \sigma_{\mathrm{p}}}{\partial t} + \frac{\partial}{\partial x}(\sigma_{\mathrm{p}} u_{\mathrm{p}}) = J_{\mathrm{p}}. \tag{2.20}$$

$$\text{Momentum}: \frac{\partial}{\partial t}(\sigma_{\mathrm{p}} u_{\mathrm{p}}) + \frac{\partial}{\partial x}(\sigma_{\mathrm{p}} u_{\mathrm{p}}^2 + p_{\mathrm{p}}) = u_{\mathrm{p}} J_{\mathrm{p}} + f_{\mathrm{p}}. \tag{2.21}$$

Energy:

$$\frac{\partial}{\partial t}\left[\sigma_{\mathrm{p}}(e_{\mathrm{p}} + u_{\mathrm{p}}^2/2)\right] + \frac{\partial}{\partial x}\left[\sigma_{\mathrm{p}} u_{\mathrm{p}}(e_{\mathrm{p}} + u_{\mathrm{p}}^2/2 + p_{\mathrm{p}}/\sigma_{\mathrm{p}})\right]$$
$$= (e_{\mathrm{p}} + u_{\mathrm{p}}^2/2) J_{\mathrm{p}} + u_{\mathrm{p}} f_{\mathrm{p}} + Q_{\mathrm{p}}. \tag{2.22}$$

$$\text{Number density}: \frac{\partial n_{\mathrm{p}}}{\partial t} + \frac{\partial}{\partial x}(n_{\mathrm{p}} u_{\mathrm{p}}) = \Psi_{\mathrm{p}}. \tag{2.23}$$

Fluid phase:

$$\text{Mass}: \frac{\partial \phi_{\mathrm{g}} \rho_{\mathrm{g}}}{\partial t} + \frac{\partial}{\partial x}(\phi_{\mathrm{g}} \rho_{\mathrm{g}} u_{\mathrm{g}}) = -J_{\mathrm{p}}. \tag{2.24}$$

$$\text{Momentum}: \frac{\partial}{\partial}(\phi_{\mathrm{g}} \rho_{\mathrm{g}} u_{\mathrm{g}}) + \frac{\partial}{\partial x}(\phi_{\mathrm{g}} \rho_{\mathrm{g}} u_{\mathrm{g}}^2 + \phi_{\mathrm{g}} p_{\mathrm{g}}) = -u_{\mathrm{p}} J_{\mathrm{p}} - f_{\mathrm{p}}. \tag{2.25}$$

Energy:

$$\frac{\partial}{\partial t}\left[\phi_{\mathrm{g}} \rho_{\mathrm{g}}(e_{\mathrm{g}} + u_{\mathrm{g}}^2/2)\right] + \frac{\partial}{\partial x}\left[\phi_{\mathrm{g}} \rho_{\mathrm{g}} u_{\mathrm{g}}\left(e_{\mathrm{g}} + u_{\mathrm{g}}^2/2 + p_{\mathrm{g}}/\rho_{\mathrm{g}}\right)\right]$$
$$= -(e_{\mathrm{p}} + u_{\mathrm{p}}^2/2) J_{\mathrm{p}} - u_{\mathrm{p}} f_{\mathrm{p}} - Q_{\mathrm{p}}, \tag{2.26}$$

$$\text{Species}: \frac{\partial}{\partial t}(\phi_{\mathrm{g}} \rho_{\mathrm{g}} Y_j) + \frac{\mathrm{d}}{\mathrm{d}x}(\phi_{\mathrm{g}} \rho_{\mathrm{g}} u_{\mathrm{g}} Y_j) = w_j, \qquad j = 1, \ldots, M, \tag{2.27}$$

where variables u, ρ, σ, e, and Y are the flow velocity, material density, partial density (or mass concentration), specific internal energy including chemical energy, and mass fraction of the gas species, respectively. The subscripts p and g refer to the solid particle and fluid phase, respectively.

The volume fractions of each phase are constrained by

$$\phi_{\mathrm{g}} + \phi_{\mathrm{p}} = 1. \tag{2.28}$$

The partial density or mass concentration of the particle flow is related to the particle material density by

$$\sigma_{\mathrm{p}} = \phi_{\mathrm{p}} \rho_{\mathrm{s}}. \tag{2.29}$$

The number density (particle number per volume) for spherical solid particles is defined by

$$n_{\mathrm{p}} = \frac{6\phi_{\mathrm{p}}}{\pi d_{\mathrm{p}}^3} = \frac{6\sigma_{\mathrm{p}}}{\pi \rho_{\mathrm{s}} d_{\mathrm{p}}^3}, \tag{2.30}$$

where d_{p} is the particle diameter. The variable p_{g} is the pressure of the gas flow occupying the partial volume ϕ_{g}, while p_{p} is the pressure of the particle flow in the control volume without interstitial gas and it is different from the material pressure inside the particle. The definition of the pressure of particle flow will be discussed in the next section. The mixture density ρ and pressure p are the sum of the partial density and the pressure, respectively:

$$\rho = \sigma_{\mathrm{g}} + \sigma_{\mathrm{p}} = \phi_{\mathrm{g}}\rho_{\mathrm{g}} + \sigma_{\mathrm{p}}, \quad p = \phi_{\mathrm{g}} p_{\mathrm{g}} + p_{\mathrm{p}}. \tag{2.31}$$

The source terms J_{p}, f_{p}, Q_{p}, and Ψ_{p} are the rates of interphase transfer functions representing the exchange for mass, momentum, energy, and the solid particle number between the phases. Evaporation and combustion of solid particles are included in the rate of mass transfer J_{p}, which adopts a negative value when particle depletion occurs. The force f_{p} and the rate of energy transfer Q_{p} have the same sign as the phase velocity difference $u_{\mathrm{g}} - u_{\mathrm{p}}$ and the phase temperature difference T_{g}–T_{p}, respectively. Solid particle agglomeration or fragmentation can be controlled through the rate of particle number change Ψ_{p}. If the reaction zone is large, the loss due to the tube wall or other lateral boundary conditions can also be included by the source terms for the rate of the momentum and heat exchange with the lateral boundaries. The source terms will be described later and various forms of transfer functions can be found in [10–12,27,28,50–56]. Since the source terms are modeled according to physical rules and empirical co-relations, the predictability of the two-phase fluid dynamics theory strongly depends on the choice of source term models for a particular flow topology [54].

The conservative form of the two-phase fluid dynamics (2.20)–(2.27) is convenient in numerical implementation of the Eulerian approaches. For problems involving multiple particle sizes or particles with a size distribution, the conservation equations for the particle flow can be rewritten in the Lagrangian derivative form that can be straightforwardly resolved in the group Lagrangian approaches. For a steady shock or detonation problem, (2.20)–(2.27) are often rewritten in a shock-attached coordinate system [$x' = x_{\mathrm{D}}(t) - x$, $t' = t$, where $x_{\mathrm{D}}(t)$ is the trajectory of the shock front]. The steady conservation equations in the shock-attached coordinate system remain the same form as (2.20)–(2.27) in which the $\partial/\partial t$ derivative terms are removed.

B Equations of State

Depending on the volume fraction occupied by the particle phase, a particular gas–particle flow topology can be classified as a dilute particle–gas flow

($\phi_p < 0.01$), a dense particle–gas flow ($0.01 \leqslant \phi_p < \phi_{p,pack}$), or a granular flow ($\phi_p \geq \phi_{p,pack}$), where $\phi_{p,pack}$ refers to the volume fraction of a packed particle system. While the conservation relations (2.20)–(2.27) are valid for any continuum two-phase flow, the equations of state for continuum fluid and discrete particles must be specified to provide closure for the solution of a specific gas–particle flow.

Equations of state for gases (e.g., ideal equation of state) can be found in textbooks. For many dense gases, liquids, and solids under high-pressure shock compression, the Grüneisen equation is applied in the form

$$p = p_H + \frac{\Gamma_s}{v}(e - e_H). \tag{2.32}$$

The subscript H denotes the shock Hugoniot state determined by the Hugoniot relations

$$p_H = \frac{c^2(v_0 - v)}{[v_0 - S(v_0 - v)]^2}, \tag{2.33}$$

$$e_H = \frac{1}{2}\left[\frac{c(v_0 - v)}{v_0 - S(v_0 - v)}\right]^2, \tag{2.34}$$

with the specific volume $v = 1/\rho$ and the Grüneisen coefficient $\Gamma_s = 2S - 1$. Γ_s is also related to the ratio of the specific heats by $\gamma = \Gamma_s + 1$. Parameters c and S are from the linear relationship of shock velocity with the flow velocity: $D = c + Su$ and can be found in the handbook of Marsh [127]. Similarly to (2.32), an equation of state for temperature can be written, from which the Hugoniot temperature can also be obtained [128].

The Jones–Wilkins–Lee equation of state is often used to model the detonation product gases and their subsequent adiabatic expansion in an explosion [129]. The constants in the Jones–Wilkins–Lee equation of state used to be the best fits to experimental measurements and hydrodynamic calculations. Now they can also be easily calculated for various energetic materials using an equilibrium code such as Cheetah [130]. Many sophisticated and specific equations of state for real gases, liquids, and solids can be found in the literature suitable for numerical solutions.

The following discussion is focused on the discrete particle flow without interstitial gas. In the limit of a dilute particle flow in which the particles are far apart from each other, the interactions between the particles are negligible. Hence, in analogy to a dilute gas molecule system, the pressure of the particle flow can be neglected: $p_p = 0$ [57]. Consequently, the sound speed of the particle flow is negligible. If the shock or detonation pressure is much lower than the particle material strength, the particle material can be considered to be incompressible: $\rho_s =$ constant.

At the other extreme limit of a granular flow where the particles are closely packed, the competition between the stresses within the particle and the external forces, exerted by the neighboring particles and the gas in voids, forms the main mechanism for the compaction, compression, and deformation of the particles. Applying the second law of thermodynamics (entropy inequality), Baer and Nunziato [52] obtained a dynamic compaction equation controlled by the mechanical nonequilibrium between the pressure inside the solid particles, p_s, and the pressure in the gas plus the pressure due to contact forces between the particles, β_s:

$$\frac{\mathrm{d}\phi_\mathrm{p}}{\mathrm{d}t} = \frac{\partial \phi_\mathrm{p}}{\partial t} + u_\mathrm{p}\frac{\partial \phi_\mathrm{p}}{\partial x} = \Pi_\mathrm{p} + \frac{J_\mathrm{p}}{\rho_\mathrm{s}} = \frac{\phi_\mathrm{p}\phi_\mathrm{g}}{\mu_\mathrm{c}}\left[p_\mathrm{s} - (p_\mathrm{g} + \beta_\mathrm{s})\right] + \frac{J_\mathrm{p}}{\rho_\mathrm{s}}. \tag{2.35}$$

This equation describes the evolution of the solid volume fraction due to the change of packing configuration, where the coefficient μ_c is interpreted as the compaction viscosity. Owing to large solid volume fractions, the pressure and sound speed of the granular flow are close to those in the solid itself. Hence, the volume-averaging assumption (2.29) can be extended to the pressure of the discrete particle flow p_p [52, 53]:

$$p_\mathrm{p} = \phi_\mathrm{p} p_\mathrm{s}. \tag{2.36}$$

Between the above two limits of particle flows, there exists a regime where the high-speed particles mostly interact through inelastic collisions and the definitions of the pressure and sound speed of the particle flow have not been well established. Difficulties are encountered when the volume-averaging assumption (2.36) is extended to dense and dilute particle flows. From (2.29) and (2.36), although the pressure of the particle flow approaches zero as the solid volume fraction approaches zero, the resultant phase-frozen sound speed of the particle flow always equals the sound speed of the solid regardless of how small the solid volume fraction becomes. This conclusion contradicts the common fact that the sound speed of the solid flow diminishes rapidly as the solid volume fraction approaches zero. Clearly, the sound speed in a discrete particle flow is different from that of the solid owing to the spaces between the particles. Though the partial pressure rule (2.36) has often been used in the classical theory of gas mixtures, distribution of the solid pressure into a large space following (2.36) may be physically unrealizable.

The analogy between the random motion of granular particles and the thermal motion of gas molecules inspired researchers to apply the gas kinetic theory to the granular and dense particle flow [131, 132]. This theory was recently applied to a planar shock wave sweeping over a bulk dust layer to lift the dust behind the shock [133]. Using the kinetic theory, one defines a concept of "granular temperature" in terms of velocity fluctuation around the mean particle flow velocity. Consequently, the pressure of the particle flow and the squared sound speed are proportional to the granular temperature. However, the transport equation for granular temperature indicates that the rate of the granular temperature change and therefore the sound speed would increase towards infinity as the solid volume fraction approaches zero. Some studies attempted to include the gas drag to explain this contradiction between the theory prediction and the experimental fact [134]. Although a macroscopic particle system and a molecular system are analogous in many ways, the two systems have fundamental differences. At the root of the unique status of dense high-speed particle flow is the existence of a large number of path-dependent inelastic particle collisions. A careful study of these fundamental differences is necessary and may lead to a new analysis of dissipative statistical mechanics. As the available computational power increases, the dissipative statistical mechanical analysis will be made easier through direct simulations of a large number of particles at the "microscopic" particle scale.

A correct definition of the sound speed in the dense particle flow is crucial for describing compressible phenomena of the flow. Owing to the lack of a sound theoretical basis, the pressure and the sound speed for the transition regime of dense particle flow were estimated using a heuristic interpolation method [54]. This is realized by applying a weighting function to the solid volume fraction between the solid limit and the dilute particle flow limit. This approximation was expected to give reasonable results for problems in which the transition regime from a granular flow to a dilute solid flow occurs in a relatively short time, such as rapid dispersal of solid particles from detonation of a condensed explosive. The heuristic model is assumed to have the form

$$p_\mathrm{p} = P(\rho_\mathrm{s}, e_\mathrm{p}, \phi_\mathrm{p}), \tag{2.37}$$

which satisfies

$$\begin{aligned} &P(\rho_\mathrm{s}, e_\mathrm{p}, \phi_\mathrm{p}) \to p_\mathrm{s}(\rho_\mathrm{s}, e_\mathrm{p}), \text{ as } \phi_\mathrm{p} \to 1, \\ &P(\rho_\mathrm{s}, e_\mathrm{p}, \phi_\mathrm{p})/\sigma_\mathrm{p} \to 0, \text{ as } \phi_\mathrm{p} \to 0 \ \& \ \rho_\mathrm{s} \to \rho_\mathrm{s0}, \end{aligned} \tag{2.38}$$

with σ_p obeying (2.29). The variable e_p is the specific internal energy defined as the internal energy per mass of the discrete particle flow, and therefore equals the specific internal energy of the solid, e_s. Defining the phase-frozen sound speed of a particle flow by

$$a_\mathrm{p}^2 = \left(\frac{\partial p_\mathrm{p}}{\partial \sigma_\mathrm{p}}\right)_{s_\mathrm{p}, \phi_\mathrm{p}} = \frac{1}{\phi_\mathrm{p}}\left(\frac{\partial p_\mathrm{p}}{\partial \rho_\mathrm{s}}\right)_{e_\mathrm{p}, \phi_\mathrm{p}} + \frac{p_\mathrm{p}}{\phi_\mathrm{p}^2 \rho_\mathrm{s}^2}\left(\frac{\partial p_\mathrm{p}}{\partial e_\mathrm{p}}\right)_{\sigma_\mathrm{p}, \phi_\mathrm{p}}, \tag{2.39}$$

where s_p denotes the specific entropy of the particle flow, the heuristic equations of state (2.37)–(2.39) yield

$$a_\mathrm{p} \to a_\mathrm{s} \text{ as } \phi_\mathrm{p} \to 1 \text{ and } a_\mathrm{p} \to 0 \text{ as } \phi_\mathrm{p} \to 0. \tag{2.40}$$

As an example, applying (2.37)–(2.40) to the Grüneisen equation of state (2.32), one can formally write a heuristic equation of state for a dense particle flow as

$$p_\mathrm{p} = g_1(\phi_\mathrm{p}) p_\mathrm{H}(\rho_\mathrm{s}) + g_2(\phi_\mathrm{p}) \Gamma_\mathrm{s} \rho_\mathrm{s} [e_\mathrm{s} - e_\mathrm{H}(\rho_\mathrm{s})], \tag{2.41}$$

with

$$g_1(\phi_\mathrm{p}) = O(\phi_\mathrm{p}^m), \ g_2(\phi_\mathrm{p}) = O(\phi_\mathrm{p}^n) \text{ with } m \geq 1 \text{ and } n > 1. \tag{2.42}$$

Here, p_H equals the value on the solid Hugoniot when $\rho_s/\rho_{s0} \geq 0$ and is assumed to be zero when $\rho_s/\rho_{s0} < 0$.

The particle temperature is computed from the internal energy:

$$T_p = \begin{cases} e_p/c_{pv} & \text{for } e_p < c_{pv}T_{pm} \\ T_{pm} & \text{for } c_{pv}T_{pm} < e_p < c_{pv}T_{pm} + L_m \\ (e_p - L_m)/c_{pv} & \text{for } c_{pv}T_{pm} + L_m < e_p < c_{pv}T_{pb} + L_m \end{cases}, \tag{2.43}$$

where the boiling point can be obtained from [135]

$$T_{pb} = \frac{L_v - 2W_p p_g/\rho_p}{R\ \ln(p^*/p_g)}. \tag{2.44}$$

Here, p^* is a constant, c_{pv} denotes the specific heat at constant volume for the particle, and L_m and L_v are the latent heat of melting and vaporization of the particle material respectively. The internal energy in (2.43) does not include chemical energy. Equation (2.44) includes the correction due to high surrounding gas pressure such as in the detonation products of condensed explosives. As $W_p p_g/\rho_p L_v \ll 1$, the second term in the numerator on the right-hand side is negligible and (2.44) is reduced to the Clausius–Clapeyron equation. For example, the boiling point for aluminum at 7×10^8 Pa is 7,315 K from (2.44) and 9,072 K from the Clausius–Clapeyron relation. A similar correction term due to high gas pressure can be introduced in the equation for the melting point T_{pm}.

C Interphase Transfers

C.1 Mass Transfer

Neglecting the influence of direct interactions between particles, the rate of mass transfer, J_p in kilograms per cubic meter per second, due to particle combustion in a diffusion-limited reaction model is expressed as [27, 136]

$$J_p = n_p \frac{dm_p}{dt} = n_p \pi d_p^2 \rho_s \frac{dr_p}{dt} = \frac{3\sigma_p}{t_b}\left(1 + 0.276\, Re^{1/2} Pr^{1/3}\right), \text{ if } T_p \geq T_{ign}, \tag{2.45}$$

otherwise $J_p = 0$. In (2.45), the particle burning time is

$$t_b = K d_{p0}^n / Y_{oxi}^{\alpha}. \tag{2.46}$$

Here, m_p, and r_p are the particle mass and radius, respectively. The parameters K, d_{p0}, Y_{oxi}, and T_{ign} are the rate coefficient, initial particle diameter, mass fraction of oxidizing gases, and particle ignition temperature respectively. $n = 2$ has often been used, which essentially assumes infinite chemical kinetics. In fact, a diffusion-limited d^2 law for the particle burning time can be derived from the liquid droplet combustion theory in a quiescent flow [99]. The model is independent of temperature and pressure and assumes a particle ignition temperature above which particles react. $n < 2$ implies the effect of finite gas-phase kinetics and possibly convective flow effects [99]. The term including the Reynolds number Re and the gas-phase Prandtl number Pr was therefore introduced in (2.45) presumably to consider the convective flow effect, where the Reynolds number is defined by the velocity difference between the two phases:

$$Re = \rho_g d_p \left| u_g - u_p \right| / \mu_g, \tag{2.47}$$

with a gas-phase dynamic viscosity $\mu_g = \mu_g(T_g)$.

For aluminum combustion, $n = 1.75$, $k = 7.3 \times 10^{-6}\,\mu\mathrm{m}^{1.75}\ \mathrm{s}^{-1}$ (with d_{p0} in micrometers), and $\alpha = 0.9$ have been used in rocket motor propulsion applications where T_{ign} is near the aluminum oxide melting point in quiescent flow conditions [136]. To model the propagation of detonation in aluminum–gas mixtures, $n = 2$, $k = 4 \times 10^{-6}\,\mu\mathrm{m}^2\ \mathrm{s}^{-1}$ (with d_{p0} in micrometers), and $\alpha = 0.9$ has been suggested where T_{ign} has a value near the melting point of aluminum (933–1,350 K) considering possible oxide coating breakup under shock conditions [14, 27].

For kinetics-limited reaction of solid particles, a surface heterogeneous reaction model can be expressed by

$$J_p = -n_p \pi d_p^2 p_{oxi}^m k_0 e^{-E/RT_p} = -\frac{6\sigma_p p_{oxi}^m}{\rho_s d_p} k_0 e^{-E/RT_p}, \tag{2.48}$$

where p_{oxi}, k_0, E, and R are the partial pressure of oxidizing gas, the rate coefficient, the activation energy, and the universal gas constant. This model is suitable for solid particles for which the shock ignition delay follows a logarithmically linear relation with $1/T_p$. Sichel et al. [111] used this model in studying the temperature distribution inside a particle during the shock ignition.

For many solid particles under shocked flow conditions, both surface chemical kinetics and diffusion reaction can take place. Field and Elperin et al. [137] applied a hybrid reaction model combining the surface oxidation and diffusion reaction to carbon particle reaction in oxygen behind a shock wave. As evidenced from the experiments reviewed in Sects. 2.3 and 2.4, micrometric and nanometric aluminum–air detonation strongly depends on the initial pressure; violent abrupt DDT suggests a highly nonlinear dependence on temperature via chemical kinetics. On the other hand, the aluminum–air detonation waves exhibit a weak transverse wave structure with a large transverse wave spacing, thus indicating slower energy release rates in the later phase of aluminum combustion that supports the detonation propagation. Hence, a kinetics–diffusion hybrid aluminum reaction model is proposed in the form [22]

$$J_p = -n_p \pi d_p^2 k_p = -n_p \pi d_p^2 \frac{\upsilon_p W_p}{\upsilon_{oxi} W_{oxi}} k = -\frac{6\sigma_p}{\rho_s d_p} \frac{\upsilon_p W_p}{\upsilon_{oxi} W_{oxi}} k, \tag{2.49}$$

with

$$k = \frac{k_d k_s}{k_d + k_s} C_{oxi}, \tag{2.50}$$

$$k_d = \frac{\nu_{oxi} W_{oxi}}{\nu_p W_p} \frac{\rho_s d_p}{2 C_{total} K d_{p0}^2} (1 + 0.276\, Re^{1/2} Pr^{1/3}), \tag{2.51}$$

and

$$k_s = k_0 e^{-E/RT_{ps}}, \quad \text{with } T_{ps} = (T + T_p)/2, \tag{2.52}$$

where T_{ps} is the particle surface temperature. k_p and k are the mass depletion rates (equaling the mass flux) of the particle and the oxidizing gas at the particle surface in the particle radial direction, respectively. k_d and k_s are the rate coefficients for diffusion and surface reaction, respectively. W, ν, C_{oxi}, and C_{total} denote the molecular weight, stoichiometric coefficient, oxidizing gas molar concentration and total gas molar concentration, respectively. The model (2.49)–(2.52) becomes surface-kinetics-limited (i.e., $k \to k_s$) when $k_s/k_d \ll 1$, and approaches the diffusion-limited reaction model (2.45) and (2.46) but without an ignition temperature assumption as $k \to k_d$ if $k_s/k_d \gg 1$.

The hybrid reaction model (2.49)–(2.52) depends on temperature and pressure (via oxidizing gas concentration) and does not need a presumed particle ignition temperature. Hence, the model is capable of describing transient combustion of particles such as an abrupt DDT via a local explosion center in an aluminum–air mixture as shown in Fig. 2.48. In this numerical computation, $K = 4 \times 10^6\,\mathrm{s\,m^{-2}}$, $k_0 = 1.2 \times 10^6\,\mathrm{kg\text{-}m\,(mol\text{-}s)^{(-1)}}$, and $E = 71.1\,\mathrm{kJ\ mol^{-1}}$

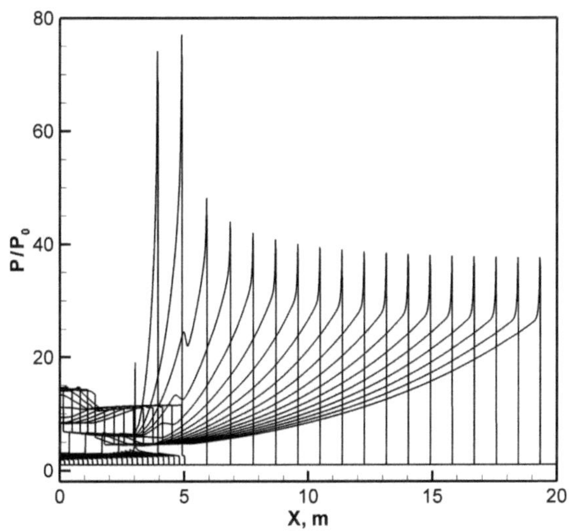

Fig. 2.48. Numerical DDT pressures in a 2-µm aluminum–air mixture at $\sigma_p = 1250\,\mathrm{g\,m^{-3}}$ and $p_0 = 2.5\,\mathrm{atm}$ [22]

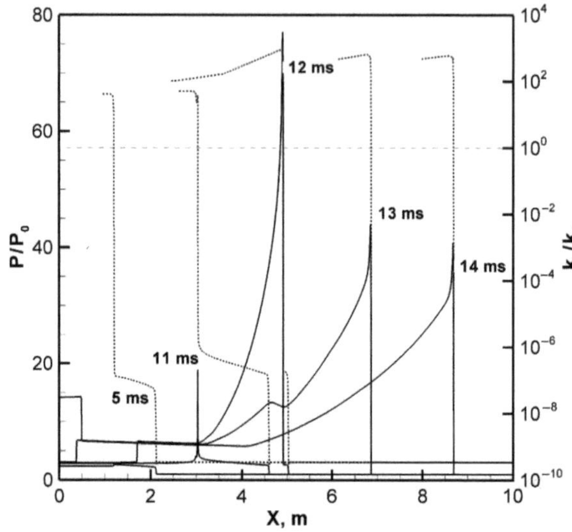

Fig. 2.49. Early process in Fig. 2.48. *Solid line* p/p_0, *dashed line* k_s/k_d [22]

were employed [22], using a 6.13 kJ m^{-2} hot spot zone (1 mm long with $p = 8.56p_0$, $T = 2{,}567$ K, and $T_p = 300$ K) for initiation at the beginning of a 2-μm aluminum–air mixture at $\sigma_p = 1{,}250$ g m^{-3} and an initial pressure of $p_0 = 2.5$ atm (the same initial conditions as for the 2.5-atm experiment displayed in Fig. 2.31). During the early DDT process, $k_s/k_d < 1$ holds behind the shock front in an induction stage that leads to local explosion, thus indicating a kinetics-limited reaction (Fig. 2.49). As the local explosion develops, the particle temperature rapidly increases and therefore results in a rapid increase in k_s/k_d. Once the detonation forms, $k_s/k_d > 1$ holds after a very short kinetics induction time behind the shock, thus showing a diffusion-limited reaction for most of the aluminum mass. The phenomena are in agreement with the experimental observation and therefore the hybrid model properly describes the detonation initiation and the abrupt DDT. Under the same initial conditions and hot spot zone, the abrupt DDT via an autoexplosion center cannot be obtained using the diffusion-limited model (2.45) and (2.46).

Noticing that all the models above are based on a single particle reaction rate multiplied by the particle number, the influence of neighboring particles on the reaction rate remains a research subject, particularly in a dense particle cloud.

C.2 Momentum Transfer

Forces exerted on the solid particles in a dilute particle–gas flow include drag, forces due to the gas pressure gradient and added gas mass on the particle surface, as well as the Basset history force [57, 138]. In the shock and detonation flow conditions, the drag is dominant; the other forces contribute less and can be incorporated into an effective drag coefficient. Neglecting the influence of direct interactions between particles and compression within a particle, one can express the drag, f_p in newtons per cubic meter, formally by

$$f_p = n_p \frac{\pi d_p^2}{4} C_d \rho_g (u_g - u_p) \left| u_g - u_p \right| /2. \tag{2.53}$$

The drag coefficient $C_D = C_D(Re)$ for a spherical particle has been expressed in a standard drag curve based on incompressible flow experiments ($M \ll 1$), and equation fits of the standard drag curve can be found in textbooks [138]. The single particle drag coefficient has been extended to $C_D = C_D(Re, M)$ including the effect of flow compressibility in a range of relative flow Mach number [139, 140]. Effective drag coefficients of gas–particle flow have also been measured under unsteady or shocked flow conditions [57,141,142]. As an example, the following drag model has been applied to both dilute and dense particle–gas flow for $Re < 10^5$ [143]:

$$\begin{aligned} C_{d1} &= \frac{24}{Re} + \frac{4.4}{Re^{0.5}} + 0.42, \text{ if } \phi_p \leqslant 0.08, \\ C_{d2} &= \frac{4}{3\phi_g}\left(1.75 + \frac{150\phi_p}{\phi_g\, Re}\right), \text{ if } \phi_p \geq 0.45, \\ C_{d3} &= [(0.92\phi_p - 0.08\phi_g)C_{d2} + (0.45\phi_g - 0.55\phi_p)C_{d1}]/0.37, \text{ if } 0.08 < \phi_p < 0.45, \end{aligned} \tag{2.54}$$

where the term $24/Re$ is the Stokes model for $Re \ll 1$.

For a granular flow, Baer and Nuziato [52] introduced a compaction-related term, $p_g \partial \phi_p / \partial x$, that can be added in (2.53) for the rate of momentum transfer.

In handling the momentum transfer in detonation of solid particles suspended in low-density gas flow, the shock interaction time in which the shock front crosses a particle is several orders of magnitude smaller than the velocity relaxation time related to the drag. Thus, a solid particle is assumed to remain stationary as the shock front crosses it [57]. In contrast, for shock and detonation in high-density gas or condensed matter containing light solid particles, the shock interaction time can be comparable to the drag-induced velocity relaxation time owing to a significant increase in the initial material density ratio of fluid to solid particles. Thus, the stationary-particle assumption during the shock interaction time is no longer valid as $\rho_g/\rho_s \rightarrow 0.1 - 1$. Under strong shock conditions, the shock compression becomes an important force to accelerate a solid particle during the shock interaction time [39].

If a shock interaction time is defined to be $\tau_s = D/d_p$, where D is the shock velocity, the velocity transmission factor of a particle after a shock front crossing it can be expressed as

$$\alpha = \frac{u_p(\tau_s)}{u_g}, \tag{2.55}$$

where u_g and u_p are the postshock gas velocity and the mass-averaged particle velocity, respectively. The momentum transfer rate or the force acting on the solid-phase control volume containing n_p particles during the shock interaction process can then be written as

$$f_p = n_p m_p \frac{du_p}{dt} \approx n_p m_p \frac{u_p(\tau_s) - u_p(0)}{\tau_s} = n_p m_p \frac{\alpha u_g - u_p(0)}{\tau_s}. \tag{2.56}$$

Thus, the momentum transfer rate during the shock interaction process can be determined by the velocity transmission factor. The function of the velocity transmission factor can be obtained from mesoscale simulations [39, 68, 69].

C.3 Heat Transfer

Neglecting the influence of direct interactions between particles and compression within a particle, one can write the rate of heat transfer, Q_p in joules per cubic meter per second, due to convection as

$$Q_p = n_p \pi d_p^2 h(T_g - T_p) = n_p \pi d_p Nu \lambda_g (T_g - T_p), \tag{2.57}$$

where parameters h and λ_g are the coefficient of convection heat transfer to particles and the gas-phase thermal conductivity, respectively. The Nusselt number, $Nu = hd_p/\lambda_g$, is defined as the ratio of the convective heat transfer to the particle to the conductive heat transfer in the gas. Curve fits of the Nusselt number $Nu = Nu(Re, Pr)$ for a spherical particle in incompressible flow ($M \ll 1$) and its extension $Nu = Nu(Re, Pr, M)$ for compressible flow can be found in textbooks [57]. Fox et al. [144] suggested a model for $0 < M < 6$ and $Re < 10^4$:

$$Nu = \frac{2\exp(-M)}{1 + 17M/Re} + 0.459\, Re^{0.55} Pr^{0.33} \frac{1 + 0.5\exp(-17M/Re)}{1.5}. \tag{2.58}$$

This model was fitted to experimental data for continuum flow and partial transition flow ($M/Re < 0.5$). It becomes a model for incompressible flow when M approaches zero, and matches compressible subsonic results and supersonic measurements in rarefied gas flow.

In general, the heat conduction inside the particle must also be considered [12]. However, when the Biot number is small (the ratio of convective heat transfer to the particle to conductive heat transfer within the particle, $Bi = hd_p/6\lambda_p < 0.1$), a uniform temperature within the particle is quickly established and can therefore be assumed to simplify the problem. The rate of convective heat transfer in the detonation flow reaches a maximum value immediately behind the shock front where the temperature difference and the relative flow velocity are maximum. At this point, large particles can have a Biot number on the order of 0.1–1. As the particle is accelerated by the flow, the relative velocity and the rate of convective heat transfer drop significantly, thereby resulting in a small Biot number for most of the processes behind a shock.

The radiative heat transfer, $n_{\rm p}\pi d_{\rm p}^2\varepsilon\sigma(T_{\rm g}^4 - T_{\rm p}^4)$, can be added to the heat transfer (2.57). Under detonation conditions, however, the radiation heat transfer is small compared with the convective heat transferred to the particle and can often be neglected. In the granular flow model of Baer and Nuziato [52], an additional term $(-p_{\rm p}\Pi_{\rm p})$ was introduced in the heat transfer (2.57) to account for the heat loss due to irreversible dissipative compaction.

Similar to the rate of momentum transfer during the shock interaction process in dense gases or condensed matter containing solid particles, a temperature transmission factor of a particle after a shock front crossing it is introduced as [68]

$$\beta = \frac{T_{\rm p}(\tau_{\rm s})}{T_{\rm g}}, \tag{2.59}$$

where $T_{\rm g}$ and $T_{\rm p}$ are the postshock gas temperature and the mass-averaged particle temperature, respectively. The rate of heat transferred to the solid-phase control volume containing $n_{\rm p}$ particles during the shock interaction process can then be written as

$$q_{\rm p} = n_{\rm p}c_{\rm p}\frac{{\rm d}T_{\rm p}}{{\rm d}t} \approx n_{\rm p}c_{\rm p}\frac{T_{\rm p}(\tau_{\rm s}) - T_{\rm p}(0)}{\tau_{\rm s}} = n_{\rm p}c_{\rm p}\frac{\beta T_{\rm g} - T_{\rm p}(0)}{\tau_{\rm s}}, \tag{2.60}$$

where $c_{\rm p}$ is the specific heat capacity for the particle. Thus, the heat transfer rate during the shock interaction process can be determined by the temperature transmission factor. The function of the temperature transmission factor can be obtained from mesoscale simulations [68, 69].

C.4 Particle Number Change

Solid particle fragmentation was observed upon high-velocity impact on a hard wall [145]. A fragmentation process can be modeled by introducing an exchange source term for the particle number density based on the following assumptions:

1. A particle–wall interaction is divided into two steps: instantaneous fragmentation and nonsliding wall reflection.
2. The fragment sizes $d_{\rm frag}$ are grouped by n particle sizes:

$$d_{\rm p} = d_{n-1}, \quad \text{if } d_{n-1} \geq d_{\rm frag} > d_n \text{ for } n = 2, \ldots n, \tag{2.61}$$

   where $d_{n-1} > d_n$, with d_1 the original particle size.
3. The mass, momentum, and energy of the solid particle flow are conserved during the fragmentation process. Thus, fragmented particles have the same material density, velocity, and temperature as the original particles.

Under the above assumptions, the rate of particle number change due to high-velocity impact fragmentation on a wall can be expressed (e.g., $n = 2$) by

$$\Psi_{\mathrm{p1}} = \begin{cases} -n_{\mathrm{p1}}(x,t)\,\delta(t - t_{\mathrm{wall}}), & \text{for } x = x_{\mathrm{wall}}, d_{\mathrm{frag}} \leqslant d_2 \\ 0, & \text{otherwise} \end{cases}, \tag{2.62}$$

$$\Psi_{\mathrm{p2}} = -\left(d_1/d_2\right)^3 \Psi_{\mathrm{p1}}, \tag{2.63}$$

where t_{wall} is the time at which the particles arrive at the wall, x_{wall}. The corresponding rate of mass transfer between the particles with different sizes is

$$J_{\mathrm{p1}} = -J_{\mathrm{p2}} = \begin{cases} -\sigma_{\mathrm{p1}}(x,t)\,\delta(t - t_{\mathrm{wall}}), & \text{for } x = x_{\mathrm{wall}},\ d_{\mathrm{frag}} \leqslant d_2 \\ 0, & \text{otherwise} \end{cases}. \tag{2.64}$$

The above formulation can be straightforwardly extended to $n > 2$. Numerical implementation in the Eulerian approaches requires n sets of conservation equations for n particle sizes. Hence, the calculation becomes more efficient in the group Lagrangian approach when n becomes large.

The fragmentation size can be obtained using a model proposed by Grady [146, 147], based on the assumption that a moving spherical particle is subjected to an internal fracturing force due to the kinetic energy associated with outward expansion motion and the particle surface tension resistance:

$$d_{\mathrm{frag}} = B\left(K_{\mathrm{IC}}/\sigma_{\mathrm{m}}\right)^2, \tag{2.65}$$

where K_{IC} is the model I critical stress intensity with units of newtons per meter to the power two thirds and σ_{m} denotes the mean stress. $B = 5$ is used in [146] and $B = 6$ is used in [147]. If the elastic energy in the particle is included in the formulation, a value of $B = 1.875$ can be obtained.

From the elastic finite-element computations of a moving particle normally impacting on a wall, the maximum mean stress obtained can be correlated with the particle normal velocity by a curve fitting to

$$\sigma_{\mathrm{m}} = A\rho_{\mathrm{s}}u_{\mathrm{p}}^2, \tag{2.66}$$

where A is a fitting coefficient. $A = 1.46$ was obtained for an aluminum particle with elastic Young's modulus $E = 7.7 \times 10^{10}$ Pa, Poisson's ratio $\nu = 0.33$, density $\rho_{\mathrm{s}} = 2{,}700\,\mathrm{kg\,m^{-3}}$, and $K_{\mathrm{IC}} = 3.0 \times 10^7\,\mathrm{N\,m^{-3/2}}$. Therefore, Grady's model (2.65) can be revised to a relationship with the particle velocity:

$$d_{\mathrm{frag}} = \frac{B}{A^2}\left(\frac{K_{\mathrm{IC}}}{\rho_{\mathrm{s}}u_{\mathrm{p}}^2}\right)^2. \tag{2.67}$$

Solid particle fragmentation will also take place as a shock or detonation wave propagates in condensed matter containing solid particles [116]. The damage and fragmentation of reactive solid particles will change the surface properties and the size distribution of the particles, thus affecting the subsequent ignition and reaction behavior of the particles. While the fragmentation process can be modeled in the same approach discussed above using a source term of particle number change, the relevant physics and criteria for mechanical damage and fragmentation of specific particles are a subject of current research efforts.

Acknowledgements

My first gratitude is to Hans Grönig for initiating my work into multiphase shock and detonation research. He sets a lasting example of how scholarly work should be done. I particularly wish to thank John Lee, Martin Sichel, and Peter Wolanski for many insightful discussions. This chapter would not have been possible without dedicated contributions from Herbert Oliver, Andrias van de Ven, Peter Greilich, Gihui Geng, Paul Thibault, Robert Ripley, Keith Gerrard, Akio Yoshinaka, Julian Lee, and Vincent Tanquay.

References

1. Strauss, W.A.: Investigation of the detonation of aluminum powder–oxygen mixtures. AIAA J. **6**, 1753–1757 (1968)
2. Nettleton, M.A., Stirling, R.: Detonations in suspensions of coal dust in oxygen. Combust. Flame **21**, 307–314 (1973)
3. Cybulski, W.B.: Detonation of coal dust. Bull. Pol. Acad. Sci., **19**, 37–41 (1971)
4. Bartknecht, W.: Explosionen, pp. 5–72. Springer, Berlin (1978)
5. Fangrat, J., Glinka, W., Wolanski, P., Wolinski, M.: Detonation structure in organic dust–oxygen mixtures. Arch. Combust. **7**, 321–332 (1987)
6. Wolanski, P.: Deflagration and detonation of dust mixtures. Prog. Astronaut. Aeronaut. **132**, 3–31 (1991)
7. Kauffman, C.W., Wolanski, P., Arisoy, A., Adams, P.R., Maker, B.N., Nicholls, J.A.: Dust, hybrid and dusty detonations. Prog. Astronaut. Aeronaut. **94**, 221–240 (1984)
8. Peraldi, O., Veyssiere, B.: Experimental study of detonations in starch particle suspensions with O_2/N_2, H_2/O_2 and C_2H_4/O_2 mixtures. Prog. Astronaut. Aeronaut. **106**, 490–504 (1986)
9. Lee, J.H.S.: Dust explosion: an overview. In: Grönig, H. (ed.) Proceedings of 16th International Symposium of Shock Waves, Aachen, Germany, pp. 31–38. VCH, Weinheim (1988)
10. Wolanski, P., Lee, D., Sichel, M., Kauffman, C.W., Nicholls, J.A.: The structure of dust detonations. Prog. Astronaut. Aeronaut. **94**, 241–263 (1984)
11. Lee, D., Sichel, M.: The Chapman–Jouguet condition and structure of detonation in dust-oxidizer mixtures. Prog. Astronaut. Aeronaut. **106**, 505–521 **(1986)**
12. Fan, B.C., Sichel, M.: A comprehensive model for the structure of dust detonations. In: Proceedings of the 22nd International Symposium on Combustion, pp. 1741–1750. The Combustion Institute, Pittsburgh (1988)
13. Tulis, A.J., Selman, J.R.: Detonation tube studies of aluminum particles dispersed in air. In: Proceedings of the 19th International Symposium on Combustion, pp. 655–663. The Combustion Institute, Pittsburgh (1982)
14. Borisov, A.A., Khasainov, B.A., Saneev, E.L., Formin, I.B., Khomik, S.V., Veyssiere, B.: On the detonation of aluminum suspensions in air and in oxygen. In: Borisov, A.A. (ed.) Dynamic Structure of Detonation in Gaseous and Dispersed Media, pp. 215–253. Kluwer, Dordrecht (1991)

15. Gardner, B.R., Winter, R.J., Moore, M.J.: Explosion development and deflagration-to-detonation transition in coal dust/air suspensions. In: Proceedings of the 21st International Symposium on Combustion, pp. 335–343. The Combustion Institute, Pittsburgh (1986)
16. Zhang, F., Grönig, H.: Spin detonation in reactive particles-oxidizing gas flow. Phys. Fluids A **3**, 1983–1990 (1991)
17. Zhang, F., Grönig, H.: Two-headed detonation in reactive particle-oxidizing gas flow, Phys. Fluids A **4**, 2308–2315 (1992)
18. Zhang, F., Greilich, P., Grönig, H.: Propagation mechanism of dust detonations. Shock Waves **2**, 81–88 (1992)
19. Grönig, H.: Dust detonations. JSME Int. J. Ser. B **40** 1–15 (1997)
20. Zhang, F., Grönig, H., van de Ven, A.: DDT and detonation waves in dust–air mixtures. Shock Waves **11**, 53–71 (2001)
21. Ingignoli, W., Veyssiere, B., Khasainov, B.A.: Study of detonation initiation in unconfined aluminum dust clouds. In: Roy, G., Frolov, S., Kailasanath, K., Smirnov, N. (eds.) Gaseous and Heterogeneous Detonations, pp. 337–350. ENAS, Moscow (1999)
22. Zhang, F., Gerrard, K., Ripley, R., Tangury, V.: Unconfined aluminum particles–air detonation. In: Hannemann, K. (ed.) Proceedings of the 26th International Symposium on Shock Waves, Goettingen (2007)
23. Veyssiere, B., Manson, N.: Sur l'existence d'un second front de détonation des mélanges biphasiques hydrogène-oxygène-azote-particules d'aluminium. Note C. R. Acad. Sci. **295 II**, 335–338 (1982)
24. Veyssiere, B.: Structure of the detonations in gaseous mixtures containing aluminium particles in suspension. Prog. Astronaut. Aeronaut. **106**, 522–544 (1986)
25. Zhang, F., Thibault, P.A., Murray, S.B.: Transition from deflagration to detonation in an end multiphase slug. Combust. Flame **114**, 13–25 (1998)
26. Afanasieva, L.A., Levin, V.A., Tunik, Y.V.: Multifront combustion of two-phase media. Prog. Astronaut. Aeronaut. **87**, 394–413 (1983)
27. Khasainov, B.A., Veyssiere, B.: Steady, plane, double-front detonations in gaseous detonable mixtures containing a suspension of aluminium particles. Prog. Astronaut. Aeronaut. **114**, 284–299 (1988)
28. Veyssiere, B., Khasainov, B.A.: Structure and multiplicity of detonation regimes in heterogeneous hybrid mixtures. Shock Waves **4**, 171–186 (1995)
29. Wolinski, M., Teodorczyk, A., Wolanski, P., Klemens, R., Sichel, M., Nettleton, M.: Hybrid detonations in oat dust clouds in methane–air mixtures. Combust. Sci. Technol. **120**, 39–53 (1996)
30. Veyssiere, B., Ingignoli, W.: Existence of the detonation cellular structure in two-phase hybrid mixtures. Shock Waves **12**, 291–299 (2003)
31. Zhang, F., Murray, S.B., Gerrard, K.B.: Hybrid detonations in aluminum dust–gas mixtures. In: Proceedings of the 19th International Colloquium on the Dynamics of Explosions and Reactive Systems, Hakone, pp. 167.1–167.4 (2003)
32. Zhang, F., Murray, S.B., Gerrard, K.B.: Hybrid detonation waves in heterogeneous explosive mixtures. In: Proceedings of the 18th International Symposium on Military Aspects of Blast and Shock, Bad Reichenhall, pp. 75.1–75.14 (2004)
33. Zhang, F., Murray, S.B., Gerrard, K.B.: Hybrid detonation waves. In: Proceedings of the 20th International Colloquium on the Dynamics of Explosions and Reactive Systems, Montreal, pp. 217.1–217.12 (2005)

34. Tulis, A.J., Sumida, W.K., Dillon, J., Comeyne, W., Heberlein, D.C.: Submicron aluminum particle size influence on detonation of dispersed fuel-oxidizer powders. In Proceedings of International Symposium on Hazards, Prevention, and Mitigation of Industrial Explosion, Schaumberg, pp. 68–76 (1998)
35. Li, Y.-C., Alexander, C.G., Wolianski, P., Kauffman, C.W., Sichel, M.: Experimental investigations of accelerating flames and transition to detonation in layered grain dust. Prog. Astronaut. Aeronaut. **154**, 170–184 (1993)
36. Vasil'ev, A.A., Gavrilenko, T.P., Topchian, M.E.: On the Chapman–Jouguet surface in multi-headed detonations. Astronaut. Acta **17**, 499–502 (1972)
37. Edwards, D.H., Jones, A.T., Phillips, D.E.: The location of the Chapman–Jouguet surface in a multiheaded detonation wave. J. Phys. D **9**, 1331–1342 (1976)
38. Weber, M., Olivier, H.: The thickness of detonation waves visualised by slight obstacles. Shock Waves **13**, 351–365 (2003)
39. Zhang, F., Thibault, P.A., Link, R.: Shock interaction with solid particles in condensed matter and related momentum transfer. Proc. R. Soc. Lond. A, **459**, 705–726 (2003)
40. Zeldovich, Ya.B., Kompaneets, A.S.: Theory of Detonation, pp. 133–205. Academic, London (1960)
41. Wood, W.W., Kirkwood, J.G.: Diameter effect in condensed explosives: the relation between velocity and radius of curvature of the detonation wave. J Chem. Phys. **22**, 1920–1924 (1954)
42. Kuznetsov, N.M.: Nonuniqueness and stability of detonation modes. Sov. Phys. JETP **25** 199–204 (1967)
43. Erpenbeck, J.J.: Stability of idealized one-reaction detonations. Phys. Fluids **7**, 684–696 (1964)
44. Fickett, W., Wood, W.W.: Flow calculations for pulsating one-dimensional detonations. Phys. Fluids **9**, 903–916 (1966)
45. Abouseif, G.E., Toong, T.Y.: Theory of unstable one-dimensional detonations. Combust. Flame **45**, 67–94 (1982)
46. Lee, H.I., Stewart, D.S.: Calculation of linear detonation instability: one-dimensional instability of planer detonations. J. Fluid Mech. **216**, 103–132 (1990)
47. He, L., Lee, J.H.S.: The dynamics limit of one-dimensional detonations. Phys. Fluids **7**, 1151–1158 (1995)
48. Sharpe, G.J.: Linear stability of pathological detonations. J. Fluid Mech. **401**, 311–338 (1991)
49. Ng, H.D., Lee, J.H.S.: Direct instability of detonation with a multi-step reaction scheme. J. Fluid Mech. **476**, 179–211 (2003)
50. Gelfand, B.E., Frolov, S.M., Nettleton, M.A.: Gaseous detonations – a selective review. Prog. Energy Combust. Sci. **17**, 327–371 (1991)
51. Powers, J.M., Stewart, D.S., Krier, H.: Theory of two-phase detonations – part II: structure. Combust. Flame **80**, 280–303 (1990)
52. Baer, M.R., Nunziato, J.W.: A two-phase mixture theory for the deflagration-to-detonation transition (DDT) in reactive granular materials. Int. J. Multiphase Flow **12**, 861–889 (1986)
53. Powers, J.M., Stewart, D.S., Krier, H.: Theory of two-phase detonations – part I: modeling. Combust. Flame **80**, 264–279 (1990)
54. Zhang, F., Frost, D.L., Thibault, P.A., Murray, S.B.: Explosion dispersal of solid particles. Shock Waves **10**, 431–443 (2001)

55. Zhang, F.: Numerical studies of detonation propagation in mixtures of combustible gases and inert dust. In: Gottlieb, J.J., Ethier C.R. (eds.) The 2nd Conference of the CFD Society of Canada, pp. 261–268 (1994)
56. Kapila, A.K., Son, S.F., Bdzil, J.B., Menikoff, R., Stewart, D.S.: Two-phase modeling of DDT, structure of the velocity-relaxation zone. Phys. Fluids A **9**, 3885–3897 (1997)
57. Rudinger, G.: Fundamentals of Gas–Particle Flow. Elsevier, Amsterdam (1980)
58. Zhang, F., Lee, J.H.S.: Friction-induced oscillatory behaviour of one-dimensional detonations. Proc. R. Soc. Lond. A **446**, 87–105 (1994)
59. Zhang, F., Chue, R.S., Frost, D.L., Lee, J.H.S., Thibault, P., Yee, C.: Effects of area change and friction on detonation stability in supersonic ducts. Proc. R. Soc. Lond. A **449**, 31–49 (1995)
60. Korobeinikov, V.: Formation of zone with high particle concentrations in dusty gases. Prog. Astronaut. Aeronaut. **132**, 287–292 (1991)
61. Fickett, W., Davis, W.C.: Detonation, pp. 153–191. University of California Press, Berkeley (1979)
62. Khasainov, B.A., Veyssiere, B., Ingignoli, W.: Numerical simulation of detonation cell structure in hydrogen–air mixture loaded by aluminum particles. In: Roy, G., Frolov, S., Netzer, D., Borisov A. (eds.) High-Speed Deflagration and Detonation: Fundamentals and Control, pp. 163–174. ELEX-KM, Moscow (2001)
63. Hayashi, A.K., Fuyuto, T., Fujiwara, T.: Triple-shock structure in dusty gas detonations. In: Sturtevant, B., Shepherd, J.E., Hornung H.G. (eds.) Proceedings of the 20th International Symposium on Shock Waves, pp. 1071–1076. World Scientific, Singapore (1996)
64. Fedorov, A.V., Khmel, T.A.: Types and stability of detonation flows of aluminum particles in oxygen. Fiz. Goreniya Vzryva **32**(2), 74–85 (1996)
65. Fedorov, A.V., Khmel, T.A.: Determination of nonideal self-sustained detonation regimes of aluminum particles in air. Fiz. Goreniya Vzryva **34**(5), 95–102 (1998)
66. Benkiewicz, K., Hayashi, A.K.: One-dimensional parametric studies of an aluminum–dust combustion model for numerical simulations of detonation waves. AIAA J. **44**, 608–619 (2006)
67. Zhang, F., Gerrard, K.B., Ripley, R.: Reaction mechanism of aluminum particles–air detonation. In: Proceedings of the 7th International Symposium on Hazards, Prevention and Mitigation of Explosion, St. Petersburg, Vol. 2, pp. 223–237 (2008)
68. Ripley, R., Zhang, F., Lien, F.S.: Shock interaction of metal particles in condensed explosive detonation. In: Furnish, M.D., Elert, M., Russel, T.P., White C.T. (eds.) Shock Compression of Condensed Matter – 2005, pp. 499–502. American Institute of Physics, Melville (2006)
69. Ripley, R., Zhang, F., Lien, F.S.: Detonation interaction with metal particles in explosives. In: Proceedings of 13th International Symposium on Detonation, pp. 214–223. Office of Naval Research ONR 351-07-01 (2006)
70. Zhang, F., Grönig, H.: Transition to detonation in cornstarch dust–oxygen and –air mixtures. Combust. Flame **86**, 21–32 (1991)
71. Urtiew, P.A., Oppenheim, A.K.: Experimental observation of the transition to detonation on an explosive gas. Proc. R. Soc. Lond. A **295**, 13–29 (1966)
72. Lee, J.H.S., Moen, I.O.: The mechanism of transition from deflagration to detonation in vapour explosions. Prog. Energy Combust. Sci. **6**, 359–389 (1980)

73. Zhang, F., Murray, S.B., Gerrard, K.B.: Aluminium dust–air detonation at elevated pressures. Shock Waves **15**, 313–324 (2006)
74. Craven, A.D., Greig, T.R.: The development of detonation over-pressures in pipelines. Chem. Eng. Symp. **25**, 41–50 (1968)
75. Kogarko, S.M.: Investigation of the pressure at the end of a tube in connection with rapid nonstationary combustion. Sov. Phys. Tech. Phys. **28**, 2041 (1958)
76. Tulis, A.J.: On the unconfined detonation of aluminum powder-air clouds. In: Wolanski, P. (ed.) 1st International Colloquium on Explosibility of Industrial Dusts, Warsaw, Poland, pp. 178–186 (1984)
77. Dove, J.E., Wagner, H.G.: A photographic investigation of the mechanisms of spinning detonation. In: Proceedings of the 8th International Symposium on Combustion, pp. 589–600. The Combustion Institute, Pittsburgh (1960)
78. Manson, N.: Propagations des detonations et des deflagrations dans les melanges gaseaux.Off. Nat. Etud. Rech. Aeronaut. Inst. Fr. Pet. Paris C. R. Acad. Sci. **222**, 46–51 (1947)
79. Denisov, Y.N., Troshin, Y.K.: Structure of gaseous detonation in tubes. Z. Tekh. Fiz. **30**, 450–459 (1960)
80. Voitsekhovskii, B.V., Mitrofanov, V.V., Topchian, M.E.: Structure of a detonation front in gases. Izdatel'stvo Sibirskogo Otdeleniya Akademiya Nauk SSSR, Novosibirsk (1963). English translation: Wright-Patterson Air Force Base Report FTD-MT-64-527 (AD0633821) (1966)
81. Schott, G.L.: Observations of the structure of spinning detonation. Phys. Fluids **8**, 850–865 (1965)
82. Mitrofanov, V.V., Soloukhin, R.I.: The diffraction of multi-front detonation waves. Sov. Phys. Dokl. **9**, 1055–1058 (1965)
83. Strehlow, R.A.: The nature of transverse waves in detonations. Astronaut. Acta **14**, 539–548 (1969)
84. Lundstrom, E.A., Oppenheim, A.K.: On the influence of nonsteadiness on the thickness of the detonation wave. Proc. R. Soc. Lond. A **310**, 463–478 (1969)
85. Edwards, D.H., Hooper, G., Job, E.M., Parry, D.J.: The behaviour of the frontal and transverse shock in gaseous detonation waves. Acta Astronaut. **15**, 323–333 (1970)
86. Strehlow, R.A., Crooker, A.J.: The structure of marginal detonation waves. Acta Astronaut. **1**, 303–315 (1974)
87. Lee, J.H.S.: Dynamic parameters of gaseous detonations. Annu. Rev. Fluid Mech. **16**, 311–336 (1984)
88. Shepherd, J.E., Pintgen, F., Austin, J.M., Eckett, C.A.: The structure of the detonation front in gases. AIAA Pap. 2002-0773 (2002)
89. Shepherd, J.E.: Detonation: a look behind the front. In: Proceedings of the 19th International Colloquium on the Dynamics of Explosions and Reactive Systems, Hakone, Japan, pp. 218.1–218.15 (2003)
90. Zeldovich, Y.B., Kogarko, S.M., Simonov, M.N.: An experimental investigation of spherical detonation of gases. Sov. Phys. Tech. Phys. **1**, 1689–1713 (1956)
91. Benedick, W.B., Guirao, C.M., Knystautas, R., Lee, J.H.: Critical charge for direct initiation of detonation in gaseous fuel–air mixtures. Prog. Astronaut. Aeronaut. **106**, 181–202 (1986)
92. Lee, J.H.S., Knystautas, R., Guirao, C.: The link between cell size, critical tube diameter, initiation energy and detonability limits. In: Lee, J.H.S., Guirao, C.M. (eds.) Fuel–Air Explosions, pp. 157–187. University of Waterloo Press, Waterloo (1982)

93. Vasil'ev, A.A.: Gaseous fuels and detonation hazards. In: Eisenreich, N. (ed.) Proceedings of 28th ICT Conference, Karlsruhe, pp. 50.1–50.14 (1997)
94. Dahab, O., Kapuscinski, M., Wolanski, P.: Influence of dust parameters on detonation velocity, structure and limits. Prog. Astronaut. Aeronaut. **133**, 356–370 (1989)
95. Zhang, F., Grönig, H.: Detonability of organic dust–air mixtures. Prog. Astronaut. Aeronaut. **154**, 195–215 (1993)
96. Kogarko, S.M., Zeldovich, Ya.B.: On detonation of gas mixtures. Dokl. Akad. Nauk SSSR **63**, 553–556 (1948)
97. Shchelkin, K.I., Troshin, Ya.K.: Gasdynamics of Combustion. Academy of Sciences, Moscow. Translated by Kuvshinoff, B.W., Holtschlag, L. Mono Book, Baltimore (1965)
98. Strehlow, R.A., Engel, C.D.: Transverse waves in detonation II: structure and spacing in H_2–O_2, C_2H_2–O_2, C_2H_4–O_2, and CH_4–O_2 systems. AIAA J. **7**, 492–496 (1969)
99. Glassman, I.: Combustion, pp. 168–193. Academic, New York (1977)
100. Friedman, R., Macek, A.: Ignition and combustion of aluminum particles in hot ambient gases. Combust. Flame **6**, 9–19 (1962)
101. Fontijn, A., Felder, W.: HTFFR Kinetic Studies of $Al + CO_2 \rightarrow AlO + CO$ from 300 to 1,800 K, a non-Arrenhius reaction. J. Chem. Phys. **67**, 1561 (1977)
102. King, M.K.: Modeling of single particle aluminum combustion in CO_2–N_2 atmospheres. In: Proceedings of the 17th International Symposium on Combustion, pp. 1317–1328. The Combustion Institute, Pittsburgh (1978)
103. Gurevich, M.A., Lapkina, K.I., Ozerov, E.S.: Ignition limit of aluminum particles. Combust. Explos. Shock Waves **6**, 172–175 (1970)
104. Davis, A.: Solid propellants: the combustion of particles of metal ingredients. Combust. Flame **7**, 227–234 (1963)
105. Pokhil, P.F., Belyayev, A.F., Frolov, Y.V., Logachev, V.S., Kotkov, A.I.: Combustion of powdered metal in active media. Nauka, Moscow (1972). US Air Force Foreign Technology Division Translation FTD-MT-24-551-73 (AD0769576) (1973)
106. Foelsche, R.O., Burton, R.I., Krier, H.: Ignition and combustion of aluminum particles in $H_2/O_2/N_2$ combustion products. J. Propuls.. Power **117**, 1001–1008 (1998)
107. Melcher, J.C., Burton, R.L., Krier, H.: Combustion of aluminum particles in solid rocket motor flows. Prog. Astronaut. Aeronaut. **185**, 723–747 (2000)
108. Melcher, J.C., Krier, H., Burton, R.L.: Burning aluminum particles inside a laboratory-scale solid rocket motor. J. Propuls.. Power **18**, 631–640 (2002)
109. Borisov, A.A., Gelfand, B.E., Timofeev, E.I., Tsyganov, S.A., Khomic, S.V.: Ignition of dust suspensions behind shock waves. Prog. Astronaut. Aeronaut. **96**, 332–339 **(1984)**
110. Boiko, V.M., Fedorov, A.V., Formin, V.M., Papyrin, A.N., Soloukhin, R.I.: Ignition of small particles behind shock waves. Prog. Astronaut. Aeronaut. **87**, 71–87 (1982)
111. Sichel, M., Baek, S.W., Kauffman, C.W., Maker, B., Nicholls, J.A., Wolanski, P.: The shock wave ignition of dusts. AIAA J. **23**, 1374–1380 (1985)
112. Geng, J.H., van de Ven, A., Zhang, F., Grönig, H.: Ignition delay of the cornstarch dust behind an incident shock wave. In: Proceedings of the 14th International Colloquium on the Dynamics of Explosions and Reactive Systems, Coimbra (1993)

113. Servaites, J., Krier, H., Melcher, J.C., Burton, R.L.: Ignition and combustion of aluminum particles in shocked $H_2O/O_2/Ar$ and $CO_2/O_2/Ar$ mixtures. Combust. Flame **125**, 1040–1054 (2001)
114. Tanguay, V., Goroshin, S., Higgins, A.J., Zhang, F.: Aluminum particle combustion in high-speed detonation products. Combust. Sci. Technol. **181**, 670–693 (2009)
115. Lee, J.J., Zhang, F.: Burning properties of aluminum in H_2O or CO_2 gas. In: Proceedings of the 20th International Colloquium on the Dynamics of Explosions and Reactive Systems, Montreal, pp. 179.1–179.7 (2005)
116. Yoshinaka, A., Zhang, F., Wilson, W.H.: Effect of shock compression on aluminum particles in condensed media. In: Elert, M., Furnish, M.D., Chau, R., Holmes, N., Nguryen J. (eds.) Shock Compression of Condensed Matter – 2007, pp. 1057–1060. APS Melville, New York (2007)
117. Fay, J.A.: Two-dimensional detonations: velocity deficits. Phys. Fluids **2**, 283–289 (1959)
118. Manson, N., Brochet, C., Brossard, J., Pujol, Y.: Vibratory phenomena and instabilities of self-sustained detonations in gases. In: Proceedings of the 9th International Symposium on Combustion, pp. 461–469. Academic, London (1963)
119. Dabora, E.K., Nicholls, J.A., Morrison, R.B.: The influence of a compressible boundary on the propagation of gaseous detonations. In: Proceedings of the 10th International Symposium on Combustion, pp. 817–830. The Combustion Institute, Pittsburgh (1965)
120. Vasiliev, A.A.: Gas detonation of a free mixture column. Lavrent'yev Institute of Hydrodynamics, Siberian Branch of the USSR Academy of Sciences, Novosibirsk, (1980)
121. Edwards, D.H., Thomas, G.O., Nettleton, M.A.: The diffraction of a planar detonation wave at an abrupt area change. J. Fluid Mech. **95**, 79–96 (1979)
122. Murray, S.B., Lee, J.H.: The influence of physical boundaries on gaseous detonation waves. Prog. Astronaut. Aeronaut. **106**, 329–355 (1986)
123. Moen, I.O., Sulmistras, A., Thomas, G.O., Bjerketvedt, D., Thibault, P.A.: Influence of cellular regularity on the behavior of gaseous detonations. Prog. Astronaut. Aeronaut. **106**, 220–243 (1986)
124. Moen, I.O., Donato, M., Knystautas, R., Lee, J.H.: The influence of confinement on the propagation of detonations near the detonability limits. In: Proceedings of the 18th International Symposium on Combustion, pp. 1615–1622. The Combustion Institute, Pittsburgh **(1981)**
125. Zhang, F.: Detonation waves in dust media: a review. AIAA Pap. 2002-0772 (2002)
126. Gogulya, M.F., Makhov, M.N., Dolgoborodov, A.Yu., Brazhnikov, M.A., Arkhipov, V.I., Shchetinin, V.G.: Mechanical sensitivity and detonation parameters of aluminized explosives. Fiz. Goreniya Vzryva **40**, 82–95 (2004)
127. Marsh, S.P.: LASL Shock Hugoniot Data. University of California Press, Berkeley (1980)
128. Walsh, J.M., Christian, R.H.: Equation of state of metals from shock wave measurements. Phys. Rev. **97**, 1544–1556 (1955)
Note to reference stylist: Please add the publisher city location. N
129. Lee, E., Hornig, H.C., Kury, J.W.: Adiabatic expansion of high explosive detonation products. Lawrence Livermore National Laboratory report UCRL-50422. Lawrence Livermore National Laboratory, Livermore, CA (1968)

130. Fried, L.E., Howard, W.M., Souers, P.C.: Cheetah 2.0 User's Manual. Lawrence Livermore National Laboratory report UCRL-MA-117541 revision 5. Lawrence Livermore National Laboratory, Livermore, CA (1998)
131. Lun, C.K.K., Savage, S.B., Jeffrey, D.J., Chepurniy, N.: Kinetic theories for granular flow: inelastic particles in couette flow and slightly inelastic particles in a general flowfield. J. Fluid Mech. **140**, 223–256 (1984)
132. Gidaspow, D.: Multiphase flow and fluidization. Academic, London (1994)
133. Fan, B.C., Chen, Z.H., Jiang, X.H., Li, H.Z.: Interaction of a shock wave with a loose dusty bulk layer. Shock Waves **16**, 179–187 (2007)
134. Blazer, G., Simonin, O., Boelle, A., Lavieville, J.: A Unifying modelling approach for the numerical prediction of dilute and dense gas-solid two phase flow. In: Proceedings of 5th International Conference on Circulating Fluidized Bath, Beijing (1996)
135. Gonor, A.: High-pressure vaporization and boiling of condensed material: a generalized Clausius–Clapeyron equation. In: Furnish, M.D., Thadhani, N.N., Horie, Y. (eds.) Shock Compression of Condensed Matter – 2001, pp. 63–66. American Institute of Physics, Melville (2002)
136. Price, E.W.: Combustion of metalized propellants. Prog. Astronaut. Aeronaut. **90**, 479–513 (1984)
137. Elperin, I., Igra, O., Ben-Dor, G.: Analysis of normal shock wave in a carbon particle-laden oxygen gas. Trans. ASME **108**, 354–359 (1986)
138. Schlichting, H.: Boundary-Layer Theory, 7th edn. McGraw-Hill, New York (1979)
139. Bailey, A.B.: Sphere drag coefficient for supersonic speeds in continuum and free-molecule flows. J. Fluid Mech. **65**, 401–410 (1974)
140. Henderson, C.B.: Drag coefficient of spheres in continuum and rarefied flows. AIAA J. **14**, 707–708 (1976)
141. Igra, O., Takayama, K.: Shock tube study of the drag coefficient of a sphere in a nonstationary flow. In: Takayama K. (ed.) Proceeding of the 18th International Symposium of Shock Waves, pp. 491–497. Springer, Berlin (1991)
142. Sommerfeld, M., Decker, S.: On the importance of the Basset history term on the particle motion induced by a plane shock wave. In: Brun, R., Dumitrescu L.Z. (eds.) Proceedings of the 19th International Symposium of Shock Waves, pp. 37–42. Springer, Berlin (1995)
143. Smirnov, N.N.: Combustion and detonation in multi-phase media, initiation of detonation in dispersed film systems behind a shock wave. Int. J. Heat Mass Transfer **31**, 779–793 (1988)
144. Fox, T.W., Rackett, C.W., Nicholls, J.A.: Shock wave ignition of magnesium powders. In: Proceedings of the 11th International Symposium on Shock Waves and Tubes, Seattle, pp. 262–268 (1978)
145. Frost, D.L., Goroshin, S., Janidlo, S., Pryszlak, J., Levine, J., Zhang, F.: Fragmentation of reactive metallic particles during impact with a plate. In: Furnish, M.D., Gupta, Y.M., Forbes, J.W. (eds.) Shock Compression of Condensed Matter – 2003, pp. 451–454. American Institute of Physics, Melville (2004)
146. Grady, D.E.: Local inertial effects in dynamic fragmentation. J. Appl. Phys. **53**, 322–325 (1981)
147. Yew, C.H., Grady, D.E., Lawrence, R.J.: A simple model for debris clouds produced by hypervelocity particle impact. Int. J. Impact Eng. **14**, 851–862 (1993)

3

Slurry Detonation

D.L. Frost and F. Zhang

3.1 Introduction

Commercial explosives and blasting agents are composed of low-density heterogeneous mixtures of fuels, oxidizers, and other components, in contrast to high-density military explosives, which are generally mixtures of molecular explosives and binders. The amount of commercial explosives sold each year, which has been estimated by the US Bureau of Mines to be two million metric tons in the USA alone, represents the vast majority of all explosives used each year [1]. Commercial explosives are used primarily for mining, tunnel construction, and other specialized applications, and their use was revolutionized in the mid-1950s with the widespread use of cost-effective ammonium nitrate fuel oil (ANFO) explosives and further in the 1970s with the development of slurry and emulsion explosives [2], which are water-resistant, safe to handle, and in most cases contain no inherently explosive ingredients.

Commercial explosives are formulated to be insensitive and are usually oxygen-balanced to minimize the amount of the poisonous gases CO and NO_x that is produced during the detonation. They can be mixed on-site and usually require the addition of sensitizers to render them detonable. Current formulations are designed to maximize the work done by the expanding detonation products, in contrast to military explosives, which have higher detonation pressures and a superior ability to accelerate metal.

The variation in the formulations of slurry and emulsion explosives is almost limitless, depending on the manufacturer and the particular application. The exact formulation of many commercial explosives is proprietary, but much information on the properties and performance of commercial explosives is available in the open literature. The primary sources used for the present review include the proceedings of the Symposia of the International Society of Explosive Engineers, the International Symposia on Detonation, and the International Pyrotechnics Symposia, information circulars from the US Bureau of Mines, textbooks such as [3], and relevant journals such as *Propellants, Explosives, Pyrotechnics* and *Combustion, Explosion, and Shock Waves.*

F. Zhang (ed.), *Shock Wave Science and Technology Reference Library: Heterogeneous Detonation*, DOI: 10.1007/978-3-540-88447-7_3,

3.1.1 Definitions of Slurry Blasting Agents and Explosives

A blasting agent is defined in the explosives industry as a mixture of a fuel and an oxidizer, for which two requirements must be satisfied: (1) none of the ingredients are classified as an explosive, and (2) it cannot be initiated unconfined by a single blasting cap (it normally needs a booster charge for reliable initiation) [3]. Blasting agents that are rendered detonable with a blasting cap (i.e., cap-sensitive) by adding solid or hollow particles or chemical sensitizers are then classified as explosives. The most common commercial blasting agent, a mixture of about 94% prilled (i.e., in pellet form) ammonium nitrate with 6% fuel oil, denoted ANFO, can be made cap-sensitive by grinding the prills (initially typically about 2 mm in diameter) into very fine particles. Slurry blasting agents can be made cap-sensitive by adding flaked (paint-grade) aluminum powder.

A slurry refers, in general, to a liquid that contains very finely dispersed solid particles. Slurry blasting agents and explosives contain a mixture of oxidizers, fuels, and sensitizers. They typically contain 5–40% water (with an average of about 15%) [4] and hence are sometimes referred to as "water–gel slurries." Gelling and cross-linking agents are added to make the slurry resistant to water, which prevents the slurry components from leaching into surrounding groundwater. The water resistance of slurry explosives is one of their principle advantages over ANFO, which cannot be used in the presence of water owing to the hygroscopic nature of ammonium nitrate prills. Slurry explosives may be mixed on-site, or my be premixed and packaged in polyethylene bags for hand loading [4]. Currently about 95% of commercial explosives are delivered to the test site using bulk methods [5].

An emulsion is, in general, a mixture of two or more immiscible liquids which is stabilized by the addition of emulsifiers. Emulsion explosives typically are in the form of a "water-in-oil emulsion" consisting of fine droplets (typically 1–10 μm) of a highly concentrated salt solution (ammonium nitrate, sodium nitrate, or calcium nitrate) separated by thin films of fuel/emulsifier. Inverse "oil-in-water emulsion" formulations are also possible. Emulsion explosives have the consistency of common emulsions such as mayonnaise and may be prepared in cartridges or in a bulk, pumpable form. The emulsion matrix has no components that are explosive and requires the addition of sensitizers, such as hollow glass microballoons, fine particulates, or gas-forming chemical ingredients, to form a detonable mixture. Owing to their inherent insensitivity to detonation, emulsion explosives are among the safest commercial explosives to handle and also have good water resistance. Emulsion explosives often contain solid particle additives, such as aluminum powder, to increase the energy density of the mixture. In this case, emulsion explosions can be considered to be a subclass of slurry explosives.

Another class of slurry explosives can be formed by adding solid (or hollow) particles to a monomolecular liquid explosive. Typically inert or reactive solid particles have been added to insensitive liquid explosives such as nitromethane

(NM; CH_3NO_2) or isopropyl nitrate (IPN; $C_3H_7NO_3$) to form an oxygen-deficient explosive. With these heterogeneous mixtures, secondary combustion of the detonation products and metal particles with the surrounding air improves the blast performance. Recently, there has been renewed interest in these "enhanced-blast" explosive formulations for military applications [14].

3.1.2 Applications of Slurry Explosives

Commercial explosives are used in the following industries (listed in decreasing order of total explosive consumption in the USA [4]): coal mining, metal mining, railways (including tunnel construction) and construction, and quarrying and nonmetal mining. Dry blasting agents such as ANFO are the most widely used type of explosive and are used primarily for coal mining, owing to the dry blasting conditions and competitive cost. Slurry explosives are used extensively in metal mining, owing to the hard rock conditions and frequent presence of water. Taconite mining, in particular, accounts for a large fraction of the total consumption of slurry explosives [4]. The recent development of the oil-shale industry is also expected to use a significant amount of slurry explosive [6]. Commercial slurry explosives are also used in a large number of specialized applications, including building demolition, seismic tests, metal-forming technology, and explosive welding [2]. Fuel-rich metalized slurry explosives are used often in confined regions where afterburning produces sustained pressure impulses. These explosives are used in mining applications to improve the "heave" of the rock and for military applications requiring enhanced blast wave strengths (in this context these explosives are sometimes referred to as "thermobaric" explosives).

3.1.3 Historical Development of Slurry Explosives

Current commercial explosives can trace their origin to a patent for a blasting compound based on ammonium nitrate applied for in 1867 [2]. In 1935 the first actual blasting agent was introduced, commercially called "Nitramon," which consisted of a mixture of ammonium nitrate and a carbonaceous sensitizer [4]. An explosive using prilled ammonium nitrate was introduced in 1955 using a solid fuel, and over the next decade the solid fuels were replaced by liquid fuel oils, forming ANFO. The lack of water resistance of ANFO prompted research into the feasibility of water–gel and slurry explosives. Melvin A. Cook of the University of Utah is credited with the first development of a water-resistant slurry explosive in 1956 consisting of water, ammonium nitrate, and aluminum [2]. The first commercial use of slurry explosives was in 1959 in the Mesabi Range in Minnesota [4]. By 1960, the development and use of commercial slurry explosives was largely responsible for the collapse of the dynamite industry [6]. At this time, the early slurry explosives used were sensitized with aluminum and/or TNT powders and TNT prill. In 1962, research at Atlas in the USA led to the development of water-in-oil and oil-in-water emulsion slurries containing neither aluminum nor explosives [6]. These early emulsion explosives contained nitric acid as well as the primary component of ammonium nitrate and utilized air bubbles as a sensitizer. The range of slurry blasting agents and slurries available commercially expanded rapidly in the 1960s and 1970s as companies produced formulations with various explosive properties, sensitivities, and fluid consistencies [7].

3.1.4 Chapter Outline

In this chapter, the particular characteristics of detonation propagation in slurry explosives will be examined. The term "slurry explosive" is used in a general sense to include water–gel slurry explosives, emulsion slurry explosives, and fuel-rich (thermobaric) slurry formulations. In Sect. 3.2, the composition of slurry explosives is described, including the procedure for generating the slurry. This is followed by a review of the explosive performance of slurry explosives. Although many different tests are used to characterize the properties and performance of explosives, emphasis in this review will be placed on the properties of the detonation wave itself, including the detonation velocity, pressure, and failure diameter. Models for detonation propagation in slurry explosives will then be reviewed, followed by a summary of the current state of the art and possible future developments in slurry explosives.

3.2 Formulations of Slurry Explosives

3.2.1 Water–Gel and Emulsion Slurry Formulations

The typical components used in water–gel slurry explosives and emulsion explosives are shown in Table 3.1, together with their function within the mixture [1]. The essential components of the emulsion matrix are (with typical weight percentages used) as follows: an oxidizing agent (85%), water (10%), and fuel oil with an emulsifier (5%) [8]. The other items listed in Table 3.1 are added to sensitize the mixture or enhance the energy density. To increase the energy density of ANFO and benefit from the water resistance of emulsion explosives, ANFO and emulsion explosives are often mixed together, forming so-called heavy ANFO.

3.2.2 Procedure for Generating Water–Gel Slurry and Emulsion Explosives

Production of a water–gel slurry follows a two-step process [1]. First, liquid components are blended to form a liquid phase that comprises 30–60 wt% of the final product. Dry nitrate salts, aluminum powder (if desired), and gelling

Table 3.1. Ingredients in slurry explosives and their function [1]

Water–gel slurry explosives Ingredient	Use
Ammonium nitrate	Oxidizer
Sodium or calcium nitrate	Oxidizer
Ammonium perchlorate	Oxidizer, sensitizer
Glass microballoons	Sensitizer
Methylamine nitrate	Sensitizer, fuel
Hexamine nitrate	Sensitizer, fuel
Aluminum (paint grade)	Sensitizer, fuel
Aluminum (coarse)	Fuel, energy enhancer
Ethylene glycol	Fuel, fluidizer
Water	Fluidizer
Guar gum	Thickener, gelling agent
Emulsion (slurry) explosives Ingredient	**Use**
Ammonium nitrate	Oxidizer
Sodium or calcium nitrate	Oxidizer
Ammonium perchlorate	Oxidizer, sensitizer
Fuel or mineral oil	Fuel
Emulsifier	Stabilizer, fuel
Aluminum (granular)	Energy enhancer
Glass bubbles/chemical gassing	Sensitizer

agents are then mixed into the liquid phase. During the mixing process, air is entrained into the mixture until the desired density is reached. The mixture is then pumped into cylindrical packages.

Water-in-oil emulsion explosives are manufactured using high-speed mixing equipment. The oxidizer solution, fuel oil, and emulsifier are heated to temperatures typically between 80 and 120℃ [1, 9] and are mixed together with a high-speed mixer. The speed (900 rpm in [9]) and time for the mixing determines the resulting size of the dispersed ammonium nitrate solution droplets that are formed. Each droplet is surrounded by the continuous oil phase. Separation of the droplets inhibits the growth of ammonium nitrate crystals as the mixture is cooled. The addition of the emulsifier stabilizes the emulsion matrix. Glass microballoons or gas-forming agents are added to the emulsion matrix with a low-speed mixer to sensitize the mixture. Fine aluminum particles or flakes can be used in emulsions as a sensitizer and energy enhancer.

3.2.3 Fuel-Rich Slurry Explosive Formulations

The pioneering work on the detonation properties of liquid explosives with added particles was carried out using fuel-rich NM as the explosive with either glass or metal particles [10–13, 34]. To vary the mass fraction of the particles, the liquid was partially gelled with the addition of several percent of either guar gum or poly(methyl methacrylate). Another insensitive liquid explosive or monopropellant that is also oxygen deficient that has been studied in the past is IPN [66, 67]. Reactive metal particles that have been added include aluminum, magnesium, titanium, and zirconium. Typical formulations contain 20–30% liquid explosive, 60–70% metal powders, and 0–5% gelling agents. Additional oxidizers such as ammonium nitrate or ammonium perchlorate, or explosive powders (e.g., HMX) have also been added to the slurry mixture. With the above-mentioned components, a wide variety of different slurry formulations may be prepared. The effect of the formulation components on the performance of the slurry explosive was investigated by Hall and Knowlton [14] in a systematic study of fuel-rich slurry formulations.

3.3 Explosive Properties of Slurry Explosives

There are many ways to characterize the performance of an explosive. For commercial explosives, both the rock-breaking capability and the work done by the expanding product gases are important to various degrees, depending on the particular application. Historically, a number of different empirical methods have been developed to characterize the relative performance of explosives. Table 3.2 gives a partial list of tests that have been used in the past. The nature and the relative importance of these tests were described in detail in [3,15] and the reader is referred to these publications for a review of this topic.

The performance of an explosive in the tests listed in Table 3.2 will be related, although not directly, to the theoretical available chemical energy in the explosive. In the literature, there are several different measures of chemical energy associated with an explosive, depending on the particular thermodynamic process that is considered. For example, the explosion energy Q_v refers to the energy released in a constant-volume explosion. The detonation energy Q_d is the heat of reaction with reference to the chemical equilibrium

Table 3.2. Explosive performance test methods [3]

Ballistic mortar
Grade strength
Brisance
Trauzl lead block test
Plate dent test
Cylinder test
Underwater detonation test
Crater test
Langefors weight strength
Breaking index from underwater detonation testing

of the combustion products at the Chapman–Jouguet detonation point. For commercial explosives, efficient conversion of the chemical energy released into mechanical or expansion work is important. The expansion work W may be defined as [3]

$$W = \int_{v_1}^{v} p \, \mathrm{d}v - \frac{u^2}{2}, \tag{3.1}$$

where p, v, and u are the fluid pressure, specific volume, and velocity, respectively, and the index 1 indicates the state at the Chapman–Jouguet detonation point. If the reaction products expand all the way to atmospheric pressure adiabatically, the expansion work will be nearly equal to the detonation energy, with the difference corresponding to the thermal energy contained in the expanded detonation products.

3.3.1 Equilibrium Thermodynamics

Equilibrium thermodynamics calculations can be carried out to determine the equilibrium Chapman–Jouguet detonation point properties. The properties of the decomposition products can then be determined as the products expand isentropically to some final pressure or volume. The utility of equilibrium calculations for characterizing slurry explosives is limited owing to the intrinsic neglect of all nonequilibrium and scale effects. Nevertheless, they provide a qualitative comparison of the properties of different explosive formulations. To illustrate the relative equilibrium predictions of detonation properties, the Cheetah 2.0 code [16] was used to compute the properties for three generic slurry explosives (an emulsion explosive, an aluminized slurry explosive, and a thermobaric explosive), in comparison with a standard dry blasting agent (ANFO) and a military explosive (HMX).

Table 3.3 shows the predicted Chapman–Jouguet detonation velocity, pressure, temperature, and the total mechanical work done by the detonation products in expanding adiabatically to atmospheric pressure. The calculations were carried out using Cheetah's version of the Becker–Kistiakowsky–Wilson (BKW) equation of state, although this equation of state has not been tested with aluminized explosives [16].

The military explosive HMX generates the highest detonation velocity and pressure, whereas the thermobaric formulation has the highest predicted detonation temperature and potential mechanical work.

Figure 3.1 shows the energy released, relative to the energy of the reactants, as a function of the volume expansion ratio. The values shown by the dashed lines on the right correspond to the total mechanical work done, W, for expansion to atmospheric pressure. The energy release from HMX occurs with the smallest relative volume increase, whereas the thermobaric explosive requires the largest degree of expansion to produce the maximum expansion work done. Note that since thermobaric explosives are fuel-rich, if afterburning with air is included, the expansion work will be larger. The emulsion explosive and ANFO have relatively similar energetic profiles.

Table 3.3. Equilibrium detonation properties for various generic explosives

Explosive	Composition	ρ (g cm^{-3})	V_{CJ} (m s^{-1})	P_{CJ} (atm)	T_{CJ} (K)	E_{Mech} (kJ cm^{-3})
Blasting agent (ANFO)	94% AN 6% fuel oil	0.90	4,900	57,065	2,984	3.554
Emulsion explosive	80% AN 15% water 5% fuel oil	1.2	5,612	88,623	2,353	3.617
Aluminized slurry explosive	30% Al 50% AN 15% water 5% fuel oil	1.2	5,483	91,138	4,539	7.328
Thermobaric explosive	30% NM 30% Al 40% HMX	1.725	6,975	190,008	5,581	13.297
Military explosive (HMX)	100% HMX	1.905	9,301	388,533	4,113	11.133

AN ammonium nitrate, *ANFO* ammonium nitrate fuel oil, *CJ* Chapman–Jouguet, *NM* nitromethane

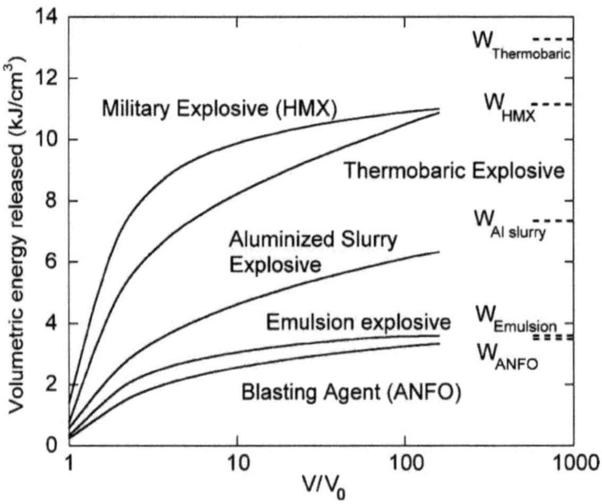

Fig. 3.1. Energy released as a function of volume expansion for a variety of generic explosives. In each case, total expansion work to atmospheric pressure W is shown as a *dashed line*. *ANFO* ammonium nitrate fuel oil

In the use of commercial explosives for rock blasting, all of the chemical energy released is not used for fracturing and displacing the rock. As the rock cracks, the hot gases escape to the surroundings. This occurs typically when the product gases have expanded by a factor of 10–20, which from

Fig. 3.1 corresponds to about 60–80% of the total mechanical work done by the explosive. For highly metalized explosives, the energy release predicted by the equilibrium calculations may also be reduced owing to the finite rate at which the aluminum oxidizes and transfers heat to the surrounding gases.

To compare the performance of different commercial explosive formulations, Persson et al. [3] have suggested that the following three values are most significant: (1) detonation energy Q_d, (2) shock wave and bubble pulse energy from underwater detonation tests, and (3) expansion work performed at $V/V_0 = 10$, 15, and 20. However, there is no agreement in the industry as to what standardized specifications should be provided. However, there is general agreement that correlating blasting performance with ideal equilibrium energy release can be misleading since other factors such as the charge diameter and the degree of confinement play an important role in the detonation properties [17]. A measure of the strength of an explosive which is useful for engineering purposes must take into account both the detonation properties of the explosive as well as the properties of the surrounding rock, and various empirical measures have been proposed [3]. More sophisticated measures of the strength of an explosive require realistic models of the detonation propagation, including models for the reaction rate of the energetic material, the equation of state of the reactants and products, and the influence of the curvature of the detonation front on the detonation velocity. Models for detonation propagation in slurry explosives are discussed in Sect. 3.3.2.

3.3.2 Detonation Velocity of Emulsion Explosives

Most commercial explosives exhibit nonideal detonation behavior in that the detonation velocity can be considerably lower than the equilibrium predicted value. In fact, the divergence of the detonation velocity from the ideal value is one measure of the nonideality of an explosive. The detonation velocity of an explosive is governed by the competing rates of chemical reaction behind the shock front and lateral expansion of the combustion products which reduces the temperature and pressure in the reaction zone. If the diameter of an explosive charge is much larger than the length of the chemical reaction zone in an explosive, and the charge is strongly confined, then the detonation velocity will be independent of the diameter. This is usually the case for military explosives, such as TNT and HMX, which have reaction zone lengths of less than 1 mm. However, commercial explosives may have reaction zone lengths greater than 10 mm [3] and hence the detonation velocity depends strongly on charge diameter, particularly for weakly confined charges.

The dependence of detonation velocity on inverse charge diameter is an important property of an explosive. Many of the reported detonation velocity data for commercial water–gel and emulsion slurry explosives are of limited utility since the explosive formulation was not reported for proprietary reasons. However, some detonation velocity data do exist for emulsion explosives.

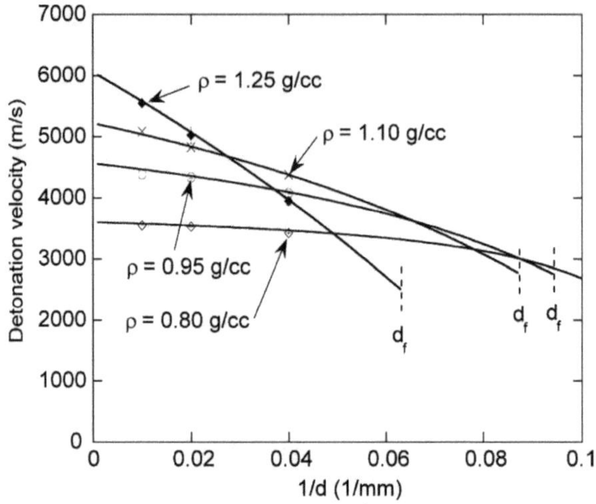

Fig. 3.2. Charge diameter effect for emulsion explosive as a function of emulsion density. *Vertical dashed lines* indicate failure diameter values [18]

For example, Lee and Persson [18] carried out a systematic study of the detonation properties of an emulsion explosive consisting of the following emulsion matrix: ammonium nitrate (66.91%), calcium nitrate (14.59%), water (12.0%), light mineral oil (5.0%), and an emulsifier (SPAN 80; 1.5%). They varied the density of the emulsion by varying the number of glass microballoons added to sensitize the mixture as well as the size of the microballoons. Figure 3.2 shows the dependence of detonation velocity on inverse diameter for this emulsion explosive with various numbers of unsieved microballoons. The velocity values were measured with ionization pins in cylindrical charges as well as with resistance probes in conical charges. Both types of data were then used to fit the data to the following function (following [20]):

$$D = D_{\mathrm{i}}\left(1 - \frac{A}{d - d_{\mathrm{c}}}\right), \tag{3.2}$$

where D is the detonation velocity, d is the charge diameter, the subscript i refers to infinite diameter, and A and d_{c} are fitting constants. As the density of the emulsion increases, the diameter effect curves become steeper, which corresponds to an increase in the fitting constant A, and the failure diameter for the explosive increases.

From dimensional reasoning, previous investigators (e.g., [21]) have speculated that the constant A in the above fitting relation is related to the 1D chemical reaction zone length, δ, of the explosive. Engelke et al. [22] noted that there cannot be a direct correspondence between A and δ since the constant A depends strongly on the confinement of the charge. However, they found that for two closely related explosives with the same confinement, the ratio of the A values is a measure of the ratio of the δ values for the two explosives. From the curve fits for the emulsion explosive, the A values varied from 0.66 mm for a density of $0.80\,\mathrm{g\,cm^{-3}}$ to 7.637 mm for a density of $1.25\,\mathrm{g\,cm^{-3}}$ [18]. Hence, from [22], the reaction zone length in the emulsion explosive increases by about an order of magnitude when the density increases by 50%, from 0.80 to $1.25\,\mathrm{g\,cm^{-3}}$.

From Fig. 3.2, for large diameters, the diameter effect curves are nearly linear and become more concave near the diameter at which the detonation fails. As the density is increased by reducing the number of microballoons, the curves become steeper and the failure diameter also increases. This is consistent with the discussion in the previous paragraph regarding the parameter A, i.e., as the density is increased, the reaction zone length increases, and hence the sensitivity is reduced and the failure diameter increases. According to the classification of explosives by Price [23], the behavior of the emulsion explosive is typical of group 2 explosives (such as ANFO) and contrasts with that of group 1 explosives (typically single-molecule explosives), in which the reaction zone length and failure diameter decrease with increasing density.

The variation of detonation velocity with density for a given charge diameter of the emulsion explosive exhibits a maximum, as illustrated in Fig. 3.3. At low densities, the effect of density on the detonation velocity dominates and the detonation velocity increases with increasing density. However, at some point, the increase in the reaction zone thickness with increasing density becomes the dominant influence on the detonation velocity and the velocity reaches a maximum and begins to decrease.

The effect of the size of the microballoons on the detonation velocity of the emulsion explosive is shown in Fig. 3.4. For low-density emulsions (i.e., high

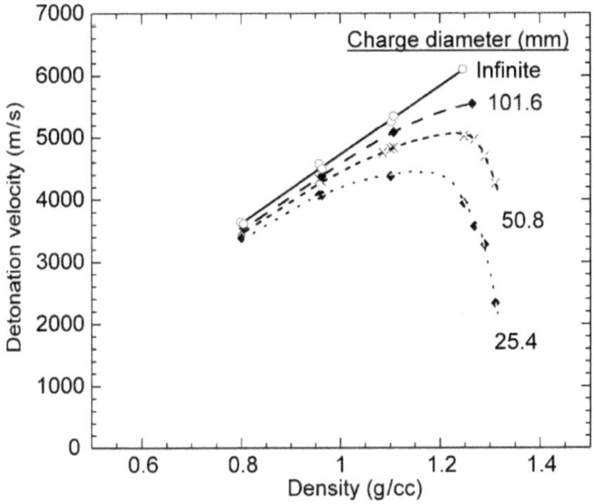

Fig. 3.3. Effect of density on detonation velocity for an emulsion explosive [18]

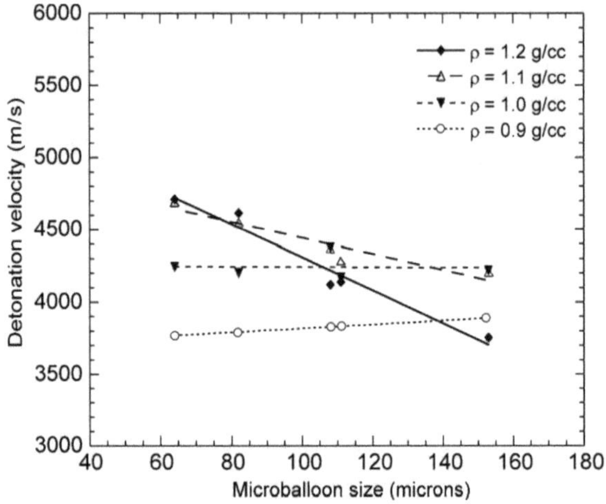

Fig. 3.4. Effect of microballoon size on detonation velocity of an emulsion explosive [18]

mass fraction of microballoons), increasing the microballoon size increases the detonation velocity; however, for high-density emulsions, the opposite trend is observed.

The composition of the oxidizer used in an emulsion explosive influences the detonation velocity. In the study described above, the oxidizer consisted of a mixture of 66.9% ammonium nitrate and 14.6% calcium nitrate. In a related study by the same research group [30], they used only ammonium nitrate (77%) as the oxidizer. In contrast, Yoshida et al. [19] studied the detonation behavior of an emulsion matrix containing 72% ammonium nitrate and 10% sodium nitrate. Lee et al. [30] compared the effect of the oxidizer composition on the detonation velocity scaled with the predicted velocity using the BKW equation of state as a function of density (the maximum density corresponds to the emulsion matrix without microballoons) (Fig. 3.5). They speculated that the lower velocities for the ammonium nitrate/calcium nitrate and ammonium nitrate/sodium nitrate mixtures are due to the slow rate of decomposition of the metallic salts calcium nitrate and sodium nitrate.

Recently, Anshits et al. [24] carried out a similar study with a different emulsion formulation consisting of ammonium nitrate (76.9%), industrial oil (6.9%), water (15.2%), and an emulsifier (1%). Rather than using commercial microballoons to sensitize the mixture, they used hollow cenospheres obtained from coal ash. The cenospheres ranged in size from 50 to 500 μm, with a peak in the distribution around 200 μm, which is larger than commercial microballoons, which typically range from 30 to 70 μm. They found a similar dependence of the detonation velocity on emulsion density as in [18], although with the larger cenospheres the critical diameter was about twice as large as with the use of glass microballoons.

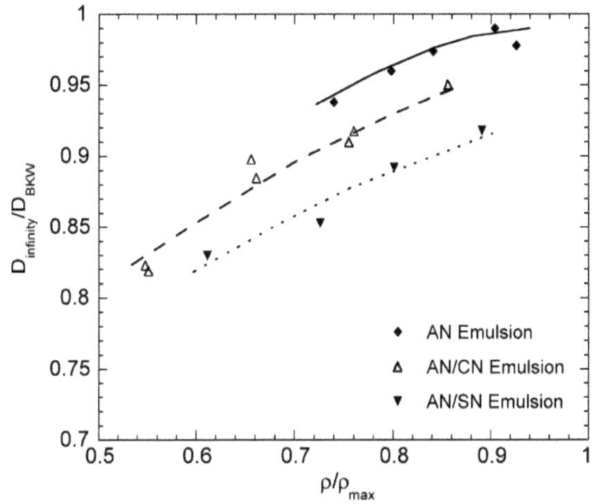

Fig. 3.5. Effect of oxidizer on velocity in emulsion explosives [30]. *AN* ammonium nitrate, *CN* calcium nitrate, *SN* sodium nitrate

3.3.3 Hot-Spot Sensitization in Emulsion Explosives

Many of the aforementioned features of emulsion explosives can be explained by examining the role of the microballoons in the detonation propagation. Since each microballoon acts as a hot spot, the number and the size of the microballoons have a strong influence on the detonation propagation. As the number of microballoons in the emulsion explosive increases, the density of hot spots increases and the explosive becomes more sensitive, corresponding to a decrease in the chemical reaction zone thickness and the failure diameter. This is consistent with the trends observed in the decrease in the slope of the diameter-effect curves (Fig. 3.2) that occurs with a reduction in the density of the emulsion. The effect of microballoon size on detonation velocity can also be interpreted in terms of hot spots [18]. For a given emulsion density, as the size of the microballoons decreases, the number density increases, but dilution of the explosive by the inert microballoon material increases (since the effective density of a small microballoon is larger than that of a large microballoon). At high emulsion densities, the effect of the decrease in the reaction zone length as the microballoon size decreases dominates, whereas at low emulsion densities, the dilution effect dominates.

Hirosaki et al. [38] carried out an extensive study on the effects of void size and volume on the detonation properties of emulsion explosives, including measurements of detonation velocity, pressure, and critical diameter. The oxygen-balanced emulsion matrix they used was composed of ammonium nitrate/sodium nitrate/water/wax and emulsifier (77.66:4.68:11.22:5.40 by weight) and had a density of 1,390 kg m^{-3}. Unlike previous studies that used glass microballoons with sizes typically smaller than 0.15 mm, their study sensitized the emulsion explosive with plastic balloons ranging from 0.05 to

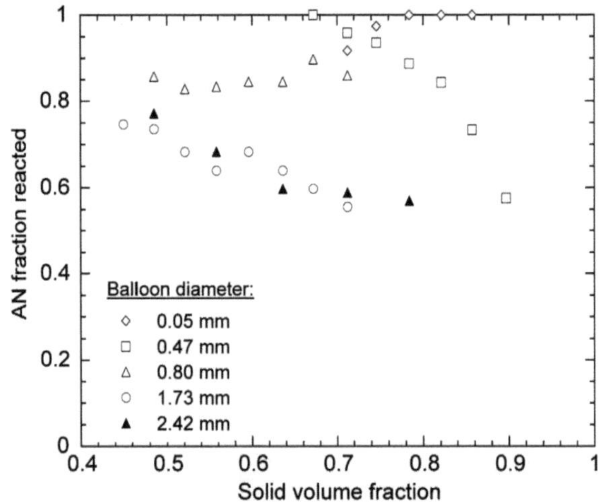

Fig. 3.6. Dependence of the fraction of AN that reacted on the solid volume fraction of an emulsion explosive [38]

2.42 mm in diameter. They compared the measured detonation velocities, extrapolated to infinite diameter, with the theoretical detonation velocities calculated with the KHT hydrothermodynamic code, with the plastic balloons treated as reactive in the calculations. From the comparison, they estimated the fraction of ammonium nitrate that reacted, assuming the other components to be completely reactive. The dependence of the fraction of ammonium nitrate that reacted on the solid volume fraction (or $1-\phi$, with ϕ being the void volume fraction) is shown in Fig. 3.6. Hirosaki et al. [38] speculated that the incomplete reaction with the large voids is due to rarefaction waves not only from the lateral expansion, but also from the void itself. They observed a linear relationship between the critical charge diameter and the thickness of bulk explosive between void interfaces. Since the intervoid distance is related to the inverse of the cube root of the void volume fraction, the critical diameter will be proportional to $1/\phi^{1/3}$, as shown in Fig. 3.7. Hirosaki et al. [38] concluded that the reaction in the emulsion explosive proceeds by hot-spot initiation and grain burning processes so that the reaction rate at the void surface and that in the bulk explosive influence the detonation propagation and failure.

3.3.4 Impact Sensitivity of Emulsion Explosives

Water–gel and emulsion slurry explosives are relatively insensitive to impact in comparison with monomolecular explosives. However, only a small number of systematic studies have been carried out on the shock sensitivity of slurry explosives. For example, Persson et al. [25] investigated the sensitivity

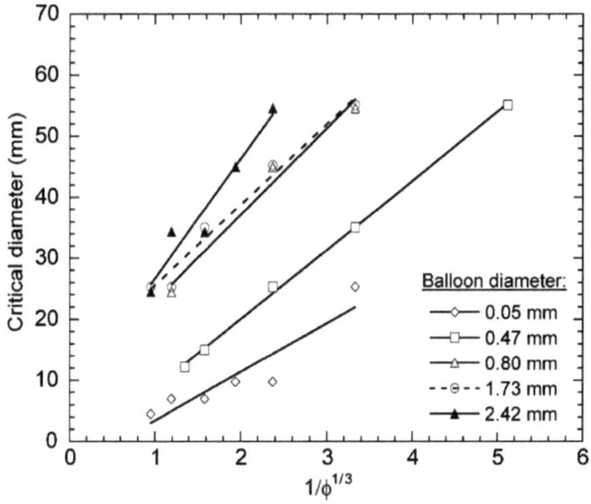

Fig. 3.7. Dependence of the critical diameter of an emulsion explosive on the inverse cube root of the volume void fraction [38]

of water–gel explosives to weak shocks. Mohan et al. [26] studied the impact of thin layers of emulsion explosives photographically. They observed crystallization effects in the layer during impact, but did not observe the initiation of an explosion. The impact of a slurry explosive was investigated numerically by Feng and Hanasaki [27]. The micromechanics of the collapse of microballoons in an emulsion explosion is closely related to detonation initiation, and this has been addressed by several investigators (including [28, 29]). Depending on the size of the microballoons and the initiating shock pressure, different phenomena may play a role, e.g., viscoplastic heating in glass and gas during symmetrical bubble collapse, asymmetric bubble collapse, and jet formation.

Lee et al. [30] carried out a study of the shock initiation properties of an emulsion explosive (consisting of 77% ammonium nitrate, 16% water, 6% oil, and 1% emulsifier) using a conventional wedge test. They found that the shock sensitivity of the emulsion explosive was less than that of cast TNT. The Pop plot (i.e., initial shock pressure versus the run distance to detonation) was quite flat, with a slight upwards concavity, so the run distance to detonation was predicted to be very long for pressures below 8 GPa for an emulsion density of 1.248 g cm^{-3}. They estimated that the emulsion matrix material would not detonate at charge diameters less than 378 mm without the addition of a sensitizing agent. Using the experimental shock and particle velocity data (U_s, u_p) with the Mie–Grüneisen equation of state, they obtained the shock Hugoniot of the emulsion matrix to be

$$U_s = 2.55 + 2.01 u_p. \tag{3.3}$$

They concluded the paper by noting that it is very difficult to find useful and reproducible data for commercial explosives. Section 3.3.5 reviews experimental data that are available for a well-defined model slurry explosive, and provides further insight into detonation propagation in heterogeneous liquid/particle mixtures.

3.3.5 Nitromethane Slurry Explosives

A particular model slurry explosive that has received considerable attention in the past consists of a heterogeneous mixture of the liquid explosive NM with the addition of either solid [10] or hollow [31] particles. Owing to the extensive research carried out on the detonation properties of pure NM, a heterogeneous mixture of NM and particles is a particularly convenient system to illustrate the relative effects of both chemical and physical sensitizers. The addition of the particles generates hot spots and this sensitization effect has been observed for a range of different particle materials and particle densities for charges confined within long glass tubes (0.6–1.3-mm thick) [32]. An example of the data obtained from the latter paper is shown in Fig. 3.8, which shows the dependence of detonation failure diameter on solid mass fraction for two different additives, 7 μm alumina particles and 1 μm tungsten powder. Addition of the particles sensitizes the mixture, reducing the failure diameter up to an addition of 30–40% particles by mass. With a further increase in solid mass fraction, the failure diameter begins to rise as the dilution effect of adding the inert solid material to the explosive counterbalances the sensitizing effect of the particles.

The significant effect of a small number of particles on the detonation properties of NM was demonstrated in the meticulous experiments of Engelke and Bdzil [11–13]. Engelke [11] added 6% silica powder (with sizes ranging

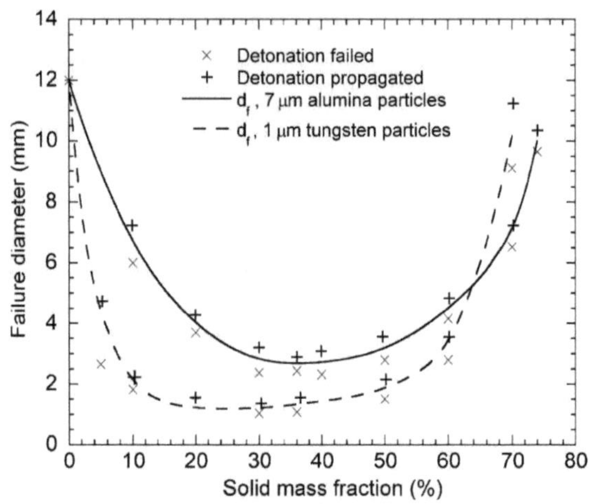

Fig. 3.8. Failure diameter as a function of solid mass fraction for mixtures of nitromethane (NM) and particles contained in long glass tubes [32]

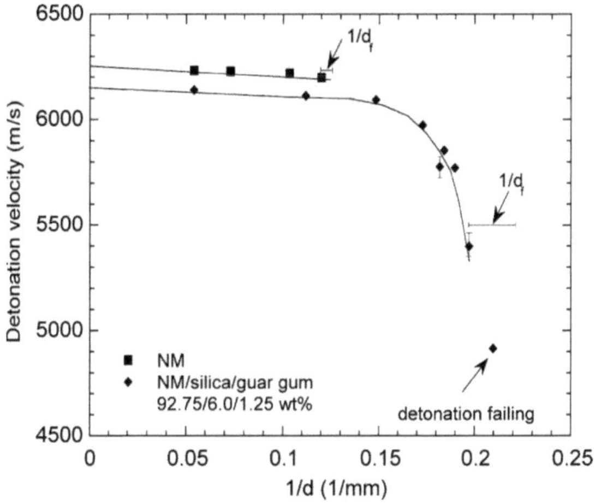

Fig. 3.9. Diameter effect curves for pure NM and NM with 6% silica powder added. The failure diameters, d_f, are noted for the two cases with the associated *error bars* obtained from half the difference of the internal diameters of the largest stick which failed and the smallest stick which propagated [11]

from 5 to 75 μm, with a peak in the distribution at 15 μm) to NM and found that the diameter-effect curve became sharply concave downwards just prior to failure of the detonation, as shown in Fig. 3.9.

Engelke attributed the change in the diameter-effect curve with the addition of particles to sensitization of the explosive by the generation of hot spots due to either shock reflections or material stagnation at the particles. Engelke and Bdzil [12] used the velocity data and solved the inverse problem to infer the chemical heat released in the reaction zone for both the homogeneous case and the heterogeneous case.

Engelke [13] also investigated the effect of the number density of heterogeneities on the failure diameter of liquid NM (with the addition of 1.25% by weight of guar gum to increase the viscosity of the NM). He dispersed small numbers of glass beads with sizes of 1–4, 35–45, and 105–125 μm in the NM/guar explosive mixture. His experimental results showed that the addition of the 35–45- and 105–125-μm-diameter heterogeneities in amounts up to 9.0 wt% produced no failure diameter reduction. However, addition of 1–4-μm-diameter beads in amounts as low as 0.5 wt% produced a failure diameter reduction. The effect was enhanced as more 1–4-μm-diameter heterogeneities were added up to, at least, 9.0 wt%. A failure diameter reduction of about 40% at both the 3.0 and the 9.0 wt% levels was observed.

Kato et al. [33–36] conducted a series of experiments to determine the detonation characteristics of aluminized NM comprising aluminum particles with a 10-μm mean diameter suspended in a mixture of 97% NM with 3% poly(methyl methacrylate) added as a gelling agent. They investigated the detonation velocity as a function of charge diameter and the effect of aluminum addition on the detonation sensitivity and on the brightness temperature of the detonation products. They found that the addition of aluminum particles decreased the detonation velocity relative to pure NM. Owing to the narrow reaction zone in NM (the most finely resolved measurements of the reaction zone in NM, by Sheffield et al. [39], indicate that early, fast reactions occur over a time of 5–10 ns) the amount of heat transfer to the aluminum particles within the reaction zone is insignificant; hence, the aluminum particles are effectively chemically inert within the detonation zone. Secondly, optical pyrometer measurements showed that the aluminum particles started to react with the gaseous products produced by the detonation of the liquid explosive after less than 1 μs. The brightness temperature at time zero (i.e., the Chapman–Jouguet temperature) decreased with increasing aluminum concentration, whereas the maximum brightness temperature increased with the aluminum concentration. They estimated that the duration of the aluminum reaction exceeded 2 μs for aluminum particles of mean diameter 10 μm. In comparison, using a conventional cylinder test, Moulard et al. [37] found that the duration of aluminum reactions was about 9 μs for the case of aluminum particles with a mean diamter of 5 μm. More recent tests by Kato et al. [48] in which they measured the detonation pressure in NM/aluminum mixtures give additional information about particle reaction times and will be described later in this section.

Lee et al. [40–42] investigated the detonation of a heterogeneous mixture consisting of sensitized liquid NM in a packed bed of inert spherical glass beads. The average propagation velocity of a detonation wave through this heterogeneous mixture is less than the detonation velocity of the liquid explosive itself but significantly in excess (approximately 50%) of the ideal equilibrium Chapman–Jouguet predictions based on full thermal, mechanical, as well as chemical equilibrium. They found that for a packed bed of inert monodisperse spherical beads saturated with sensitized NM, the dependence of failure diameter on bead size exhibited two distinct regimes, with an abrupt transition between the regimes, as shown in Fig. 3.10. In regime I, for bead diameters of 1 mm or greater [for NM sensitized with 15% diethylenetriamine (DETA)], the failure diameter decreases (i.e., the explosive becomes more sensitive) as the bead size increases. For small beads, in regime II, the opposite behavior occurs and the mixture becomes more sensitive as the bead size decreases. Figure 3.10 also shows that the λ-shaped dependence of the failure diameter with bead size shifts if the chemical sensitivity of the NM is altered from 10 to 15% sensitization levels of NM with DETA.

The dependence of detonation velocity on inverse charge diameter was linear for all bead sizes, except for bead diameters of 0.93 and 1.02 mm, which are near the bead diameter of 1.1 mm ± 0.1 mm at which there is the abrupt transition in failure diameter (see Fig. 3.10). In this case, the velocity varied between 3.0 and 4.5 km s^{-1} from one trial to the next, with little correlation with charge diameter [42]. These two types of behavior are illustrated in Fig. 3.11, which shows the detonation velocity as a function of inverse diameter for bead diameters of 0.34 and 1.02 mm.

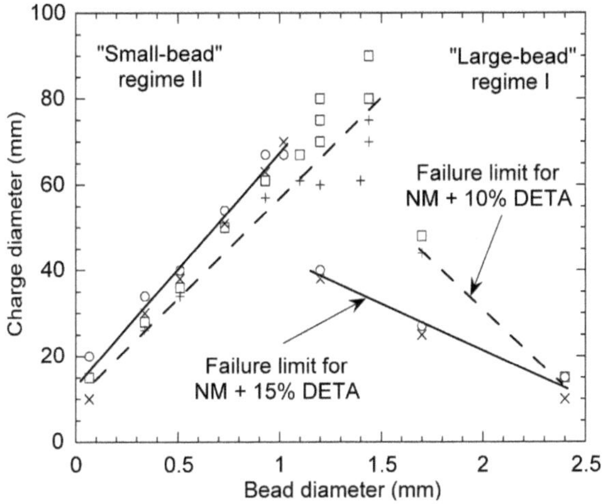

Fig. 3.10. Effect of degree of chemical sensitization on the failure dependence for NM/glass bead mixtures. *DETA* diethylenetriamine. (Data from [41] and D.L. Frost, H. Kleine, and S. Janidlo, unpublished results, 2005)

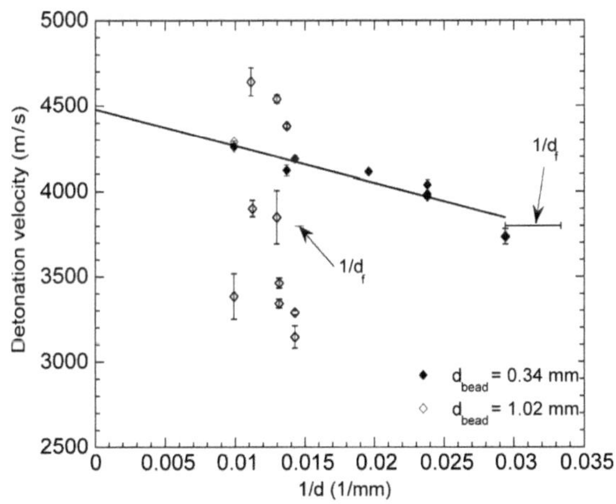

Fig. 3.11. Diameter-effect data for NM + 15% diethylenetriamine for glass beads with diameters of 0.34 and 1.02 mm [42]

As noted above, except for the case of bead diameters near the transition point (1.1 mm), the diameter-effect curves for the heterogeneous explosive exhibit a linear decrease in velocity to the failure point. Figure 3.12 shows linear curve fits to the diameter-effect data for bead sizes ranging from 0.066 to 1.7 mm, plotted in nondimensional coordinates after scaling with the charge failure diameter d_f and extrapolated infinite charge diameter velocity D_∞. The dimensional values of the extrapolated detonation velocities at infinite

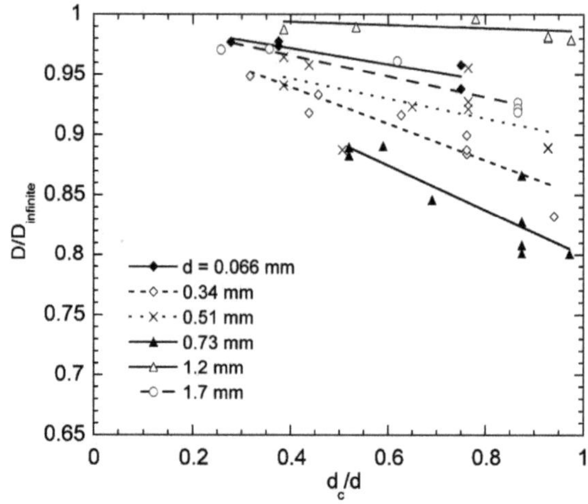

Fig. 3.12. Diameter-effect data for NM + 15% DETA with glass beads ranging in diameter from 66 µm to 1.7 mm [42]

Table 3.4. Extrapolated detonation velocity at infinite diameter and at failure for NM + 15% DETA with glass beads [42]

Bead size (mm)	D_∞ ($\mathrm{km\,s^{-1}}$)	D_f ($\mathrm{km\,s^{-1}}$)	Velocity deficit (%)
0.066	4.24 ± 0.04	3.91 ± 0.22	8
0.34	4.49 ± 0.07	3.81 ± 0.18	15
0.51	4.30 ± 0.09	3.85 ± 0.21	10
0.73	4.61 ± 0.08	3.69 ± 0.17	19
0.93	–	–	–
1.02	–	–	–
1.2	4.49 ± 0.08	4.43 ± 0.10	1
1.7	4.63 ± 0.05	4.23 ± 0.11	9
2.4	4.53 ± 0.01	–	–

diameter (D_∞) and at failure (D_f) are shown in Table 3.4 as a function of bead diameter. Although the diameter-effect curves are nearly linear as for an ideal explosive, the detonation velocity deficit at failure varies from 1 to 19%, which is considerably higher than the velocity deficit at failure for pure NM, which is on the order of 2%. The linear diameter effect near detonation failure differs from the sharp drop in detonation velocity near failure observed by Engelke [11] for small numbers of fine (1–4-µm) glass particles added to NM. Lee [42] speculated that this linear velocity–diameter behavior for relatively large beads may be due to a relatively weak hot-spot effect for the large glass beads.

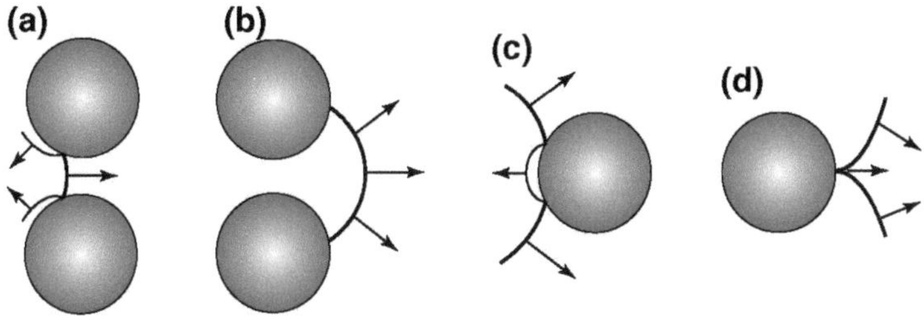

Fig. 3.13. Local detonation wavelet propagation in the "large-bead" regime, illustrating wave diffraction between beads (**a**, **b**), wave–particle interaction (**c**), and wave collision (**d**) (adapted from [42])

On the basis of their experimental data, Lee et al. [40] hypothesized that two distinct propagation mechanisms exist in the liquid/particle mixture, depending on the particle size. The heterogeneous explosive is most insensitive (i.e., the failure diameter reaches a maximum) when the bead diameter is a critical value, which is on the order of the failure diameter for the sensitized NM (which is approximately 1 mm). For larger beads, they postulated that detonation wavelets can successfully propagate in the liquid explosive between the beads, shown schematically in Fig. 3.13 [42]. For pore sizes on the order of the failure diameter, there may also be local failure and reinitiation events due to shock/bead interactions.

The global detonation front thus consists of a series of wavelets that propagate in winding paths through the geometric irregularities of the pores. In this "large-bead" regime, as the bead size increases, diffraction effects become less severe and hence the failure diameter of the mixture decreases.

For bead diameters less than the critical value, the detonation fails to propagate around the beads, but instead the shock propagation through the beads is sufficient to initiate the explosion of the liquid in the interstitial pores. In turn, the explosions in the liquid explosive generate more shocks in the neighboring inert beads ahead. This results in sustained "sympathetic" detonation propagation through the heterogeneous explosive medium, shown schematically in Fig. 3.14 [42].

In this "small-bead" regime, as the bead size decreases, the density of artificial hot spots associated with the beads increases and the failure diameter decreases. The average detonation velocity in the large-bead regime is typically larger than that for the small-bead regime.

The above phenomenological description of two propagation regimes is consistent with the dependence of failure diameter on the sensitivity of the liquid explosive. Card gap tests have shown that the sensitivity of NM is maximized with the addition of 15% DETA [51]. In the large-bead regime, it is expected that the failure of the wave will be dominated by local diffraction effects. When the sensitivity of the NM is increased by increasing the amount of DETA from 10 to 15%, the chemical reaction zone thickness decreases and

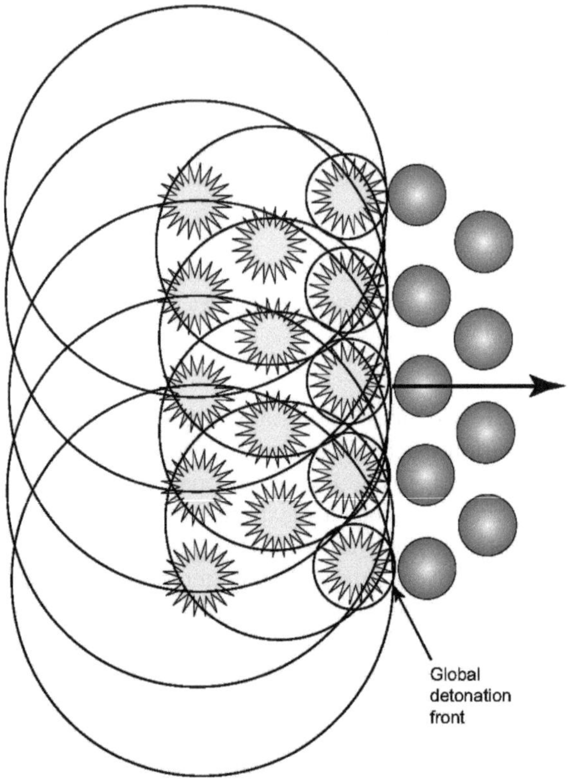

Fig. 3.14. Sympathetic wave propagation in the "small-bead" regime, in which the detonation wave is formed from the locus of shock waves from explosions of individual pockets of liquid explosive between beads (adapted from [42])

hence more severe diffraction will be required to fail the wave, implying that the failure curve shifts down. In the small-bead regime, the propagation is dominated by hot spots generated by the beads and failure of the detonation will occur owing to global curvature of the wavefront. In this regime, the amount of sensitizer plays the role of a diluent and does not directly control the failure mechanism. Hence, increasing the amount of sensitizer dilutes the NM and actually causes the failure curve to shift up slightly, indicating a small decrease in overall sensitivity of the mixture.

The particle material influences the failure diameter in both of the propagation regimes. Figure 3.15 shows the λ-shaped failure diameter curves for both glass and steel beads. In this case, the tests were carried out using the less sensitive explosive mixture NM + 10% triethylamine (TEA), which has a failure diameter of about 2.6 mm with glass confinement as compared with about 1 mm for NM + 15% DETA [43]. In this trial series (D.L. Frost, H. Kleine, and S. Janidlo, unpublished results, 2005) about 70 experiments

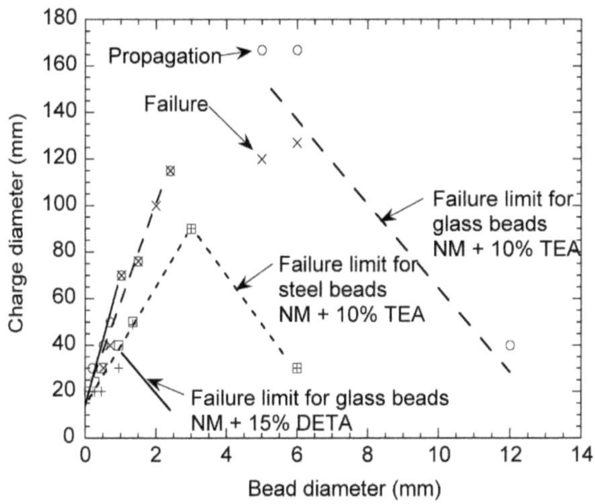

Fig. 3.15. Effect of bead material on failure diameter for a NM + 10% triethylamine (*TEA*) mixture

were carried out; only the cases immediately above and below detonation failure are shown in Fig. 3.15 for clarity. For glass beads, the transition between regime I and regime II shifts to a larger bead diameter in comparison with the results for NM + 15% DETA, consistent with the increase in failure diameter of the liquid alone. For steel beads, the failure diameter is smaller for both propagation regimes. Changing the bead material from glass to steel increases the shock impedance mismatch between the liquid explosive and the beads, which reduces the diffraction losses for propagation in regime I. The steel beads also provide a stronger effective confinement of the multiphase mixture, which sensitizes the mixture in propagation regime II. To resolve the effect of bead material on the wave interactions within the heterogeneous explosive, detailed multidimensional mesoscale modeling is necessary, which will be discussed further in Sect. 4.4.

Haskins et al. [52] reported experimental results on the detonation characteristics of *neat* NM containing high volume percentages of small spherical glass beads and aluminum particles. These mixes were found to detonate at diameters less than that of the pure liquid explosive partly owing to the hot-spot sensitization effect of solid particles. The detonation velocity of such mixtures again was less than that of NM alone, but was higher than would be expected if the particles and detonation products were always in equilibrium in the detonation process. The authors suggested that the shock velocity within the solid particles also plays a role in determining the detonation velocity.

In general, for the particle sizes that were typically used (10–100 μm) in the above-mentioned experiments, the time scale of the ignition and combustion of the aluminum particles was much longer than that of the explosive itself [35]. Therefore, within the detonation zone of the explosive, the transfer of heat and momentum to the particles actually serves to reduce the detonation velocity and pressure. However, the later burning of the particles in the expanding detonation products increases the temperature of the product gases and the energy available to do expansion work, which is important for applications such as the commercial blasting of rock and underwater explosions.

The characteristics of the particle reaction in NM/magnesium and NM/aluminum slurry explosives in cylindrical charges were investigated by Frost et al. [44, 45]. They observed three particle reaction regimes: no reaction, delayed reaction, and prompt reaction. The charge diameter above which prompt particle reaction occurred, denoted "critical diameter for particle ignition" (CDPI), depends on the competition between particle heating from the detonation products and expansion cooling of the products. This critical diameter depends on the particle diameter, d_p, as well as the particle material. For magnesium particles, the critical diameter was found experimentally to be a strong function of the particle diameter, i.e., CDPI $\sim {d_\mathrm{p}}^{1.75}$. A similar dependence was obtained by Tanguay et al. [46] using a simple analytical model. For aluminum particles, a more complex dependence of particle reaction on particle and charge diameter is observed [45]. For a given size of aluminum particle, the transition from no particle reaction to prompt particle reaction occurs as the charge diameter is increased. A transitional regime occurs in which ignition of the particles occurs at isolated spots or rings within the conically expanding particle cloud. Examples of the subcritical (i.e., no ignition), critical, and supercritical regimes are shown in Fig. 3.16. The particle reaction regime is also a function of the cylinder material, which determines the confinement of the mixture and rate of expansion of the products [47]. The reaction rate of the particles together with the particle dynamics have a strong effect on the strength of the blast wave from the heterogeneous explosive [50].

The distance between the bright detonation front and the start of the bright luminosity in the particle cloud is a measure of the apparent ignition delay of the particles. For small particles, this gap is quite short. For example, Fig. 3.17 shows photographs (the time between frames is 50 μs) of the propagation of a detonation through a mixture of nanoscale aluminum powder(Alex aluminum from Argonide, with a particle size of 100–200 nm) and NM (sensitized with 10% TEA) in a 41 mm inner diameter glass tube. The short gap between the bright detonation front and the start of the luminosity in the particle cloud is about 2.6 cm. To estimate the ignition delay time corresponding to this gap, we need an estimate of the aluminum particle velocity behind the detonation front. Given the small size of the particles, it is likely that the aluminum velocity is close to the fluid particle velocity behind the front, which can be estimated from an equilibrium Chapman–Jouguet calculation of the detonation properties. Using the Cheetah 2.0 code [16] for a mixture of 25% aluminum (inert) and 75% NM (sensitized with 10% TEA) gives a detonation velocity of $V_\mathrm{CJ} = 5.78\,\mathrm{km\,s^{-1}}$ and a particle velocity of $u_\mathrm{p} = 1.55\,\mathrm{km\,s^{-1}}$. The ignition delay time, t_ignition, will then be given by $L/(V_\mathrm{CJ} - u_\mathrm{p})$, or about 6 μs.

Fig. 3.16. Ignition regimes for cylindrical charges containing sensitized NM and aluminum particles. Conditions from *top* to *bottom*: subcritical case for the dispersion of H-95 particles (114 ± 40 µm) in a 19 mm inner diameter (id) tube, 114 µs between frames; critical case for H-15 particles (20 ± 10 µm) in a 48.5 mm id tube, 86 µs between frames; critical case for H-50 particles (54 ± 21 µm) in a 34 mm id tube, 57 µs between frames; supercritical case for H-50 particles in a 74 mm id tube, 80 µs between frames

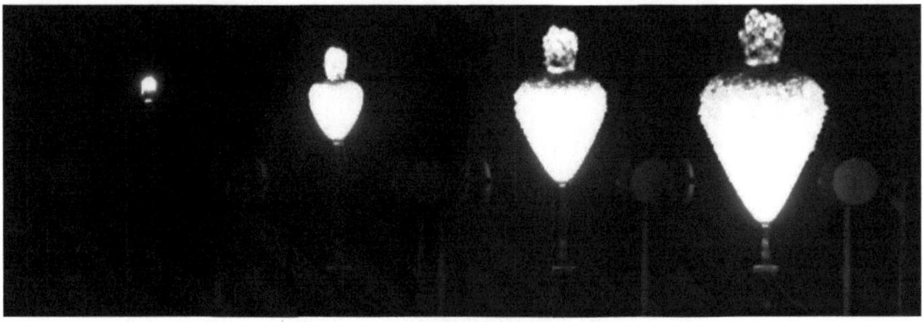

Fig. 3.17. Detonation propagation in a mixture of Alex aluminum (100–200-nm-diameter particles) in a 41-mm-id tube with NM + 10% TEA (mixture 25% aluminum by weight). Time between frames, 50 μs. The distance between the detonation front and the luminous particle cloud of about 2.6 cm corresponds to a delay time for particle reaction of about 6 μs

Fig. 3.18. Detonation propagation in mixture of H-2 aluminum (3-μm-diameter particles) in a 64-mm-id tube with NM + 10% TEA (mixture 65% aluminum by weight). Time between frames, 50 μs. The distance between the detonation front and the luminous particle cloud of about 6.5 cm corresponds to a reaction onset time of about 15 μs

This visual ignition delay will overestimate the actual ignition delay time of the particles if they start to react in the center of the charge and the visibility of the burning particles is obscured by soot (or glass fragments) at the edge of the particle cloud.

As the particle size increases, the visual distance between the detonation front and the initiation of particle reaction increases. For example, Fig. 3.18 shows the detonation propagation in a 64 mm inner diameter glass tube containing H-2 aluminum particles (3.0 ± 1.5 μm) with sensitized NM. In this case, with a similar calculation as above, the average distance between the detonation front and the start of the luminosity in the conically shaped particle cloud (approximately 6.5 cm) corresponds to a delay time of about 15 μs.

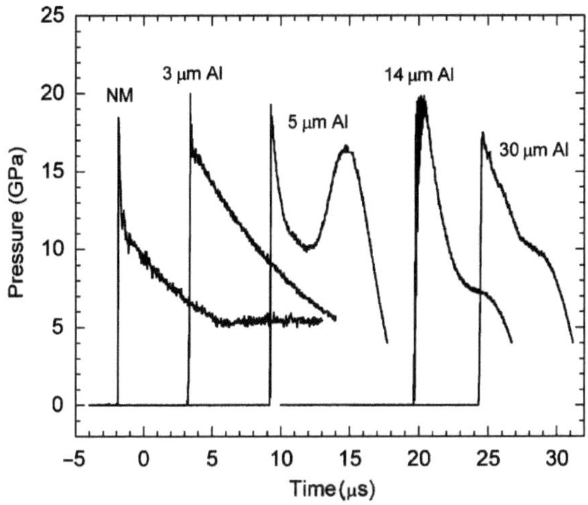

Fig. 3.19. Pressure–time profile measured with a poly(vinylidene difluoride) pressure gauge at the end of a steel-cased cylindrical charge containing NM alone (on the *left*) or NM with aluminum particles with diameters of 3, 5, 14, and 30 μm (the aluminum mass fraction in the NM/aluminum mixtures was 57, 60, 73, and 75%, respectively) [49]

For H-50 particles (54 μm), the distance between the detonation front and the luminous particle zone (approximately 10 cm) from the last frame in Fig. 3.16 corresponds to an upper bound for the delay time for onset of particle reaction of about 25 μs.

In a study of the detonation properties of packed beds of aluminum particles saturated with NM in poly(vinyl chloride) and steel tubes, Kato et al. [48,49] measured the pressure history at the end of the cylindrical charges with the use of a poly(vinylidene difluoride) pressure gauge. Figure 3.19 shows the pressure recorded for charges containing NM alone or NM with aluminum particles with diameters of 3, 5, 14, and 30 μm. The charges were confined in steel tubes (38 mm inner diameter, 49 mm outer diameter, 140 mm long). For the 3-μm particles, immediately after the von Neumann spike, the pressure is elevated in comparison with that for pure NM, presumably owing to the reaction of the aluminum. For the 5-μm particles, the pressure decays after the shock, then abruptly rises after about 2.5 μm owing to particle reaction. For the 14-μm particles, only a slight rise in pressure is observed about 4 μs after the shock wave. The pressure signal recorded for a mixture containing 30-μm particles suggests that in this case the particle reaction was initiated after a time longer than 7 μs. For the largest particle sizes, a distinct von Neumann spike structure is not present.

3.4 Models for Detonation Propagation in Slurry Explosives

3.4.1 Equilibrium Model

The simplest model to predict the detonation properties of an explosive, or Chapman–Jouguet model, is a steady solution based on equilibrium thermodynamics in which the downstream flow velocity is sonic relative to the wave. The Chapman–Jouguet detonation solution may diverge considerably from experimental data for slurry explosives, depending on the importance of nonequilibrium effects. For emulsion explosives, with a very finely structured emulsion matrix giving intimate contact between the fuel and the oxidizer, experimental velocity data approach closely the predicted Chapman–Jouguet values, particularly for high emulsion densities [19]. However, for slurry explosives, particularly those containing large metal particles, the assumption of thermal and mechanical equilibrium at the Chapman–Jouguet plane is not valid and predicted velocity values may be considerably lower than experimental values. For example, for a slurry explosive consisting of sensitized NM with millimeter-sized steel particles, the predicted Chapman–Jouguet velocity value is about 40% less than the extrapolated experimental infinite-diameter result [53].

3.4.2 Zeldovich–von Neumann–Döring Model

The next level of sophistication in modeling a detonation wave is the so-called Zeldovich–von Neumann–Döring model, which assumes a steady structure of a shock followed by a coupled reaction zone, which is described in Chap. 2 in this volume. For a multiphase mixture, the model consists of the 1D steady Euler equations for each component of the mixture. The Euler equations contain source terms of mass, momentum, and energy, which represent the exchange of these quantities between the components. With the presence of the source terms, the classical Chapman–Jouguet criterion can no longer be used to select the unique detonation solution. Rather, an eigenvalue solution may be defined, which represents a balance between the rate of exothermic chemical energy deposition in the gas flow and the rates of energy losses including endothermic reactions and momentum and heat transfer to the particle phases as well as to the boundaries (lateral expansion of the flow and friction and heat loss to the wall). It is normally assumed that the chemical reaction zone thickness is much less than the charge diameter and the lateral expansion of the multidimensional detonation can be considered to be a perturbation to a planar, 1D detonation front. To complete the formulation of the problem, appropriate equations of state are required for the reactants and detonation products, and kinetic rate laws must be specified.

One reaction rate model that has been used extensively for commercial slurry explosives was developed as part of the CPeX code [54,55] and contains three terms that correspond to the hot spots, matrix material, and solid phases within the explosive, i.e.,

$$\frac{\mathrm{d}\lambda}{\mathrm{d}t} = (1-\lambda)\left(\frac{a_\mathrm{h}(p-p_\mathrm{h})^{b_\mathrm{h}}}{t_\mathrm{h}} + \frac{a_\mathrm{l}p^{b_\mathrm{l}}}{t_\mathrm{l}} + \frac{a_\mathrm{s}p^{b_\mathrm{s}}}{t_\mathrm{s}}\right), \tag{3.4}$$

where λ is the degree of reaction, and the subscripts h, l, and s refer to "hot-spot," "emulsion matrix," and "included solid phases" within the explosive mixture, respectively. The a_i factors are given form functions, dependent only on the explosive formulation and the degree of reaction of the explosive. The b_i pressure exponents are normally set to unity and the three t_i time constants are fitting parameters. The critical pressure for activation of the hot spot, p_h, is a fourth fitting constant. Knowledge of either the kinetic behavior or the diameter-effect curve can be used to determine the other. Of course, with four fitting parameters, the utility of the model is dependent on the availability of high-quality experimental data on the diameter effect. From various forms of the diameter-effect curves, Leiper and Cooper [55] obtained the corresponding reaction rate form. They identified two classes of reaction: "thermal explosion" behavior characteristic of monomolecular military explosives and "heterogeneous-type" behavior associated with commercial slurry explosives.

This ignition and growth model was refined by Kirby and Chan [56] and has the form

$$\frac{\mathrm{d}\lambda}{\mathrm{d}t} = \frac{1-\lambda^{0.67}}{\tau_\mathrm{h}}\left(a_\mathrm{h}\left(P-P_\mathrm{crit}\right) + \frac{1-a_\mathrm{h}}{\tau_\mathrm{s}}P\right), \tag{3.5}$$

where

$$a_\mathrm{h} = \exp\left[-\left(\frac{\lambda}{P_\mathrm{s}W_\mathrm{h}}\right)^3\right]. \tag{3.6}$$

The critical pressure P_crit specifies the pressure below which the hot-spot reaction stops. The parameter τ_s corresponds to a relative time constant for the pressure-dependent burn of the matrix material. A new feature with this reaction model is inclusion of the term $P_\mathrm{s}W_\mathrm{h}$ which specifies the fraction of the material reacting in the hot spot to be shock pressure dependent. The model determines the four kinetic parameters by fitting the predicted diameter-effect curve to experimental data. Additional particle velocity data were also used in the fitting procedure.

3.4.3 Detonation Shock Dynamics

Another approach that has been taken in modeling detonation propagation in condensed explosives is the detonation shock dynamics (DSD) theory. Developed by Bdzil et al. [57,58] (see also the review by Bdzil and Stewart [59]),

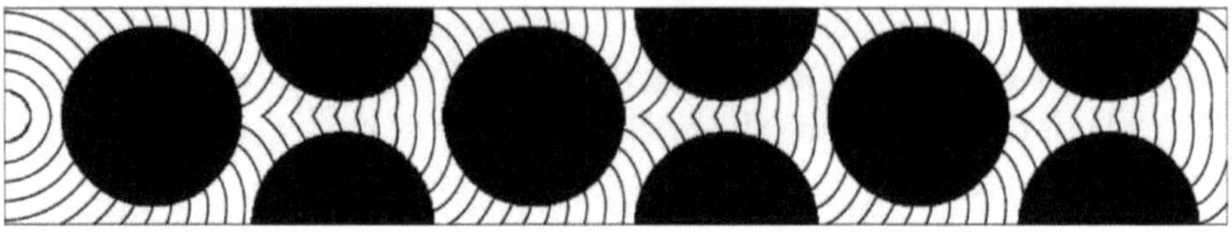

Fig. 3.20. Detonation wavelets propagating through a staggered array of cylindrical obstacles computed with the detonation shock dynamics model [53]

it is based on the idea that the detonation velocity normal to the shock, D_n, is only a function of the shock curvature, κ. With the specification of the $D_n(\kappa)$ relation, which requires accurate experimental front-curvature data, the subsequent detonation propagation can be tracked readily through complex geometries with the use of the level-set method [60, 61]. They have also developed a higher-order DSD theory, which is also tailored for explosives with highly state sensitive reaction rates [62]. Kennedy [63] applied the DSD theory to modeling detonation propagation in heavy ANFO (a mixture of 77% ANFO with 23% ammonium nitrate based emulsion explosive). He noted the difficulties in determining shock curvature from experimental breakout traces and found that no single-valued $D_n(\kappa)$ relationship existed for heavy ANFO. Hence, he concluded that the original form of the DSD theory is not suitable for this highly nonideal commercial explosive. Further refinements to the DSD theory have been presented by Bdzil et al. [64, 65], with a calibration developed for the nonideal explosive ANFO [64]. The DSD model has not been used for conventional commercial emulsion explosives, for which commercial codes such as CPeX remain the primary tool for computing detonation properties.

Application of the DSD model to the propagation of detonation wavelets in the "large-bead" regime discussed earlier was carried out by Frost et al. [53], as shown in Fig. 3.20. With the Huygens assumption, the average detonation velocity through the particle array is less than that for the liquid explosive, owing to the increased path length. When the detonation velocity is assumed to depend on front curvature, the average detonation velocity through the particles is reduced further by a smaller amount, with the reduction being dependent on the particle material.

3.4.4 Mesoscale Continuum Modeling

Much of the current computational modeling of heterogeneous material describes macroscale behavior using ensemble-averaged continuum theory. While there is general agreement on the overall conservation laws in the formulation of multiphase continuum theory [68, 69], disagreement centers on the deviation of interactions associated with the exchange of mass, momentum, and energy between phases needed to provide closure of the theory [70, 71]. Physically, these interactions occur at the mesoscale associated with the discrete nature of the mixture and they have therefore been mostly obtained on the basis of intuition or experimental guidance. While experimental techniques at the mesoscale crystal/particle level are currently in development, mesoscale continuum modeling, with the development of modern parallel computer technology, exhibits the time and space resolution necessary to gain insight into the mechanical, thermal, and chemical behavior of local explosive material or particles under shock and detonation conditions. Hence, it becomes a powerful means to understand the mechanism for the detonation initiation and propagation in multiphase explosives. In the mesoscale continuum models, both explosive and individual metal particles have been treated as continua using their own fluid or solid governing equations accompanied with equations of state and constitutive equations. Interactions between the explosive and the particles and interactions between the particles are described by appropriate boundary conditions. The choice of the governing equations and mesoscale boundary conditions depends on the sophistication of the model and problem to be solved. Mader and Kershner [72, 73] were among the first to use a 3D Eulerian mesoscale continuum model in studying shock initiation of heterogeneous energetic materials. Baer [74] has provided a comprehensive overview of the current capabilities of mesoscale modeling of nonequilibrium multiphase mixtures of "crystals" and binder, and has gained insight into and understanding of the process of mechanical initiation of granular energetic materials using the 3D Eulerian CTH shock physics code in combination with statistical and probabilistic descriptions. Milne [75] employed a 2D Eulerian code to simulate detonation propagation in liquid NM filling the gap between large (with respect to the small detonation zone) spherical metal particles. Zhang et al. [71] used a 2D mesoscale continuum model and later Ripley et al. [76, 77] used a 3D Eulerian model in investigating the interactions associated with the momentum and energy transfer between phases during the shock or detonation propagation in an explosive with metal particles. These references are the primary source for the following overview with the intent to provide an introduction to the mesoscale modeling and applications based on simplistic models.

3.4.4.1 Shock and Detonation Interaction with Particles

Considering the ratio of the characteristic particle size, d_{p}, to the detonation reaction zone thickness, L, one can identify three interaction cases of multiphase detonation: (1) $d_{\mathrm{p}}/L \ll 1$, where the initial detonation-particle interaction can be treated as the interaction with a frozen heavy-side step shock; (2) $d_{\mathrm{p}}/L \sim 1$, for which the particle interacts with the detonation through a shock followed by an expanding reacting flow; and (3) $d_{\mathrm{p}}/L \gg 1$, where the interaction is controlled by the diffraction of the thin detonation front followed by the expansion flow subjected to the boundary conditions of curved particle surfaces and the rear flow.

Ripley et al. [76,77] used an inviscid Eulerian multiple material mesoscale continuum model in a finite-volume framework with Cartesian grids as the simplest model in studying the three interaction cases described above. In this approach, the condensed explosive and solid particles are treated as a continuum, where the proportion of each is accounted for with advection of mass fraction scalars. Depending on which material is present in a computational cell, appropriate equation-of-state parameters are applied. In cells containing multiple materials, a pressure equilibrium approach is applied. A metalized slurry explosive consisting of a mixture of liquid NM and aluminum particles is used as the prototypical multiphase explosive. The particles are assumed to be inert within the detonation front because of the larger characteristic time associated with particle ignition and reaction. Detonation of NM is modeled using a one-step Arrhenius reaction law and a combination of the Mie–Grüneisen equation of state for the unreacted explosive and the Jones–Wilkens–Lee equation of state for the detonation products. The deformable metal particles are also modeled using the Mie–Grüneisen equation of state incorporated with the linear approximation of the unreacted shock Hugoniot, $D = C + Su$, where D is the shock velocity and u is the fluid velocity. Material strength has been neglected and only volumetric strain is assumed to occur in the particles. Temperatures are determined using the fitting technique of Walsh and Christian [78]. A second-order accurate 3D unstructured mesh approach is employed in a parallel computing framework. The Harten–Lax–van Leer with contact correction (HLLC) approximate Riemann solver is used to solve for the fluxes and provides high resolution particularly for resolving shock and material interfaces (or slip lines) in a multimaterial mixture [12]. The inviscid continuum model is also justified by comparing its results with those from a fully viscous continuum modeling, with minor differences resulting, due to the magnitude of the intrinsic numerical cell viscosities.

A closely packed particle matrix is selected as illustrated in Fig. 3.21 for the discussions here as it provides the highest solid volume fraction (ϕ_{packed}

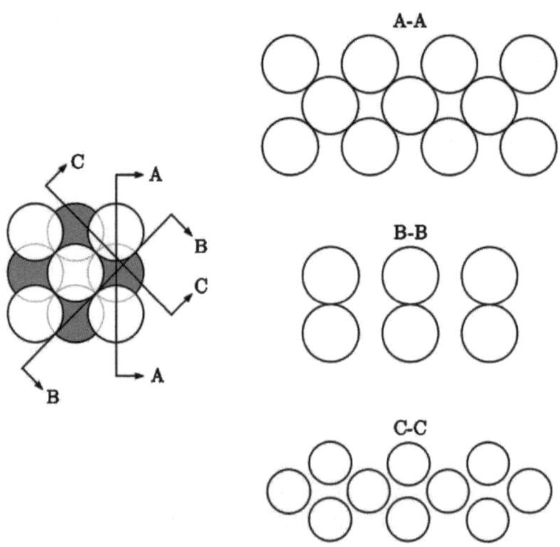

Fig. 3.21. A closely packed particle matrix

= 0.74) among various packing configurations studied in [76,77]. The discrete solid volume fraction is varied by increasing the spacing, s, between particles.

3.4.4.2 Case 1: $d_p/L \ll 1$

For a frozen or inert shock wave crossing a solid particle, the inviscid governing equations and rate-independent material response models do not introduce any additional length scales or time scales except for the particle size. Therefore, at any given time, t_{d_2}, the surrounding flow field and the dynamic response of a particle with diameter d_2 can be geometrically scaled to the same state as for a particle with diameter d_1 at a time t_{d_1} [71]:

$$\frac{t_{d_2}}{t_{d_1}} = \frac{d_2}{d_1}. \tag{3.7}$$

Consequently, calculations need only be conducted for one particle diameter, and the results can be scaled to any other particle diameter. Hence, in the following discussion, the particle diameter is arbitrarily chosen as 10 µm. Figure 3.22 shows the computational result of the aluminum particle deformation in a mixture of NM and 3D closely packed particles ($s = 0.2d_p$) subjected

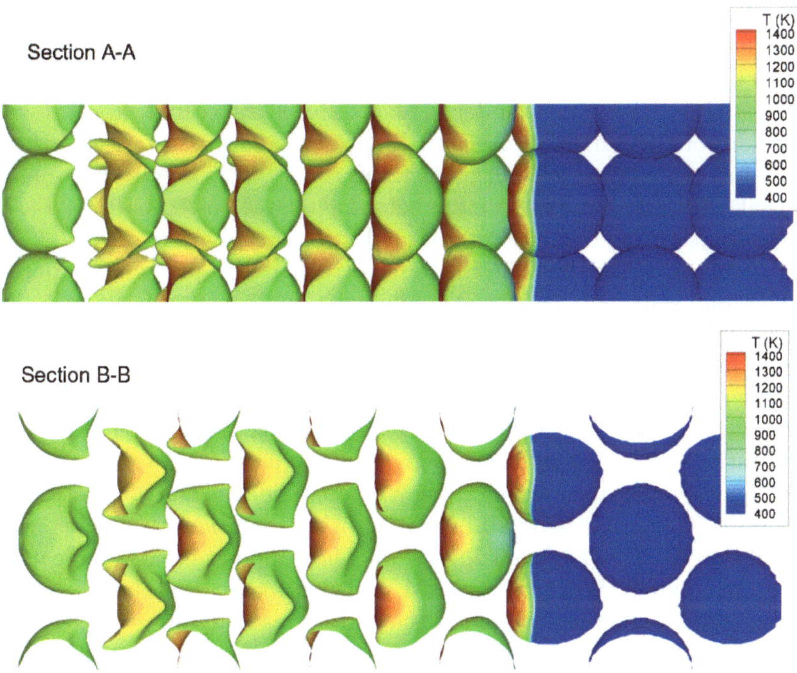

Fig. 3.22. Particle deformation and temperature contour in a closely packed aluminum/NM mixture ($s = 0.2d_p$) subjected to an inert shock [77]

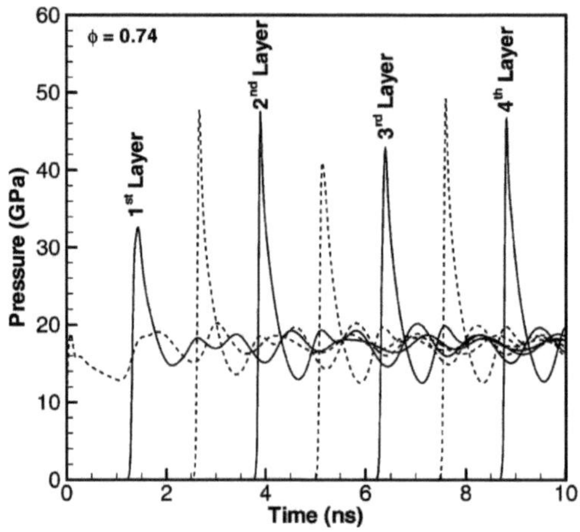

Fig. 3.23. Particle leading edge pressures in a closely packed aluminum/NM mixture ($s = 0$) [76]

to a 10.1-GPa inert step shock traveling from left to right. While the first layer of particles in the model receives a planar shock, subsequent layers are subject to nonuniform loading induced by neighboring particles and complex reflected waves. A quasi-steady shock propagation is achieved after passing several layers of particles. The deformed particles resemble a saddle shape, strongly influenced by the complex shock reflections from upstream and neighboring particles. If a failure model were introduced, one could expect damage or breakup of the particles to occur. Figure 3.23 displays pressure histories observed at the leading edge of particles in the $s = 0$ closely packed matrix. The peak pressures exhibit periodic behavior related to the particle packing configuration. Owing to multiple shock reflections and focusing, the peak pressures are much higher than the theoretical transmission pressure of 20.9 GPa for a 1D solid slab. The peak pressure decays rapidly as the shock crosses the particle and then oscillates about a quasi-steady value. Corresponding to the pressure history, the mass-centered particle velocity, shown in Fig. 3.24, also increases rapidly as the shock crosses the particle and then oscillates about a mean value above the theoretical 1D transmission value of 1.110 mm μs^{-1}. Computations show that the shock velocity through the mixture and the transmitted quasi-steady pressure increase with solid volume fraction between no particles (4.769 mm μs^{-1} and 10.1 GPa) and the theoretical transmission for a 1D solid slab (7.0 mm μs^{-1} and 20.9 GPa). The temperature field indicates localized hot spots in the interstitial NM voids between particles in the matrix. The peak fluid temperature found at the particle leading edge is 3,828 K in the close-packing configuration, while it reaches only 2,383 K for the single cubic packing. This local hot-spot behavior will significantly affect the detonation initiation process.

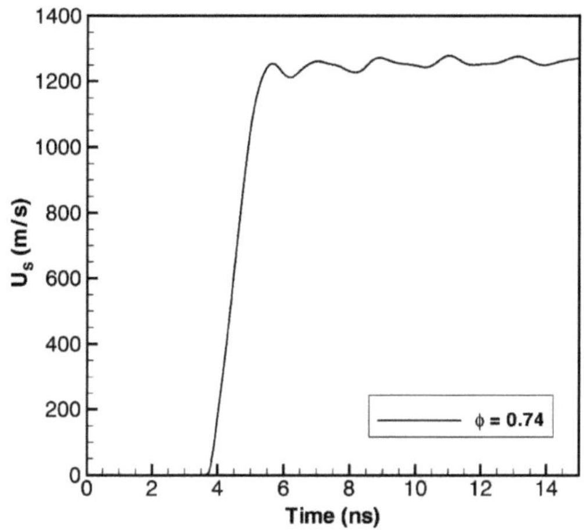

Fig. 3.24. Mass-centered particle velocity in a closely packed aluminum/NM mixture ($s = 0$) [76]

3.4.4.3 Case 2: $d_p/L \sim 1$

The detonation case has two length scales (the reaction zone length and the particle size); hence, the simple similarity relation (3.7) becomes invalid and the problem depends on the choice of d_p at a given L. Estimates in the literature for the 1D reaction zone length in NM detonation range from a few microns to 100 µm. Mader [73] simulated reaction zone lengths ranging from 0.24 to 70.5 µm; Engelke and Bdzil [12] predicted a NM reaction time of 6 µs on the basis of the detonation front curvature. Recently, Sheffield et al. [39] used VISAR interferometry to measure a reaction time of about 5–10 ns for a rapid reaction zone and 50 ns to what is assumed to be the sonic point with a particle velocity of about 2 mm µs^{-1} (corresponding to a reaction zone length on the order of 10–100 µm). To study the detonation interaction with a packed particle array, Ripley et al. [77] used a NM detonation wave with an L = 1.0 µm rapid reaction zone length (measured from the von Neumann spike to the end of the rapid pressure change) as the initial condition upstream of 20 layers of particles. The stable detonation had a propagation velocity of 6.690 mm µs^{-1}, for which the reaction was 99.9% complete at the Chapman–Jouguet point with the gaseous products of 1.551 g cm^{-3} density, 13.80-GPa pressure, 3,657-K temperature and 1.827 mm µs^{-1} flow velocity.

As an illustrative example, $d_p = 1$ µm or $d_p/L = 1$ is chosen for the mesoscale modeling. The 3D domain features quarter-symmetric sphere segments arranged in a dense packed configuration using a mesh resolution of 10 nm in each direction with 16 million total cells in the mesh. All boundaries are reflective, which models a semi-infinite and periodic bed of packed particles. Figure 3.25 illustrates the particle deformation in a mixture of NM and 3D closely packed aluminum particles ($s = 0.2d_p$) subjected to the detonation

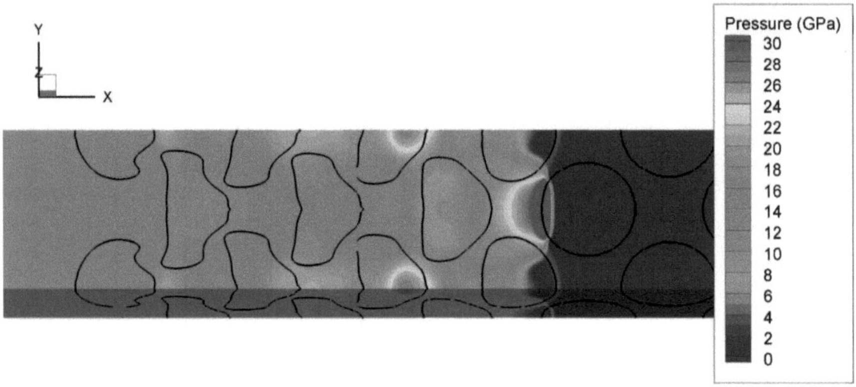

Fig. 3.25. Material deformation and pressure contour in a closely packed aluminum/NM mixture ($s = 0.2d_\mathrm{p}$) subjected to a detonation wave [77]

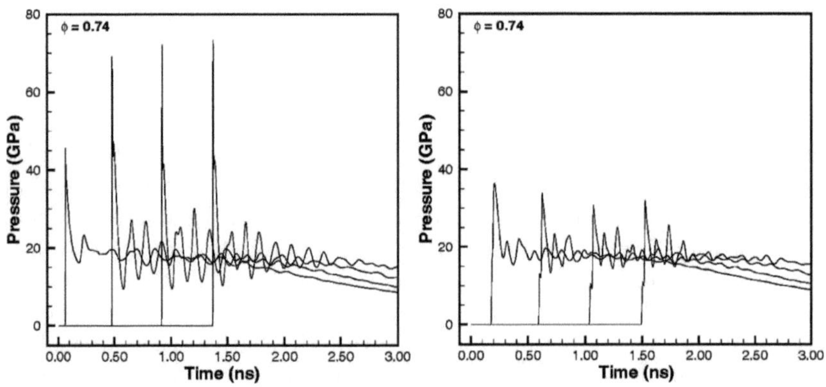

Fig. 3.26. Pressures in particle leading edges (*left*) and in NM voids (*right*), corresponding to the first, third, fifth and seventh layer in a closely packed aluminum/NM mixture ($s = 0$) [77]

wave traveling from left to right. Pressure histories at the particle leading edges are displayed in Fig. 3.26 (left) for the mixture of NM and $s = 0$ closely packed particles. The first trace shows a peak pressure, resulting from the reflected von Neumann spike, consistent with 1D wave transmission theory (47.9 GPa). Subsequent waves feature a shock front pressure of 65–85 GPa due to reflection and focusing where the diffracted Mach stems from four spherical particles collide on the centerline upstream of the particle leading edge. NM reaction follows the shock front and the pressure decays and oscillates around a mean value before further decaying owing to the Taylor expansion dictated by the rear boundary. Pressure histories for the NM in the interstitial regions of the matrix show a similar behavior with a leading shock pressure higher than the von Neumann spike followed by an oscillating detonation wave (Fig. 3.26, right). Both the particle leading edge and the void histories show a pressure oscillation about a mean value of approximately 17.5 GPa, higher than the Chapman–Jouguet pressure in pure NM. While the reaction zone length prior to entering the aluminum/NM mixture was 1.0 μm, the reaction rate in the mixture increases and the reaction zone length decreases to about 0.3 μm, owing to the resultant high temperatures in the mixture. For the various packing volume fractions ($s = 0$–$10d_{\mathrm{p}}$), maximum reflected shock temperatures at the particle leading edges range from 8,100 to 7,400 K, but quickly decay after the shock front. In comparison, temperatures within the voids are sustained above 3,000–3,700 K for the duration of the simulation.

The propagation of the detonation is initially transient but becomes quasi-steady after a propagation distance of a few particle layers, where the detonation velocity exhibits a deficit with respect to the entering detonation velocity of $D = 6.690\,\mathrm{mm\,\mu s^{-1}}$. Figure 3.27 shows the detonation velocity deficit obtained as a function of aluminum mass fraction in comparison with the equilibrium detonation velocity of the mixture calculated using Cheetah 2.0 [16]. The equilibrium results do not account for the particle size and show a monotonic decrease with the aluminum mass fraction up to a maximum velocity deficit at 65% aluminum mass fraction. In contrast, the mesoscale nonequilibrium results indicate that the detonation velocity decreases with aluminum mass fraction up to 40%, above which the detonation velocity approaches a constant, owing to the wave transmission in the dense particle layers. Currently there are not sufficient experimental data at different aluminum mass fractions to validate the trend of the mesoscale results shown in Fig. 3.27. Most velocity data are for packed particle mixtures, in which the solid mass fraction depends on the particle size and morphology (e.g., Kato

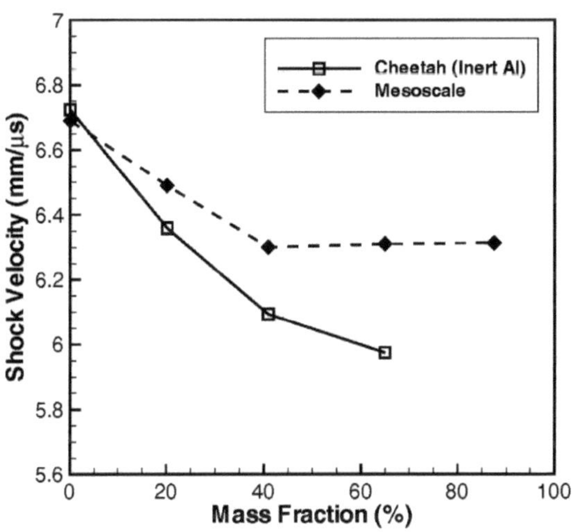

Fig. 3.27. Detonation velocity as a function of aluminum mass fraction [77]

and Murata [49] obtained solid mass fractions ranging from 57 to 75% for aluminum particles ranging in size from 3 to 30 μm).

The above-mentioned work was based on simplistic material descriptions. In reality, the morphology of the particle matrix is of a stochastic nature where the particles are of a different size and shape and the packing is random with an irregular geometrical configuration. More advanced material and failure models must be considered to deal with particle damage and fragmentation under shock and detonation conditions. The particle damage and fragmentation are important in understanding the subsequent combustion of the metal particles. Figure 3.28 displays more sophisticated mesoscale modeling for a mixture of liquid IPN, RDX grains, and aluminum flakes by Baer et al. [74,79], using the CTH code with appropriate equations of state and constitutive models. The RDX crystals are represented as a distribution of cubes with 50- and 100-μm sizes that are randomly mixed with 200 μm × 200 μm × 10 μm aluminum platelets to create a packing configuration that conforms to a given volume fraction, saturated with liquid IPN in the remaining interstitial regions. A booster explosive is used at the bottom of the mixture to initiate

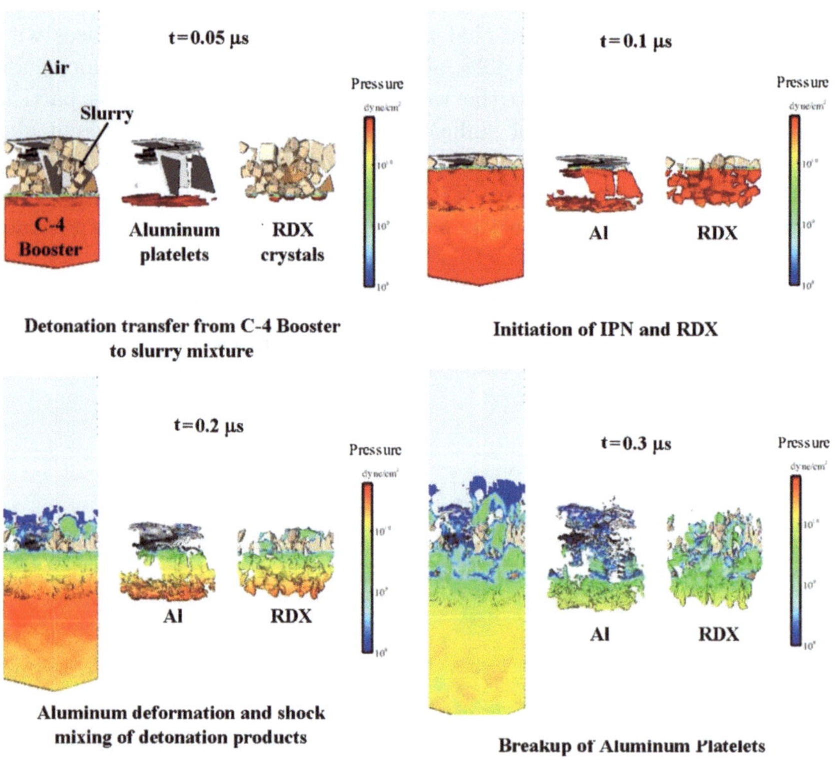

Fig. 3.28. Material deformation and pressure contours in a packed aluminum/RDX/isopropyl nitrate mixture subjected to detonation initiation [79]

the RDX and IPN, while the aluminum is treated as inert. As shown in Fig. 3.28, after material deformation during the shock loading, high shear forces between the materials exist that modify the surface characteristics of the dispersed aluminum flakes and cause them to break up behind the detonation wave. Subsequently, the ignition and combustion characteristics of the dispersed aluminum require an analysis of the nonpassivated highly strained metal additive that is different from the initial unstressed material. The severe particle deformation and damage indicated in [77, 79] during the shock and detonation processes may further lead to a removal of the thin aluminum oxide coating, exposing hot aluminum to the oxidizing gases and leading to rapid reaction of the aluminum. The high pressure and temperature environment under detonation conditions may also cause particle agglomeration, thereby increasing the effective particle diameter and altering the mass, momentum, and heat transfer characteristics. These topics require development of more advanced mesoscale models.

3.4.4.4 Macroscopic Interphase Transfer Functions

For dense and packed matrices of particles in an explosive, the exchanges of mass, momentum, and energy between phases are complicated functions, subjected to the interactions with neighboring particles under shock or detonation conditions. An important application of mesoscale modeling is to use it to perform mesoscale "numerical experiments" from which the macroscopic "empirical" exchange functions such as momentum and heat transfer between the phases during the shock or detonation process can be derived. These functions can then be used as the interphase transfer source terms applied to the macroscopic continuum modeling for detonation of a multiphase explosive and subsequent dense reactive particle-gas flow.

As described in Chap. 2, the momentum transfer rate acting on the particle-phase control volume containing n_p particles during the shock interaction process can be written as

$$f_\mathrm{p} = n_\mathrm{p} m_\mathrm{p} \frac{\mathrm{d}u_\mathrm{p}}{\mathrm{d}t} \approx n_\mathrm{p} m_\mathrm{p} \frac{\alpha u_1 - u_\mathrm{p}(0)}{\tau_\mathrm{s}}, \tag{3.8}$$

and the heat transfer rate for the control volume can be obtained from

$$q_\mathrm{p} = n_\mathrm{p} c_\mathrm{p} \frac{\mathrm{d}T_\mathrm{p}}{\mathrm{d}t} \approx n_\mathrm{p} c_\mathrm{p} \frac{\beta T_1 - T_\mathrm{p}(0)}{\tau_\mathrm{s}}, \tag{3.9}$$

where m_p and c_p are the particle mass and specific heat capacity, respectively, and τ_s is the shock interaction time defined as the shock velocity divided by the particle diameter. A velocity transmission factor, α, and a temperature transmission factor, β, of a particle are defined as follows:

$$\alpha = \frac{u_\mathrm{p}(\tau_\mathrm{s})}{u_1}; \beta = \frac{T_\mathrm{p}(\tau_\mathrm{s})}{T_1}, \tag{3.10}$$

where u_1 and T_1 are the postshock flow velocity and temperature, and u_p and T_p are the mass-averaged particle velocity and temperature. Thus, the momentum and heat transfer rate during the shock interaction process can be determined by the transmission factors that can be obtained from the mesoscale simulations.

For a step shock wave in condensed matter passing a single spherical particle, the velocity transmission factor was studied for a particle of magnesium, beryllium, aluminum, nickel, uranium, and tungsten in NM or RDX with initial densities of 1.4–1.8 g cm^{-3} subjected to a shock of 5.07–20.3 GPa [71]. The particle velocity after the shock interaction is found to strongly depend on the initial density ratio of explosive to metal, as summarized in Fig. 3.29. The influence of other parameters, such as particle acoustic impedance, shock strength, and bulk unreacted explosive's shock Hugoniot is implicitly included in the effect of the initial density ratio. Figure 3.29 indicates that the momentum transferred to light-metal particles is significant and the particle velocity for aluminum, beryllium, and magnesium achieves 60–100% of the value of the shocked explosive's velocity. A curve fit of the mesoscale numerical values of α suggests the following correlation [71]:

$$\alpha = \frac{1}{a+b}\left(a + b\frac{\rho_0}{\rho_{s0}}\right)\frac{\rho_0}{\rho_{s0}}, \tag{3.11}$$

where $a = 3.947$ and $b = -1.951$.

For a shock or detonation wave propagating in an explosive with a matrix of particles, the velocity and temperature transmission factors can be calculated using the inviscid 3D mesoscale model described above [76, 77]. Figure 3.30 shows the time-dependent behavior of mass-averaged velocity and

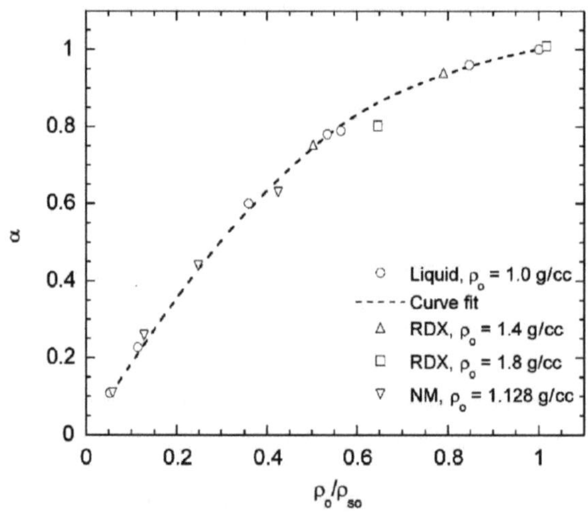

Fig. 3.29. Velocity transmission factor of a single metal particle in an explosive subjected to an inert shock [71]

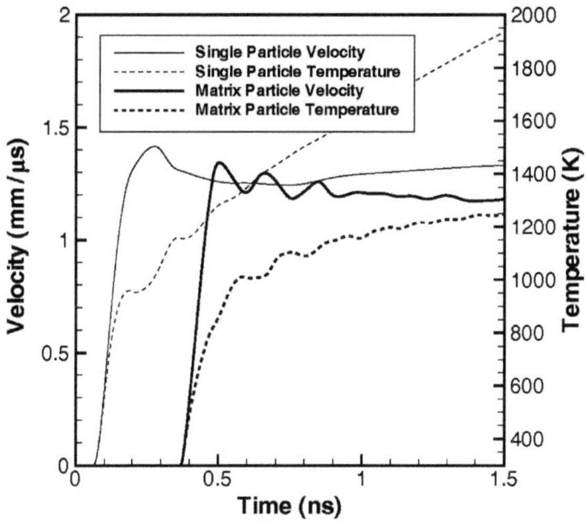

Fig. 3.30. Mass-averaged particle velocity and temperature history for a single Al particle and a closely packed Al in NM subjected to NM detonation [77]

temperature of a particle within a closely packed aluminum/NM mixture ($s = 0$) subjected to the NM detonation with $d_{\mathrm{p}}/L = 1$ as compared with that for the same detonation wave crossing a single particle. The different shock arrival times in Fig. 3.30 for the single particle and the particle within a packed matrix are due to the additional distance that the detonation travels prior to interacting with the subject particle. During the shock interaction time ($\tau_{\mathrm{s}} = 0.15\,\mathrm{ns}$), a rapid increase in velocity and temperature is observed. This is consistent with results for inert shock interaction. Following the leading shock crossing the particle, a decrease in velocity is seen corresponding to the change in conditions in the expanding gas products flow behind the shock. For the single-particle case, the particle velocity becomes steady, whereas for the particle in a packed matrix, the particle velocity oscillates owing to multiple shock reflections from neighboring particles. The increase in mass-averaged temperature after the shock interaction time is partially due to numerical diffusion between the particle material and the hot detonation product gases.

Figure 3.31 summarizes the velocity and temperature transmission factors as a function of metal volume fraction, ϕ, for closely packed aluminum/NM mixtures in three interaction cases [77]: (1) $d_{\mathrm{p}}/L \ll 1$, inert step shock ($D = 4.769\,\mathrm{mm}\,\mu\mathrm{s}^{-1}$, $p_1 = 10.1\,\mathrm{GPa}$) crossing particles with $d_{\mathrm{p}} = 10\mu\mathrm{m}$; (2) $d_{\mathrm{p}}/L = 1$, detonation ($D = 6.69\,\mathrm{mm}\,\mu\mathrm{s}^{-1}$, $p_1 = p_{\mathrm{vN}} = 22.8\,\mathrm{GPa}$) crossing particles with $d_{\mathrm{p}} = 1.0\,\mu\mathrm{m}$; and (3) $d_{\mathrm{p}}/L = 30$, detonation ($D = 6.69\,\mathrm{mm}\,\mu\mathrm{s}^{-1}$, $p_1 = p_{\mathrm{vN}} = 22.8\,\mathrm{GPa}$) crossing particles with $d_{\mathrm{p}} = 30\,\mu\mathrm{m}$. In Fig. 3.31, the $\phi \to 0$ case is simulated by interaction with a single particle, and $\phi = 1$ corresponds to transmission into a 1D slab of aluminum. In general, the transmission factors for detonation conditions are smaller than those obtained for a step

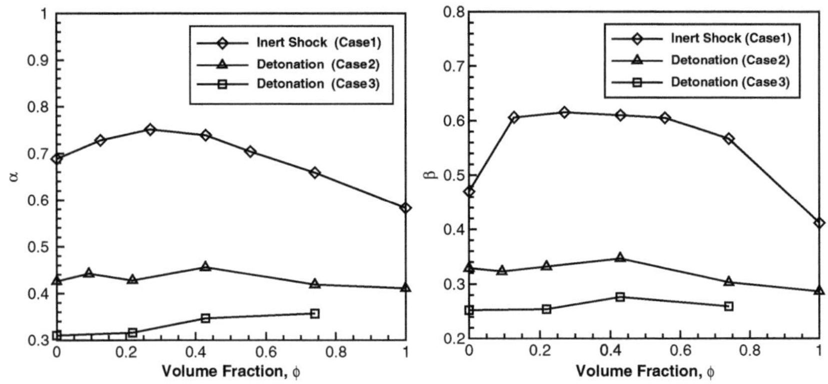

Fig. 3.31. Velocity (*left*) and temperature (*right*) transmission factors for shock and detonation conditions [77]

shock, primarily owing to the expansion following the shock wave front in the detonation cases. The expansion surrounding a particle reduces the local particle velocity and temperature when the shock front crosses the particle and therefore reduces the transmission factors that are based on the mass-average over the entire particle. The transmission factors for detonation cases with various ratios of d_p/L are bounded within the limits of $d_p/L \ll 1$ and $d_p/L \gg 1$ and decrease with increasing d_p/L, owing to longer expanding flow surrounding the particle as the shock front passes the particle. Note that the transmission factors for $d_p/L \ll 1$ are independent of the particle size chosen according to the scaling rule (3.7) in the frame of the current inviscid mesoscale model. The scaling for the other cases depends on reaction zone length and particle size with appropriately scaled rear boundary conditions.

3.5 Concluding Remarks

Although ANFO is still the predominant commercial explosive, slurry (including emulsion) explosives are increasingly used owing to their inherent reliability and safety. Future improvements to the sensitivity, strength, and handling characteristics of slurry explosives will allow them to compete favorably with other commercial explosives such as dynamite and ANFO. Increasing energy demand and shortages of fuel oil will also increase the consumption of slurry explosives in the future [6].

Further developments in fuel-rich metalized slurry explosives for military applications are aimed at improving their blast performance through better control of the time and position over which the metal particles release their chemical energy. Important factors include the oxygen balance of the explosive, which determines the availability of oxidizers within the detonation products; the particle size and density, which influences the particle dynamics and the degree to which they mix with the surrounding air; and the particle material, which determines the particle reaction mechanism. A particularly challenging problem is to determine the reaction rate of the metal particles during the highly transient particle dispersal process as a function of the ambient flow conditions. To increase the surface area available for particle reaction, a number of investigators are considering the use of sub-micron-sized metallic particles (or multilayer metal foils) mixed with explosives.

Current models for the propagation of detonations in heterogeneous explosives depend on the use of a number of empirical factors that describe the interaction of a shock wave with heterogeneities and the formation of hot spots and subsequent reaction of the explosive and heterogeneities. Improvements to models for detonation propagation in metalized slurry explosives require a better understanding of the physical processes that occur at the mesoscale. For example, the shock/particle interactions will cause deformation of the metal particles which will influence the subsequent transfer of momentum and energy to the particles. The refined mesoscale models for the constitutive physics then need to be incorporated into hydrocodes for determining the performance of the heterogeneous explosive at larger scales. Further developments in our understanding of the detonation of slurry explosives will continue to depend on the close coupling between highly resolved numerical models and well-characterized experimental results and should lead to new commercial and military applications.

Acknowledgements

The authors gratefully acknowledge the contributions of data and figures included in this chapter by Y. Kato, R. Ripley, M. Baer, and M. Cooper.

References

1. Gilion, J.B., Eck, G., Machacek, O.: Excess energetic materials as ingredients in commercial explosives. ISEE Proc. 1994R, 163–174 (1994)
2. Tinggenberg, O., Mathieu, J.: Commercial high explosives. Chimia **58**, 390–393 (2004)
3. Persson, P.-A., Holmberg, R., Lee, J.: Rock Blasting and Explosives Engineering. CRC, Boca Raton (1994)
4. Dick, R.A.: The impact of blasting agents and slurries on explosives technology. U.S. Bureau of Mines information circular IC 8560, US Department of the Interior, Pittsburgh, PA (1985)
5. Hopler, R.B.: History of the development and use of bulk loaded explosives, from black powder to emulsions. ISEE Gen. Proc. 1993G, 177–198, (1993)
6. Gehrig, N.E.: The future of slurry explosives. ISEE Gen. Proc. 1982G, 223–233 (1982)

7. Ingram, C.B., Gehrig, N.E.: Gulf slurry products and their uses. ISEE Gen. Proc. 1977G, 216–219 (1977)
8. Sumiya, F., Tokita, K., Nakano, M., Ogata, Y., Wada, Y., Seto, M., Katsuyama, K., Itoh, S.: Experimental study on the channel effect in emulsion explosives. J. Mater. Proc. Tech. **85**, 25–29 (1999)
9. Cudzilo, S., Kohlicek, P., Trzcinski, V.A., Zeman, S.: Performance of emulsion explosives. Combust. Explos. Shock Waves **38**(4), 463–469 (2002)
10. Campbell, A.W., Davis, W.C., Travis, J.R.: Shock initiation of detonation in liquid explosives. Phys. Fluids **4**, 495–510 (1961)
11. Engelke, R.: Effect of a physical inhomogeneity on steady-state detonation velocity. Phys. Fluids **22**(9), 1623–1630 (1979)
12. Engelke, R., Bdzil, J.B.: A study of the steady-state reaction zone structure of homogeneous and heterogeneous explosives. Phys. Fluids **26**(5), 1210–1221 (1983)
13. Engelke, R.: Effect of the number density of heterogeneities on the critical diameter of condensed explosives. Phys. Fluids **26**(9), 2420–2424 (1983)
14. Hall, S., Knowlton, D.: Development, characterization and testing of high blast thermobaric compositions. Proc. Int. Pyrotech. Semin, Fort Collins, CO, July 11–16, 663–678 (2004)
15. Suceska, M.: Test Methods for Explosives. Springer, Berlin (1995)
16. Fried, L.E., Howard, W.M., Souers, P.C.: Cheetah 2.0 user's manual. LLNL UCRL-MA-117541 revision 5. University of California (1998)
17. Mohanty, B.: Parameters affecting explosive energy release – a case for realistic rating of commercial explosives. ISEE Gen. Proc. 1988G, 304–314 (1988)
18. Lee, J., Persson, P.-A.: Detonation behavior of emulsion explosives. Propellants Explos. Pyrotech. **15**, 208–216 (1990)
19. Yoshida, M., Iida, M., Tanaka, K., Fujiwara, S., Kusakabe, M., Shiino, K.: Detonation behavior of emulsion explosives containing glass microballoons. In: Proceedings of Eighth International Detonation Symposium, Albuquerque, NM, Naval Surface Weapons Center, White Oak, MD, pp. 993–1000 (1985)
20. Campbell, A.W., Engelke, R.: The diameter effect in high density heterogeneous explosive. In: Proceedings of Sixth International Symposium on Detonation, Coronado, CA, Office of Naval Research, Arlington, VA, pp. 642–652 (1976)
21. Taylor, J.: Detonation in Condensed Explosives, p. 152. Clarendon, Oxford (1952)
22. Engelke, R., Sheffield, S.A., Stacy, H.L.: Chemical-reaction-zone lengths in condensed-phase explosives. Phys. Fluids **16**(11), 4143–4149 (2004)
23. Price, D.: Contrasting patterns in the behavior of high explosives. In: Proceedings of Eleventh Symposium (International) on Combustion, pp. 693–702. The Combustion Institute, Pittsburgh (1967)
24. Anshits, A.G., Anshits, N.N., Deribas, A.A., Karakhanov, S.M., Kasatkina, N.S., Plastinin, A.V., Reshetnyak, A.Yu., and Sil'vestrov, V.V.: Detonation velocity of emulsion explosives containing cenospheres. Combust. Explos. Shock Waves **41**(5), 591–598 (2005)
25. Persson, A., Jerberyd, L., Almgren, L.-Å.: Testing the sensitivity of watergel explosives to weak shocks. Propellants Explos. **5**, 45–48 (1980)
26. Krishna Mohan, V., Field, J.E., Swallowe, G.M.: High-speed photographic studies of impact on thin layers of emulsion explosive. Propellants Explos. Pyrotech. **9**, 77–81 (1984)

27. Feng, K.K., Hanasaki, K.: A numerical study of the behaviour of liquid and slurry explosives under projectile impact. Propellants Explos. **6**, 74–87 (1981)
28. Chaudhri, M.M., Almgren, L.-Å, Persson, A.: Detonation behaviour of 'water-in-oil' type emulsion explosive containing glass microballoons of selected sizes. In: Proceedings of Tenth International Detonation Symposium, Boston, MA, Office of Naval Research, Arlington, VA, pp. 741–748 (1993)
29. Khasainov, B.A., Ermolaev, B.S., Presles, H.N.: Effect of glass microballoons on shock wave sensitivity and detonation critical diameter of condensed explosives. In: Proceedings of Tenth International Detonation Symposium, Boston, MA, Office of Naval Research, Arlington, VA, pp. 749–757 (1993)
30. Lee, J., Sandstrom, F.W., Craig, B.G., Persson, P.-A.: Detonation and shock initiation properties of emulsion explosives. In: Proceedings of Ninth International Detonation Symposium, Portland, OR, Office of Naval Research, Arlington, VA, pp. 573–584 (1989)
31. Presles, H.N., Campos, J., Heuze, O., Bauer, P.: Effects of microballoons concentration on the detonation characteristics of nitromethane-pmma mixtures. In: Proceedings of Ninth International Detonation Symposium, Portland, OR, Office of Naval Research, Arlington, VA, pp. 925–929 (1989)
32. Kurbangalina, R.Kh.: Critical diameter of liquid explosives as a function of powder content. Zh. Prikl. Mekh. Tekn. Fiz. **10**(4), 133–136 (1969)
33. Kato, Y., Bauer, B., Brochet, C., Bouriannes, B.: Brightness temperature of detonation wave in nitromethane-tetranitromethane mixture and in gaseous mixtures at a high initial pressure. In: Proceedings of Seventh International Detonation Symposium, Annapolis, MD Naval Surface Weapons Center, White Oak, MD, pp. 768–774 (1981)
34. Kato, Y., Brochet, C.: Cellular structure of detonation in nitromethane containing aluminum particles. In: Proceedings of Sixth International Detonation Symposium, Coronado, CA, Office of Naval Research, Arlington, VA, pp. 124–132 (1976)
35. Kato, Y., Brochet, C.: Detonation temperatures of nitromethane aluminum gels. Prog. Astronaut. Aeronaut. **94**, 416–426 (1984)
36. Kato, Y., Mori, N., Sakai, H., Tanaka, K., Sakurai, T., Hikita, T.: Detonation temperature of nitromethane and some solid high explosives. In: Proceedings of Eighth International Detonation Symposium, Albuquerque, NM, Naval Surface Weapons Center, White Oak, MD, pp. 558–566 (1985)
37. Moulard, H., Fauquignon, C., Lichtenberger, M., Lombard, J.M.: Détonation de mélange nitrométhane-Al-NO_2. In: Proceedings of HDP Symposium, pp. 293–307. Commissariat a l'Energie Atomique, Paris (1979)
38. Hirosaki, Y., Murata, K., Kato, Y., Itoh, S.: Detonation characteristics of emulsion explosives as functions of void size and volume. In: Proceedings of Twelfth International Detonation Symposium, San Diego, CA, Office of Naval Research, Arlington, VA, pp. 263–270 (2002)
39. Sheffield, S.A., Engelke, R., Alcon, R.R., Gustavsen, R.L., Robbins, D.L., Stahl, D.B., Stacy, H.L., Whitehead, M.C.: Particle velocity measurements of the reaction zone in nitromethane. In: Proceedings of Twelfth International Detonation Symposium, San Diego, CA, Office of Naval Research, Arlington, VA, pp. 159–166 (2002)
40. Lee, J.J., Frost, D.L., Lee, J.H.S., Dremin, A.: Propagation of nitromethane detonation in porous media. Shock Waves **5**(1/2), 115–120 (1995)

41. Lee, J.J., Brouillette, M., Frost, D.L., Lee, J.H.S.: Effect of diethylenetriamine sensitization on detonation of nitromethane in porous media. Combust. Flame **100**, 292–300 (1995)
42. Lee, J.J.: Detonation mechanisms in a condensed-phase porous explosive. Ph.D. thesis, Université de Sherbrooke (1997)
43. Choong, K.H.: A comparison of the sensitization effects between triethylamine and diethylenetriamine on nitromethane. Honours thesis, McGill University (1998)
44. Frost, D.L., Zhang, F., Murray, S., McCahan, S.: Critical conditions for ignition of metal particles in a condensed explosive. In: Proceedings of Twelfth International Detonation Symposium, San Diego, CA, Office of Naval Research, Arlington, VA, pp. 693–701 (2002)
45. Frost, D.L., Goroshin, S., Levine, J., Ripley, R., Zhang, F.: Critical conditions for ignition of aluminum particles in cylindrical charges. AIP Conf. Proc. **845**, 972–975 (2005)
46. Tanguay, V., Higgins, A.J., Zhang, F.: A simple analytical model for reactive particle ignition in explosives. Propellants Explos. Pyrotech. **32**(5), 371–384 (2007)
47. Zhang, F., Yoshinaka, A., Frost, D.L., Ripley, R., Kim, K., Wilson, W.: Casing influence on ignition and reaction of aluminum particles in an explosive. In: Proceedings of Thirteenth International Detonation Symposium, Norfolk, VA, Office of Naval Research, Arlington, VA, pp. 233–244 (2006)
48. Kato, Y., Murata, K., Itoh, S.: Detonation characteristics of packed beds of aluminum saturated with nitromethane. In: Proceedings of Thirteenth International Detonation Symposium, Norfolk, VA, Office of Naval Research, Arlington, VA, pp. 187–195 (2006)
49. Kato, Y., Murata, K.: Detonation in packed beds of metal particles saturated with nitromethane. In: Proceedings of Europyro 2007 – 34th International Pyrotechnics Seminar, Beaune (2007)
50. Frost, D.L., Zhang, F.: The nature of heterogeneous blast explosives. In: Proceedings of the 19th International Symposium on Military Aspects of Blast and Shock, Calgary, AB, Oct. 1–6 (2006)
51. Trocino, J.L. and Associates: Techniques for sensitizing and detonating nitromethane-based explosive systems. Technical report bulletin JLTN-2. Trocino and Associates, Sherman Oaks (1974)
52. Haskins, P.J., Cook, M.D., Pilgrim, A., Briggs, R.I.: The effect of additives on the detonation characteristics of a liquid explosive. AIP Conf. Proc. **620**, 890–893 (2001)
53. Frost, D.L., Aslam, T., Hill, L.G.: Application of detonation shock dynamics to the propagation of detonation in nitromethane in a packed inert particle bed. AIP Conf. Proc. **505**, 821–824 (1999)
54. Kirby, I.J., Leiper, G.A.: A small divergent detonation theory for intermolecular explosives. In: Proceedings of Eighth International Detonation Symposium, Albuquerque, NM, Naval Surface Weapons Center, White Oak, MD, pp. 176–186 (1985)
55. Leiper, G.A., Cooper, J.: Reaction rates and the charge diameter effect in heterogeneous explosives. In: Proceedings of Ninth International Detonation Symposium, Portland, OR, Office of Naval Research, Arlington, VA, pp. 197–208 (1989)

56. Kirby, I.J., Chan, S.K.: Analysis of VOD-diameter data using an analytical two-dimensional non-ideal detonation model. AIP Conf. Proc. **845**, 453–456 (2005)
57. Bdzil, J.B.: Steady-state two-dimensional detonation. J. Fluid Mech. **108**, 195–226 (1981)
58. Bdzil, J.B., Fickett,W., Stewart, D.S.: Detonation shock dynamics: a new approach to modeling multi-dimensional detonation waves. In: Proceedings of Ninth International Detonation Symposium, Portland, OR, Office of Naval Research, Arlington, VA, pp. 730–742 (1989)
59. Bdzil, J.B., Stewart, D.S.: The dynamics of detonation in explosive systems. Annu. Rev. Fluid Mech. **39**, 263–292 (2007)
60. Osher, S., Sethian, J.A.: Fronts propagating with curvature-dependent speed: algorithms based on Hamilton–Jacobi formulations. J. Comput. Phys. **79**, 12–49 (1988)
61. Bdzil, J.B., Aslam, T.D., Stewart, D.S.: Curved detonation fronts in solid explosives: collisions and boundary interactions. In: Sturtevant, B., Shepherd, J.E., Hornung, H.G. (eds.) Proceedings of the 20th International Symposium on Shock Waves, Pasadena, CA, pp. 97–106. World Scientific, River Edge (1995)
62. Aslam, T.D., Bdzil, J.B., Hill, L.G.: Extensions to DSD theory: analysis of PBX 9502 rate stick data. In: Proceedings of Eleventh International Detonation Symposium, Snowmass, CO, Office of Naval Research, Arlington, VA, pp. 21–29 (1998)
63. Kennedy, D.L.: Multi-valued normal shock velocity versus curvature relationships for highly non-ideal explosives. In: Proceedings of the Eleventh International Detonation Symposium, Snowmass, CO, Office of Naval Research, Arlington, VA, pp. 181–192 (1998)
64. Bdzil, J.B., Aslam, T.D., Catanach, R.A., Hill, L.G., Short, M.: DSD front models: nonideal explosive detonation in ANFO. In: Proceedings of Twelfth International Detonation Symposium, San Diego, CA, Office of Naval Research, Arlington, VA, pp. 409–417 (2002)
65. Bdzil, J.B., Short, M., Aslam, T.D., Quirk, J.J.: Higher-order DSD for detonation propagation: DSD for detonation driven by multi-step chemistry models with disparate rates. In: Proceedings of Thirteenth International Detonation Symposium, Norfolk, VA, Office of Naval Research, Arlington, VA, pp. 726–736 (2006)
66. Brochet, C.: Monopropellant detonation: isopropyl nitrate. Astronaut. Acta **15**, 419–425 (1970)
67. Zhang, F., Yoshinaka, A., Murray, S.B., Higgins, A.: Shock initiation and detonability of isopropyl nitrate. In: Proceedings of Twelfth International Detonation Symposium, San Diego, CA, Office of Naval Research, Arlington, VA, pp. 781–790 (2002)
68. Drumheller, D.S: Introduction to wave propagation in nonlinear fluids and solids. Cambridge University Press, Cambridge (1998)
69. Baer, M.R., Nunziato, J.W.: A two-phase mixture theory for the deflagration-to-detonation transition (DDT) in reactive granular materials. Int. J. Multiphase Flow **12**, 861–889 (1986)
70. Zhang, F., Frost, D.L., Thibault, P.A., Murray, S.B.: Explosive dispersal of solid particles. Shock Waves **10**, 431–443 (2001)
71. Zhang, F., Thibault, P.A., Link, R.: Shock interaction with solid particles in condensed matter and related momentum transfer. Proc. R. Soc. Lond. A **459**, 705–726 (2003)

72. Mader, C.L., Kershner, J.D.: The heterogeneous explosive reaction zone. In: Proceedings of the Ninth International Detonation Symposium, Portland, OR, Office of Naval Research, Arlington, VA, pp. 693–700 (1989)
73. Mader, C.L.: Numerical modeling of explosives and propellants, 2nd edn. CRC, Boca Raton (1998)
74. Baer, M.R.: Modeling heterogeneous energetic materials at the mesoscale. Thermochim. Acta **384**, 351–367 (2002)
75. Milne, A.M.: Detonation in heterogeneous mixtures of liquids and particles. Shock Waves **10**, 351–362 (2000)
76. Ripley, R.C., Zhang, F., Lien, F.-S.: Shock interaction of metal particles in condensed explosive detonation. AIP Conf. Proc. **845**, 499–502 (2005)
77. Ripley, R.C., Zhang, F., Lien, F.-S.: Detonation interaction with metal particles in condensed explosives. In: Proceedings of the Thirteenth International Detonation Symposium, Norfolk, VA, Office of Naval Research, Arlington, VA, pp. 214–223 (2002)
78. Walsh, J.M., Christian, R.H.: Equation of state of metals from shock wave measurements. Phys. Rev. **97**(6), 1544–1556 (1955)
79. Cooper, M.A., Baer, M.R., Schmitt, R.G., Kaneshige, M.J., Pahl, R.J., Desjardin, P.E.: Understanding enhanced blast explosives: a multi-scale challenge. In: Proceedings of the 19th International Symposium on Military Aspects of Blast and Shock, Calgary, AB, Oct. 1–6 (2006)

4

Detonation of Metalized Composite Explosives

M.F. Gogulya and M.A. Brazhnikov

4.1 Introduction

From the end of the nineteenth century, readily oxidized ingredients and metals (Mg, Zn, and Al among them) came into use as a component of explosives to increase their fugacity. The first patent on Al incorporation into explosive formulations was given in 1899. Al came into wide use during the First World War as an ingredient of ammonals, in which its content varied from 7.5 to 23% [1]. The studies performed by Kast and actual practice showed that the enhanced properties of the explosives were not the same in all cases; moreover, the increase in fugacity was not observed for all explosives. At the same time, it was found that the addition of Al was accompanied by an increase in both mechanical sensitivity and the sensitivity to heating. According to Kast, grained Al powder, and in some cases Al flakes, offered advantages over fine powders. Reactions of Al with detonation products (DPs) were considered as secondary effects with respect to the explosive decomposition of the base high explosive (HE) [1].

A wealth of experimental data on the detonation performance of Al-containing formulations based on the different types of HE have been acquired over the course of the past century [1–27]. In general, the studies performed do not rule out the inferences of the pioneer works. The investigations are focused on formulations, conditions for their use, and various performance evaluations. The basic scientific questions are whether Al interacts with DPs and what reaction mechanisms occur at the different stages of the expanding products.

Metal-containing explosives are classified into two types. The compositions with a large amount of a HE having a high shock sensitivity and a small critical diameter (d_{cr}) are the first type. The second type are "fuel plus explosive oxidizer" mixtures [e.g., ammonium nitrate (AN) or ammonium perchlorate (AP) with a fuel]. The latter are characterized by relatively low detonation performance, low shock wave sensitivity, large infinite detonation diameter (d_{inf}), and large d_{cr}. Such "commercial" or "industrial" explosives are particularly

F. Zhang (ed.), *Shock Wave Science and Technology Reference Library: Heterogeneous Detonation*, DOI: 10.1007/978-3-540-88447-7_4,

nonideal. Their performance has been collated and analyzed in a number of handbooks [10, 14, 22, 26]. Commercial explosives are widely used; however, the study of the macroscopic kinetics of their explosion decomposition is a rather complicated problem because the decomposition proceeds as a multiphase process.

Binary and ternary metal-containing explosives are based, generally, on HE with low d_{cr}, which allows tests to be performed in a laboratory-scale explosion chamber and many experimental investigations of this type have been conducted. The studies usually include measurements of detonation velocity (DV), pressure history or particle velocity history, temperature, acceleration ability, and heat of explosion. Mechanical and shock wave sensitivity, critical diameter, brisance, and fugacity can also be measured. In doing so, the role of the factors controlling the effect produced by incorporating Al into explosives was discovered. The role of the following parameters was examined:

- Metal content
- Metal particle size and particle morphology
- Size of explosive grains
- Density and diameter of the charge
- The origin and oxygen balance (OB) of the base explosive

Progress is being made towards the numerical simulation of the detonation performance of Al-containing explosives. Advantages have been gained in understanding the mechanisms and regularities of the detonation in such materials.

On the basis of the results presented in a number of monographs and reviews [14,15,22,26], one can make some inferences on the Al effect. Incorporation of Al into the formulations (up to 20%) results in an increase in the heat of explosion and fugacity. Brisance is the same as it is for the base explosive (or slightly less). A moderate gain in acceleration ability is observed, while the DV reduces as the concentration of the metal particles decreases. The addition of Al involves a decrease of both pressures and particle velocities. Moreover, in some cases, detonation parameters are reduced more than that with the addition of inert ingredients such as LiF. Most of the aforementioned inferences are based on the data obtained for micrometric Al ingredients, while recently some new techniques for the production of energetic materials with uniformly distributed nanometric components, such as nanoscale Al in an explosive matrix, have been developed. Consequently, there is great motivation to investigate formulations of nanometric Al and HE.

This chapter mainly considers some results on the detonation performance of binary and ternary formulations, based on powerful HEs, including bis(2,2,2-trinitroethyl)nitramine (BTNEN) and ammonium dinitramide (ADN) – organic and inorganic oxidizers, with micrometric and nanometric Al powders. Particular attention has been given to nanocomposite explosives.

Since the literature on the subject is quite voluminous, this review is mainly restricted to the results reported in Russia, which may not be available to other nations. The exclusion of works published in Europe and America is recognized. The selection of the data may also be caused by the authors' scientific interests.

4.2 Detonation Velocity

DV is the parameter most sensitive to the transformations which take place in a zone of the wave closest to the shock front (or detonation zone). The majority of the work which has been carried out shows that the addition of Al to a mixture of a common HE (with negative or close to zero OB) results in a decrease in DV [4,12,17,22,28].

4.2.1 DV in Metalized Explosives

The DVs of metal-containing formulations have been the subject of many investigations. Some experimental velocities (D) and calculated velocities (D_{calc}) are given in Table 4.1 [28]. The latter were calculated assuming a formulation as a mixture of explosive grains with the spaces between grains filled by a nonexplosive admixture. The admixtures were assumed to neither react nor heat up in the course of detonation decomposition of the basic HE.

The calculation procedure employed is an empirical scheme suitable for engineering purposes only [28]. It does not determine whether the Al reacts or not. There are two viable explanations for the reduction in DV. First, the metal particles have not begun to react during the HE detonation but act as inert particles; energy released at detonation is consumed by the compression

Table 4.1. Detonation velocities in metalized explosives

Formulation (wt%)	Metal particle size (μm)	ρ_0[a] (g cm^{-3})	D (km s^{-1})	D_{calc} (km s^{-1})
PETN–Mg (85:15)	2	1.42	6.87	6.97
PETN–polysiloxane–Mg (70:16:14)	20	1.53	7.12	7.08
PETN–polysiloxane–Al (70:5:25)	30	2.20	6.82	6.77
PETN–polysiloxane–Ni (70:5:25)	2	2.02	7.05	6.98
PETN–polysiloxane–Ti (70:5:25)	2	1.85	7.16	7.18
RDX–polysiloxane–W (65:5:30)	2	3.77	4.18	4.20
RDX–Al (50:50)	10	2.10	7.75	7.80

PETN pentaerythritol tetranitrate
[a]Charge density

(and possible heating) of the metal particles. Second, the metal, e.g., Al, partially reacts; however, the expansion of the DP caused by the energy release of metal oxidation does not compensate for the molar loss in gaseous products due to formation of higher oxides.

Mg and Al are the metals which have been widely used as admixtures in explosives from the very beginning of the twentieth century [1]. Trademarks of Al and Mg used in the majority of the works performed in Russia are given in Table 4.2. In this chapter, metal powders are denoted by the type of the metal with the mean particle size written in parentheses, or the metal is denoted by its trademark, as reported in the original works.

In [3, 4], RDX-based formulations with MPF-3 and ASD-1 powders were studied. The samples, 40 mm in diameter, had the same RDX density: 1.14 or 1.65 g cm^{-3}. The DVs in high-density, Mg-containing explosives were also calculated using the scheme [28]. The data are presented in Fig. 4.1. The general trend is a decrease in DV with an increase in the metal content. In [3], the authors noted that the effect of Al content on detonation performance, in particular DV, was different for low-density and high-density charges. The reduction in DV is much less in pressed charges than in loose-packed mixtures. The DV in low-density Al-containing charges is higher than in Mg-containing charges. Charges with finer Al, but the same metal content, showed a decrease in DV. This effect is more evident in the case of low-density charges (see Fig. 4.1).

Aniskin and Shvedov [3, 4] considered whether Al had partially reacted in the detonation zone. On the one hand, DVs measured experimentally were higher than those calculated. On the other hand, considering the fact that the

Table 4.2. Al and Mg powders [9, 14, 101, 112]

Particle form	Trademark	Mean size (μm)	Metal content (wt%)
Al			
Irregular	PA-3[a]	150	≥98
Spherical	ASD-1	15	≥99
	ASD-4	7	≥98
	ASD-6	3.6	≥96
	Sub-micron[d]	0.5	≥73
	Ultrafine[e]	0.1	≥87
	Alex	0.1–0.2	≥85–88
Flaked	PAP-2[b]	1	≥85
	PP-1[c]	8	≥97
Mg			
Spherical	MPF-3	130	≥99.5

[a]Specially sieved fraction
[b]With mean particle dimensions $1 \times 20 \times 50\,\mu m^3$
[c]Size 160 μm or less
[d]Laboratory produced by plasma technique
[e]Laboratory produced by levitation Gen-Miller technique

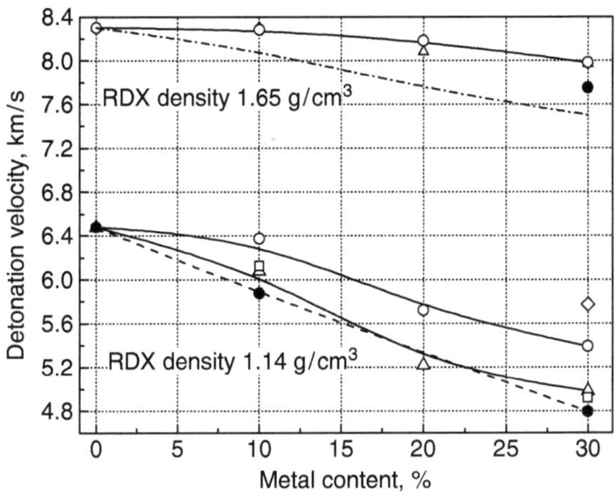

Fig. 4.1. Detonation velocity versus metal content in RDX-based formulations [3,4]. Experimental points: *triangles* Mg(50–200 µm), *open circles* Al(1–50 µm), *diamonds* Al(100–400 µm), *squares* Al(1–10 µm), *filled circles* Al (1–5 µm). The *dash-dotted line* is the calculation according to [28]

same DVs had been measured for Al- or Mg-containing high-density charges, the authors concluded that Al had to be inert within the detonation zone [4]. Finally, using the Arpege simulation program [29], they calculated the DV for an RDX–Al (85:15) formulation and compared the results with the experimental point at a charge density of 1.68 g cm^{-3}. The data obtained were as follows: 7.42 km s^{-1} (reactive Al), 7.43 km s^{-1} (inert Al), 7.48 km s^{-1} (experiment). Thus, the efforts to clarify the question of Al reactivity in mixtures with RDX based on the DV data alone have not been successful.

In [30], the formulations studied contained 20, 30, 40, or 50 wt% Mg powder with a purity over 98.5% (mean particle size 60–300 µm) and commercial HMX with 5 wt% wax [HMX(5w)]. The DVs measured in unconfined charges (d = 30 mm) and in copper tubes of different internal diameters ($d_{\rm i}$) are presented in Fig. 4.2. The introduction of Mg results in a decrease in the DV compared with that of HMX(5w).

Calculations of the detonation parameters were performed using the TIGER program [21] with the set of parameters in the Becker–Kistiakowsky–Wilson equation of state recommended in [31]. The DVs were calculated assuming the chemical activity of Mg occurred in the reaction zone and full thermochemical equilibrium of the compounds (see line 5 in Fig. 4.2). The velocities were also calculated assuming inert Mg (see line 4 in Fig. 4.2). In [32], it was shown that, in the reaction zone, the temperature between the DPs and the inert additive is not equalized for particles greater than 1–2 µm in diameter. Thus, the DV was calculated with the assumption of no heat exchange between the components. The specific volume and internal energy of the components, as well as complete mechanical equilibrium, were assumed.

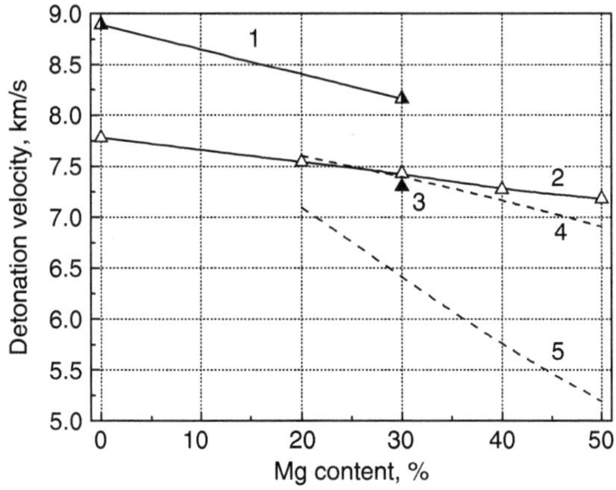

Fig. 4.2. Detonation velocity versus Mg content in HMX with 5 wt% wax [HMX(5w)]-based formulations [30]. Experimental points: *1* copper tubes d_i = 25 mm, $\rho_0 = 1.74$–$1.78\,g\,cm^{-3}$; *2* unconfined charge, $d = 30$ mm, $\rho_0 = 1.60\,g\,cm^{-3}$; *3* copper tubes $d_i = 15$ mm, $\rho_0 = 1.60\,g\,cm^{-3}$. Calculated points: *4* inert Mg; *5* reactive Mg

The DPs were described by the Jones–Wilkins–Lee [33] equation of state. For the additives and nonreacted explosive, the Grüneisen equation of state, based on the shock adiabate [34], was used.

The DVs obtained for Mg as an active chemical additive are considerably lower than the DVs calculated assuming Mg as an inert additive. Cudzilo and Trzcinski [30] believe that in fact some additional heat is released during the reaction between Mg and oxygen; but, on the other hand, MgO occurs as a condensed phase and it does not contribute to the pressure of the DP. Furthermore, Mg fixes oxygen from HMX(w) products, preventing part of the carbon from transforming into gaseous carbon oxides. The increase in the pressure of the DP owing to the heat of exothermic reactions with Mg is lower than the reduction in the pressure resulting from the reduced number of moles of the gaseous products.

The DVs in aluminized HE based on RDX and HMX with 5.5% wax [RDX(5.5w) and HMX(5.5w), respectively] were reported in [17] and are reproduced here as Figs. 4.3 and 4.4. The charges were 20 mm in diameter and about 160 mm long. The charge density varied between 89 and 97% of theoretical maximum density (TMD).

In Fig. 4.3, the D–ρ_0 dependencies are approximated by straight lines. The lowest correlation coefficient occurred for the largest Al content (20%). The DV measured in an HMX(5w)–Mg (80:20) mixture [30], normalized to the same explosive density (see the Appendix) as in HMX(5.5w)–Al, lies on the line constructed for 20% Al. This agrees with the inferences of Aniskin and Shvedov [3,4], where no difference in the DV for dense Mg- and Al-containing charges was observed.

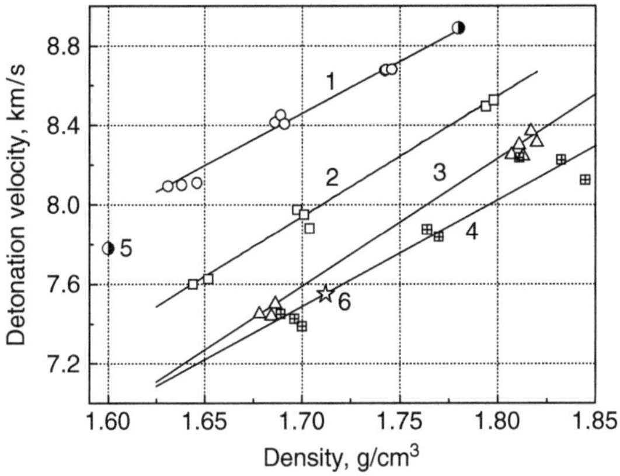

Fig. 4.3. Experimental detonation velocity versus charge density in HMX-with-wax-based formulations. ASD-1 Al content [17]: *1* 0 wt%, *2* 10 wt%, *3* 15 wt%, *4* 20 wt%. Mg content [30]: *5* 0 wt%, *6* 20 wt%

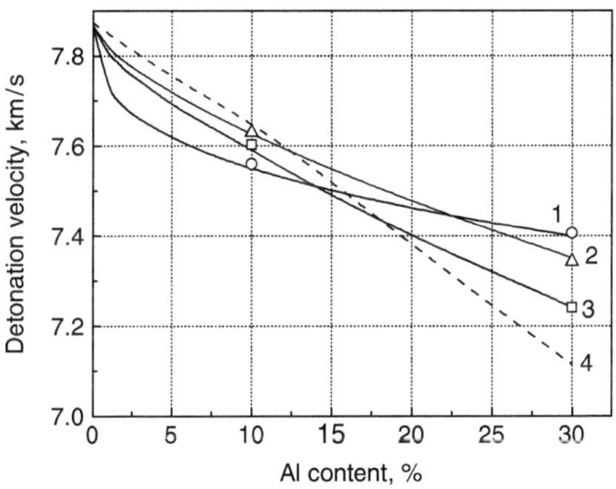

Fig. 4.4. Detonation velocity versus Al content in RDX(5.5w)-based formulations (90% of the theoretical maximum density) [17]. Experimental data: *1* ASD-4, *2* PP-1, *3* ASD-1. Calculations: *4* made according to [28]

In [17], the effect of Al particle size and particle shape on the DV was examined at different metal contents, as shown in Fig. 4.4. Spherical ASD-1 and ASD-4, and flaked PP-1 were used. For RDX(5.5w) formulations, the authors came to the following conclusions:

- The DV depends not only on the Al particle size but also on the particle shape.
- For an Al content less than 15–20%, the decrease in DV is greater for spherical Al particles (ASD-1, ASD-4) than for flaked Al [Al(fl)] (PP-1). In addition, smaller Al particles reduced the DV further.

- For an Al content more than 20%, a larger velocity reduction is observed for larger particle sizes; the decrease is independent of particle shape.

The latter conclusion is in conflict with the data from [3,4] shown in Fig. 4.1. This contradiction can be related to the difference in the base explosive [RDX and RDX(5.5w)] and the charge diameter used. For HMX(5w) formulations, the effect of charge diameter was observed in [30]. DVs measured in copper casings with $d_i = 15\,\text{mm}$ and $d_i = 30\,\text{mm}$ differed slightly (by $140\,\text{m}\,\text{s}^{-1}$) at 30% Mg content [30]. Thus, we may assume that for formulations based on RDX(5.5w) or HMX(5w) with a high metal content, $d_{\text{inf}} > 20\,\text{mm}$. The larger spread of the velocities in the case of HMX(5.5w)-based formulations with 20% Al [17] can be explained in the same manner.

4.2.2 DV in Aluminized Explosives

A comprehensive treatise of aluminized formulations based on different explosives (including those with different negative and positive OBs) was undertaken in [35–41]. The base explosives were HMX, nitroguanidine (NQ), and BTNEN. The characteristics of the Al powders are given in Table 4.2. Two classes of commercial HMX were used: HMX(A), with a mean particle size 200–300 μm, and HMX(B), with a mean particle size 10–20 μm. The accuracy of the DV measurement was less than $\pm 50\,\text{m}\,\text{s}^{-1}$.

4.2.2.1 HMX-Based Explosives

HMX–Al charges, 20 mm in diameter and about 100 mm long, were tested. The measured DVs were normalized to the density of pure HMX (see the Appendix), $1.808\,\text{g}\,\text{cm}^{-3}$ (see Figs. 4.5, 4.6). The experimental curves of DV

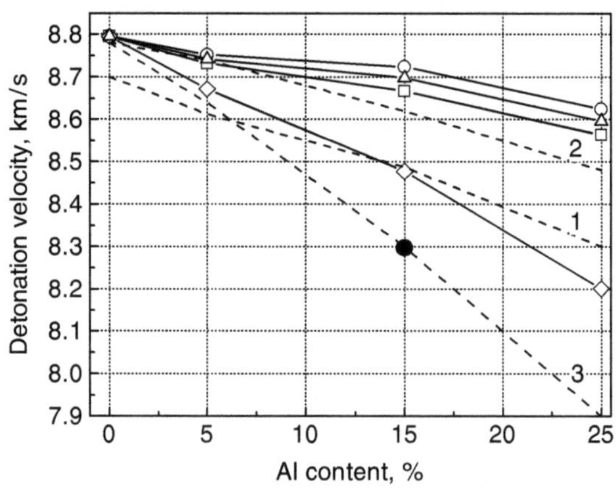

Fig. 4.5. Normalized detonation velocity versus Al content in HMX(A)-based formulations. Experiments [37, 43]: *open circles* Al(150 μm), *triangles* Al(15 μm), *squares* Al(7 μm), *diamonds* Al(0.5 μm), *filled circles* Al(0.1 μm). Calculations: *1* reactive Al [42]; *2* inert Al, *3* reactive Al [36]

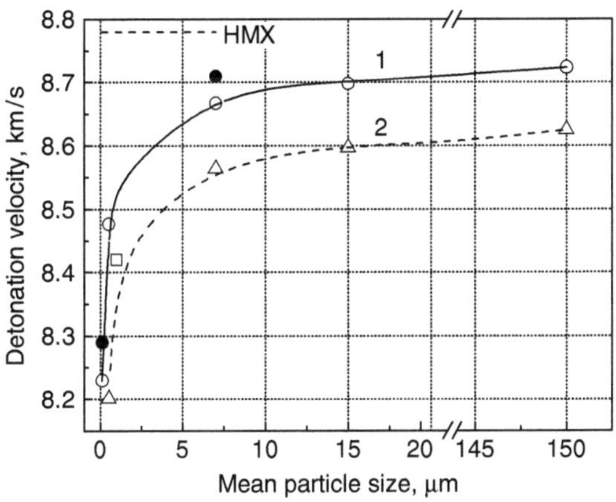

Fig. 4.6. Normalized detonation velocity versus Al particle size in HMX-based formulations [37,43]. *1* HMX(B)–Al (85:15), *open circles* spherical Al, *square flaked* Al [Al(fl)], *filled circles* HMX(A). *2* HMX(A)–Al (75:25), *triangles* spherical Al. The *horizontal line* is the detonation velocity of pure HMX at $\rho_0 = 1.808\,\mathrm{g\,cm^{-3}}$

versus Al content for micron-sized Al (i.e., diameter greater than 1 μm) are approximately linear and have the same gradient (i.e., decreasing DV with increasing Al content) (see Fig. 4.5). The largest gradient for the DV was observed for sub-micron-sized Al powders and the gradient increases with increasing Al content.

The Wood–Kirkwood kinetic detonation model has been applied to the analysis of Al-containing explosives [42]. A kinetic rate law was developed for Al combustion in condensed detonation. The combustion rate that best fits the experimental data in [43] depended on the pressure, concentration of water, and the surface area. The inclusion of the surface area in the model seems to be of great importance for a better understanding of the actual processes occurring in the detonation wave. Unfortunately, simple surface area scaling of the rate did not replicate the experimental particle-size dependencies and the only reasonable correlation was found for 15-μm Al-containing formulations. The calculated velocities are indicated as curve 1 in Fig. 4.5 in comparison with the experimental data. The difference between the calculated and experimental data can be reduced by taking into account the binder, as the authors did for HMX in [35]. For HMX(5w), Mader [21] gives a value $8.73\,\mathrm{km\,s^{-1}}$ (at $1.776\,\mathrm{g\,cm^{-3}}$), which correlates with the calculated DV that can be slightly above the one presented in Fig. 4.5. Other detonation models describing aluminized explosives were proposed in [36] (see Sect. 4.3.1). DVs calculated according to that model are represented by lines 2 (Al inert) and 3 (Al reactive) in Fig. 4.5.

DV dependencies versus Al particle size are shown in Fig. 4.6. The curves constructed for 15 and 25% Al content are very similar. The reduction in the DV with decreasing Al particle size in HMX-based formulations agrees well with the data for RDX-based explosives [3,4].

Rapid velocity decreases take place when the Al particle size drops below about 10 μm. The DVs in the mixtures with Al(fl) and Al(0.5 μm) are identical, within experimental error [37]. The powders have a similar mass content of pure Al, but have different particle shapes. The velocities measured for HMX(A)- and HMX(B)-based compositions with Al(0.1 μm) and Al(7 μm) were very close to each other. The DVs for the finer HMX base are a little lower than those for the coarser HMX base. The difference in Al particle shape or HMX grains used (at the same Al content) results in different initial charge structures (uniformity of Al distribution). We believe that the structure has a minor effect on the DV in HMX-based charges in comparison with the mass content of metal additive and the mean size of the particles.

Thus, the general trend for high-density HMX-, pentaerythritol tetranitrate (PETN)-, or RDX-based explosives containing metal additives is a reduction in DV with increasing mass of metal. However, [38] provides experimental results showing an increase in DV for RDX-based formulations with Alex. The charges were 20 mm in diameter and 70 mm long. The RDX density in the charge was $0.90\,g\,cm^{-3}$. The DV decreased from 5.4 to $4.7\,km\,s^{-1}$ at an Alex mass content below 30%. However, a very fast rise in DV was observed with increasing Alex content (up to $7.1\,km\,s^{-1}$ at 50% Alex). This result was explained by the special features of Alex powder and the additional energy restored [38]. The authors believe that the additional energy is released owing to particle structure ordering when the Al density in the charge exceeds some threshold; this liberated energy promotes Al interaction with the DP, which, in turn, results in a sharp increase in DV. However, [39] shows that the amount of any structurally bound energy in Alex, if it exists, is very small. The value of the DV given, $7.1\,km\,s^{-1}$, measured for the low-density formulation is comparable to the value $7.75\,km\,s^{-1}$ reported in [28] for high-density charges with micron-sized Al (see Table 4.1). At 50% Alex content, the charge density is $1.33\,g\,cm^{-3}$, but pure RDX of the same density would have practically the same velocity of $7.16\,km\,s^{-1}$. This means that interaction of Alex with the DPs should give the same effect on detonation velocity as an increase in RDX density by $0.43\,g\,cm^{-3}$ (as its density in the composite charge was only $0.9\,g\,cm^{-3}$). The experimental difference in the solid explosion products of the formulations with micron-sized and nanosized Al at the same metal content is displayed in Table 4.3 [40]. Nevertheless, the reported DV increase

Table 4.3. Explosion products of RDX–Al formulation

Al (wt%)	Surface equivalent particle size (μm)	Weight content of solid products (%)		
		AlN	Al_2O_3	Al
10	10	0.7	98.1	1.2
	0.23	1.2	97.5	1.3
50	10	49.3	48.6	2.1
	0.23	57.1	41.0	1.9

at 50% Alex content in the RDX-based mixture remains unexplained and calls for further investigation.

The reduction in DV can be caused by energy consumption on shock compression, heating of Al particles and involving them in the flow, and reduction of the amount of gaseous products in the course of Al reaction with the DPs within the detonation zone. The data given in the following sections can partially clarify the aforementioned problem. Two series of experiments were performed with different base HEs: NQ – an explosive with a high hydrogen content – and BTNEN – a powerful organic oxidizer. The DP of NQ contains a large amount of water, and its interaction with Al does not reduce the total number of moles of gaseous products (unlike the interaction with carbon dioxide). The DP of BTNEN contains free oxygen.

4.2.2.2 NQ-Based Explosives

NQ ($CH_4O_2N_4$) is a group II explosive with an OB of −30.8% [41]. The DV was measured in pressed charges of 40-mm diameter containing 15% Al, and a total charge length of about 50 mm. The data, shown in Fig. 4.7, were normalized to the NQ density of the charge, $1.635\,g\,cm^{-3}$. The addition of 15% spherical Al(0.1 μm) to NQ did not effect the DV, while the addition of Al(fl) slightly reduced the velocity by about $100\,m\,s^{-1}$.

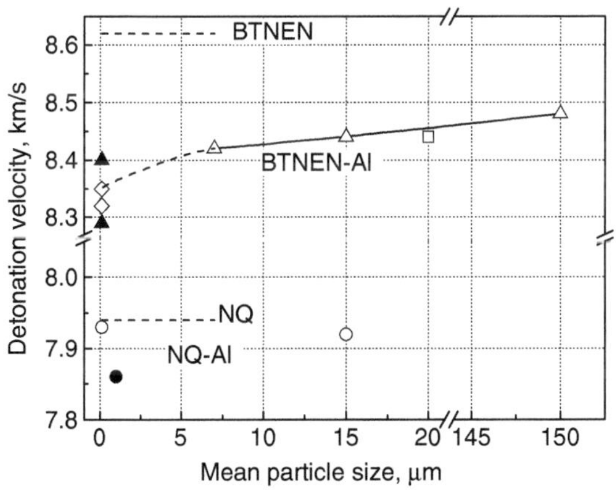

Fig. 4.7. Normalized detonation velocity versus Al particle size in nitroguanidine (*NQ*)- and bis(2,2,2-trinitroethyl)nitramine (*BTNEN*)-based formulations with 15% Al content. Experiments [37, 74, 75, 77]: NQ *open circles* Al(0.1 μm) and Al(15 μm), *filled circles* Al(fl); BTNEN *open triangles* and *filled triangles* charge $d = 20$ mm with and without copper casing, respectively, *diamonds* unconfined charge $d = 40$ mm, *squares* mixture containing LiF, unconfined charge $d = 20$ mm. The *horizontal lines* are the detonation velocity in NQ and BTNEN at $\rho_0 = 1.635$ and $1.900\,g\,cm^{-3}$, respectively

4.2.2.3 BTNEN-Based Explosives

BTNEN ($C_4H_4O_{14}N_8$) is an explosive with a positive OB of 16.5%. The compositions tested contained 15% Al. The data on DV versus particle size are shown in Fig. 4.7. In formulations with micron-sized Al, the DV was measured in copper tubes with d_i = 20 mm and a length of 180 mm (by the T-20 test method described in Sect. 4.3). For Al(0.1 μm), tests were performed with charges 20 and 40 mm in diameter and 100 and 60 mm long, respectively. The difference in DVs lies within the experimental error.

The decrease in the velocity for the compositions with micron-sized Al is comparable to the effect produced by the addition of LiF. The largest decrease was obtained with Al(0.1 μm). For 85:15 HMX–Al(0.1 μm), the relative velocity decrease (referred to pure HMX) is about 6%, while the decrease is only about 3% for the similar BTNEN-based formulation. It can be assumed that Al(0.1 μm) reacts with the free oxygen in the DP with heat release which far exceeds that produced by Al interaction with H_2O, CO_2, and CO (the main oxygen-containing DPs of explosives with negative OB).

Detonation of composite explosives based on HMX, RDX, or BTNEN was simulated using the thermodynamic code TDS [44] and the resultant DV data are given in [45]. The calculations were based on the statistical mechanics equation of state for fluid DPs and the semiempirical equation of state for nanosized carbon particles and liquid phases of Al and Al_2O_3. Two possible modes of Al behavior in the reaction zone were considered: (1) Al was inert but able to melt; and (2) Al reacted with the DP. In the latter case, the possible formation of liquid and solid phases of Al_2O_3 was taken into account. In the context of the model, relationships between the DV and explosive density in the charge (ρ_{HE}) were calculated. Linear relationships were observed for most individual HEs. The model draws a more complicated picture for Al-containing explosives. The melting of Al_2O_3 results in a deceleration of the DV growth with increasing explosive density. Moreover, with increasing Al content (up to 25%) and completeness of its oxidation, the charge density at which the deviation from a linear dependence takes place becomes greater and is accompanied with a greater transition area (see Fig. 4.8). One can estimate the DVs in RDX–Al (75:25) compositions from the experimental data given in [3, 4] as 5.6 km s^{-1} at ρ_{HE} = 1.14 g cm^{-3} and 8.1 km s^{-1} at ρ_{HE} = 1.65 g cm^{-3}. The values are inconsistent with those shown in Fig. 4.8.

The aforementioned numerical simulation was further conducted and compared with experiments reported in [46] with a 75:25 BTNEN–Al formulation of Al(3.6 μm), for which one could expect the greatest effect. The charges were 20 mm in diameter and about 60 mm long. The charge density varied from 1.01 to 2.05 g cm^{-3}. The mixtures with densities less than 1.32 g cm^{-3} were placed in Plexiglas casings with 0.65-mm-thick walls. The charges of higher density were unconfined. The resulting experimental data are shown in Fig. 4.9.

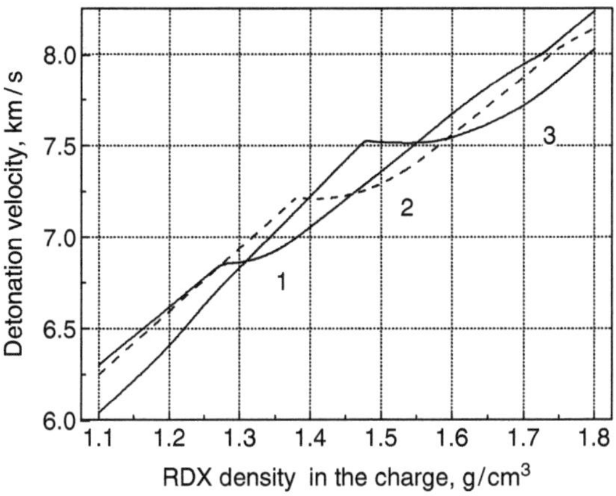

Fig. 4.8. Calculated detonation velocities in RDX–Al (75:25) formulations versus RDX density of the charge [45]. Completeness of Al oxidation: *1* 50%, *2* 75%, *3* 100%

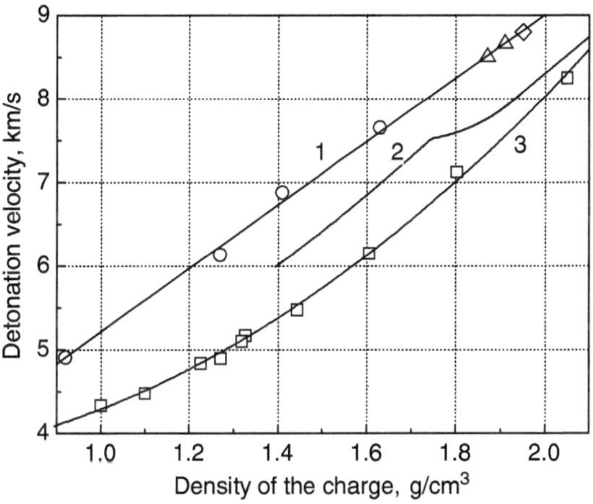

Fig. 4.9. Detonation velocities in BTNEN-based formulations versus charge density. BTNEN: *1* experimental points, *circles* [46], *triangles* [37], *diamonds* [110] and straight line approximation. BTNEN–Al: *2* calculation assuming 50% Al oxidation [45], *3* experimental points *(squares)* and the parabola approximation [46]

With practically the same accuracy, the experimental D versus ρ_0 data [46] can be fit to either a single parabola or two linear segments, as follows:

$$D(\rho_0) = 3.934 - 1.348\rho_0 + 1.696\rho_0^2 \tag{4.1}$$

or

$$D(\rho_0) = 1.770 + 2.511\rho_0, \text{ at } \rho_0 = 1.00 - 1.42\,\text{g cm}^{-3} \tag{4.2}$$

and

$$D(\rho_0) = -1.210 + 4.612\rho_0, \text{ at } \rho_0 > 1.42\,\text{g cm}^{-3}. \tag{4.3}$$

Nonlinear $D(\rho_0)$ curves were also reported in [47] for RDX–TNT–Al (45:30:25) and TNT–Al (80:20) mixtures. In Cook et al.'s [47] view, the nonideal character of the explosives and the influence of the initial charge density on the ratio between solid Al_2O_3 and gaseous Al_2O in the DP explain such behavior.

It is likely that the changes in the gradient of the $D(\rho_0)$ dependence for BTNEN-based formulations are caused by changes in the qualitative composition of the DPs of aluminized explosives or their phases in the corresponding density regions. In the context of the assumptions made in [45], one can explain the changes in gradient by phase changes in Al_2O_3. If $\rho_0 < 1.42\,\mathrm{g\,cm^{-3}}$, the oxide can be considered as a liquid and above this value, as a solid. In [45], it was pointed out that "... if the experimental research did not detect breaks in the HE slope in D–ρ_0 curve, this would not, however, mean that the Al remains inert during the detonation. There could be a variety of reasons for the failure of the agreement with experiments including the case that our EOS [equation of state] of liquid Al_2O_3 is possibly too rough due to the lack of high pressure data for this material."

The data on the DV in BTNEN–Al (90:10) placed in steel tubes, $d_{\mathrm{i}} = 12\,\mathrm{mm}$, reported in [11] are described by a linear $D(\rho_0)$ dependence (about $7.3\,\mathrm{km\,s^{-1}}$ at $\rho_0 = 1.68\,\mathrm{g\,cm^{-3}}$ and about $8.4\,\mathrm{km\,s^{-1}}$ at $\rho_0 = 2.0\,\mathrm{g\,cm^{-3}}$), but quantitatively they are inconsistent with the data in [46]. While the possibility of the effect of the diameter on the $D(\rho_0)$ dependence found in [46] may not be ruled out, such an influence at 20-mm diameter seems to be unlikely for the highly detonable BTNEN-based formulations.

Elucidation of the proper mechanism controlling the DV and its dependence on the density can be performed in the course of developing an adequate model describing the detonation of aluminized explosives.

4.2.2.4 TNT-Based Explosives

TNT is the most widely used basic HE, with an OB more negative than that of RDX and HMX, and a critical diameter larger than that of these two explosives (27.5 mm for melt-cast TNT [48]). DVs in melt-cast TNT-based formulations containing Al, including nanometric Al, were measured in a number of works [9,49,50]. Some of these data are shown in Fig. 4.10.

The advantageous effects of Alex on DV were not observed for ternary TNT–RDX–Al mixtures, nor for Al-containing HMX- or RDX-based plastic-bonded explosives, nor for RDX-based plastic-bonded explosives with AP [9,49,50]. The DV versus Al particle size relationship in composition B with 10% Al is shown in Fig. 4.10a, as an example. It is similar to the dependence constructed for HMX-based formulations (see Fig. 4.6); however, the DV of composition B-based mixtures containing micron-sized Al and Alex particles differs by only $150\,\mathrm{m\,s^{-1}}$.

It is seen that the Al particle size affects the measured DV; the difference in DV is larger at small charge diameters and Brousseau et al. [9]

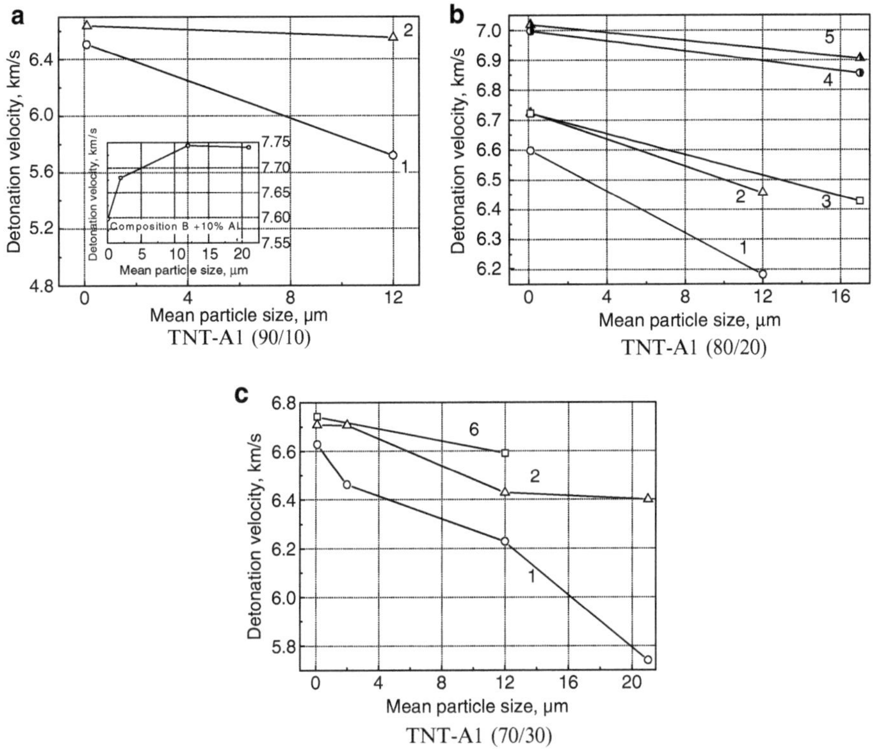

Fig. 4.10. Detonation velocity in melt-cast TNT charges versus Al particle size [9,49,50]. Charge diameter: *1* 25.4 mm, *2* 31.8 mm, *3* 41.1 mm, *4* 74.8 mm, *5* 81.8 mm, *6* 57.9 mm

believe that this diameter effect will be eliminated at larger (infinite) diameters. For 80:20 compositions, the velocity in Alex-containing formulations is close to that in pure TNT, which has the same density as the aluminized charge of 81.8-mm diameter (7.04 km s^{-1} at 1.63 g cm^{-3} [26]). A possible explanation for the observed results is a sensitizing effect when Al is incorporated into TNT-based charges, which leads to a reduction of d_{cr}. Replacement of conventional Al with Alex (100–200-nm mean size) reduces the d_{cr} of Tritonal-type formulations (80:20 TNT–Al) by nearly a factor of 2 [9]: for high-density formulations containing Alex, d_{cr} was estimated to be less than 10 mm.

4.2.2.5 AN-Based Explosives

It is well known that the addition of Al powder to commercial-grade AN results in an increase in DV [47,51], as shown in Fig. 4.11. For comparison, the DV measured in pure unconfined AN, at $d = 460$ mm and $\rho_0 = 1.04$ g cm^{-3}, was less than 3 km s^{-1}.

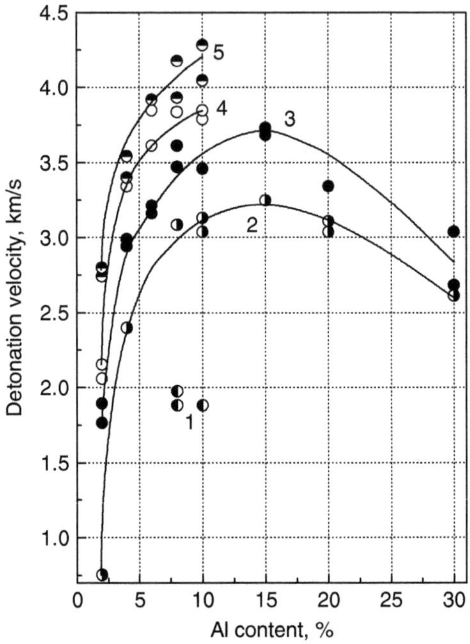

Fig. 4.11. Detonation velocity versus percentage of Al in ammonium nitrate (AN)–Al mixtures at average $\rho_0 = 1.05\,\mathrm{g\,cm^{-3}}$ [47]. Diameters of unconfined charges: *1* 25 mm, *2* 50 mm, *3* 75 mm, *4* 99 mm, *5* 127 mm

It was also found that d_{cr} of AN-based explosives containing Al was drastically reduced in comparison to that of pure AN. As reported by Cook et al. [47], at 4–15% Al content, $d_{\mathrm{cr}} \approx 25\,\mathrm{mm}$ and at 20–30% Al content, $d_{\mathrm{cr}} = 50\text{–}75\,\mathrm{mm}$. The dependence of d_{cr} on Al content exhibits a minimum near a content of 8%. For unconfined charges ($\rho_0 = 1.05\,\mathrm{g\,cm^{-3}}$), the highest DV was measured at 15% Al content (see Fig. 4.11). At the same time, Cook [51] concluded that "low-density AN–Al mixtures are in general nonideal over a broad range of diameters and particle sizes of both AN and Al." From the experimental data, it followed that for pure AN at $\rho_0 = 1.04\,\mathrm{g\,cm^{-3}}$, d_{inf} was greater than 460 mm, while for an AN-based mixture with 12% Al content at $\rho_0 = 1.05\,\mathrm{g\,cm^{-3}}$ [51], the estimated "ideal" DV is about $4.5\,\mathrm{km\,s^{-1}}$ and $d_{\mathrm{inf}} \approx 300\,\mathrm{mm}$.

Recently, DVs were measured in loose-packed AN–Al mixtures ($\rho_0 = 0.92\text{–}0.98\,\mathrm{g\,cm^{-3}}$) manually tapped into carbon steel tubes with $d_{\mathrm{i}} = 16.1$ or 21.5 mm and a wall thickness of 2.6 or 2.7 mm, respectively, and a length of 300 mm (see Figs. 4.12, 4.13) [52]. DVs were measured in the 25.1-mm-diameter charges at various contents of Al(fl) (see Fig. 4.12). The DV at 10% Al content was found to be the highest. At 10% Al content, two types of powder, atomized and flaked, were tested (see Fig. 4.13). In both cases, the results showed that the velocity had increased with a decrease in the Al particle size.

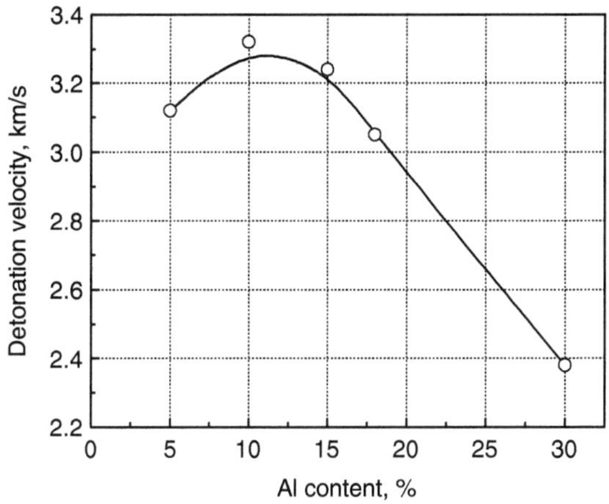

Fig. 4.12. Effect of Al content on detonation velocity of AN-based mixtures with 75-μm Al(fl) at $d_\mathrm{i} = 21.5\,\mathrm{mm}$

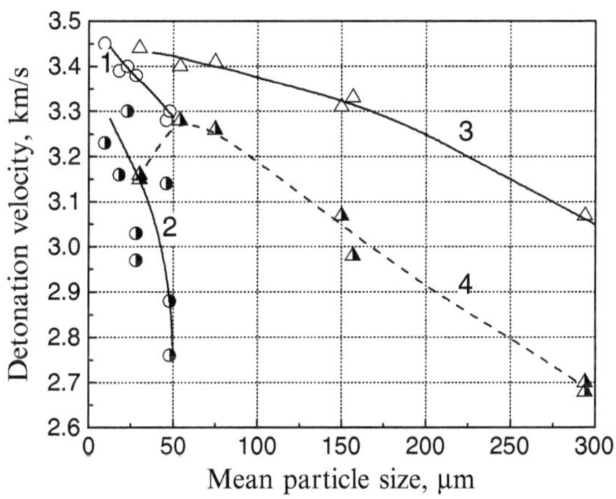

Fig. 4.13. Effect of Al particle size on detonation velocity of an AN–Al (90:10) mixture. *1, 2* atomized Al, *3, 4* Al(fl). *1, 3* $d_\mathrm{i} = 25.1\,\mathrm{mm}$, *2, 4* $d_\mathrm{i} = 16.1\,\mathrm{mm}$

The DV in loose-packed pure AN ($\rho_0 = 0.69\,\mathrm{g\,cm^{-3}}$) placed in steel tubes, $d_\mathrm{i} = 15$ or 21 mm and wall thickness about 3 mm, was measured as about 1.17 and $1.22\,\mathrm{km\,s^{-1}}$, respectively [53]. In summary, it seems that the addition of Al results in a significant increase in the DV of AN-based explosives. Moreover, the trend in the DV behavior of AN-based explosives, when Al particle size is decreased, is the opposite of that observed in powerful HEs.

Nevertheless, to estimate the role of Al powder and its contribution to the detonation from possible reactions of Al with the DP of AN, one should take into account that the diameter of the steel tubes used in the work reported in [52] was close to or less than the critical diameter for AN-based mixtures with Al.

If the experiments were performed with a charge diameter greater than an infinite detonation diameter d_{inf}, it would be easier to determine whether Al plays a chemically active or passive role in the detonation zone. The DV in AN ($\rho_0 = 0.85\,\mathrm{g\,cm^{-3}}$) placed in a steel tube, $d_{\mathrm{i}} = 100\,\mathrm{mm}$, is influenced by the wall thickness when it is varied from 5 to 20 mm (the DV increases from about 1.7 to about $3.0\,\mathrm{km\,s^{-1}}$, respectively). Moreover, the DV measured in 10-mm-thick steel tubes ($d_{\mathrm{i}} = 300\,\mathrm{mm}$) is about $3.95\,\mathrm{km\,s^{-1}}$ [54]. Since there are no experimental data on the ideal DV in AN, a calculated value of the ideal DV in AN has to be used for comparison, e.g., $4.61\,\mathrm{km\,s^{-1}}$ at $\rho_0 = 1.0\,\mathrm{g\,cm^{-3}}$ estimated in [55]. Comparing the data for AN–Al (90:10) mixtures, we find that the DVs measured in steel tubes are less than the velocities obtained for unconfined AN–Al charges of larger diameters and are also less than that calculated for pure AN. Summarizing the discussion above, we can assume that the benefits to the DV of AN–Al mixtures over pure AN are observed for charges with diameters far from the infinite diameter.

The peculiarities in the behavior of Al-containing HEs near the critical diameter d_{cr} can be illustrated with the data obtained for AN mixed with coarse TNT (mean size 0.8 mm), where the TNT was partially replaced with Al [56]. Two types of Al powder, flaked (160 μm) and spherical (less than 300 μm), were used. Several experiments with AN–TNT (80:20) and AN–TNT–Al (80:10:10) mixtures in poly(vinyl chloride) (PVC) casings and copper tubes ($d_{\mathrm{i}} = 25\,\mathrm{mm}$) were performed. For unconfined AN–TNT mixtures with a density $1.05\,\mathrm{g\,cm^{-3}}$, d_{cr} was estimated to be 35 mm. The data on the DV are presented in Fig. 4.14. The results show that it is more effective to use Al(fl) rather than TNT as the additive to AN for unconfined charges with diameters near d_{cr}. The reason for this effect is that the incorporation of more rigid Al particles into the mixture changes its structure and its dynamic impedance. It is possible to assume that the introduction of Al causes a reduction in d_{cr}. The data show that Al has no advantage over AN when the charge diameters are much greater than d_{cr} or when the charges are heavily confined. For the compositions tested, $d_{\mathrm{inf}} > 100\,\mathrm{mm}$ and the effective diameter of the charges placed into the copper tubes with $d_{\mathrm{i}} \approx d_{\mathrm{cr}}$ is much less than d_{inf}, as shown in Fig. 4.14.

The last example shows that an increase in the DV at the cost of using finer Al particles is observed not only for 1-μm and sub-micron-sized particles, but also for much coarser particles incorporated into charges with large critical diameters (for AN, $d_{\mathrm{cr}} = 127$–$160\,\mathrm{mm}$ at $\rho_0 = 1.04\,\mathrm{g\,cm^{-3}}$ [47]). Thus, a gain in DV was observed for charges with large d_{cr}, such as melt-cast TNT and AN. The addition of Al to these base explosives causes a decrease in d_{cr}. In general, the critical diameter may decrease or increase, depending on the competing effects of the sensitization due to the shock–particle interaction and desensitization from the momentum and heat transfer, or explosive dilution. The question is whether Al reacts with the DPs in the course of shock–particle interaction or, in other words, whether the observed decrease in d_{cr} is caused by additional energy release (Al reaction with the DPs) in

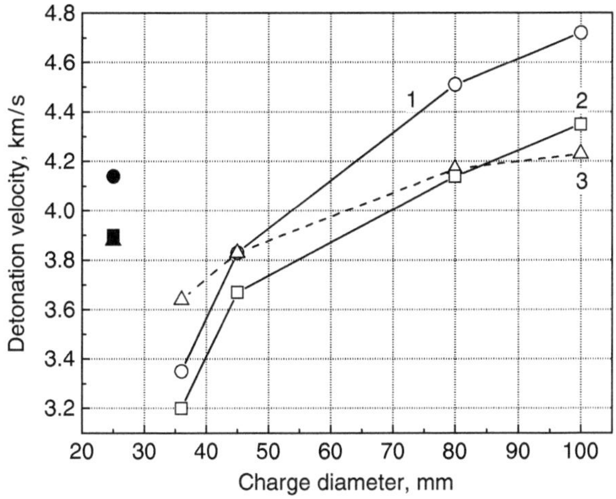

Fig. 4.14. Detonation velocity in AN-based explosives [56]. *1* AN–TNT (80:20), *2, 3* AN–TNT–Al (80:10:10), Al spherical less than 300 μm and 160-μm Al(fl). All in poly(vinyl chloride) cases. *Filled symbols* are mixtures 1, 2, and 3 in copper tubes

the detonation zone. The time and the location of Al ignition in a composite explosive are of great importance for detonation propagation itself, but they are also important for afterburning processes in the blast wave that have been investigated by many, e.g., see [57].

4.3 Pressure and Temperature Time Histories

Experimental data on pressure and temperature time histories in Al-containing explosives were reported in [37, 43, 63, 74, 75, 77, 111]. Pressed charges 40 mm in diameter and 45 mm long for HMX- and BTNEN-based formulations and 50 mm long for NQ-based formulations were tested. In the majority of the tests, detonation was initiated by a plane-wave generator made of RDX(w) and RDX booster of 1.7 g cm^{-3} and 10 mm thick. The records were obtained with the help of a two-channel optical pyrometer with a time resolution about 10 ns.

Pressure histories were recorded using an indicator technique [60], in which bromoform placed on the end of the charge was used as the indicator. Variations in pressure occurring at the DP–bromoform interface result in adequate changes in pressure at the shock wave front that spreads in the bromoform. However, the changes at the front proceed approximately 2 times more slowly. The pressure histories are presented in the DP nearby the interface with bromoform. The accuracy of the determination of absolute pressure values is about 3%.

The temperature, in the case of condensed reagents, is the most sensitive parameter which indicates the exothermic reaction. However, brightness temperature is a value averaged over the observing area. In the case of composite explosives, the luminosity measured by an optical pyrometer from the end surface of the charge is an integral characteristic of the emitting area, which consists of the spots, strongly differing in temperature within the whole recording time. At the very beginning of the history, the temperature reflects, to a large extent, the interface processes and the processes that occur at the Al particles–DP surfaces, while the pressure profiles characterize the processes which take place in the volume of the DP. The time of pressure equalization is less than that of the temperature. Moreover, the temperature, measured in the frame of the so-called window technique, is determined in the reflected wave (shock wave or rarefaction) [61]. Therefore, it is incorrect to compare pressure and temperature histories directly; only qualitative agreement or disagreement between the profiles can be examined. In the papers reviewed, brightness temperatures were measured at the DP–LiF interface. The accuracy of the measurements can be estimated as 4–5%.

4.3.1 HMX-Based Explosives

Pressure profiles measured in the DP of HMX-based formulations containing Al are shown in Figs. 4.15–4.17. Initial charge densities and detonation pressures are given in Tables 4.4–4.6.

The distinctive feature of the pressure time histories is the presence of a two-peak profile for most of the compositions tested with Al content not less than 15% (see Fig. 4.15). Maximum and minimum pressures are given in Table 4.4. Different trends for different Al admixtures are seen in Fig. 4.15. One can observe the decrease in pressure with Al content growth for the most coarse Al used, Al(150 μm), used, while the opposite tendency is seen for Al(0.5 μm); the pressure of the first peak remains practically unchanged for Al(15 μm) (see Table 4.4).

Two sets of experiments were performed to clarify this complicated problem. The results of the first one are given in Fig. 4.16 and Table 4.5. Two-peak profiles appear in HMX(A)–Al(15 μm) at an Al content of 5% and above; some profile transformations are already seen at a minor Al content of 3%. Averaging the data on the first peak for Al(15 μm) admixtures of different mass content (see Table 4.5), one can estimate the pressure as 34.7 ± 0.6 GPa. The value lies within the experimental error.

In the second series, the formulations were of the same admixture content of 15%, but differed in the batch of the Al used and HMX grain size. Pressure profiles are shown in Fig. 4.17 (some data are collected in Table 4.6.). Also tested were formulations with Al and LiF, whose mean particle size was about 20 μm; the value is comparable with that of Al(15 μm). There is no evidence of any profile transformation in LiF-containing charges, while a two-peak profile is observed at the same content of Al(15 μm) (see Figs. 4.15, 4.17). The pressure in the charges with LiF is higher than that in the mixtures with Al

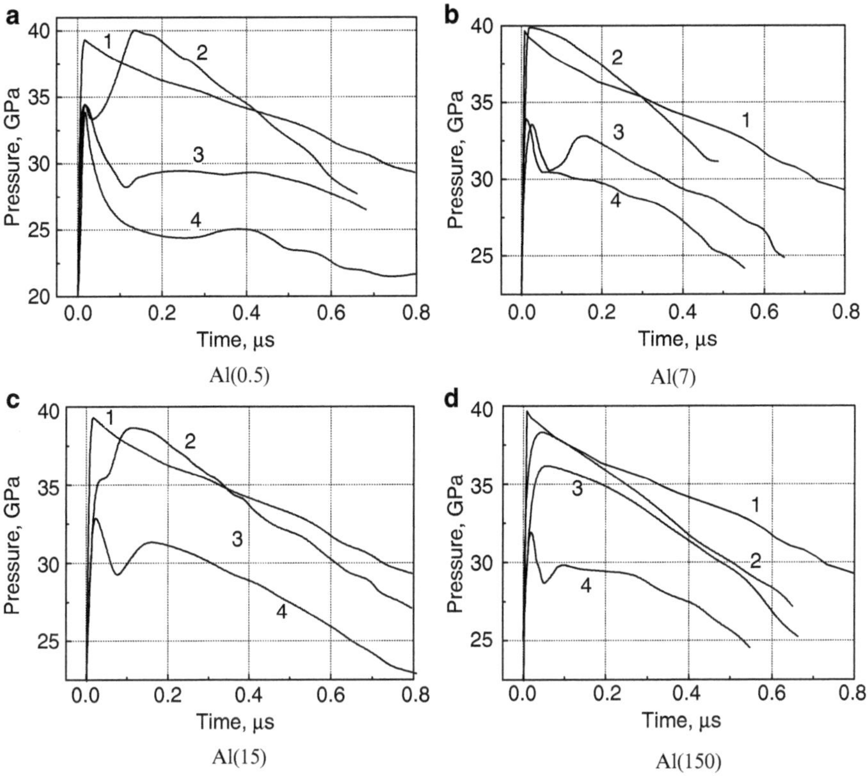

Fig. 4.15. Pressure profiles in the detonation products of HMX(A)–Al formulations. Al content: *1* 0%, *2* 5%, *3* 15%, *4* 25%

within the interpeak time (about 0.3 μs), while the inverse picture is observed for longer times (see curve 3 in Fig. 4.15c and curve 2 in Fig. 4.17).

The charges of the same Al content differed in the structure: the most uniform Al distribution was in HMX(B)–Al(0.1 μm); the least uniform was in HMX(B)–Al(fl). Thus, degeneration of the two-peak profile is observed for the former formulation, while the longest and the deepest pressure decrease takes place in the latter (see Fig. 4.17).

Variously shaped pressure profiles, which were measured for charges differing in initial density, cannot be compared directly to extract the information on the role of Al in the admixture. To perform the comparison, first both peak pressures were normalized to the HMX density in the charge (see the Appendix) and then the same was done with the values of the pressures averaged over the first 0.5 and 0.75 μs (see Table 4.6) behind the front.

Normalized second peak pressures turned out to be practically the same for micron-sized Al, while the value for nanometric admixtures is slightly higher. The role of Al in HMX can be characterized by the average pressures

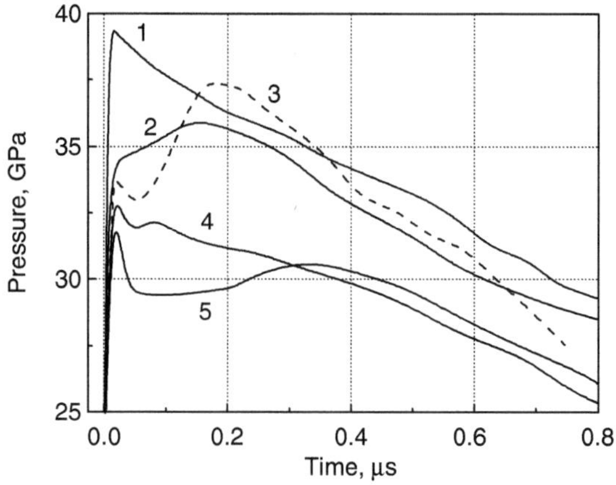

Fig. 4.16. Pressure profiles of the detonation products HMX(A)–Al(15). Al content: *1* 0%, *2* 3%, *3* 7.5%, *4* 4.5%, *5* 10%

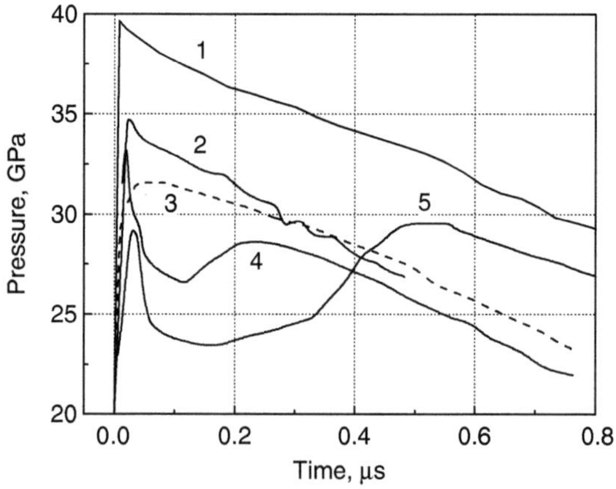

Fig. 4.17. Pressure profiles of the detonation products of HMX-based formulations (85:15) for different admixtures used. *1* HMX(A), *2* HMX(A)–LiF, *3* HMX(B)–Al(0.1 μm), *4* HMX(A)–Al(0.1 μm), *5* HMX(B)–Al(fl)

more correctly. Inert LiF reduces the pressure within the first 0.5 μs to a lesser extent than any Al admixture. Only minor advantages with regard to pressure of Al(0.1) mixed with HMX(B) over micrometric Al are observed, while the pressure is still below that value in the DP of HMX. Thus, a negative role of the two-peak profile configuration, which reduces the average pressures near the detonation front, is clearly seen. The difference in the pressures measured for micron-sized Al powders reduces with time. The role of Al in the expanding DP of HMX for longer times can be seen in the tests on acceleration ability.

Table 4.4. Detonation pressures in HMX(A)–Al formulations

Al	β (%)	ρ_0 (g cm^{-3})	P_1 (GPa)	$P^*{}_1$ (GPa)	$P^*{}_{min}$ (GPa)	$P^*{}_2$ (GPa)
Al(150 µm)	5	1.835	38.79	38.97	–	–
	15	1.885	36.29	37.18	–	–
	25	1.952	32.10	33.00	29.33	30.76
Al(15 µm)	5	1.835	35.07	35.24	–	38.84
	15	1.881	33.82	34.84	28.53	31.50
	25	1.930	33.41	35.41	30.90	33.32
Al(7 µm)	5	1.835	39.90	40.09	–	–
	15	1.870	33.62	35.15	31.90	34.34
	25	1.920	33.96	36.52	32.75	Slow decrease
Al(0.5 µm)	5	1.835	34.66	34.83	33.44	40.38
	15	1.867	34.48	36.23	29.18	30.93
	25	1.910	34.27	37.46	26.63	27.39

β Al mass content, ρ_0 initial charge density, P_1 pressure of the first peak, corresponding to the initial density, P_1^*, P_2^*, $P_{\min}^*$ pressures of the first and the second peaks, and minimum pressure normalized to the same HMX density in the charge, 1.808 g cm^{-3}(see the Appendix)

Table 4.5. Detonation pressures in HMX(A)–Al(15 µm) formulations

β (%)	ρ_0 (g cm^{-3})	P_1 (GPa)	$P^*{}_1$ (GPa)	$P^*{}_{min}$ (GPa)	$P^*{}_2$ (GPa)
0	1.808	39.67	–	–	–
3	1.816	34.39	34.86	Growth	36.41
4.5	1.834	33.81	33.88	32.74	37.52
5	1.835	35.07	35.24	–	38.84
7.5	1.844	33.87	34.34	32.28	32.84
10	1.850	33.32	34.23	30.19	31.49
15	1.881	33.82	34.84	28.53	31.50
25	1.930	33.41	35.41	30.90	33.32

Table 4.6. Detonation pressures in HMX-based formulations containing 15% of an additive *Al(fl)* flaked Al

Formulation	ρ_0 (g cm^{-3})	P_1 (GPa)	$P^*{}_1$ (GPa)	$P^*{}_{min}$ (GPa)	$P^*{}_2$ (GPa)	$P^*{}_{0.5}$ (GPa)	$P^*{}_{0.75}$ (GPa)
HMX(A)–Al(0.1 µm)	1.796	34.14	39.70	30.85	33.30	32.3	30.6
HMX(B)–Al(0.1 µm)	1.831	31.57	34.94	–	–	33.0	31.1
HMX(B)–Al(fl)	1.836	29.20	32.04	25.61	32.41	27.9	29.2
HMX(A)–Al(15 µm)	1.881	33.82	34.84	28.53	31.50	30.1	29.5
HMX(A)–LiF	1.832	34.85	38.34	–	–	33.4	31.0
HMX(B)	1.808	39.67	–	–	–	35.8	34.3

From the tests, it follows that two-peak profiles appear in HMX-based formulations with Al and the phenomenon depends on Al content, type of Al used, and the charge structure. Practically all charges tested were of the same porosity (94–95% of the TMD) and had the same length (45 mm); thus, the role of these factors was not studied experimentally.

When the two-peak pressure profiles are analyzed, it is important to take into account the times characterizing the behavior of the multiphase reactive flow of DPs, such as the duration of the detonation zone, the burn time of Al particles, and the times of particle velocity and thermal equilibration. Unfortunately, there is no exhaustive information; nevertheless, we can identify some results which can be useful for analysis.

The duration of the detonation zone of high-density RDX- and HMX-based explosives has been studied in a number of works [65–67]. For pure HMX with a charge density of 1.85 g cm^{-3}, the time was measured as 0.04 μs [65,66], while for a plastic-bonded HMX-based explosive with a binder content up to 9%, it was estimated as 0.015–0.025 μs [67] at $\rho_0 = 1.84\,\text{g cm}^{-3}$. Despite the difference in the data, the duration time of the detonation zone evaluated for high-density HMX-based explosives is, as a rule, less than the interpeak time of the pressure profiles observed for aluminized HMX-based explosives. It is interesting that information on the burn rate of HMX has been published recently. For instance, the reaction propagation rate for ultrafine HMX particles with a grain size of about 3 μm was measured as 257 ± 26 and $641 \pm 70\,\text{m s}^{-1}$ at 25 and 35 GPa, respectively [68].

There is a wealth of literature on Al combustion at relatively low pressures but there are limited data on combustion in DPs. The data obtained in cylinder tests with Al-loaded explosives were analyzed using a numerical two-phase flow model to estimate the burn rate of Al particles. In [69], two burn laws were mentioned that could be incorporated in numerical models: constant radial burn velocity, and constant rate of area change. Application of the former law to the description of nitromethane (NM)–Al mixtures with 40% of 5-μm Al resulted in a burn time of 24 μs best fitting the experimental data on the cylinder wall expansion. The usage of the latter law gave practically the same value, 25 μs. "There appears to be little dependence on the nature of the law, only the speed. If the burn time were much longer this may not remain true [69]." It was stated that the relationship between burn time and particle diameter had to be more complex than a simple square law. Nevertheless, with use of a d^2 law and the burn time of 5-μm Al, the burn time of 100-nm Al was estimated as 9.6 ns, and this gave a good fit to the experiment.

In [15], a model was proposed to calculate an average rate of Al consumption due to diffusion-controlled oxidation by the HE decomposition products in a detonation wave. The basic element of the model is a spherical cell reproducing the average heterogeneous characteristics of the explosive mixture. The cell comprises a spherical Al particle and HE decomposition products surrounding it in the form of a spherical layer. In the frame of the model, the total dimensionless time of Al depletion in HMX mixtures with Al content ranging from 5 to 25% was calculated; the time was found to depend slightly on Al content. For 10-μm particles, the total burn time was estimated as a few microseconds, assuming the effective diffusion coefficient to be about $0.1\,\mathrm{cm^2\,s^{-1}}$.

In the course of the detonation propagation through condensed composite explosives, plastic deformation of Al particles, caused by shock compression, can take place. The deformation can result in particle instability with consequent fragmentation and formation of very chemically active "fresh" surfaces. The particle fragmentation, if it exists, would control the mass transfer (effective diffusion coefficient), the burn rate, and the reaction proceeding as a whole, but there are still great difficulties to incorporate particle fragmentation mechanisms into numerical models of Al burning.

For detonation in a condensed explosive containing metal particles of 0.1–1 μm, the shock–particle interaction time is about the same order as or 1 order of magnitude less than the velocity relaxation time. Therefore, momentum transfer during the shock–particle interaction together with that behind the shock front could influence the detonation initiation and structure. In [58], numerical and theoretical calculations were reported in liquid explosives and RDX with various compressible metal particles under conditions of detonation pressure. The study indicated that the momentum transfer from the condensed explosive to the metal particles is significant during the particle crossing of the shock front. According to the study, the particle velocity after the shock–particle interaction strongly depends on the initial density ratio of explosive to metal, but is relatively insensitive to the other parameters, such as the particle acoustic impedance, shock strength, and bulk explosive shock Hugoniot. The transmitted particle velocity decreases with an increase in the particle volume fraction. The results showed that immediately behind the shock front the velocity of particles such as Al and Mg can achieve 60–94% of the value of the shock velocity of the explosive [58].

Indirect information on the time of pressure equilibration in a multiphase reactive flow can be extracted from experimental data obtained for inert high-density mixtures of different compressibilities shocked by impact of Al flyers [59, 70]. Several series of experiments were performed, recording pressure time histories of mixtures such as Al–KI (50:50) and tungsten–paraffin (66.2:33.8). It was found that the characteristic time of the pressure equilibration for the Al–KI mixture, with components of similar compressibility (at $P \approx 30\,\mathrm{GPa}$), could be estimated as about 0.05 μs for ASD-1 and about 0.1 μs for the coarser Al powder PP-1. "Frontolysis" of the shock wave front has been observed for the tungsten–paraffin mixture, with components of vastly different compressibility. Nevertheless, a two-peak profile was not found in any of the nonreactive mixtures tested.

The two-peak phenomena, like those seen in HMX–Al formulations, have been observed in low-density analogous systems of Al particle–gas flow [71]. As stated in [71], apart from the endothermic reaction or mole number reduction, the particle–fluid system possesses its own unique features, namely, the momentum and heat loss from the fluid to the particles within the detonation zone of the fluid. This provides a mechanism to satisfy the necessary conditions for a weak detonation. Various weak detonation solutions including the two-peak solutions can be realized by different late particle energy releases to match the rear flow conditions. Secondary pressure rises were observed in hydrogen–oxygen mixtures in which the products of partial combustion of Al-containing mixtures with oxidizers (NH_4ClO_4 or NH_4NO_3) had been injected. The pressure rises were caused by Al particle burning, and it was assumed that the second pressure rise would be able to catch the leading front under certain conditions [62].

In [37, 43, 63], several possible reasons were considered to explain the two-peak profiles experimentally observed in HMX-based mixtures:

- Al interaction with DPs, the intensity of which depends on the effective surface area of the powder and pure Al content
- Possible changes in macroscopic kinetics of HMX decomposition, grains of which are in contact with relatively cold Al particles
- Nonequilibrium of the particle velocity, particularly owing to inclusion of Al particles in the flow at the initial stage of the wave formation

The aforementioned assumptions need further experimental and theoretical study, but the role of the first factor, i.e., Al interaction with DPs, should be inspected more closely.

In [36], the effect of Al addition on detonation front parameters as well as on the structure of "behind-front" flows in the detonation wave was studied. The questions of macroscopic kinetics were considered in connection with thermodynamic peculiarities of the DP behavior and the gas-dynamic mechanism of a self-sustaining detonation regime in composite explosives. The thermodynamic properties of the explosion products of metalized HE with negative OB were identified as abnormal. This means that the process of Al oxidation to the highest oxide, as the metal enters into the reaction, is followed by additional heat release simultaneously with a decrease in the total mole number of gaseous components and their degradation. As a result, contrary to what one could expect from the exothermic character of Al oxidation, the pressure of a high-density explosive decreases in the p–V plane at constant volume.

Thermodynamic analysis of the DP shock Hugoniots of aluminized HE in the p–V plane performed in [22, 64] shows that the Chapman–Jouguet (C–J) parameters corresponding to the equilibrium adiabat of the final DP with condensed Al_2O_3 are lower than those corresponding to the partial equilibrium Hugoniot calculated assuming Al to be inert. However, the isentrope, which passes through the C–J point at the nonequilibrium adiabat, is steeper in the p–V plane. It intersects the equilibrium adiabat in the area corresponding to the 1.5–2-fold DP expansion. Thus, the main part of the energy of Al reaction with DP is releasing while the products are expanding up to relatively lower pressures (about several gigapascals) [72]. The detonation behavior of composite aluminized explosives differs from that of a molecular explosive, for which any Hugoniot of the partially reacted (nonequilibrium) products is situated between the Hugoniot of the explosive and the shock adiabat of the final DP. Thus, it is often assumed that the nonequilibrium adiabats describing the detonating explosive "moves" monotonically from the Hugoniot of the nonreactive explosive to that of the final DP in the course of detonation decomposition. For an aluminized explosive, according to [36], the "moving" is of nonmonotonic character, the main result of which is the principal possibility for such an explosive to detonate in a self-sustaining undercompressed regime [22,64]. In this mode, the DV is higher, and the particle velocity and pressure in the DP are lower than those in normal mode determined by the point of tangency of the equilibrium adiabat of the final DP and the Rayleigh straight line. The aforementioned reasoning is supported by the results of both numerical thermodynamic simulations and 1D gas-dynamics calculations. In [36], for aluminized HMX-based formulations, the numerical results of the particle velocity, pressure, and temperature varying with charge length are presented. In the variants of the computations, some results of which are given below, it was assumed that an HMX booster (1.9 g cm^{-3} and 30 mm long) initiated the compositions studied.

Calculated DVs versus metal content assuming Al to be inert or reactive (curves 2 and 3 in Fig. 4.5, respectively) were given in Sect. 4.2.2.1. For micron-sized Al, the experimental points are close to those calculated for inert Al, while the experimental point for Al(0.1 μm) lies at the curve constructed for the reactive metal. At the same time, the Wood–Kirkwood model [35] predicts the velocities for micron-sized Al, assuming it reacts, well. Thus, there is a particular interest to analyze the results of both models and experimental data on pressure and temperature.

To compare the experimental and calculated pressure profiles, the latter were brought to the same scale (dividing the length by the value of the DV calculated assuming Al to be reactive) [36]. Of course, the procedure is not absolutely correct because the assumption of moving any point of the profile with the DV is an approximation. The rescaled profiles are shown in Fig. 4.18. If fast Al burning within the first microsecond is assumed, the calculations can result in the formation of a "plateau" expanding with charge length, which is peculiar to the undercompressed detonation mode (see Fig. 4.18a). The plateau, depending on the conditions of the detonation initiation, can transform into a two-peak profile, whose duration changes with charge length (see Fig. 4.18b).

The most important outcome of the model developed in [36] is that the undercompressed detonation mode, assuming Al to be reactive, can result in a two-peak profile, whose configuration changes with charge length. The first-peak pressures calculated are about twice as large as the experimentally measured ones, while the second-peak pressures are similar. The model is independent of Al particle shape and size.

Temperature time histories are shown in Figs. 4.19 and 4.20; the initial densities of the charges and temperatures measured at about 1 μs are given

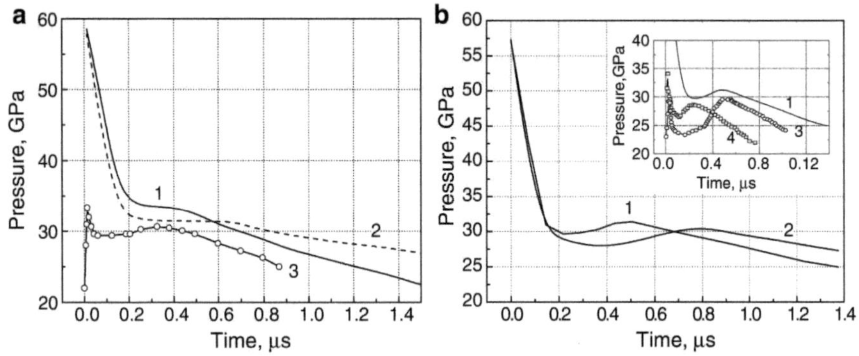

Fig. 4.18. Calculated [36] and experimental [37, 43] pressure profiles in the detonation products of HMX–Al formulations. **a** HMX–Al (90:10). Calculations: charge length *1* about 27 mm, *2* about 55 mm. Experiment: *3* HMX(A)–Al(15 μm). **b** HMX–Al (85:15). Calculations: charge length *1* 26 mm; *2* about 53 mm. Experiment: *3* HMX(B)–Al(fl), *4* HMX(A)–Al(0.1 μm)

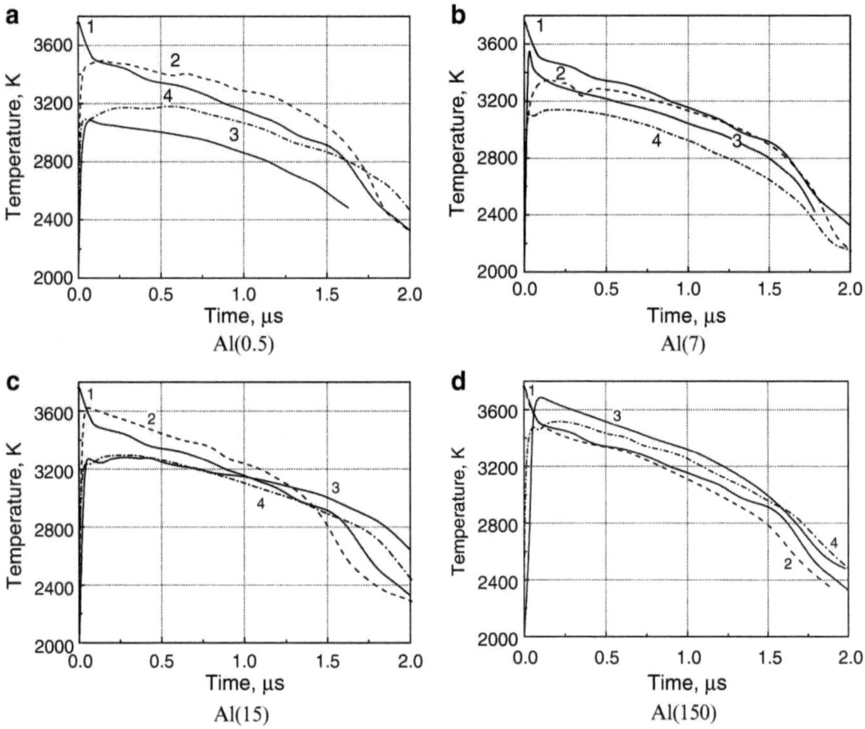

Fig. 4.19. Brightness temperature time histories of the detonation products of HMX(A)–Al formulations. Al content: *1* 0%, *2* 5%, *3* 15%, *4* 25%

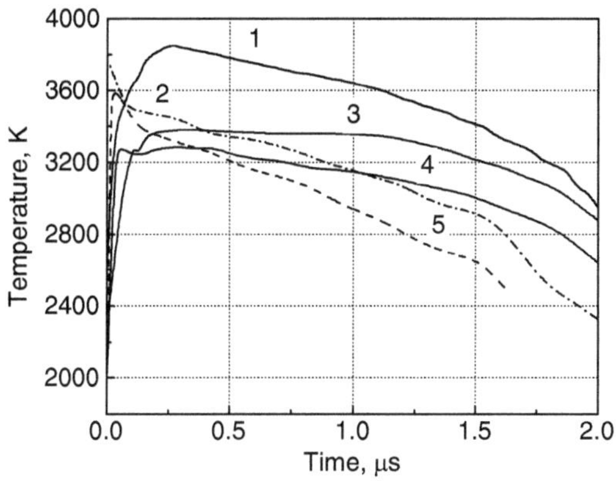

Fig. 4.20. Brightness temperature time histories of the detonation products of HMX-based formulations with 15% of different admixtures. *1* HMX(B)–Al(fl), *2* HMX(A), *3* HMX(B)–Al(0.1), *4* HMX(A)–Al(15 μm), *5* HMX(A)–LiF(20 μm)

Table 4.7. Brightness temperatures of HMX-based formulations

Formulation	β (%)	ρ_0 (g cm^{-3})	T (K) (1 μs)	Formulation	β (%)	ρ_0 (g cm^{-3})	T (K) (1 μs)
HMX(A)–Al(150 μm)	5	1.835	3,100	HMX(A)–Al(7 μm)	5	1.832	3,130
	15	1.891	3,320		15	1.873	3,040
	25	1.951	3,250		25	1.917	2,920
HMX(A)–Al(15 μm)	5	1.821	3,240	HMX(A)–Al(0.5 μm)	5	1.835	3,290
	10	1.841	3,360		15	1.867	2,860
	15	1.871	3,150		25	1.905	3,070
	25	1.921	3,100				
HMX(B)–Al(0.1 μm)	15	1.826	3,350	HMX(A)	0	1.802	3,160
HMX(B)–Al(fl)	15	1.860	3,640	HMX(A)–LiF(20 μm)	15	1.833	2,940

in Table 4.7. The entrance of the rarefaction to the DP–LiF interface results in more rapid temperature decrease, which is observed at about 1.5 μs (see Fig. 4.19).

Analyzing the data for HMX(A)-based formulations with different content of micron-sized and sub-micron-sized Al (Fig. 4.19), one can see that all the data, including those for pure HMX, are close to each other within experimental error. A larger reduction in temperature is seen for HMX(A)–Al(0.5 μm) (75:25); the high content of aluminum oxide in the powder can be one of the reasons. Nevertheless, the fact that replacement of some of the explosive with the metal additive does not significantly affect the measured temperature favors the assumption of Al reacting with the DP near the detonation front. The finer the Al used, the slower the temperature decrease observed in the DP compressed by the reflected wave. This effect can be explained by the reaction growth with time, but this explanation is not the only one. The same trend was observed for HEs with a few percent of carbon black [73].

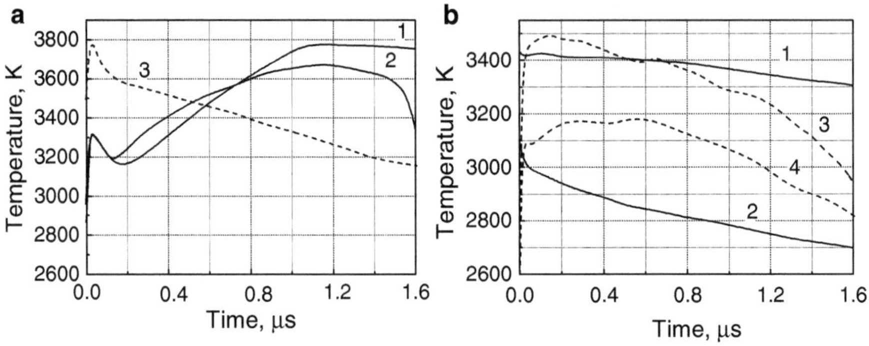

Fig. 4.21. Calculated and experimental temperatures in HMX–Al formulations. **a** HMX–Al (90:10). Calculations [36]: charge length: *1* about 55 mm; *2* about 27 mm. Experiment [111]: *3* HMX(A)–Al(15 μm). **b** Calculations [35]: *1* HMX–Al (95:5), *2* HMX–Al (75:25). Experiment [43]: *3* HMX(A)–Al(0.5 μm) (95:5), *4* HMX(A)–Al(0.5 μm) (75:25)

The role of Al is seen more clearly in Fig. 4.20, in which temperature histories are given for 15% of the admixture differing in particle size. The histories measured for pure HMX and HMX with inert LiF are similar, but the latter lies lower. The first temperature peak has a complicated origin. It was assumed that the luminosity of the detonation complex in HMX within the first 0.1–0.2 μs is of a nonthermal character [61,73]. Nontransparent Al screens that luminosity, and after the first tenths of a microsecond, the difference in the temperature measured for inert LiF(20 μm) and Al(15 μm) increases with time. The temperature profile for finer Al(0.1 μm) lies higher than that for Al(15 μm). The highest temperatures have been measured for nonspherical particles Al(fl), for which the pressure profile is the most highly nonmonotonic.

Comparison of experimental brightness temperatures with those calculated is shown in Fig. 4.21.

The temperature data calculated in [36] are rescaled in the same manner as described above for pressure profiles. The experimental and calculated temperature histories are different as shown in Fig. 4.21a. One of the reasons is that the brightness temperature is measured not in the direct wave, but in the secondary compression wave reflected from the window; however, the difference in the compressibility of the DP and LiF is small. Another possible reason is that the calculation of the temperature was performed assuming complete Al oxidation within about 1 μs [36]. Actually, this does not take place. We believe Al(15 μm) to be neither inert nor completely oxidized in the detonation zone. The tendency for the temperature to increase within the first microsecond is observed for finer Al powder and larger metal content in the formulation (see Figs. 4.20, 4.21b). The temperatures calculated in [35] for 5 and 25% Al content are neither similar to those calculated in [36] nor similar to the experimentally measured values. Better agreement is achieved for 5% Al content.

The analysis of the data suggests that the reaction between Al and the DP of HMX takes place in the detonation zone or in its immediate vicinity. However, Al oxidation is incomplete within the 3-μs interval, which covers both the reaction zone behind the detonation wave and part of the rarefaction wave adjacent to it. A fine structure of detonation waves in HMX mixtures with Al manifests itself in various temperature histories and pressure profiles at different contents and particle sizes of Al and provides evidence of a complicated nature of physicochemical interaction between Al and products of the explosive decomposition.

4.3.2 NQ-Based Explosives

Pressure time histories measured in the DP of NQ-based formulations are shown in Fig. 4.22. No two-peak pressure profile is observed. For pure NQ of 1.635 g cm^{-3}, the C–J pressure was estimated as 22.1–21.6 GPa [74, 75]. The C–J pressure range corresponds to the reaction zone from 0.71 mm (90 ns) to 1.00 mm (125 ns) and a polytropic index of 3.66–3.77. NQ, as a HE with a large hydrogen content, is characterized by relatively high DVs and low C–J pressures. The addition of Al does not change the character of the measured pressure histories in comparison with that for pure NQ, though the charges differed in the initial density and it is difficult to compare the profiles accurately. It seems that the highest pressure corresponds to the NQ–Al(0.1 μm) formulation, but its charge density is also the largest (see Fig. 4.22). For a comparison, one can normalize the pressures to the same density of the base explosive in the charge (1.635 g cm^{-3}) (see the Appendix). In doing so, it

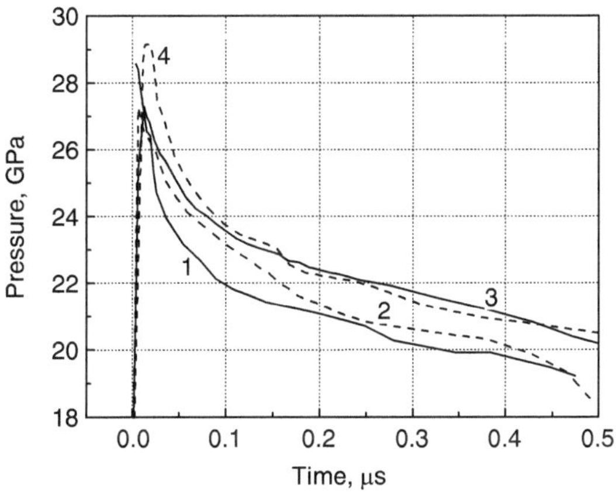

Fig. 4.22. Pressure profiles of the detonation products NQ–Al (85:15). *1* NQ, $\rho_0 = 1.635$ g cm^{-3}; *2* NQ–Al(15 μm), $\rho_0 = 1.742$ g cm^{-3}; *3* NQ–Al(fl), $\rho_0 = 1.733$ g cm^{-3}; *4* NQ–Al(0.1 μm), $\rho_0 = 1.775$ g cm^{-3}

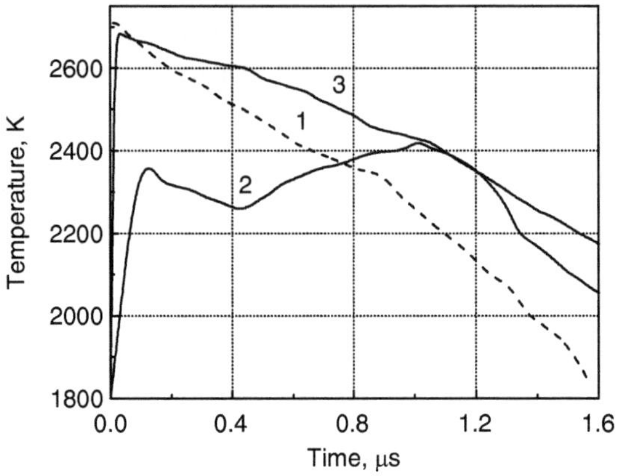

Fig. 4.23. Brightness temperature time histories of the detonation products of NQ–Al (85:15). *1* NQ, $\rho_0 = 1.649\,\mathrm{g\,cm^{-3}}$; *2* NQ–Al(fl), $\rho_0 = 1.728\,\mathrm{g\,cm^{-3}}$; *3* NQ–Al(0.1 μm), $\rho_0 = 1.778\,\mathrm{g\,cm^{-3}}$

would appear that Al addition slightly decreased the peak pressure values (actually, it would be 28.6 GPa in pure NQ and about 27.5 GPa in both Al-containing formulations). For a longer time, the normalized pressures in NQ and NQ–Al(0.1 μm) formulations are the same, but the addition of Al(fl) slows down the pressure decrease. Thus, in 0.5 μs, the normalized pressures are the following: 19.0, 19.1, and 20.4 GPa in NQ, NQ–Al(0.1 μm), and NQ–Al(fl) formulations, respectively. One can assume that both Al(0.1 μm) and Al(fl) partially react within the recorded time.

The brightness temperatures, shown in Fig. 4.23, are much lower than those determined for HMX-based formulations. The histories obtained for pure NQ and the NQ–Al(0.1 μm) mixture are similar, but the temperature decreases more slowly in the latter case (the same effect as in HMX–Al mixtures). It seems that the observed slowing down of the temperature decrease can be caused by the partial chemical reaction between Al and the DP.

The temperature history measured for NQ–Al(fl) has a complicated character, and it does not seem to correlate with the pressure profile. At the beginning of the record, the nonuniformity in the temperature distribution is greatest; moreover, one can assume that it depends on the initial charge structure, which, in turn, depends on many factors, such as volume fraction, the various sizes of the particles in the components, and the particle mean sizes and shapes. The needlelike grains of NQ mixed with Al(fl) give the most complicated structure [as previously stated for HMX–Al(fl) formulations]. Moreover, the flakes are able to cover the grains of an explosive totally, and in such a way flakes can screen the radiation of hot DPs within the first tenth of a microsecond (the lower temperatures within this time in comparison with those for spherical Al can be explained by a larger "screen" effect of the flakes). The decrease in pressure within the first 0.5 μs and that in temperature within the first 0.4 μs can be compared qualitatively only. It would be incorrect to call the time of the first 0.4 μs the "ignition" time in the case of NQ–Al(fl) formulations as there is not enough information; nevertheless, we believe that the observed further temperature growth is evidence of the reaction, which starts in close vicinity to the detonation front. This agrees with the higher normalized pressure determined for NQ–Al(fl) mixtures. Unfortunately, there is no simple procedure to normalize measured temperatures to the same initial charge density. In spite of the difference in ρ_0, the temperatures of aluminized formulations appear to be the same at 1.0–1.2 μs (further parts of the records are influenced by rarefactions, the entrance of which at the interface depends on LiF thickness), while the normalized pressures are higher in the case of NQ–Al(fl). Both pressure and temperature profiles show evidence for the possible implication of Al in the reaction with the DP of NQ.

4.3.3 BTNEN-Based Explosives

Pressure time histories measured in the DP of BTNEN-based formulations are shown in Fig. 4.24. The records show the existence of a small "stepwise" part at the beginning of the pressure profile. We believe that these steps reflect the peculiarity of BTNEN detonation decomposition – namely, it occurs in two stages. Possibly, the first stage is slightly endothermic. Recently, similar complicated particle velocity histories have been observed for some common HE charges (e.g., RDX) under certain conditions [76]. Still, there is no proper explanation for the phenomenon.

These "steps" are smoothed out with decreasing Al particle size, and the pressure rise at the leading front becomes slower (see Fig. 4.24). The pressure

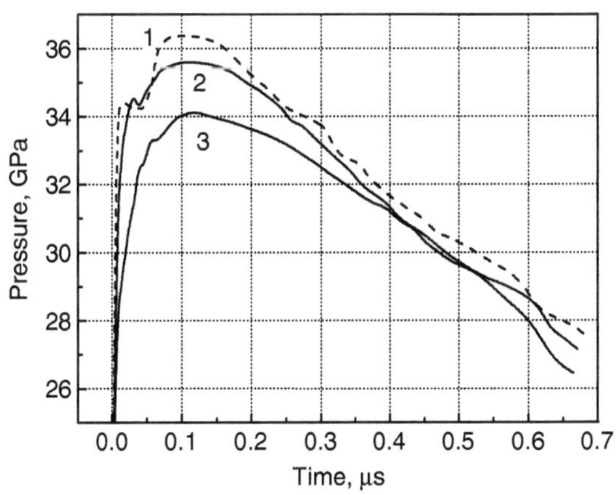

Fig. 4.24. Pressure profiles in the detonation products BTNEN–Al (85:15). *1* BTNEN, $\rho_0 = 1.881\,\mathrm{g\,cm^{-3}}$; *2* BTNEN–Al(15), $\rho_0 = 1.967\,\mathrm{g\,cm^{-3}}$; *3* BTNEN–Al(0.1), $\rho_0 = 1.924\,\mathrm{g\,cm^{-3}}$

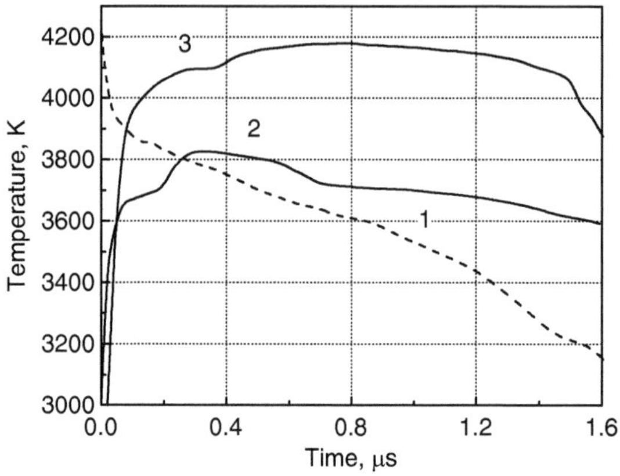

Fig. 4.25. Brightness temperature time histories of the detonation products of BTNEN–Al (85:15). *1* BTNEN, $\rho_0 = 1.881\,\mathrm{g\,cm^{-3}}$; *2* BTNEN–Al(15 μm), $\rho_0 = 1.975\,\mathrm{g\,cm^{-3}}$; *3* BTNEN–Al(0.1 μm), $\rho_0 = 1.922\,\mathrm{g\,cm^{-3}}$

profiles given for the charges differed slightly in initial density. For comparison, maximum pressures were normalized to the same BTNEN density in the charge of $1.900\,\mathrm{g\,cm^{-3}}$ (see the Appendix), and the pressures were the same within experimental accuracy [37.4, 36.7, and 37.5 GPa for BTNEN; BTNEN–Al(15 μm), and BTNEN–Al(0.1 μm), respectively]. The normalized pressures at 0.5 μs can be estimated as follows: 31.1, 30.7, and 32.5 GPa for BTNEN, BTNEN–Al(15 μm), and BTNEN–Al(0.1 μm), correspondingly. The highest pressure occurs in BTNEN–Al(0.1 μm), but the difference in pressure between pure HE and the mixture with nanosized Al is only slightly above the experimental error.

Brightness temperature histories measured in BTNEN-based formulations are shown in Fig. 4.25. For pure BTNEN, a peak of luminosity occurred at the instant the shock wave entered the DP–LiF interface and the peak could be caused by the luminescence (observed also in HMX and RDX [75, 77]). It has already been discussed that pressure and temperature profiles cannot be compared directly; nevertheless, for BTNEN–Al formulations one can see a little plateau region at the initial part of the temperature history, after which temperature increases again. The appearance of the plateau corresponds to the same features observed in the pressure records. Temperatures in aluminized explosives become higher than that of pure BTNEN after a few tenths of a microsecond. Flattened temperature profiles of Al-containing formulations can be considered as evidence of a chemical reaction between Al and the DP; moreover, the temperatures measured for BTNEN–Al(0.1 μm) are extremely high (above 4,000 K) and they exceed those of BTNEN–Al(15 μm). This means that nanoscale Al can undergo more intensive oxidation immediately behind the wave front.

Table 4.8. Temperatures of Al-containing compositions *AP* ammonium perchlorate

Composition (wt%)	Additives	ρ_0 (g cm^{-3})	T (K) (1 µs)
RDX–AP–Al–wax (20:43:25:12)	Al(0.1 µm)	1.805	3,490
	Al(5 µm)	1.793	2,582
RDX–AP–Al–wax (50:24:12:14)	Al(0.1 µm)	1.664	3,037
	Al(5 µm)	1.654	2,947
	LiF(5 µm)	1.659	2,453

4.3.4 RDX-AP-Based Explosives

Optical pyrometry is applied for temperature measurement in a variety of Al-containing energetic materials. In [78, 79], RDX–AP–Al–wax formulations, with different component ratios, containing 5-µm and 100-nm Al were investigated. The measurements were performed with the help of a six-channel pyrometer with an effective wavelength from 500 to 1,510 nm; LiF was used for the window. Temperature time histories are given in [78]. On the basis of the analysis of the emission spectra and the heat conduction at the DP–LiF interface, the temperatures at 1 µs after the shock entrance inside the LiF window were determined (see Table 4.8) [79].

It is seen that replacement of Al by inert LiF results in a significant temperature decrease. The temperatures measured in the formulations containing nanoscale Al are higher than those with micron-sized Al.

4.4 Acceleration Ability

The acceleration ability of the explosive formulations is an important characteristic. In Russia, it is studied, mainly, using two techniques: plate acceleration "M-60" and its analogues [12, 22] and "T-20" [22] (a variant of the "cylinder test" [16, 80]). The results on the acceleration ability allow one to consider the conversion of the energy of Al oxidation to the kinetic energy of the DP and the accelerated body within the characteristic times: up to about 20 µs for plate acceleration and 10 µs for cylinder expansion.

The acceleration ability of the Al-containing formulations based on HMX, NQ, and BTNEN with component mass ratio 85:15 HE–Al was studied in [75, 77, 81]. The different types of Al powder were tested.

4.4.1 Plate Acceleration

According to the technique, a steel plate 4 mm thick is accelerated from the end of a cylindrical charge (40 mm long by 40 mm in diameter) in the channel of a steel shell. The times at which the flyer plate reaches the contact needles, located parallel to the shell axis, are detected using an oscilloscope.

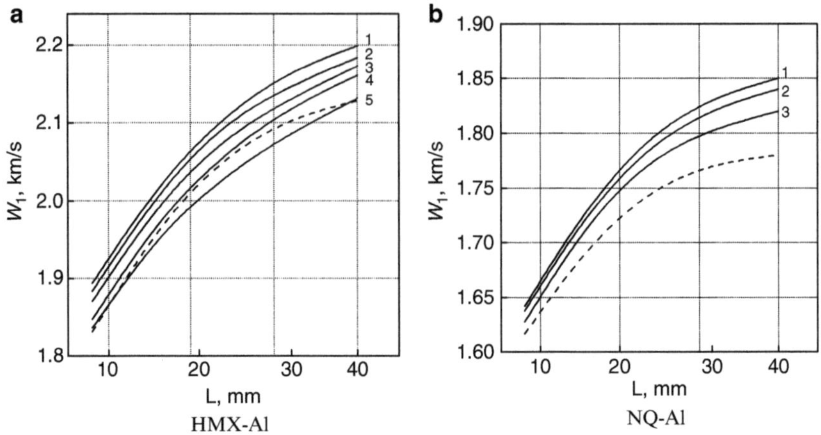

Fig. 4.26. Plate velocity versus flight length. *1* Al(fl), *2* Al(0.1 μm), *3* Al(15 μm), *4* Al(7 μm), *5* Al(150 μm). *Dashed lines* correspond to the pure high explosives. **a** HMX(A) was used in the formulation with Al(7 μm), the rest of the compositions contained HMX(B). The curve obtained for HMX(B)–Al(7 μm) coincides with that for HMX(B)–Al(fl)

The data obtained are used to determine the dependence of the plate velocity on the flight distance. The 40-mm distance from the initial position of the plate is adopted as the standard length for comparing acceleration ability. Measured velocities of the steel plate (W_1) versus flight length (L) are presented in Figs. 4.26 and 4.27. In [81], the curves given in Figs. 4.26 and 4.27 and in Figs. 4.28 and 4.29 were normalized to the following HE densities in the charge: 1.808 g cm^{-3} for HMX; 1.635 g cm^{-3} for NQ, and 1.90 g cm^{-3} for BTNEN.

The addition of Al to HMX increases the final W_1 of the plate except for Al(150 μm) (see Fig. 4.26a). The $W_1(L)$ curves constructed for HMX(B)–Al(7 μm) and HMX(B)–Al(fl) practically coincide and that for HMX(B)–Al(0.1 μm) is slightly lower. For compositions with Al(fl), Al(0.1 μm), and Al(7 μm), the difference in final W_1 lies within the experimental error, namely, $W_1(40\,\text{mm}) = 2.19 \pm 0.02\,\text{km}\,\text{s}^{-1}$. Thus, the gain in velocity relative to pure HMX is about 3%. The replacement of HMX(B) by HMX(A) in formulations with Al(7μm) results in a minor decrease of the final W_1 (from 2.19 to 2.16 km s^{-1}) and can be explained by a less uniform structure of the charges with coarser HMX.

NQ and its mixtures were initiated using an RDX booster charge (10 mm thick); the height of the charges tested was 35 mm. Although the OB of NQ is more negative than that of HMX, the addition of Al to NQ leads to an identical gain in the final W_1 (3.4%). The addition of Al to NQ increases W_1 predominantly on the final legs of flight (see Fig. 4.26b). For NQ–Al(fl) and NQ–Al(0.1 μm), the final W_1 values also coincide; the average velocity is $1.84 \pm 0.02\,\text{km}\,\text{s}^{-1}$.

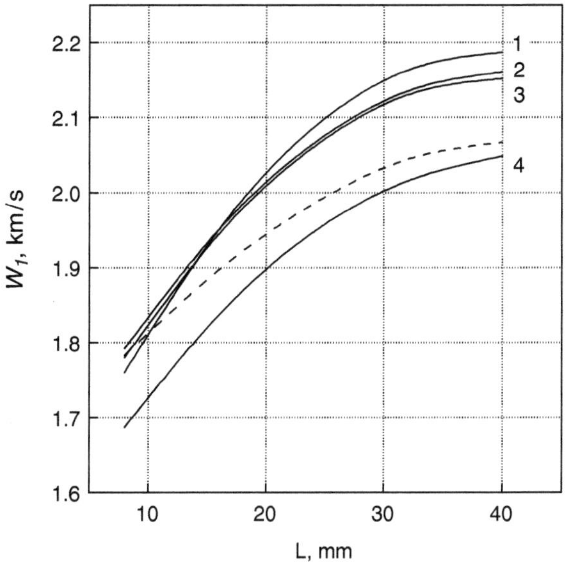

Fig. 4.27. Plate velocity versus flight length for BTNEN–Al. *1* Al(7 μm), *2* Al(15 μm), *3* Al(0.1 μm), *4* Al(150 μm). The *dashed line* corresponds to BTNEN

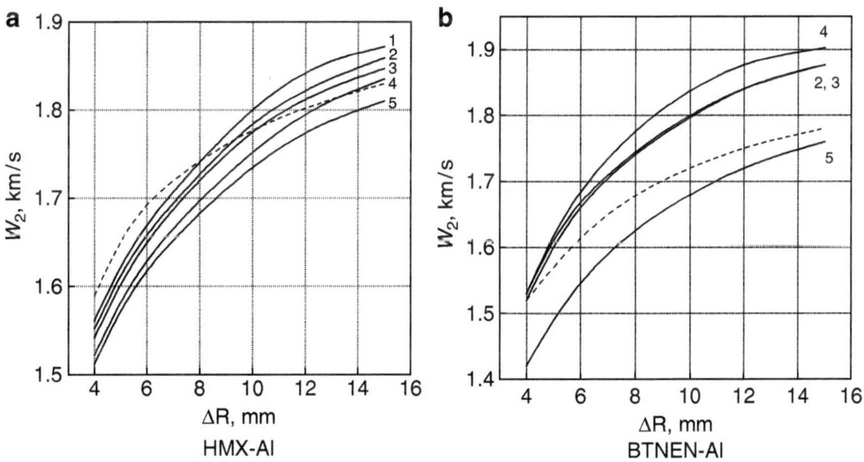

Fig. 4.28. Cylinder shell velocity versus radius increment. *1* Al(fl), *2* Al(0.1 μm), *3* Al(15 μm), *4* Al(7 μm), *5* Al(150 μm). *Dashed lines* correspond to the pure HEs. **a** HMX(A) was used in the formulation with Al(7 μm), the rest of the compositions contained HMX(B). The curve obtained for HMX(B)–Al(7 μm) coincides with that for HMX(B)–Al(fl)

The largest increase in the final W_1 is observed for BTNEN–Al mixtures (see Fig. 4.27). The addition of 15% Al(7 μm) increases the velocity by about 6%. This indicates more active oxidation of Al by the free oxygen of the BTNEN DPs and a more considerable energy release due to this process. However, because the acceleration ability of pure BTNEN is lower than that of HMX, the BTNEN–Al(7 μm) composition exhibits the same final $W_1 (2.19 \pm 0.02\,\mathrm{km\,s^{-1}})$ as HMX(B)–Al(7 μm). The maximum W_1 at the final flight distance was obtained for BTNEN–Al(7 μm) and the minimum W_1 was obtained for BTNEN–Al(150 μm) (see Fig. 4.27). Unlike the compositions with HMX and NQ, BTNEN–Al(0.1 μm) has no advantages with regard to W_1 at the main legs of the flight even over BTNEN–Al(15 μm). For the mixtures with Al(0.1 μm) and Al(15 μm), the average final W_1 is $2.15 \pm 0.02\,\mathrm{km\,s^{-1}}$. This can be explained by the fact that at a high completeness of Al oxidation by the DP of BTNEN, the heat effect is determined primarily by the content of the pure metal in the Al powder. The oxide content in Al(0.1 μm) is larger than that in Al(7 μm) and Al(15 μm).

In the case of BTNEN-based formulations (see Fig. 4.27), the velocity gradients for pure HE and mixtures become equal at the final legs of the flight [except for BTNEN–Al(150 μm) owing to lower specific surface of Al(150 μm)]. Apparently, this can be explained by active Al oxidization in the DP at the initial stage of the product expansion, and a considerable part of the reaction heat is expended before the final legs of the flight.

With use of a similar experimental set-up, the plate acceleration ability of HMX(5.5w)- and BTNEN-based formulations with Al (PP-1 and ASD-4) was tested [12]. The data are given in Table 4.9. The values are consistent with those presented above; the BTNEN density in aluminized charges was 1.91–1.92 $\mathrm{g\,cm^{-3}}$.

Table 4.9. Plate acceleration ability of aluminized high explosives (HEs)

Formulation	Component ratio	ρ_0 ($\mathrm{g\,cm^{-3}}$)	W_1 ($\mathrm{km\,s^{-1}}$)		
			10 mm	20 mm	38 mm
HMX(5.5w)	–	1.78	1.80	1.96	2.07
HMX(5.5w)–PP-1	90:10	1.80	1.77	1.98	2.11
	80:20	1.87	1.72	1.90	2.06
HMX(5.5w)–ASD-4	80:20	1.87	1.67	1.91	2.07
BTNEN	–	1.90	1.68	2.00	2.08
	90:10	1.98	–	–	2.19
BTNEN–ASD-4	85:15	2.00	–	–	2.25
	80:20	2.03	1.73	2.07	2.21

W_1 measured for bis(2,2,2-trinitroethyl)nitramine (*BTNEN*)–ASD-4 (85:15) is greater than that for pure BTNEN over the whole flight distance. The relative increment in the final W_1 for BTNEN–ASD-4 formulations is 5.3, 8.2, and 6.3% for 10, 15, and 20% Al, respectively. The dependence of W_1 on charge density of BTNEN–ASD-4 obtained can be approximated by a straight line.

4.4.2 Cylinder Test

The radial velocities of the outer surface of the copper cylindrical shell (W_2) for various radius increments (ΔR) are shown in Fig. 4.28. For HMX- and BTNEN-based compositions, the cylinder expansion velocities agree qualitatively with the results on acceleration of steel plates. Thus, for some of the compositions, the rates of increment of the cylinder shell velocity and the final velocities are higher than those for individual HEs. These results indicate that the energy of Al oxidation transforms into the kinetic energy of the DP and the cylinder shell in the course of product expansion. The curve for the composition with Al(150 μm) is below the curves constructed for other mixtures (see Fig. 4.28). For the HMX-based formulations with Al(fl), Al(0.1 μm), and Al(7 μm), W_2 is $1.87 \pm 0.02\,\mathrm{km\,s^{-1}}$ within the experimental error, and for BTNEN with Al(15 μm), Al(0.1 μm), and Al(7 μm), the velocity is $1.89 \pm 0.02\,\mathrm{km\,s^{-1}}$.

The data on HMX formulations with coarse Al(150 μm) qualitatively correlate with the results of cylinder tests of HMX(5w)–Mg mixtures (70:30) (Mg particles were 60–300 μm in size) [30]. The tests were performed in two types of copper tube with internal to external diameter ratios of 15:18 and 25:30. The primary results of the tests were presented in the form of the shapes of the external boundary of the copper tubes; final expansion of the tubes corresponded to a sevenfold to tenfold volume increase. For both types of tube, the X-ray recorded shapes exhibited less expansion for Mg-containing formulations than for HMX(5w). Nevertheless, by calculating the final Gurney energy per unit mass of the explosive component based on the experimental data, Cudzilo et al. [30] confirmed that reaction of magnesium with the expanding DPs releases an additional heat.

4.5 Heat of Explosion

The heat of explosion (Q) is a basic parameter which controls the potential ability of HEs to perform work in the course of the expansion of the explosion products. The greatest effect resulting from introduction of Al into the explosives is observed by measuring Q [5, 75, 77, 81–83]. The actual values of Q in metal-containing explosives depend on the conditions of the experiment. The completeness of metal oxidation and that of energy release depend on the total sizes of the charge, charge confinement and its properties, and the properties of the surrounding medium in which explosion products expand. For a TNT–Al(15 μm) formulation (80:20), the Q measured in an unconfined charge 10 mm in diameter is $4,980\,\mathrm{kJ\,kg^{-1}}$, while for the same formulation placed in a copper tube of the same diameter with 5-mm wall thickness, the value is $6{,}070\,\mathrm{kJ\,kg^{-1}}$. The Q of a TNT–Al(200 μm) mixture (80:20) was measured, respectively, as 5,250 and $5{,}650\,\mathrm{kJ\,kg^{-1}}$ in copper tubes with $d_\mathrm{i} = 10$ and 5-mm wall thickness and in copper tubes having $d_\mathrm{i} = 20$ and 7-mm wall thickness [83]. Therefore, Q determined in a bomb calorimeter should not be considered as an absolute criterion for the energy content of a HE with metal admixtures. However, it can be used as a relative measure to compare the effect reported in [83].

In tests with the calorimetric bomb, the time of the Al interaction with the products exceeded tens of microseconds. The considerable time provides for a high degree of metal oxidation and release of large amounts of additional energy. However, complete oxidation of Al does not occur even under such conditions. The results of Q measurements agree qualitatively with the data on the plate and cylinder velocities.

Q was measured in a 5-l bomb calorimeter [84]. Many of the experimental data given in Table 4.10 were obtained for charges placed into a copper casing with $d_i = 20$ and 7-mm wall thickness [83]. The values of Q correspond to the condition with H_2O as a gas in the products. The composition of the DP was calculated using the semiempirical methods proposed in [82, 85, 86].

Analyzing the data in Table 4.10, Makhov [83] concluded that, as expected, the effect of incorporation of Al into explosives rises as the oxygen content in the HE increases. The largest gain in Q (in comparison with the base explosive) is observed for BTNEN. This considerable increase in Q is attained not only because of high completeness of Al oxidation, but also owing to the highest energy effect of metal oxidation by free oxygen in the DP of BTNEN. Explosion products of both TNT and RDX contain CO, which is less chemically active with Al in comparison to H_2O and CO_2. Nevertheless,

Table 4.10. Heat of explosion in Al-containing formulations [83]

β (%)	Al particle size (μm)	ρ_0 ($\mathrm{g\,cm^{-3}}$)	Q ($\mathrm{kJ\,kg^{-1}}$)	β (%)	Al particle size (μm)	ρ_0 ($\mathrm{g\,cm^{-3}}$)	Q ($\mathrm{kJ\,kg^{-1}}$)
TNT–Al				RDX–Al			
0	–	1.60	4,390	0	–	1.70	5,610
10	7	1.66	5,270	10		1.75	6,570
20	7	1.71	6,200	20		1.80	7,280
	15	1.72	6,070	30	7	1.84	7,980
	15[a]	1.70	5,330	35		1.87	7,980
	200	1.72	5,250	40		1.89	7,850
30		1.76	6,660	45		1.90	7,710
35	7	1.78	6,920	PETN–Al			
40		1.80	6,820	0	–	1.70	5,900
BTNEN–Al				20		1.79	7,850
0		1.90	5,230	35		1.85	9,210
20		1.98	9,310	40	7	1.88	9,480
30	7	2.00	9,890	45		1.92	9,440
40		2.01	10,580	50		1.95	9,320
45		2.03	10,420				

[a]Coarse TNT particle 1.5 mm in size; in other cases the mean size is 50–100 μm

Table 4.11. Heat of explosion in 85:15 HE–Al formulations [81]

Al	HMX(B)		NQ		BTNEN	
	ρ_0 ($g\,cm^{-3}$)	Q ($kJ\,kg^{-1}$)	ρ_0 ($g\,cm^{-3}$)	Q ($kJ\,kg^{-1}$)	ρ_0 ($g\,cm^{-3}$)	Q ($kJ\,kg^{-1}$)
–	1.81	5,610	1.63	3,480	1.89	5,230
Al(0.1 µm)	1.83	6,960	1.78	4,960	1.90	8,350
Al(fl)	1.84	6,960	1.71	4,960	–	–
Al(7 µm)	1.87	7,000	–	–	1.94	8,600
Al(15 µm)	1.89	6,920	1.74	4,820	1.95	8,500
Al(150 µm)	1.89	6,690	–	–	1.94	8,280
Al(7 µm)[a]	1.88	6,780	–	–	–	–

[a]HMX(A)

the larger CO content in the explosion products of TNT compared with RDX leads to larger gain in Q for TNT.

For TNT–Al (80:20), the role of both explosive grain size and Al particle size has been studied. It is seen that the use of coarser particles reduces Q. The data also illustrate the dependence of Q on the Al content in the mixture, which has a maximum near 40% Al.

Table 4.11 gives the results for formulations with 15% Al powder of different particle sizes. Of particular interest are the data for nanoscale Al in comparison with micrometric Al. At low oxide-film content in the Al powder, the maximum Q is attained for the smallest Al particle size. The addition of 15% Al(7 µm) to HMX(B) results in an approximately 25% increase in Q. The absence of any advantages with regard to Q for the mixtures of HMX(B) with Al(0.1 µm) and Al(fl) over the compositions containing Al(7 µm) is due to the fact that an increase in specific surface area of the metal does not compensate for the losses related to the larger oxide-film content. As in the case of TNT, the use of coarse HMX in the formulation decreases Q.

From the data it follows that the addition of Al increases Q considerably more for NQ-based formulations than for mixtures with HMX, when relative values are compared. The addition of 15% Al(0.1 µm) and Al(fl) to NQ gives a more than 40% increase.

In spite of the fact that Q of pure BTNEN is lower than that of HMX, BTNEN–Al mixtures, on average, have a 20% gain in Q over HMX–Al compositions. The results obtained indicate a high degree of Al oxidation in the course of the reaction with the DP of BTNEN. As in the case of HMX, the largest Q is observed for BTNEN–Al(7 µm). The Q values measured for Al(0.1 µm) and Al(15 µm) admixtures are similar.

The Q value of the most individual HEs was considered as a linear function of the initial charge density; however, more careful study indicates that $Q(\rho_0)$ dependencies for explosives with strongly negative OB have an S-like shape [86]. The influence of the initial charge density on the Q value of Al-containing

Table 4.12. The effect of the HE charge density in Al-containing formulations

Formulation	ρ_0 ($g\,cm^{-3}$)	Q ($kJ\,kg^{-1}$)	Formulation	ρ_0 ($g\,cm^{-3}$)	Q ($kJ\,kg^{-1}$)	ΔQ ($kJ\,kg^{-1}$)
BTNEN	1.75	5,230	BTNEN–Al(7 µm) (60:40)	2.03	10,580	5,350
	0.80	5,230		1.11	10,370	5,140
TNT	1.47	4,240	TNT–Al(7 µm) (60:40)	1.80	6,820	2,580
	0.87	3,530		1.19	6,460	2,930
	1.57	4,360	TNT–Al(15 µm) (80:20)	1.72	6,070	1,710
	1.20	3,930		1.35	5,730	1,800

formulations based on HEs with different OB was studied in [83, 86, 87]. The data are given in Table 4.12.

The densities in the HE charges correspond to those given for the pure base explosive. It is seen that charge density affects the gain (ΔQ) in the heat of explosion caused by Al addition. It was reported that "the density influences the completeness of Al oxidation" [86]. In contrast to BTNEN, a HE with a positive OB, for HEs with a strongly negative OB, such as TNT, the addition of Al attenuates the density effect on ΔQ. It was stated that the reduction in the TNT density in the charge would result in increase of CO content in the explosion products, whose heat of reaction according to Al_2O_3 formation is larger in comparison with those with CO_2 and H_2O [86]. This effect is clearly seen for large Al content in TNT only.

The influence of the OB of the base explosive on Q in formulations with Al(15 µm) and Al(0.1 µm) at different admixture contents was examined. The HEs tested were BTNEN (+16.5% OB), ADN (+25.8% OB), HMX (−21.6% OB), and TNT (−74.0% OB). The experimental values of Q (H_2O–gas) for the pressed charges normalized to the same explosive density in the mixture (Q^*), as given in [81, 87], are presented in Fig. 4.29. For PETN (−10.1% OB) with Al(7 µm), the curve fitted to the experimental points given in Table 4.10 is also constructed.

All the curves have a maximum corresponding to an Al content of about 40%, which slightly depends on the type of Al and the base explosive used. Increase of Q with Al content is characterized by a larger gradient for base explosives with a positive OB, while the gradient of the curve constructed for PETN is moderate. The lowest gradients are observed for TNT- and HMX-containing formulations.

The Q^* curve for ADN–Al(15 µm) lies higher than that for ADN–Al(0.1 µm). The completeness of oxidation of Al is high when the HE molecule has an excess of oxygen. The advantages of Al(0.1 µm) over Al(15 µm) in Q^* are observed for HMX and TNT mixtures. At 40% Al, –Al(0.1 µm) is slightly (by about 3%) superior to HMX–Al(15 µm) (Fig. 4.29). At 15% Al, the values

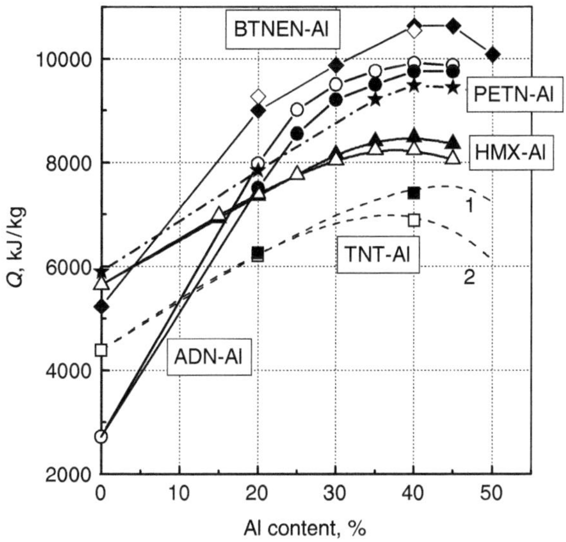

Fig. 4.29. Heat of explosion versus Al content. *Diamonds* BTNEN ($1.86\,\mathrm{g\,cm^{-3}}$), *circles* ammonium dinitramide *(ADN)* ($1.70\,\mathrm{g\,cm^{-3}}$), *triangles* HMX ($1.81\,\mathrm{g\,cm^{-3}}$), *squares* TNT ($1.59\,\mathrm{g\,cm^{-3}}$). The explosive density in the formulations is given in parentheses. *Filled symbols* Al(0.1 μm), *open symbols* Al(15 μm). *Stars* PETN–Al(7 μm) (experimental points at ρ_0). *Dashed lines* calculated curves of TNT-based mixtures: *1* Al(0.1 μm), *2* Al(15 μm) [87]

of Q^* for HMX mixtures with Al(0.1 μm) and Al(15 μm) are nearly equal. For BTNEN-based mixtures at 20% Al, BTNEN–Al(0.1 μm) is inferior to BTNEN–Al(15 μm) with regard to Q^*. The compositions containing BTNEN and 40% Al of both types have similar Q^* values. As a whole, the addition of 40–45% Al to BTNEN and to HMX leads to an approximately 100% and an approximately 50% increase in Q^*, respectively.

All the data given above were obtained under the same conditions and can be analyzed jointly. Obviously, a reduction in the Al particle size owing to an increase in the specific surface area of the metal should result in an increase in Q. An increase in the aluminum oxide content with reduction in the Al particle size lowers Q. This latter effect is a controlling factor if the completeness of oxidation of the pure metal is close to 100% (a high OB of the HE and a low Al content in the mixture). In this case, the replacement of the coarse Al powder with a finer one can lower Q. Such a situation is observed for BTNEN–Al (80:20) mixtures, where the composition with Al(0.1 μm) has a lower Q than that with Al(15 μm). If the completeness of oxidation of the metal is much lower than 100%, the role of the specific surface area increases; so the use of ultrafine Al increases Q. This was observed for the 60:40 HMX–Al(0.1 μm) and TNT–Al(0.1 μm) formulations. The fact that the maximum Q corresponds to a similar Al content in HEs, despite the considerable difference in their OB (see Fig. 4.29), becomes clear if one takes into account that an increase in the

Table 4.13. Heat of detonation of Al-containing formulations [49, 88]

β (%)	Al particle size (μm)	ρ_0 (g cm^{-3})	Q (kJ kg^{-1})	β (%)	Al particle size (μm)	ρ_0 (g cm^{-3})	Q (kJ kg^{-1})
TNT–Al				Composition B–Al			
10	12	1.67	5,307	0	–	1.69	5,389
	0.1–0.2	1.67	5,479	10	21	1.74	5,621
20	17	1.71	5,912		12		5,816
	0.1–0.2	1.76	6,021		2		6,004
30	21	1.83	6,184		0.1–0.2		5,927
	12	1.82	6,519	Plastic-bonded explosive[a]			
	2	1.83	6,660	20	12	1.72	4,632
	0.1–0.2	1.82	6,749		0.1–0.2	1.73	4,661

[a]Hydroxyl-terminated polybutadiene–dioctyl adipate–TDI–HMX–Al (9.6:5.6:0.8:64:20), where TDI is toluene diisocyanate

OB leads, in addition, to an increase in the completeness of Al oxidation, and to increase in consumption of the oxygen in the HE.

In [49, 88], data on the heat of detonation measured for TNT-based compositions were reported, and some of the data are given in Table 4.13.

The experimental set-up differs from that used in the aforementioned works. This can explain the difference in the "heat of explosion" given above and the "heat of detonation" presented below for similar formulations. Nevertheless, the trends in behavior of Al-containing formulations found by the authors seem to be very similar. Moreover, the aforementioned analysis [83] can be applied to the regularities found in [49, 88].

For binary TNT–Al formulations, one can see the trend for Q to increase with a decrease in Al particle size. The differences in Q measured for micrometric and nanoscale Al slightly increase with Al content growth. In [49] it was stated that "…the amount of heat was also greater for small particles, probably because detonation calorimeter charges were small and detonated more easily with smaller aluminium. The very small aluminium is also likely to react more completely than regular in those oxygen deficient mixes." Analyzing the data for formulations based on composition B, Brousseau et al. [88] (see Table 4.13) believe that "the reaction may occur rapidly enough behind the detonation front that the performance of these explosives is affected in the detonation calorimeter configuration." One can see that nanoscale Alex (100–200 nm) has no advantages with regard to Q over 2-μm powder as an ingredient of composition B mixes, at 10% content. This agrees with the data on HMX-based mixtures with 15% Al (see Table 4.11). For PBX formulations containing 64% HMX and 20% Al, the measured Q was practically the same for 12-μm Al powder and Alex [88]. One of the possible reasons for the absence of the expected advantages of nanopowders over conventional Al is the lower content of pure Al in nanometric powders (for Alex it is about 85%).

The effect of Al on the acceleration ability and the heat of explosion of a HE depends on the nature of the HE, the component ratio, the particle size, and the oxide-film content in the Al powder. The results of studies using the cylinder and the plate acceleration tests showed that the addition of Al to HEs could slightly enhance the acceleration ability. The comparison of the data for HMX- and NQ-based compositions indicates that for HEs with a negative OB the largest increase in the acceleration ability can be expected in the case of explosives with high hydrogen content. Because of the positive OB of BTNEN, the addition of Al to this HE provides for a more notable increase in the acceleration ability than addition of Al to HMX and NQ. However, the absolute value of the acceleration ability for Al mixtures with BTNEN exceeds that for pure HMX only slightly. The absence of advantages of compositions with Al(0.1 μm) over compositions with Al(7 μm) is due primarily to the high oxide-film content in Al(0.1 μm).

Because of the considerable time of interaction of Al with the DP under conditions of an explosion in a calorimetric bomb, the addition of Al to HEs significantly increases the heat of explosive decomposition. Mixtures with BTNEN have the largest values.

4.6 Nanocomposite Explosives

In recent years, there has been great interest in the development of new methods of production of nanoscale energetic materials, such as HE with nanoscale Al, as well as investigations of their properties. The results of some works in relation to the study of ADN- and HMX-based formulations are collected in this part. Two types of nanoscale formulation have been examined. The first type, called a "mechanical mixture" according to the procedure of its production is a common mixing of the components (it is denoted through the whole text as "HE–Al"). In the second type, called a "nanoscale composite," or simply "nanocomposite," nanometric Al particles are bound to the explosive more tightly; thus the nanoscale formulations can be considered as nanometric Al particles in an explosive matrix (it is denoted in the following as "HE/Al"). Nanocomposites can be produced in different ways.

4.6.1 ADN and Nanometric Al Formulations

ADN [$NH_4N(NO_2)_2$] is an explosive with an experimentally measured value of crystal density varying from 1.81 to 1.84 g cm^{-3} [89–91, 96]. The ADN enthalpy of formation is 150 kJ mol^{-1} [90]. Besides crystalline ADN [ADN(c)], granulated explosive [prilled ADN, ADN(p)] is widely used. The scientific literature contains two main granulation technologies and the properties of granules produced by each technology are different [92]. The first granulation method, sometimes referred to as the "prilling tower," was elaborated by Thiokol [93]; it is the same as technology as used for AN granulation. The second method is named the "technology of oil immersion" or the "technology of melt mixing." This technology was first developed in Russia, and then in the USA and in Sweden [91].

4.6.1.1 Mechanical Sensitivity

Formulations manufactured by mechanical mixing of ADN(c) or ADN(p) (produced by oil immersion technology) with nanometric Al posses high mechanical sensitivity. Data on the sensitivity of ADN and ADN–Al (75:25) mixtures [94] obtained with the use of the method of "disintegrative tubes" [95] are given in Table 4.14. The technique determines the critical pressure of explosion initiation (P_{cr}) of the primary and secondary HEs. A high P_{cr} indicates a low sensitivity of the explosive to mechanical loading.

Addition of Al(15 μm) to ADN(c) or to ADN(p) resulted in a decrease in P_{cr} in comparison with that for pure ADN. The decrease in P_{cr} of ADN(p)–Al mixtures is more pronounced. This can be partially explained by the difference in P_{cr} values of the base explosives. Replacing Al(15 μm) with Al(0.1 μm) in the ADN-based mixtures resulted in a sharp increase of mechanical sensitivity. The sensitivity of Al(0.1 μm)-containing compositions was comparable to that of the primary HE. This can be explained by chemical interaction of the products of ADN thermal decomposition and the fuel in the course of hot spot formation under explosive charge disintegration in the shot. The intensity of this interaction would depend on Al reactivity, which is higher for Al(0.1 μm) owing to its larger specific surface area.

The sensitivity of ADN(c) and ADN(p) (produced by the "technology of melt mixing") was also tested in [91] using a BAM drop weight apparatus with a 2-kg drop weight. The results were based on the tests on both sides of the 50% probability level using an up-and-down method. The corresponding drop height ($h_{50\%}$) is given in Table 4.14.

All ADN-containing formulations are moisture-absorbing mixes of high mechanical sensitivity. This is of importance for compositions with Al because of its possible oxidation in the acid medium of dissolved ADN when stored. Thus, ADN-based formulations usually contain some binder to prevent

Table 4.14. Critical pressure of explosion initiation [91, 94] *ADN(c)* crystalline ammonium dinitramide (ADN), *ADN(p)* prilled ADN

Formulation	P_{cr} (GPa)	Formulation	P_{cr} (GPa)	$h_{50\%}$ (cm)
Fulminate of mercury	0.20 ± 0.01	ADN(c)	0.80 ± 0.05	31
ADN(c)–Al(0.1 μm)	0.23 ± 0.03	ADN(p)–Al(15 μm)	0.85 ± 0.03	–
ADN(c)–Al(0.1 μm)	0.25 ± 0.03	HMX	1.00 ± 0.03	–
Lead azide	0.38 ± 0.03	ADN(p)	1.12 ± 0.04	59
ADN(c)–Al(15 μm)	0.75 ± 0.04	RDX	1.15 ± 0.03	38

moisture absorption and to reduce their sensitivity. In [96], it was reported that ADN/Al nanocomposites were prepared by adding a small amount of binder (Viton or Kel-f) to the formulations.

4.6.1.2 Detonation Velocity

In [96], it was reported that the incorporation of nanometric Al into ADN-containing charges resulted in a DV increase in comparison with the mixtures with micron-sized Al and pure ADN. To analyze this information, it is necessary to take into account the data on the DV in pure ADN and its failure diameter (bearing in mind that the positive effect in the DV was observed for TNT with the nanoscale Al additive near the failure diameter of TNT).

The DV in the ADN formulations not containing Al was measured in a number of works [91, 94, 96–98] and the data are collected in Table 4.15.

It is possible to assume that ADN belongs to the group-2 explosives according to Donna Price's classification [41]. In [96], the infinite DV for ADN was estimated as only 5.2 km s^{-1}, while the maximum velocity measured in the charges of 40-mm diameter at 1.6 g cm^{-3} is approximately 1 km s^{-1} higher [94]. Recently, a comprehensive treatise of melt-cast ADN was undertaken [98] (see Table 4.15). It was shown that the infinite DV was about 6.3 km s^{-1}. This agrees with the data for unconfined charges of pressed ADN(p).

For unconfined charges, d_{cr} of ADN lies within the range 4–8 mm [99] (3.9 km s^{-1} – the corresponding DV); the range 3–7 mm was reported for ADN (1.1 g cm^{-3}) with crystals of 100–400-μm size [100]. For pressed ADN(p), detonation failure was observed in unconfined charges of 10-mm diameter [91]. For melt-cast ADN with 0.5% ZnO, detonation deteriorated in 12.7-mm Plexiglas casings [97] and the authors stated that "...the failure diameter of ADN given by Pak [99] is considerably smaller than that found in this work for melt-cast ADN. From the data presented here, one may conclude that the unconfined failure diameter of melt-cast ADN is greater than 25 mm. The steady detonations observed by Miller et al. [96] were performed in a heavily confined test, and unconfined failure diameter cannot be inferred from those results" [97]. The data on the DV in melt-cast ADN [97, 98] correlate well. The infinite unconfined diameter for pressed ADN(p) is above 40 mm (see Table 4.15). For melt-cast ADN, the infinite diameter is larger than 100 mm [98].

Analyzing the wealth of the data on ADN presented above, one should keep in mind that at the same density of the charge, the infinite DVs have to be independent of the manufacturing procedure, while the critical diameters can differ by a factor of 10, e.g., for melt-cast and pressed TNT [26]. The latter can influence the DV measured experimentally for different types of ADN.

In [94], the DV versus charge density in pressed ADN(p) was reported (see Fig. 4.30, Table 4.15). Charges with densities less than 1.45 g cm^{-3} were placed into Plexiglas casings of 40-mm internal diameter and 44-mm outer diameter. Detonation was initiated with the aid of a booster charge of RDX through the

Table 4.15. Detonation velocities in ADN formulations

Formulation	ρ_0 (g cm^{-3})	D (km s^{-1})	d (mm)	l (mm)	Confinement
ADN(p) pressed	1.201	5.46	40	66	Plexiglas [94]
	1.452	6.30			
	1.600	6.20		14.45	Unconfined [94]
	1.600	6.17		42.29	
	1.602	6.23		42.51	
	1.613	6.17		56.28	
	1.761	5.47		37.09	
	1.780	4.95		45.68	
	1.785	4.88		44.04	
ADN/MgO (99:1) melt-cast	1.72	Failing	25	~60	Plexiglas [98]
		4.250	40		
		5.430	60		
		5.990	100		
		5.605	51.95	300	Copper [98]
ADN	1.696	4.239	6.4	~60	Brass or steel [96]
	1.604	4.186			
ADN(p) pressed	1.568	5.013	25.1	150	Plexiglas [91]
	1.658	5.260	43.9	240	
ADN/Viton (97:3)	1.752	4.254	6.4	~60	Brass or steel [96]
	1.735	4.056			
	1.689	4.376	12.7	~120	
	–	4.9	25.4	–	Unspecified [96]
ADN/ZnO (99.5:0.5) melt-cast	1.71	Failing	12.7	76.2	Plexiglas [97]
	1.70	3.42			Plexiglas/ brass[a] [97]
ADN/AN/ZnO (95.5:4:0.5) melt-cast	1.63	4.54	13.2	76.2	Steel [97]
	1.63	3.91	25.9	126.8	Plexiglas [97]

l charge length, AN ammonium nitrate
[a]Plexiglas tube is confined by a brass one

4-mm-thick intermediate Plexiglas plate. Charges with larger densities were unconfined and initiated by plane-wave generators made of RDX(5.5w).

Comparing the data reported on the DV in ADN, one can see the complicated character of the detonation behavior. The $D(\rho_0)$ dependence has a maximum near 1.5 g cm^{-3}. Moreover, the detonation seemed to deteriorate at a density above 1.70–1.75 g cm^{-3} (see Fig. 4.30).

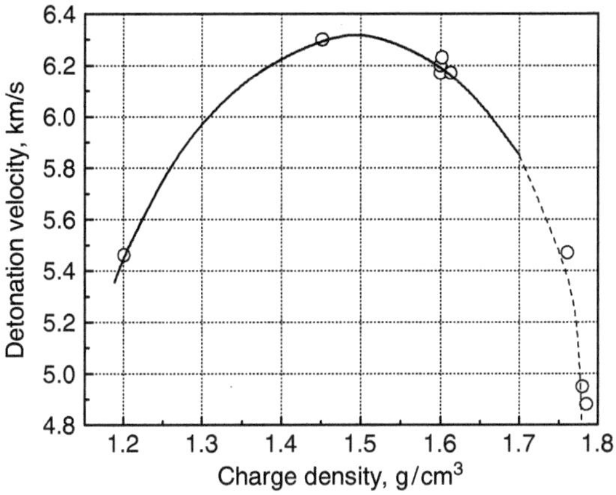

Fig. 4.30. Prilled ADN detonation velocity versus charge density [94]

Table 4.16. Detonation velocities in ADN-with-Al formulations

Formulation	Al content in the metal powder (%)	ρ_0 (g cm^{-3})	D (km s^{-1})	d (mm)	l (mm)	Confinement
ADN(p)–Al(15 µm) (75:25) pressed	98	1.801	5.86	40	28.25	Unconfined [94]
		1.803	5.78		56.4	
ADN–Al–MgO (64:35:1) melt-cast	Unspecified		4.036	60	∼60	Plexiglas [98]
ADN/Al(3 µm) (75:25) composite	∼99	1.794	4.104	6.35	∼60	Brass or steel [96]
ADN/Al(3 µm)/Viton (73:24:3) composite		1.744	4.086	12.7	∼120	
ADN/Al(60 µm)/Viton (73:24:3) composite	>99	1.835	4.323			
ADN/Al(0.15 µm)/Viton (73:24:3) composite	80–85	1.841	5.066			
ADN/Al(0.05 µm)/Viton (73:24:3) composite	65–70	1.714	5.028			
		1.804	6.0–6.6			

Thus, the hypothesis on ADN belonging to group 2 [94, 97] should be verified by testing long charges differing in density and diameter. It seems that the detonation performance of ADN can be affected by its method of manufacture: crystalline, prilled, or melt, at $d < d_{\text{inf}}$.

The detonation performance of ADN-with-Al formulations was studied in [94, 96]. The data on the DV are given in Table 4.16. In [96], the effect of

Al particle size on the DV was studied. The Al particles used in that study were spherical or spherical-like: H-60 – 60 μm, H-3 and H-5 –3 and 5 μm, Alex –150 nm; and LANL Al – 50 nm. In Table 4.16, the mean size of Al particles is given in parentheses. Most of the formulations tested contained Viton as a binder, which slightly decreases the DV. It is difficult to compare the reported data as they differed not only in Al particle size, but also in charge diameters and ADN density in the charge. It is thought that the use of a heavy casing increases the effective charge diameter by 3–5 times [22]; thus, the effective diameter of the charge placed into the 6.35-mm-diameter brass tube is still less than the infinite diameter of pure ADN. A similar effect was observed for an AN–TNT–Al mixture (see Fig. 4.14). This can explain the difference between the velocities measured in formulations with Al(60 μm) and Al(3 μm). As for nanoscale Al, it was stated that "The effect of the aluminium particle size on the DVs is clearly established for particle sizes of 150 nm or less. The aluminium described as 'Siberian' or 'Alex' with particle size of ∼150 nm gave an increase of about 25% in velocity. The aluminium obtained from LANL, ∼50 nm, also produced an increase of 25% in velocity in the 1/4 inch test, but gave an increase of over 50% in the 1/2 inch test configuration. The measured velocity was increasing throughout the experiment and the final velocity was unknown. The 60 and 3 μm particle size aluminium demonstrated no effect on the observed velocities" [96].

Different explanations can be given for the aforementioned fact; one is that incorporation of nanometric Al (especially with a high content of nearly incompressible aluminum oxide) into insensitive explosives, with a relatively large critical diameter, results in a reduction in the DV. Moreover, the DV does not actually exceed the maximum ADN velocity reported in [94]. The role of nanometric Al in ADN charges can be similar to that in TNT formulations, in spite of the difference in the OB. The DVs measured in ADN–Al(15 μm) are lower than those reported for pure ADN [94,98], but they are higher than those presented for micrometric Al in [96]. The DV in melt-cast ADN-based compositions with Al was about 1 km s^{-1} less than that in a pure explosive at the same charge diameter [98].

In [97], the effect of nanometric diamond additions on the DV of melt-cast ADN was studied (see Table 4.17). It is seen that the addition increases the velocity, and reduces d_{cr} of the charges placed into a Plexiglas tube. Both factors, diamond reaction in the DP and the reduction of the critical diameter, can lead to increase of the DV.

The data on the DV obtained for nanoscale Al and diamond in ADN mixtures are of great importance and interest to clarify the role of admixtures.

4.6.1.3 Acceleration Ability

The acceleration ability of ADN-based mixtures was estimated in so-called plate-push tests [96]. A baseline for the plate-push test was established by replacing the energetic samples with an inert material to account for the RP-2

Table 4.17. Detonation velocities in ADN-with-diamond formulations [97]

Melt-cast formulation (%)				ρ_0 ($g\,cm^{-3}$)	d (mm)	Confinement	l (mm)	D ($km\,s^{-1}$)
ADN	AN	ZnO	Nanodiamond					
92.37	0	0.48	7.15	1.31	12.7	Plexiglas	76.2	4.83
92.37	0	0.48	7.15	1.32	12.7			4.61
90.61	0	0.5	8.89	1.39	12.7		76.1	4.51
88.0	4.0	0.5	7.5	1.67	13.2	Steel	76.1	4.06

Table 4.18. Plate-push test results [96]

Formulation	ρ_0 ($g\,cm^{-3}$)	W ($m\,s^{-1}$)	Energy (mbar cm^3) cm^{-3}	Calculated energy, (mbar cm^3) cm^{-3}
PBXN-5	1.82	1,030	0.10	0.10
ADN	1.73	745	0.051	0.051
ADN/Al(0.150 μm)/Viton (74:23:3)	1.94	812	0.064	0.065–18% Al reaction
	1.86	813	0.063	
ADN/Al(3 μm)/Viton (74:23:3)	1.86	855	0.066	
ADN/Al(60 μm)/Viton (74:23:3)	1.95	712	0.044	0.46–0% Al reaction
RP-2 detonator/inert binder		372		

detonator's (32 mg PETN, 18 mg RDX) energy contribution to accelerating a steel plate, which was 2 mm thick and 6.35 mm in diameter. An average velocity (W) was obtained from the plate moving a distance of 63.5 mm, which corresponded to a time of 30–40 μs depending on the ultimate plate velocity (see Table 4.18). The recorded velocities were converted to kinetic energy after correction for the detonator energy. The results were normalized to the PBXN-5 energy content. The kinetic energies, experimentally determined and calculated, were compared to extract information on the completeness of Al oxidation (see Table 4.18). The results indicated that Al(60 μm) contributed little or nothing to the kinetic energy of the plate, while the Al(3 μm), Al(0.15 μm), and Al(0.05 μm) compositions all produced a similar increase in the estimated plate kinetic energy above that for ADN alone. Miler at al. [96] pointed out that even though Al(3 μm) produced no apparent effect on the DV, it did react sufficiently fast behind the detonation front to affect the plate velocity. They also mentioned that the effect of increased alumina content on the detonation properties as the particle size was reduced also needed consideration.

The data presented need some comments. First, it should be mentioned that the diameter of the tubes used in the tests was close to the critical diameter, so the effect of nanometric Al on d_{cr} should be taken into the account. Second, the formulations differed in the density; possible belonging of ADN to the group-2 explosives can explain the absence of an increase in W for Al(0.15 μm)-containing compositions of higher density. Thus, the lack of an advantage of nanometric Al over Al(3 μm) can be explained in two ways. First, by the high content of alumina (as was done for BTNEN–Al mixtures). Second, the higher density of ADN in the charge resulted in a decrease of its detonation performance as a group-2 explosive.

4.6.2 HMX and Nanometric Al Formulations

Nanometric Al, with a mean particle size varying from 40 to 140 nm, produced by the technique described in [101] has been used for manufacturing nanoscale composites. Several types of coating were synthesized to prevent agglomeration of Al particles and their "cold" welding [101, 102] (see Table 4.19).

Nanocomposites were produced on a laboratory scale using the technique of atomization drying of Al suspended in an HMX solution [103–105]. Solid particles of the composite containing nanometric Al surrounded by an HMX matrix were formed in the course of the gas jet moving. The conditions of the composite production depend on the type of coating used. The images of nanoscale composites made in characteristic Al emission show a relatively uniform Al distribution (see Fig. 4.31). The loose-packed density of the nanoscale composite powder containing 15% Al was about 0.10–0.25 g cm^{-3}. Ultrafine micrometric HMX (denoted by HMX(u-f)), with a mean particle size of 1.38 μm, was produced in the same manner.

4.6.2.1 Mechanical Sensitivity

The mechanical sensitivity of nanoscale HMX-based composites and mechanical mixtures with the same components was studied in [105, 106] using the method of "disintegrative tubes" [95]. The results obtained can be summarized as follows:

Table 4.19. Performance of nanosized Al powder

Al powder	Coating reactant	Mean particle size (nm)	Metallic Al content (%)
Al(1-1S)	Atmospheric air	94	86
Al(2-1S)		38	77
Al(1-6S)	Hexamethyldisilazane vapors	143	86
Al(2-6S)		86	82
Al(1-Ox)	Dry oxygen	123	83
Al(2-Ox)		65	70

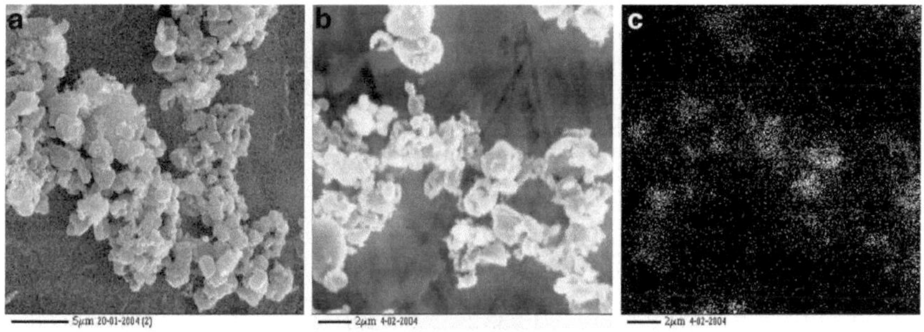

Fig. 4.31. Different structures of nanocomposites containing HMX/Al(1-6S) **a** A mixture of irregular crystallites with a 1.1 micrometer mean particle size **b** Fragments of microscopic hollow spheres with a 0.2–0.4 micrometer shell thickness **c** Images of the spheres (see **b**) in characteristic radiation of Al

- An increase of Al content (up to 25%) in nanoscale composites and mechanically prepared mixtures results in an increase in sensitivity.
- The sensitivity of HMX–Al mixtures increases with a decrease in metal particle size. Mixtures of HMX(B) with Al(0.04–0.14 μm) posses the highest sensitivity compared with mixtures containing Al(7 μm).
- Nanocomposites containing Al passivated in open air are characterized by lower sensitivity in comparison with similar mechanical mixtures of the same component mass ratio.
- The sensitivity of nanocomposites is practically independent of Al particle size in the range of their variation with the same type of coating.
- The sensitivity of both nanocomposites and mechanically prepared mixtures drastically increases with a decrease in Al "age." Metalized HE containing "junior" Al possesses a higher sensitivity compared with that of primary explosives (lead azide).
- HMX particle size does not influence the sensitivity of mixtures either with Al(0.1 μm) passivated in open air or with encapsulated Al. Formulations based on HMX(B) are more sensitive than those containing HMX(u-f).

4.6.2.2 Detonation Velocity

The testing of HMX-based formulations containing 15% Al was reported in [87, 103]. Explosive charges with $d = 20$ mm were initiated by an RDX(5.5w) booster. A pellet of the tested explosive ($l = 1.5d$) was placed between the booster and the base charge ($l \geq 2d$). Experimental DVs were normalized to the charge density of 1.760 g cm^{-3}. The experimental error was less than 30 m s^{-1}. The normalized velocities, D^*, are given in Table 4.20.

Neither the mechanical mixture with nanometric Al nor the nanocomposite has advantages with regard to the DV over the HMX(B)–Al(7 μm) mixture. The DV measured for both nanoscale composites and mixtures was lower than that in pure HMX. The DV in HMX(u-f) exceeded that in HMX(B); the

Table 4.20. Detonation velocities

Formulation	Al	ρ_0 (g cm^{-3})	D (km s^{-1})	D^* (km s^{-1})	ΔD (km s^{-1})
Mechanical mixture	Al(1-6S)	1.760	7.81	7.81	+0.15
Nanoscale composite		1.775	8.02	7.96	
Mechanical mixture	Al(1-Ox)	1.752	7.85	7.88	+0.21
Nanoscale composite		1.807	8.25	8.09	
Mechanical mixture	Al(2-6S)	1.688	7.63	7.88	−0.16
Nanoscale composite		1.770	7.76	7.72	
Mechanical mixture	Al(2-Ox)	1.733	7.69	7.79	+0.08
Nanoscale composite		1.712	7.70	7.87	
Mechanical mixture	Al(7)	1.870	8.55	8.18	–
HMX(B)	–	1.720	8.48	8.27	+0.12
		1.812	8.79		
HMX(u-f)	–	1.794	8.84	8.39	

ΔD is the difference in detonation velocity between the composite and the mixture

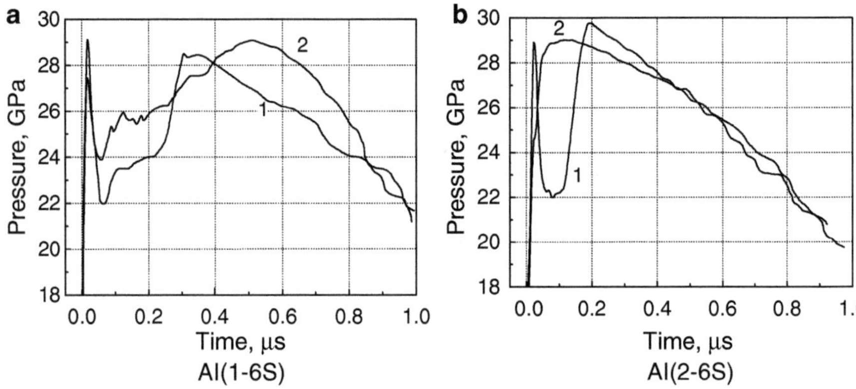

Fig. 4.32. Pressure histories in detonation products of HMX-based formulations. *1* mechanical mixture with HMX(B), *2* nanoscale composite

same effect was observed in [107]. A possible explanation is that HMX(u-f) is of higher purity (less RDX content in the powder) owing to the atomization drying technique.

Preliminary investigations with CL-20 and HMX compositions reported in [96] showed no measurable positive effect of a nanometer-sized powder admixture on the DV.

4.6.2.3 Pressure Histories

Pressure histories for nanoscale formulations were measured using an indicator technique [60]. The profiles determined in the DP are shown in Fig. 4.32 [87,103]. Experimental data are also given in Table 4.21.

Table 4.21. Pressures in the detonation products of HMX-based formulations with 15% Al

Parameters	Al(1-6S)				Al(2-6S)			
	Mixture		Composite		Mixture		Composite	
	Experiment	Normalized	Experiment	Normalized	Experiment	Normalized	Experiment	Normalized
ρ_0 (g cm^{-3})	1.800	1.760	1.816	1.760	1.746	1.760	1.710	1.760
P_1 (GPa)	29.2	27.6	27.5	25.4	29.0	29.6	24.6[a]	26.5[a]
P_2 (GPa)	28.6	26.9	29.1	26.9	29.8	30.4	29.0[a]	31.3[a]
P_{min} (GPa)	22.0	20.7	24.0	22.1	22.0	22.4	–	–
$<P>$ (GPa)	25.3	23.9	26.4	24.4	25.4	25.9	25.7	27.6

P_1, P_2, P_{min} pressure values in the detonation product of the first and the second peaks, and of the minimum pressure in-between, $<P>$ pressure averaged over the time history

[a]Monotonic pressure growth

A two-peak pressure history was observed for all explosives tested except the composite HMX/Al(2-6S), which was assumed to be the most uniform. The same trend was found for HMX-based mixtures with micrometric and sub-micron-sized Al. The pressure drop between the peaks is less both in terms of time and magnitude for the mechanical mixture with finer Al.

The pressure values were also normalized to the charge density of $1.760\,\mathrm{g\,cm^{-3}}$ (see Table 4.21). For a given Al, the normalized second-peak pressures were the same (within experimental error) in the composite and the corresponding mechanical mixture, while it was higher for the composite with finer Al(2-6S). Second-peak pressures were achieved at different times and this fact resulted in different averaged pressures.

Normalizing pressures to the same explosive density in the charges, we assume the same gradient in $D(\rho_0)$ dependence as occurs for pure HMX; the procedure is more correct for cases where there is less difference in density between formulations. This is the reason why the mixtures with micrometric Al were normalized to one magnitude of HMX density in the charge ($1.808\,\mathrm{g\,cm^{-3}}$), as it was closer to the experimental densities, and the nanoscale formulations were normalized to quite another value $-1.657\,\mathrm{g\,cm^{-3}}$. The data given in Table 4.22 illustrate the detonation behavior of the formulations when they were pressed to the same HMX density in the charge $-1.808\,\mathrm{g\,cm^{-3}}$; comparing the values, one should keep in mind that these data have lower accuracy. However, the procedure seems to be adequate as a first approximation. The experimental accuracy of 3% means that the pressures averaged within the first 0.5, 0.75, and 1.0 µs ($P^*_{0.5}$, $P^*_{0.75}$, $P^*_{1.0}$, respectively) are estimated within ± 0.9 GPa.

It is seen that the mixture with Al(fl) is inferior to all formulations in terms of average pressure within the first 0.5 µs, but at longer times no difference is observed for both mechanical mixtures and the formulations with Al(1-6S). The HMX/Al(2-6S) composite offers minor advantages over the mixture, but actually it has no advantage over HMX. Nevertheless, the reaction in the DP of the composite compensates for the losses on heating and compression of

Table 4.22. Average pressures in HMX-based formulations with 15% Al

Type of formulation	Admixture (15%)	Average pressure		
		$P^*_{0.5}$ (GPa)	$P^*_{0.75}$ (GPa)	$P^*_{1.0}$ (GPa)
Mixture	Al(15 µm)	30.1	29.5	28.5
	Al(fl)	27.9	29.2	28.9
	Al(1-6S)	29.9	29.9	29.3
Composite		30.2	30.7	29.8
Mixture	Al(2-6S)	33.6	32.6	31.7
Composite		36.8	35.5	33.9
HMX(B)	–	35.8	34.3	33.2

the Al particles and those caused by reduction in the amount of HMX in the formulation and the mole number of gaseous products within the first microsecond.

4.6.2.4 Temperature Histories

Temperature histories were obtained by recording radiation emitted from the DP–LiF interface (see Fig. 4.33). Temperatures of the first peak and the second peak, as well as the temperature measured at 1 μs, are given in Table 4.23. The pressure and temperature histories cannot be compared directly, but they correlate qualitatively. The temperature rise exhibits a more or less prominent "step" (see Fig. 4.33). The smoothest temperature history was measured for the most uniform composite, HMX/Al(2-6S). This inference correlates with the results obtained for micrometric Al. Simultaneously obtained temperature histories illustrate one more peculiarity of explosive formulations. The temperature histories reflect the process of Al reacting near the detonation front to a larger extent than the pressure profiles.

Several factors play an important role in this process: the effective specific surface area of Al, the relative movement of the components, and the content

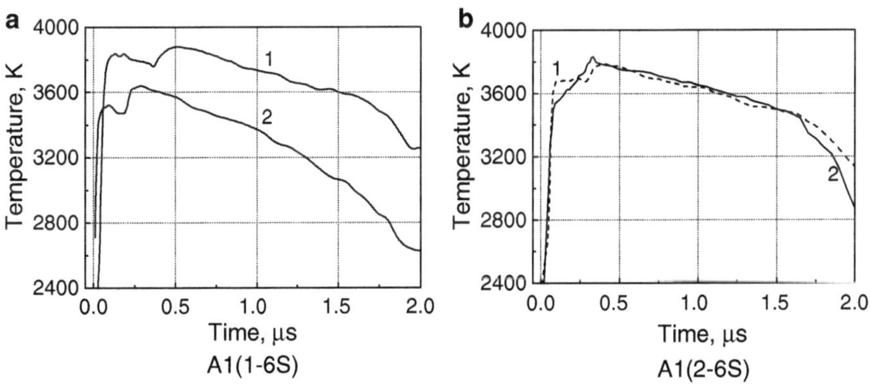

Fig. 4.33. Brightness temperature histories of the detonation products of HMX-based formulations. *1* mechanical mixture with HMX(B), *2* nanoscale composite

Table 4.23. Brightness temperature measured at the detonation products–LiF interface

	HMX(B)–Al(1-6S)	HMX/Al(1-6S)	HMX(B)–Al(2-6S)	HMX/Al(2-6S)
ρ_0 (g cm^{-3})	1.800	1.775	1.749	1.732
T_1 (K)	3,840	3,520	3,665	3,540[a]
T_2 (K)	3,885	3,645	3,795	3,835
T (K) (1 μs)	3,740	3,370	3,655	3,640

T_1 temperature of the first peak, T_2 temperature of the second peak
[a] Monotonic temperature growth

of pure metal in the Al powder. It can be assumed that the relative movement of the components is more intensive for the case of a nonmonotonic pressure history. This means that for explosives with such pressure profiles, all things being equal, the reaction would proceed with greater completeness and would be followed by higher temperatures. For the mechanical mixtures considered above, the content of pure Al was practically the same in Al(0.1 μm) and Al(fl) powders, but their pressure profiles differed the most. We believe that relative movement of the components is the most intensive in the case of a mechanical mixture with Al(fl), and the measured temperatures turned to be the highest. Another factor is an effective specific surface area, which was larger in mixtures with Al(0.1 μm) than, e.g., in explosives containing Al(0.5 μm); thus, the measured temperatures were higher for Al(0.1 μm).

The same regularities were found for the explosives tested: high temperatures were measured in HMX(B)–Al(1-6S) – a mechanical mixture, in which a two-peak structure appears to be most evident, and in HMX/Al(2-6S) – the most uniform nanoscale composite with the smallest particles and the highest specific surface area. Unfortunately, the differences in initial density cannot be taken in account accurately. At the same time, the observed difference in brightness temperatures lies within the experimental error and the inferred conclusions hold only within the times of interaction considered and under the given conditions of the experiments.

4.6.2.5 Acceleration Ability

One could expect, on the basis of the aforementioned results, that the averaged values of pressure obtained for HMX-based nanoscale composites would have some advantage with regard to acceleration ability over formulations with micrometric Al. Nevertheless, the acceleration tests with a steel plate [over a distance of about 150 mm by an HMX–Al–Viton (75:20:5) pellet 12.7 mm thick] showed that nanometric Al(0.15 μm) had no advantage with regard to terminal plate velocities over micrometric Al(5 μm). Moreover, both formulations exhibited velocities lower than the HMX–Viton (95:5) formulation [96]. The results correlate with those reported for HMX–Al mixtures [81].

4.6.2.6 Heat of Explosion

To analyze the heat effect produced by the incorporation of nanometric Al, Gogulya et al. [87, 103] presented the results of comparative studies of aluminized mechanical mixtures and nanocomposites. The Q values were measured using two calorimetric set-ups with bombs of 2 and 5 l in volume [94]. The experimental data for the explosives with 15% Al are given in Table 4.24, where Q is the heat of explosion at an experimental density and Q^* is the heat normalized to a charge density of $1.760\,\mathrm{g\,cm^{-3}}$. The table also gives the values characterizing the oxidation completeness, γ_1 and γ_2. The former was calculated taking into account only the mass content of pure Al in the metal

Table 4.24. Heat of explosion of the compositions

Composition	ρ (g cm^{-3})	Q (kJ kg^{-1})	Q^* (kJ kg^{-1})	γ_1	γ_2
HMX	1.81	5,610	5,500	–	–
HMX–Al(150 µm)	1.86	6,600	6,560	0.707	0.700
HMX–Al(15 µm)	1.89	6,980	6,930	0.852	0.835
HMX–Al(3.6 µm)	1.85	7,080	7,040	0.910	0.874
HMX–Al(0.1 µm)	1.83	6,930	6,900	0.958	0.824
HMX–Al(1-6S)	1.84	7,020	6,990	0.995	0.856
HMX/Al(1-6S)	1.75	6,960	6,970	0.986	0.848
HMX–Al(2-6S)	1.76	6,840	6,840	0.980	0.804
HMX/Al(2-6S)	1.72	6,770	6,780	0.960	0.787

Q heat of explosion at the experimental density, Q^* heat normalized to a charge density of 1.760 g cm^{-3}

powder; the latter took into account the total mass of the Al including alumina in the powder; γ_1 and γ_2 values correspond to Q^*.

From the results, it follows that Q^* values of the explosives with the same nanometric Al are closely related, irrespective of the manufacturing process (mechanical mixture or nanoscale composite). The reason is that nanometric Al undergoes practically complete oxidation during the explosion (γ_1, Table 4.24). In this case, Q^* depends mainly on the content of pure Al. Thus, the high degree of uniformity of the composites provides no additional increase in Q^*. The alumina content is higher in explosives with Al(2-6S), so the Q^* obtained is below the values measured in formulations with Al(1-6S).

Explosives with nanometric Al have a lower Q^* than mixtures with micron-sized Al owing to the large content of the oxide film; the mixture with Al(3.6 µm) has the highest Q^*. From Table 4.24, γ_1 is always higher in the case of nanosized Al; however, Q^* is greater for larger γ_2 values.

4.7 Concluding Remarks

At the end of this chapter, we would like to give some remarks on the results obtained for base explosivse containing nanometric Al, including nanocomposites.

The experimental techniques reported in the chapter let us examine the interaction of Al with the DP within different time scales. The DV is affected by the processes occurring in close proximity to the detonation front within the first 0.1 µs for powerful explosives. Pressure and temperature histories illustrate the processes taking place in the first microsecond, while the data on acceleration ability as well as on the heats of explosion provide information on the interaction of Al with the expanding DP within the first tens of microseconds.

Incorporation of nanoscale Al into an explosive has an ambiguous effect on the DV. For HEs with a small d_{cr}, the DV decreases, but the difference in the velocity measured for nanocomposites and corresponding mechanical mixtures lies in the range 150–200 $m\,s^{-1}$ at a DV of approximately 8 $km\,s^{-1}$.

For explosives with larger d_{cr} (e.g., AN, ADN, melt-cast TNT), the addition of nanometric Al can result in velocity growth compared with micrometric Al, but the effect is more prominent close to the critical diameter. A similar effect was found in melt-cast ADN with nanometric diamonds. The effect can be explained both by a sensitizing effect of a rigid admixture (nanometric Al contains a large amount of Al_2O_3) and by the Al reacting in the products. The latter explanation was offered for NQ–Al formulations, in which a decrease in the velocity was not observed with Al incorporation.

Similar effects are observed for liquid base explosives, NM or isopropyl nitrate (IPN), containing Al [108, 109]. NM and IPN have vastly different d_{cr}, so the behavior of nanoscale Al is assumed to be different.

For small packed beds of small spherical Al and glass particles saturated by NM, a reduction in both DV and d_{cr} was found. A higher DV was observed in a NM mixture with Alex compared with one containing Al(10 μm). Nevertheless, it was noted that "whilst, in principle, aluminium is a reactive additive it is not believed that there is time for any significant reaction of even the Alex material within the reaction zone of NM" [108]. It was also observed that the propagation velocities in mixtures of NM with Al or glass are higher than would be expected if the particles and the DP were in equilibrium within the reaction zone. With regard to possible detonation mechanisms, small particles were considered as the source of hot spots generated through shock interaction; a correlation between sound speed in solid particles and the measured DVs can also be seen, so the shock velocity in the particles could play a role determining detonation propagation [108].

The addition of nanometric Al into IPN results in a DV deficit with respect to the pure IPN, but an increase in detonation pressure and temperature as well as a significant decrease in critical diameter (from $d_{cr} > 30\,cm$ for pure IPN to $d_{cr} = 4.8\,cm$ for IPN/Alex in PVC tubes) [109]. Near d_{cr}, an increase in Al mass content from 25 to 32% results in an increase in the DV of 130 $m\,s^{-1}$. In contrast, it was found that a detonation could not be sustained in IPN/Al (3 μm or flakes) mixtures with large Al mass fractions in a 12.5-cm-diameter PVC tube. It was suggested that a large reaction zone of the IPN detonation would provide an opportunity for nanometric grades of Al to burn within the detonation reaction zone when these particles were mixed with IPN [109].

Of course, it is not known whether the aforementioned ideas on composite metalized liquid explosives can be extended to the solid ones directly. Nevertheless, we believe that the comments given by Zhang et al. [109] for composite liquid explosives are also true for solid ones: "Variation of the detonation velocity depends on the interplay between the total energy release from the explosives with particles and the momentum and heat transfer between them. In order to determine the main cause for the reduction of the critical diameter, i.e., due to shock-particles interactions or reactions of the particles, inert nanometric particles with approximately the same size, material density and mass fraction of packed particles as with those of Alex must be chosen and tested."

HMX-based nanocomposites exhibit high detonation temperatures and are characterized by higher pressures averaged over the first microsecond in comparison with mechanical mixtures. The observed two-peak pressure profiles, which are less pronounced in the case of nanocomposites, can be explained in terms of the detonation theory developed for Al-containing formulations.

In spite of the fact that the average pressures are found to be slightly higher for nanocomposites, preliminary data on the acceleration ability show no advantages of HMX-based formulations with nanometric Al over explosives with Al(10 μm).

No advantages with regard to the heat of explosion were found for formulations with nanometric Al (both mixtures and composites) over explosives containing fine micrometric Al. However, the data obtained for HMX-based nanocomposites displayed evidence of practically complete Al oxidation in the DP.

On the basis of the data reviewed, we may conclude that nanoscale Al enters into the reaction earlier, reacts faster and more completely than microscale Al, but its advantages are seriously restricted by the higher content of Al_2O_3.

The actual conditions and properties of the surrounding medium have an escalating effect on the work of Al-containing explosives at the later stages of expansion of the products, while the influence of the initial charge structure persists. However, in a certain sense, to react faster does not mean to produce a larger work of explosion, or action. The nonideal behavior of explosives originates from the retardation of energy release. This includes slow or multiple reactions which cause energy to be released later. Therefore, a large part of the energy, liberated beyond the influence zone, does not contribute to detonation wave propagation and must wait to catch up with the blast wave that propagates into the air. Nonideal characteristics of the explosives reduce the irreversible losses to the air close to the explosive charge by reducing the peak pressure and, therefore, the temperature compression heating of the air. Fewer irreversible losses to the air mean more energy propagates to a greater distance [113].

We began the chapter citing the fundamental work of Kast, who was a pioneer in the field of aluminized explosives. We have to conclude that the general inferences made almost a century ago on Al usage in explosive formulations hold true today. At the same time, scientific investigations have advanced the understanding of the detailed mechanism of Al reacting with DPs; a significant amount of progress has been made in developing new explosive formulations with a given detonation performance.

Appendix

Mixtures containing the same explosive with metal powders of particles of different size exhibit different compaction abilities; thus, the charges at the same metal content being manufactured in the same manner differed in ρ_0 slightly (90–95% of the TMD). Experimentally measured DVs (D) were normalized to the same density of the explosive in the charge (${\rho_{\mathrm{HE}}}^*$) to correctly compare the results. Such normalization presumes that decomposition of the base explosive plays the leading role in detonation of metalized explosives. In normalization, we used linear ρ_{HE} dependencies of DV for the base explosives. The last assumption can be justified owing to the small ρ_{HE} regions of extrapolation. Normalized DVs, $D^*({\rho_{\mathrm{HE}}}^*)$, were calculated using the relationship

$$D^* = D - b(\rho_{\mathrm{HE}} - {\rho_{\mathrm{HE}}}^*), \tag{4.4}$$

where ρ_{HE} is the density of the base explosive in the charge tested and b is the slope of the linear dependence of the base explosive. The ρ_{HE} values were calculated according to the equation

$$\rho_{\mathrm{HE}} = \alpha\rho_0\rho_{\mathrm{met}}/[\rho_{\mathrm{met}} - (1-\alpha)\rho_0], \tag{4.5}$$

where α is the explosive mass fraction, ρ_0 is the initial charge density, and ρ_{met} is the metal density. $D(\rho_{\mathrm{HE}})$ dependencies for base explosives were taken from [37].

Experimental values of the pressure, P, in the DP of composite explosive charges, differing in ρ_0, were normalized to the same charge density:

$$\rho_0^* = (\alpha/{\rho_{\mathrm{HE}}}^* + (1-\alpha)/\rho_{\mathrm{met}})^{-1}. \tag{4.6}$$

The detonation pressure can be calculated according to the formula

$$P = \rho_0 D^2/(n+1), \tag{4.7}$$

where n is the polytropic index of the DP. We assume that the similarity of pressure profiles is retained and the composition of the DP does not change (this is equivalent to the assumption $n = \mathrm{const}$) when the initial charge density is varied from ρ_0 to ${\rho_0}^*$. In doing so, normalized pressure values (P^*) can be calculated according to the formula

$$P^* = (D^*/D)^2 P({\rho_0}^*/\rho_0). \tag{4.8}$$

Acknowledgements

M.N. Makhov and A.Yu. Dolgoborodov are gratefully acknowledged for their cooperation in performing the majority of the experimental work.

References

1. Kast, H.: Spreng- und Zundstoffe. Vieweg-Verlag. Braunschweig (1921)
2. Afanasenkov, .N., Bogomolov, V.M., Voskoboinikov, I.M.: Calculation of detonation wave parameters for mixtures of explosives with inert additives. Combust. Explos. Shock Waves **6**(2), 182–186 (1970)
3. Aniskin, A.I., Shvedov, K.K.: Effect of aluminium and magnesium admixtures on detonation characteristics of mixtures with RDX. In: Detonation Critical Phenomenon. Physical–Chemical Transformations in Shock Waves, pp. 26–30. Nauka, Chernogolovka (1978)
4. Aniskin, A.I.: Detonation of composite high explosives containing aluminium. In: Proceedings of the VIII All-Union Symposium on Combustion and Explosion, pp. 26–32. Nauka, Chernogolovka (1986)
5. Arkhipov, V.I., Makhov, M.N., Pepekin, V.I., Shchetinin, V.G.: Investigations into detonation of aluminized high explosives. Chem. Phys. Rep. **18**(12), 2329–2337 (2000)
6. Baudin, G., Lafrancois, A., Bergues, D., Bigot, J., Champion, Y.: Combustion of nanophase aluminium in the detonation products of nitromethane. In: Proceedings of the 11th International Detonation Symposium, pp. 989–997, ONR 33300–5, Snowmass, (2000)
7. Belyaev, A.F.: Combustion, Detonation and Explosion Work of Condensed Systems. Nauka, Moscow (1968)
8. Bjarnholt, G.: Effects of aluminum and eithium flouride admixtures on metal acceleration ability of cmp . In: Proceedings of the 6th Symposium (International) on Detonation, pp. 510–521, ONR, ACR-221, Coronado (1976)
9. Brousseau, P., Dorsett, H., Cliff, M., Anderson, C.J.: Detonation properties of explosives containing nanometric aluminum powder. In: Proceedings of the 12th International Detonation Symposium, pp. 11–21, ONR 333-05-2, San Diego (2002)
10. Cook, M.A.: The Science of Industrial Explosives. IRECO Chemicals, Salt Lake City (1974)
11. Davydov, V.Yu., Grishkin, A.M., Feodoritov, I.I.: Experimental-theoretical study of the process of aluminium oxidation in detonation wave. Combust. Explos. Shock Waves **28**(5), 124–128 (1992)
12. Davydov, V.Yu., Grishkin, A.M., Murashov, E.Yu.: Gasodynamic conditions effect on the extent of secondary reactions energy transformation into explosive propelling action. Combust. Explos. Shock Waves **29**(2), 109–114 (1993)
13. Dremin, A.N., Pokhil, P.F., Arifov, M.I.: Effect of aluminium on the trotyl detonation constants. Rep. Acad. Sci. USSR **131**(5), 1140–1142 (1960)
14. Dubnov, L.V., Bakharevich, N.S., Romanov, A.I.: Commercial Explosives. Nedra, Moscow (1988)
15. Ermolaev, B.S., Khasainov, B.A., Baudin, G., Presles, H.-N.: Behavior of aluminum in detonation of high explosives. Surprises and interpretations. Chem. Phys. Rep. **18**(6), 1121–1140 (2000)
16. Finger, M., Hornig, H.C., Lee, E.L., Kury, J.W.: Metal acceleration by composite explosives. In: Proceedings of the 5th Symposium (International) on Detonation, pp. 55–63, ACR-184, Annapolis (1970)
17. Grishkin, A.I., Dubnov, L.V., Davydov, V.Yu., Levshina, Yu.A., Mikhailova, T.N.: The influence of admixtures of powder-like aluminium on detonation

parameters of high-order explosive systems. Combust. Explos. Shock Waves **29**(2), 115–117 (1993)
18. Hobbs, M.L., Baer, M.R.: Calibrating the BKW-EOS with a large product species database and measured C–J properties. In: Proceedings of the 10th International Detonation Symposium, pp. 409–418, ONR 33395–12, Boston (1995)
19. Howard, W.M., Sourers, P.C., Fried, L.E.: Kinetic calculations of explosives with slow burning constituents. In: Schmidt, S.C., Dandekar, D.P., Forbes, J.W. (eds.) Shock Compression of Condensed Matter – 1997, pp. 349–352. American Institute of Physics, New York (1998)
20. Johnson, J.N., Mader, C.L., Goldstein, S.: Performance properties of commercial explosives. Propellants Explos. Pyrotech. **8**(4), 8–18 (1983)
21. Mader, C.L.: Numerical Modeling of Detonation. University of California Press, Berkeley (1979)
22. Orlenko, L.P. (ed.): Physics of Explosion, 3rd edn. Fizmatlit, Moscow (2002)
23. Pokhil, P.F., Belyaev, A.F., Frolov, Yu.V., Logachev, V.S., Korotkov, A.I.: Burning of Powdered Metals in Active Media. Nauka, Moscow (1972)
24. Selesenev, A.A., Kreknin, D.A., Lashkov, V.N., Lobanov, V.N., Fedorov, A.V., Imkhovik, N.A.: Comparative analysis of the effects of aluminum and aluminum hydride on the detonation parameters and performance of mixed explosives. In: Proceedings of the 11th International Detonation Symposium, pp. 231–236, ONR 33300–5, Snowmass (2000)
25. Sosnova, G.S.: On boron and aluminium burning to their highest oxides at high pressures and temperatures. In: Proceedings of the Third All-Union Symposium on Combustion and Explosion, pp. 455–458. Nauka, Moscow (1972)
26. Stanukovich, K.P. (ed.): Physics of Explosion, 2nd edn. Nauka, Moscow (1975)
27. Voskoboinikov, I.M., Voskoboinikova, N.F.: Evaluation of projectile ability of explosives. In: Proceedings of the 2nd USSR Conference on Detonation, vol. 2, pp. 64–67. Nauka, Chernogolovka (1981)
28. Voskoboinikov, I.M., Kotomin, A.A.: Calculations of detonation parameters of explosive-inert addition. Combust. Explos. Shock Waves **21**(5), 93–97 (1985)
29. Cheret, R.: Le code ARPEGE: application a l'etude d'um explosif a l'aluminium. Acta Astronaut. **1**, 893–898 (1974)
30. Cudzilo, S., Trzcinski, W.A.: Studies of HMX-based explosives containing magnesium and polytetrafluoroethylene. Khim. Fiz. **22**(1), 82–89 (2003)
31. Crawford, P.: Getting started with TIGER. Lawrence Livermore National Laboratory, Livermore (1986)
32. Trebinski, R., Trzcinski, W., Wlodarczyk, E., Widkowski, E.: Analysis of the influence of the polymorphous transformation of the detonation process of explosive mixtures with a substances undergoing a transformation. J. Tech. Phys. **31**(2), 129–154 (1990)
33. Lee, E.L., Horning, H.C., Kury, J.W.: Report UCRL-50422. Lawrence Livermore National Laboratory, Livermore (1988)
34. McQueen, R.G., Marsh, S.P., Taylor, J.W., Fritz, J.N., Carter, W.J.: On equation of state of solids. In: Kinslow, R. (ed.) High-velocity impact phenomena. Academic, New York (1970)
35. Howard, W.M., Fried, L.E., Souer, P.C.: Kinetic modelling of non-ideal explosives with cheetah. In: Proceedings of the 11th International Detonation Symposium, pp. 998–1006, ONR 33300–5, Snowmass (2000)

36. Imkhovik, N.A.: Parameters and regimes of detonation wave propagation in explosive mixtures with different metals. In: Mikhailov, A.L. (ed.) Proceedings International Conference "VII Khariton's Topical Scientific Readings", pp. 48–53. VNIIEF, Sarov (2005)
37. Gogulya, M.F., Makhov, M.N., Dolgoborodov, A.Yu., Brazhnikov, M.A., Arkhipov, V.I., Shchetinin, V.G.: Mechanical sensitivity and detonation parameters of aluminized explosives. Combust. Explos. Shock Waves **40**(4), 445–457 (2004)
38. Reshetov, A.A., Schneider, V.B., Yavorovsky, N.A.: Influence of ultra-dispersive aluminum the speed of detonations of hexamethly-3-nitroamine. In: Proceedings of the First All-Union Symposium on Macroscopic Kinetics and Chemical Gas Dynamics vol. 1, pp. 56–57. Nauka, Chernogolovka (1984)
39. Sanden, R.: Characterization of electro-exploded aluminium (Alex). In: Proceedings of the 29th International Conference of ICT. Energetic Materials. Productions, Processing and Characterization, pp. 77.1–77.10. ICT, Karlsruhe (1998)
40. Il'in, A.P., Reshetov, A.A.: Composition of condensed products from the detonation of composite explosives. Combust. Explos. Shock Waves **35**(4), 92–94 (1999)
41. Price, D.: Contrasting patterns in the behaviour of high explosives. In: Proceedings of the 11th Symposium (International) on Combustion, pp. 693–702. The Combustion Institute, Pittsburgh (1967)
42. Howard, W.M., Fried, L.E., Souer, P.C.: Modeling of non-ideal aluminized explosives. In: Furnish, M.D., Chhabildas, L.C., Hixson, R.S. (eds.) Shock Compression of Condensed Matter – 1999, pp. 389–392. American Institute of Physics, Melville (2000)
43. Gogulya, M.F., Dolgoborodov, A.Yu., Brazhnikov, M.A., Baudin, G.: Detonation waves in HMX/Al mixtures (pressure and temperature measurements). In: Proceedings of the 11th International Detonation Symposium, pp. 979–988, ONR 33300–5, Snowmass (2000)
44. Victorov, S.B., Gubin, S.A., Maklashova, I.V., Revyakin, I.I.: Thermodynamic TDS code: Application to detonation properties of condensed explosives. In: Proceedings of the 32nd Annual Conference of ICT. Energetic Materials. Ignition, Combustion and Detonation, pp. 69.1–69.15. ICT, Karlsruhe (2001)
45. Victorov, S.B.: The effect of Al_2O_3 phase transitions on detonation properties of aluminized explosives. In: Proceedings of the 12th International Detonation Symposium, pp. 369–376, ONR 333–05–2, San Diego (2002)
46. Gogulya, M.F., Makhov, M.N., Dolgoborodov, A.Y., Brazhnikov, M.A.: Detonation velocity of BTNEN/Al. Combust. Explos. Shock Waves **42**(4), 480–485 (2006)
47. Cook, M.A., Filler, A.S., Keyes, R.T., Partridge, W.S., Ursenbach, W.O.: Aluminized explosives. J. Phys. Chem. **61**(2), 189–196 (1957)
48. Dremin, A.N., Savrov, C.D., Trofimov, V.C., Shvedov, K.K.: Detonation Waves in Condensed Media. Nauka, Moscow (1970)
49. Brousseau, P., Cliff, M.: The effect of ultrafine aluminium powder on the detonation properties of various explosives. In: Proceedings of the 32nd Annual Conference of ICT. Energetic Materials. Ignition, Combustion and Detonation, pp. 37.1–37.14. ICT, Karlsruhe (2001)
50. Cliff, M.D., Dorsett, H.E. and Lu, J.P.: Combustion of nanometric aluminium in detonating solid explosives. Paper presented to the TTCP WPN Group

TP-4, technical workshop on the usage and performance of mano-materials held at JHU-APL, Laurel (2002)
51. Cook, M.A.: The Science of High Explosives. Reinhold, New York (1959)
52. Arai, H., Nakamura, J.: Influence of aluminium powder on the detonation of ammonium nitrate and aluminium mixture. In: Schelling, F.J. (ed.) Proceedings of the Thirty-Third International Pyrotechnics Seminar, pp. 405–411. IPSUSA Seminars, Fort Collins (2006)
53. Ermolaev, B.S., Khasainov, B.A., Presles, H-N., Vidal, P., Sulimov, A.A.: Low-velocity detonation in ammonium nitrate and its mixtures. In: The 13th Symposium on Combustion and Explosion. http://ore13.rsl.ru/nettext/russian/gor_i_vzr/content/ermolaev (2005)
54. Miyke, A., Van der Steen, A.C., Kodde, H.H.: Detonation velocity and pressure of the non-ideal explosive ammonium nitrate. In: Proceedings of the 9th Symposium (International) on Detonation, pp. 560–565, OCNR 113391–7, Portland (1989)
55. Afanasenkov, A.N., Kotova, L.I., Sergeev, S.S., Shvedov, K.K.: Calculation of detonation parameters of commercial explosives. Preprint. Ukrainian Academy of Sciences, Subbotin Institute of Geophysics, Department of Explosion Geodynamics, Kiev (1993)
56. Maranda, A., Cudzilo, S., Trzcinski, W.: Behaviour of aluminium and TNT in the detonation waves of ammonium nitrate explosives. In: Proceedings of the 27th International Annual Conference of ICT. Energetic Materials. Technology Manufacturing and Processing, pp. 43.1–43.10. ICT, Karlsruhe (1996)
57. Zhang, F., Yoshinaka, A., Frost, D., Ripley, R., Kim, K., Wilson, W.: Casing influence on ignition and reaction of aluminium particles in an explosive. In: The 13th International Detonation Symposium. http://www.intdetsymp.org/detsymp2006 (2006)
58. Zhang, F., Thibault, P.A., Link, R., Gonor, A.L.: Momentum transfer during shock interaction with metals particles in condensed explosives. In: Furnish, M.D., Thadhani, N.N., Horie, Y. (eds.) Shock Compression of Condensed Matter – 2001, pp. 934–937. American Institute of Physics, Melville (2002)
59. Dolgoborodov, A.Yu., Voskoboinikov, I.M., Tolstov, I.K., Sudarikov, A.V.: Characteristic properties of shock wave propagation in mixtures. Combust. Explos. Shock Waves **28**(3), 106–112 (1992)
60. Gogulya, M.F., Dolgoborodov, A.Yu.: Indicator technique for investigations of shock and detonation waves. Chem. Phys. Rep. **13**(12), 2059–2069 (1995)
61. Gogulya, M.F., Brazhnikov, M.A.: Radiation of condensed explosives and its interpretation (temperature measurements). In: Proceedings of the 10th International Symposium on Detonation, pp. 542–548, ONR 33395–12, Boston (1995)
62. Borisov, A.A., Khasainov, B.A., Komissarov, P.V., Mailkov, A.E., Khomik, S.V., Silakova, M.A., El'shin, R.N.: Blast waves in systems "a suspension or a layer of an energetic materials in air". Khim. Fiz. **23**(4), 95–108 (2004)
63. Gogulya, M.F., Dolgoborodov, A.Yu., Brazhnikov, M.A.: Fine structure of detonations waves in HMX-Al mixtures. Chem. Phys. Rep. **17**(1–2), 51–54 (1998)
64. Imkhovik, N.A.: Modelling of detonation features of metalized high explosives as systems with non-monotonous energy release. In: International Workshop. New Model Hydrocodes for Shock Wave Processes in Condensed Matter, pp. 75–80. High Pressure Center, Moscow (2002)

65. Lubyatinsky, S.N., Loboiko, B.G.: Density effect on detonation reaction zone length in solid explosives. In: Schmidt, S.C., Dandekar, D.P., Forbes, J.W. (eds.) Shock Compression of Condensed Matter-97, pp. 743–746. American Institute of Physics, Woodbury (1998)
66. Lubyatinsky, S.N., Loboiko, B.G.: Detonation reaction zone of solid explosives. In: Short, J. (ed.) Proceedings of the 11th International Detonation Symposium, pp. 836–844, ONR 33300–5, Snowmass (2000)
67. Gustavsen, R.L., Sheffield, S.A., Alcon, R.R.: Detonation wave profiles in HMX based explosives. In: Schmidt, S.C., Dandekar, D.P., Forbes, J.W. (eds.) Shock Compression of Condensed Matter-97, pp. 739–742. American Institute of Physics, Woodbury (1998)
68. Farber, D.L., Esposito, A.P., Zaug, J.M., Reaugh, J.E., Aracne, C.M.: First results of reaction propagation rates in HMX at high pressure. In: Furnish, M.D., Thadhani, N.N., Horie, Y. (eds.) Shock Compression of Condensed Matter – 2001, pp. 1015–1018. American Institute of Physics, Melville (2002)
69. Evans, D.J., Milne, A.M., Softley, I.: The burning rate of aluminium particles in cylinder tests. In: Furnish, M.D., Thadhani, N.N., Horie, Y. (eds.) Shock Compression of Condensed Matter – 2001, pp. 1011–1014. American Institute of Physics, Melville (2002)
70. Dolgoborodov, A.Yu., Voskoboinikov, I.M.: On "Frontolysis" of shock-wave front in mixtures. Preprint. Ukrainian Academy of Sciences, Subbotin Institute of Geophysics, Department of Explosion Geodynamics. ONTI IPP AN Ukrainy, Kiev (1993)
71. Zhang, F.: Detonation in reactive solid particle-gas flow. J. Propuls. Power **22**, 1289–1309 (2006)
72. Imkhovik, N.A., Soloviev, V.S.: Mathematical simulation of the acceleration dynamic of the elements projected by the detonation products of aluminium containing HE. Def. Tech. **1**, 5–13 (1995)
73. Gogulya, M.F., Brazhnikov, M.A.: On the characteristic times of chemical reactions in heterogeneous systems under dynamic loads. Chem. Phys. Rep. **13**(11), 1887–1901 (1995)
74. Gogulya, M.F., Dolgoborodov, A.Yu., Brazhnikov, M.A., Makhov, M.N., Arkhipov, V.I.: Explosive performance of aluminised NQ-based compositions. In: Proceedings of the 32nd International Conference of ICT. Energetic Materials. Ignition, Combustion and Detonation, pp. 77.1–77.14. ICT, Karlsruhe (2001)
75. Gogulya, M.F., Dolgoborodov, A.Yu., Brazhnikov, M.A., Makhov, M.N., Arkhipov, V.I.: Aluminised explosive compositions based on NQ and BTNEN. In: Furnish, M.D., Thadhani, N.N., Horie, Ya. (eds.) Shock Compression of Condensed Matter – 2001, pp. 962–965. American Institute of Physics, New York (2002)
76. Utkin, A.V., Kolesnikov, S.A., Fortov, V.E.: Steady-state detonation wave structure in pressed RDX. Rep. Russ. Acad. Sci. **381**(6), 760–762 (2001)
77. Gogulya, M.F., Dolgoborodov, A.Yu., Mashov, M.N., Brazhnikov, M.A., Shchetinin, V.G.: Detonation performance of aluminized compositions based on BTNEN. In: Proceedings of the 12th International Detonation Symposium, pp. 249–255, ONR 333–05–2, San Diego (2002)
78. Lefrançois, A., Le Gallic, C.: Expertise of nanometric aluminum powder on the detonation efficiency of explosives. In: Proceedings of the 32nd International

Conference of ICT. Energetic Materials. Ignition, Combustion and Detonation, pp. 36.1–36.14. ICT, Karlsruhe (2001)
79. Lefrançois, A., Baudin, G., Le Gallic, C., Boyce, P., Coudoing, J-P.: Nanometric aluminum powder influence on the detonation efficiency of explosives. In: Proceedings of the 12th International Detonation Symposium, pp. 22–32, ONR 333–05–2, San Diego (2002)
80. Short, J.M., Helm, F.H., Finger, M., Kamlet, M.J.: The chemistry of detonations. VII. A simplified method for predicting explosive performance in the cylinder test. Combust. Flame **43**(1), 99–109 (1981)
81. Makhov, M.N., Gogulya, M.F., Dolgoborodov, A.Yu., Brazhnikov, M.A., Arkhipov, V.I., Pepekin, V.I.: Acceleration ability and heat of explosive decomposition of aluminized explosives. Combust. Explos. Shock Waves **40**(4), 458–466 (2004)
82. Makhov, M.N.: The heat and products of explosion of aluminized high explosives. In: Proceedings of the 31st International Annual Conference of ICT. Energetic Materials. Analysis Diagnostics and Testing, pp. 42.1–42.11. ICT, Karlsruhe (2000)
83. Makhov, M.N.: Explosion heat of aluminized high explosives. Chem. Phys. Rep. **19**(9), 1723–1731 (2001)
84. Pepekin, V.I., Makhov, M.N., Lebedev, Yu.A.: Heat of explosive decomposition of individual explosives. Rep. Acad. Sci. USSR **232**(4), 852–855 (1977)
85. Makhov, M.N., Pepekin, V.I.: Calculation of chemical composition of detonation products. Pol. J. Chem. **55**(6), 1381–1385 (1981)
86. Makhov, M.N.: The effect of charge density on the explosion heat of high explosives. In: Proceedings of the 33rd International Annual Conference of ICT. Energetic Materials. Synthesis, Production and Application, pp. 73.1–73.13. ICT, Karlsruhe (2002)
87. Gogulya, M.F., Makhov, M.N., Dolgoborodov, A.Yu., Brazhnikov, M.A., Leipunsky, I.O., Jigatch, A.N., Kuskov, M.L., Laritchev, M.N.: Aluminised nanocomposites based on HMX. In: Proceedings of the 35th International Annual Conference of ICT & 32nd International Pyrotechnics Seminar. Energetic Materials. Performance and Safety, pp. 65.1–65.11. ICT, Karlsruhe (2005)
88. Brousseau, P., Dorsett, H., Cliff, M., Anderson, C.J.: Detonation properties of explosives containing nanometric aluminum powder. In: Proceedings of the 12th International Detonation Symposium, pp. 11–21, ONR 333–05–2, San Diego (2002)
89. Gidaspov, B.V., Tselinskii, I.V., Mel'nikov, V.V., Margolis, N.V., Grigor'eva, N.V.: Crystalline, molecular, structure of the salts of dinitramide and its acid-basic properties. J. Gen. Chem. **65**(6), 995–1002 (1995)
90. Golfier, M., Graindorge, H., Longevialle, Y., Mace, H.: New energetic molecules and their applications in energetic materials. In: Proceedings of the 29th International Annual Conference of ICT. Energetic Materials. Productions, Processing and Characterization, pp. 3.1–3.18. ICT, Karlsruhe (1998)
91. Karlsson, S., Ostmark, H.: Sensitivity and performance characterization of ammonium dinitramide (ADN). In: Short, J. (ed.) Proceedings of the 11th International Detonation Symposium, pp. 801–806, ONR 33300–5, Snowmass (2000)
92. Ramaswami, A.L.: Energetic material combustion experiments on propellant formulations containing prilled ADN. Combust. Explos. Shock Waves **36**(1), 131–137 (2000)

93. Highsmith, T.K., McLeod, C., Wardle, R.B., Schmitt, R., Bottaro, J., Penwell, P., Bomberger, D., Brough, J.: ADN manufacturing technology. In: Proceedings of the 29th International Annual Conference of ICT. Energetic Materials. Productions, Processing and Characterization, pp. 20.1–20.14. ICT, Karlsruhe (1998)
94. Gogulya, M.F., Dolgoborodov, A.Yu., Mashov, M.N., Brazhnikov, M.A.: Detonation Performance of ADN and its mixtures with Al. In: Proceedings of the EuroPyro 2003 (8eme Congres Intern. de Pyrotechnie du "Groupe de travail de pyrotechnie"), pp. 18–27, Saint Malo (2003)
95. Schetinin, V.G.: Estimation of the mechanical sensitivity of solids explosives by the method of destructible shell. Combust. Explos. Shock Waves **35**(5), 570–575 (1999)
96. Miller, P.J., Bedford, C.D., Davis, J.J.: Effect of metal particle size on the detonation properties of various metalized explosives. In: Short, J. (ed.) Proceedings of the 11th International Detonation Symposium, pp. 214–220, ONR 33300–5, Snowmass (2000)
97. Doherty, R.M., Forbes, J.W., Lawrence, G.W., Deiter, J.S., Baker, R.N., Ashwell, K.D., Stherland, G.T.: Detonation velocity of melt-cast and ADN/nano-diamond cylinders. In: Furnish, M.D., Chhabildas, L.C., Hixson, R.S. (eds.) Shock Compression of Condensed Matter – 1999, pp. 833–836. American Institute of Physics, Melville (2000)
98. Ostmark, H., Helte, A., Karlsson, S., Hahma, A., Edvinsson, H.: Detonation properties and reaction rate modeling of melt cast ammonium dinitramide (ADN). In: Proceedings 12th International Detonation Symposium, pp. 775–780, ONR 333–05–2, San Diego (2002)
99. Pak, Z.P.: Some ways to higher environmental safety of solid rocket propellant application. In: The 29th Joint Propulsion Conference, paper 93–1755. American Institute of Aeronautics and Astronautics, Reston (1993)
100. Teselkin, V.A.: Impact sensitivity and detonation ability of ADN/PMMA compositions. In: Proceedings of the 35th International Annual Conference of ICT & 32nd International Pyrotechnics Seminar. Energetic Materials. Performance and Safety, p. 80.1. ICT, Karlsruhe (2005)
101. Jigatch, A.N., Leipunsky, I.O., Kuskov, M.L., Stoenko, N.I., Storozhev, V.B.: An apparatus for the production and study of metal nanoparticles. Instrum. Exp. Tech. **43**(6), 839–845 (2000)
102. Jigatch, A.N., Leipunsky, I.O., Kuskov, M.L., Pshechenkov, P.A., Berezkina, N.G., Larichev, M.A., Krasovsky, V.G.: Synthesis of coatings on ultra-fine aluminum particles surface. Khim. Fiz. **21**(4), 72–78 (2002)
103. Gogulya, M.F., Makhov, M.N., Dolgoborodov, A.Yu., Brazhnikov, M.A., Leipunsky, I.O., Jigatch, A.N., Kuskov, M.L., Laritchev, M.N.: Aluminized nanocomposites based on HMX. Manufacturing and properties. In: Mikhailov, A.L. (ed.) Proceedings of the International Conference "VII Khariton's Topical Scientific Readings", pp. 33–38. VNIIEF, Sarov (2005)
104. Jigatch, A.N., Leipunsky, I.O., Kuskov, M.L., Pshechenkov, P.A., M.A., Larichev, M.A., Krasovsky, V.G., Gogulya, M.F.: A technique to prepare aluminized nanosized explosives. In: Schelling, F.J. (ed.) Proceedings of the Twenty-Ninth International Pyrotechnics Seminar, pp. 583–589. IPSUSA, Westmister (2002)
105. Leipunsky, I.O., Jigatch, A.N., Kuskov, M.L., Pshechenkov, P.A., Larichev, M.A., Gogulya, M.F., Teselkin, V.A.: Aluminized nanocomposite explosive –

production and properties. In: Mikhailov, A.L. (ed.) Proceedings of the International Conference "V Khariton's Topical Scientific Readings", pp. 219–223. VNIIEF, Sarov (2003)
106. Teselkin, V.A.: Mechanical Sensitivity of Nano-Scaled Explosives. In: Mikhailov, A.L. (ed.) Proceedings of the International Conference "VII Khariton's Topical Scientific Readings", pp. 61–67. VNIIEF, Sarov (2005)
107. Shunai, Z., Jingliu, Z., Jingyu, W.: Preparation and properties of sub-micron nitramine explosives. In: Proceedings of the 35th International Annual Conference of ICT. Energetic Materials. Structure and Properties, pp. 100.1–100.4. ICT, Karlsruhe (2004)
108. Haskins, P.J., Cook, M.D., Briggs, R.I., The effect of additives on the detonation characteristic of a liquid explosive. In: Furnish, M.D., Thadhani, N.N., Horie, Y. (eds.) Shock Compression of Condensed Matter – 2001, pp. 890–893. American Institute of Physics, Melville (2002)
109. Zhang, F., Yoshinaka, A., Murray, S.B., Higgins, A.: Shock initiation and detonability of isopropyl nitrate. In: Proceedings of the 12th International Detonation Symposium, pp. 781–790, ONR 333–05–2, San Diego (2002)
110. Zhukov, B.P. (ed.): Energetic Condensed Materials. Janus-K, Moscow (2000)
111. Gogulya, M.F., Brazhnikov, M.A.: On the temperature of detonation of metalized high explosives. Chem. Phys. Rep. **15**(12), 1817–1820 (1996)
112. Naboichenko, S.S. (ed.): Nonferrous metal powders, handbook. Metallurgiya, Moscow (1997)
113. Romero, V., Williams, Ph.E.: Blast Waves from Non-ideal Explosives. In: Schmidt, S.C., Dandekar, D.P., Forbers, J.W. (eds.) Shock Compression of Condensed Matter – 1997, pp. 763–767. American Institute of Physics, Woodbury, New York (1998)

5

Shock-Induced Solid–Solid Reactions and Detonations

Yu.A. Gordopolov, S.S. Batsanov, and V.S. Trofimov

5.1 Introduction

In this chapter, we consider the theoretical and practical aspects of shock wave processes in condensed media, including solid–solid detonations (SSDs), i.e., conversion of solid-phase reactants to solid-phase products [1]. Numerous experimental data imply that shock processing may be used to induce very fast chemical reactions in compacted reactive powder mixtures.

It is known [2] that a shock wave process is a kind of motion in a continuous medium which is accompanied by propagation of special waves (shocks) at a hypersound velocity. A shock (sudden change) represents a thin, relatively stable zone within which elementary volumes of matter spasmodically change their velocity and density. Depending on the properties of the medium, either compression or rarefaction shocks can be formed. Since rarefaction shocks are encountered infrequently, in further discussion we will deal only with compression shocks.

Within the shock, the medium may undergo various physicochemical transformations (chemical reaction, phase transition, collapse of pores, etc.). Shocks without transformations and shocks in chemically inert porous media are normally termed "shock waves." The shocks accompanied by physicochemical transformations are termed either "shock waves" or "detonation waves," depending on the type of transformation. The difference between shock waves and detonation waves will be discussed later. Now let us only note that a leading shock in a self-propagating shock wave process in condensed explosives or reactive gaseous mixtures can be classified as a detonation wave. Concerning other shock wave processes, the difference between shock and detonation waves is a subject of controversy and argument.

The width of shocks without transformation of matter is comparable to the free path of molecules in gases or to intermolecular distance in condensed matter [2]. As a mathematical image of such a shock, the notion of a traveling finite discontinuity surface can be used. The width of a shock accompanied by transformations of matter is greater by several orders of magnitude. For

F. Zhang (ed.), *Shock Wave Science and Technology Reference Library: Heterogeneous Detonation*, DOI: 10.1007/978-3-540-88447-7_5,

instance, in powder mixtures the shock width is comparable to the particle size. Nevertheless, in some cases the shock with transformation can also be modeled as a surface of finite discontinuity.

It is believed that a self-sustained shock wave process (detonation) may develop only in condensed explosives or reactive gaseous mixtures where the reaction is accompanied by vigorous gas evolution. The possibility of detonation in systems that react without evolution of gases (so-called gas-free detonation) has been predicted theoretically [3], and a quantitative thermodynamic criterion for this to occur in any condensed media was suggested in [3] and then specified in [4] (see Sect. 5.4.4).

5.2 Shock-Induced Solid–Solid Reactions

5.2.1 Experimental Observations

For the first time, the occurrence of shock-induced solid–solid reactions (SSRs) was detected in experiments with recovery fixtures [5]. A key point here is the duration of the SSR. When the reaction is accomplished in microseconds (i.e., within the high-pressure zone), one can expect nontrivial results. Otherwise, the reaction will proceed upon unloading as a result of heating. In this case, we deal with conventional thermal reactions. The reaction time can be measured by the kinetic method suggested in [6]. But since such shock wave experiments are difficult to carry out even in well-equipped laboratories, we have to seek other indirect ways to resolve the problem.

The reaction time can be measured directly or indirectly, by the presence/absence of high-pressure phases in synthesized products. Here we will analyze the available experimental data with special emphasis on the mechanism of ultrafast transport phenomena taking place during SSRs within the shock wave.

5.2.2 Temperature Measurements

Our temperature measurements [7–11] for the Sn–S system allowed us to gradually decrease the inertia of experiments from 10^{-1} s to below 10^{-4} s through the use of thinner and thinner wires and foils. These measurements were conducted in recovery ampoules rigidly fixed to a massive steel plate.

Shock experiments were carried out under similar conditions [standard steel ampoule, high explosive (HE) RDX, porosity of samples about 30%] with powders of Sn, S, SnS (nonreactive powders), and Sn–S (reactive mixture). The residual temperature (in 10^{-1} s after explosion) was found to be 110, 120, 130, and 1,110°C, respectively.

For the complete conversion Sn + S → SnS, the product temperature is expected to be 1,960°C ($\Delta H_\mathrm{r} = 110.2\,\mathrm{kJ\,mol^{-1}}$, $c_\mathrm{p} = 49.3\,\mathrm{J\,kmol^{-1}}$). This implies that the degree of conversion was around 0.56 (obtained as a ratio of the measured temperature of 1,110–1,960°C, corresponding to 100% conversion). The recovered sample data showed that the reaction occurred largely within the axial region of the cylinder (in the so-called Mach stem). The conversion degree and product composition (SnS or SnS_2) were found to depend on the particle size of the original powder, which is indicative of SSRs. The above conclusion was also supported by calorimetric data.

Further improvement in the time resolution of temperature measurements was achieved by using optical pyrometry. For the systems Al–Fe_2O_3 [12], Ni–Al [13, 14], Sn–S [15], and Sn–Te [16], taken as examples, the reaction was completed within 10^{-7} s. In 10 ns, the temperature of the Sn–S mixture was found to reach 1,300°C [15]. The bell-shaped dependence of ΔT on p (curve 3 in Fig. 5.1) implies that the reaction may proceed only within a limited range of p: with increasing p, the formation of SnS (accompanied by an increase in v) is promoted, while ΔH_r tends to decrease [17].

The temperature profile for shock-induced SSRs of Mg, Al, and Ti with S was measured in real-time experiments [18, 19]. In the above-mentioned mixtures, SSRs proceed within 50 ns behind the shock front at a conversion

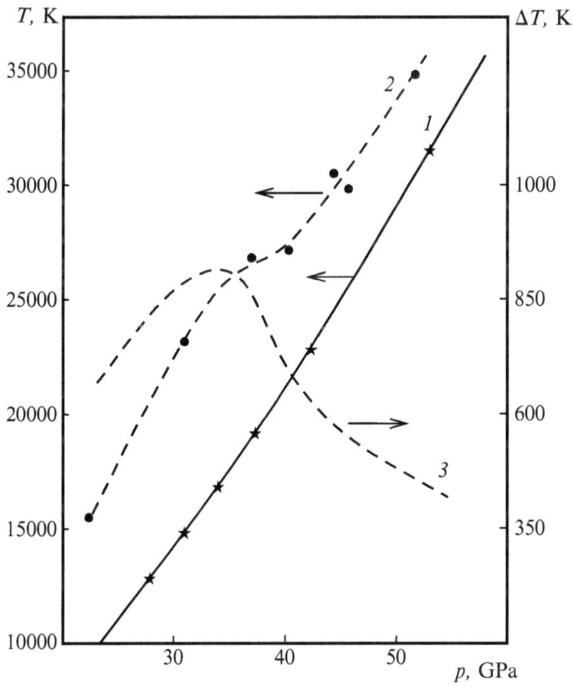

Fig. 5.1. The pressure dependence of T and ΔT for the Sn + S mixture. *1* inert mixture (T_1), *2* reactive mixture (T_2), and *3* $\Delta T = T_1 - T_2$ [15]

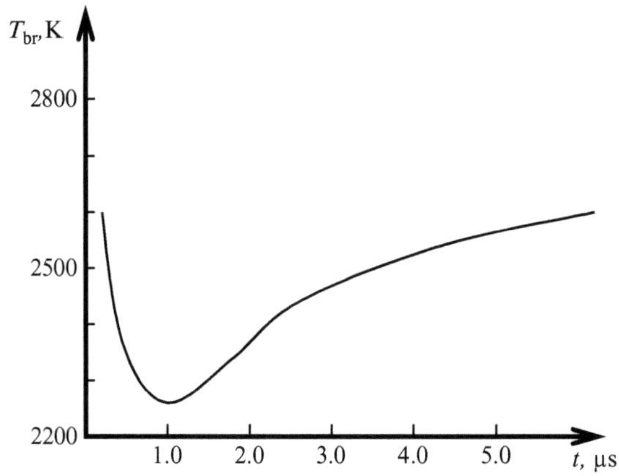

Fig. 5.2. Brightness temperature T_{br} versus time t for the shock-induced solid–solid reaction of Mg with S [18, 19]

degree of 0.2 for Mg–S and 0.5 for Al–S. Optical measurements with the Al–S system have shown that the reaction kinetics depends on the size of the reactive particles, which is also indicative of SSRs.

In the case of the Mg–S system loaded with a cast TNT–RDX explosive [18, 19], the primary peak was found to be followed by a subsequent gradual rise in temperature caused by the reaction taking place after the drop of pressure (Fig. 5.2). Such behavior is similar to that in the recovery ampoule.

5.2.3 Kinematic Measurements

A well-known example of ultrafast reactions is the detonation of HEs: this involves an intramolecular process which is not controlled by diffusion. It has been established experimentally that shock-induced solid–gas reactions can be accomplished in microseconds [20, 21]. A similar situation was observed under shock compression of the $Pb(NO_3)_2$–Al system. The curves of shock compression for $Pb(NO_3)_2$ and $Pb(NO_3)_2 + 5\%$ Al systems [22] show that, for $p > 3.5$ GPa, the reaction is accompanied by an increase in p.

The diffusion rate in solids is known to be exceedingly slow: 0.1–1.0 mm s^{-1} under normal conditions and even slower under high pressure [2]. For this reason, an ultrafast SSR within the shock appears unlikely.

In view of this, the observed [23] shift of the curve of shock compression for the Sn–S mixture for $p \geq 15$ GPa toward greater v and/or p seemed unexpected. The kinematic measurements [23] gave a conversion degree of 0.27 ± 0.1 in 0.5 μs after explosion, which agreed qualitatively with temperature measurement data (see Sect. 5.2.2). Later, the kinematic measurement data confirmed the occurrence of SSRs in the Sn–Te system for $p \geq 45$ GPa [16] and in the Ti–C mixture for $p \geq 7.5$ GPa [24] (Fig. 5.3).

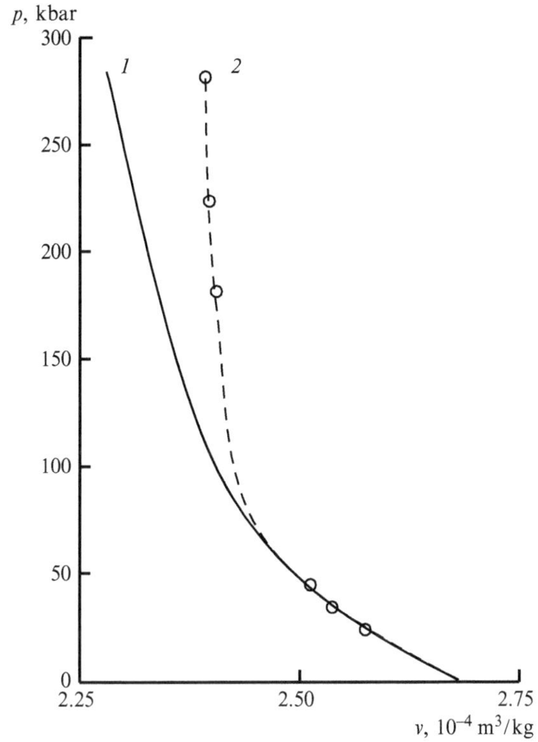

Fig. 5.3. Shock adiabat (*1*) and shock compression curve (*2*) for the Ti–C system (relative density 0.7, grain size 20 μm for Ti and 5 μm for C) [24]

The measured shock velocity in the Mn–S mixture ($2.3 \pm 0.5\,\mathrm{km\,s^{-1}}$) was associated with the action of HE [25]. Preliminary heating of the Zn–Te system up to 150°C was found [26] to increase the velocity of shock from 2.3 to $3.3\,\mathrm{km\,s^{-1}}$ owing to the occurrence of a SSR. Recently, Xu and Thadhani [27] reported on the shock-induced SSR of Ti with Ni accompanied by a volume increase and shock acceleration at 3.2 GPa.

Since 1993, some shock-induced microsecond-scale chemical reactions have been detected with piezoelectric gauges [28–30].

5.2.4 Mechanical Consequences in Recovery Ampoules

The aforementioned experimental techniques are expensive and labor-consuming. For this reason, they are unsuitable for express evaluation of the reaction time in recovery ampoules. Since SSRs take place largely within the axial part of cylindrical ampoules (in the Mach stem, whose diameter depends on the type of energy conversion in a shock), the occurrence/failure of SSRs can be readily inferred from the diameter d and thickness b of the so-called spall plate formed upon explosion [31].

The experiments were carried out as follows. Cylindrical ampoules containing a sample (surrounded by a cylindrical charge of RDX) were placed on a 1-mm-thick steel plate. After explosion, the diameter (d) and the depth (b_{sp}) of the spall on the bottom of the recovery ampoule were measured.

For instance, the shock compression of Sn–S ($D_0 = 6.2\,\mathrm{km\,s^{-1}}$) gave $d = 2.5\,\mathrm{mm}$ and $b_{sp} = 1.5\,\mathrm{mm}$, while that of SnS (an inert compound) gave $d = 1\,\mathrm{mm}$ and $b_{sp} = 0.5\,\mathrm{mm}$ (the velocity of the spall plate being $4.2\,\mathrm{km\,s^{-1}}$ in both experiments). According to our estimates, the kinetic energy of the spall plate (per mole of SnS) attained a value of 60 and $3\,\mathrm{kJ\,mol^{-1}}$, respectively. The difference ($57\,\mathrm{kJ\,mol^{-1}}$, or $0.3\Delta H_r$) is input to the power of the Mach wave. Since the yield of SnS in the Mach stem is also about 0.3 (see Sects. 5.2.2, 5.2.3), the technique described above can be used as a method for express evaluation.

Similar experiments with the Ti–C ($D_0 = 6.2\,\mathrm{km\,s^{-1}}$) and Zn–S ($D_0 = 7.2\,\mathrm{km\,s^{-1}}$) systems [32] have led to formation of TiC and ZnS in yields of 7 and 90%, respectively.

5.2.5 Solid–Solid Syntheses

The formation of high-pressure phases can be regarded as evidence for the occurrence of chemical reactions within the high-pressure zone. The known SSRs can be subdivided into the reactions of decomposition and synthesis. Decomposition reactions require no mass transport and can proceed exceedingly fast. In contrast, synthesis reactions need some time for the dispersion of matter, intermixing, and growth of product grains.

Since water under high pressure is known to acquire the properties of acid, it can be expected to dissolve the metals preceding hydrogen in the electrochemical series. We carried out [33,34] the shock compression of a frozen (with liquid nitrogen) suspension of Zn powder in water in a cylindrical recovery ampoule ($D_0 \geq 6.2\,\mathrm{km\,s^{-1}}$). Analysis of the recovered product showed formation of ZnO in the reaction $Zn + H_2O \rightarrow ZnO + H_2$.

Similar behavior was exhibited by other acid-soluble metals (B, Al, Ga, Ge, Ti, Zr, Nb, Cr, Mo, W, Mn) [33, 34]. To reduce the residual temperature, the experiments [35] were carried out as shown in Fig. 5.4: products of shock compression were scattered into a big container. In this geometry, we synthesized the cubic (high-pressure) phase of ZrO_2 [35].

In the same year, Sekine [36] synthesized hexagonal diamond (lonsdaleite) through the shock-induced reaction $MgCO_3 + Fe \rightarrow MgO + FeO + C$.

The data given above demonstrate that shock-induced SSRs can actually occur on a microsecond time scale.

Let us now consider additional evidence for the shock wave character of SSRs. Shock wave experiments can also be used to enter the region of negative pressures (stretching material until it fails). In this region, phase transformations may lead to the formation of loose material and structures that cannot

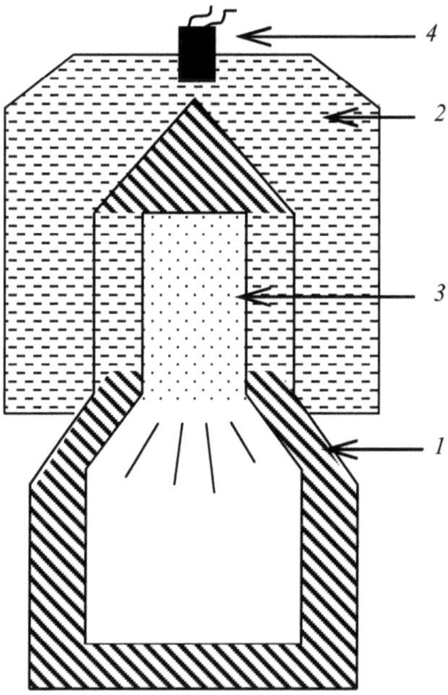

Fig. 5.4. Shock experiments with quenching of products upon scattering. *1* recovery ampoule, *2* high explosive, *3* mixture under study, *4* detonator [33,34]

form at $p \geq 0$. Back in 1965, we carried out the shock compression of turbostratic BN to obtain a new modification of BN termed the "E phase" (from "explosion") [37]. Later, the synthesis of E-BN was reproduced by other workers [38–44].

$$\text{E-BN}: \ a = 11.14, \ b = 8.06, \ c = 7.40\,\text{Å}, \ \rho = 2.50\,\text{g}\,\text{cm}^{-3}$$
$$C_{60}: \ a = 11.16, \ b = 8.17, \ c = 7.58\,\text{Å}, \ \rho = 2.50\,\text{g}\,\text{cm}^{-3}$$

The cell parameters and the density of E-BN are close to those of the fullerene C_{60} synthesized 20 years later [45]. The stabilization of E-BN requires the presence of several percent boron oxide [46].

The shock compression of Nd_2O_3 led to the synthesis of E-Nd_2O_3 [47]. The spectral data show the presence of SiO_2 (approximately 30 wt%) originating from the spalls split from the bottom and walls of the recovery ampoule. The density of E-Nd_2O_3 is as low as $1.6\,\text{g}\,\text{cm}^{-3}$ (cf. $7.42\,\text{g}\,\text{cm}^{-3}$ for Nd_2O_3 and $2.65\,\text{g}\,\text{cm}^{-3}$ for quartz). A very small number of single crystals were also isolated. They had a monoclinic unit cell ($a = 7.5$, $b = 8.7$, $c = 10.3$ Å, $\beta = 104°$) and very low refractive indices ($n_g = 1.57$, $n_m = 1.56$, $n_p = 1.54$; cf. $n = 2.10$ for Nd_2O_3). This material is insoluble in acids and alkalis and is highly heat resistant (heating is accompanied by reversible thermochromism: from white to violet). The attempts to obtain this compound by shock compression of

Nd_2O_3 mixtures with Si, SiO, or SiO_2 were unsuccessful: at any equivalence ratio, the reaction only yielded neodymium silicates with high ρ and n.

Shock compression of the Zr + 2S mixture in a steel recovery ampoule yielded ZrS_2 with low lattice parameters. Analysis showed that this product is present as the $Zr_{1-x}Fe_xS_2$ solid solution, iron originating from the walls of the recovery ampoule. However, shock compression of the ZrS_2 + Fe mixture did not yield the above-mentioned solid solution [48]. Furthermore, such solid solutions cannot be prepared by heating (upon heating, $Zr_{1-x}Fe_xS_2$ undergoes decomposition into ZrS_2 and FeS).

Shock-induced compression can also be used to synthesize new compounds from reagents with close electronegativity ($\Delta H_r \cong 0$). Kikuchi et al. [49] observed the formation of Ta_2O_5 under shock compression of SiO_2 in a recovery ampoule with tantalum walls, although under normal conditions the redox reaction Ta + SiO_2 is thermodynamically unfavorable.

It is noteworthy that E phases in shock-compressed BN and Nd_2O_3 are only formed in the axial part of a cylindrical recovery ampoule, provided that voids are formed within this area. The latter is a result of tensile stresses arising upon irregular impact of shock waves. Having noticed this fact, we undertook [50, 51] special syntheses of germanium halcogenides in cylindrical ampoules equipped with a large container for scattering shock-compression products (Fig. 5.4). All shock-synthesized compounds (GeSSe, GeSTe, GeSeTe) were found to have ρ values lower than those of thermally synthesized products (and even the original mixtures). Upon heating, these compounds underwent an exothermal transition into normal (higher-density) phases [50,51]. In our laboratory, we succeeded in synthesizing numerous other loosely packed modifications, which opened new horizons for shock chemistry at negative pressures.

Yet another type of "loose" material (foams) can be prepared by shock compression in very strong recovery ampoules. A mixture of the substance under study and a small amount of HE were placed into a cylindrical ampoule. Upon explosion of the outer charge and compression, the inner charge detonated and formed a strong highly porous material. The foam density was found to depend on the substance to HE equivalence ratio [52]. Foams are formed owing to contact melting (gluing) of grains; therefore, this technique is not applicable to the synthesis of high-melting materials (e.g., $MoSi_2$). Nevertheless, for the Mo + 2Si + HE mixture, the ΔH_r value turned out to be sufficient for formation of the $MoSi_2$ foam.

The data given above unequivocally show the feasibility of a SSR within the zone of high pressure or rarefaction on a microsecond time scale. This is also supported by the observation that the exothermic ($\Delta H_r > 0$) reaction within the axial part of the recovery ampoule does not spread over the entire shock-compressed sample. The conversion degree within the Mach stem is about 30%. Had this transformation happened upon unloading, the high temperature attained would have been sufficient for initiation of the reaction over the entire volume of the ampoule. According to [53], this does not happen owing to the extremely high pressure in the Mach stem, which leads to a decrease in temperature and, accordingly, to a low density of shock-synthesized product.

5.2.6 Mechanism of Ultrafast Diffusion

The aforementioned data require a nontrivial explanation because the velocity of conventional diffusion in solids is several orders of magnitude lower than that which ensures chemical reactions take place on a microsecond time scale. Numerous suggestions have been made regarding possible mechanisms for acceleration of diffusion in shock-compressed materials: large defect concentrations and plastic deformation, heterogeneous heating of components, shear strain, plastic flow to hot spots (as in explosives), and phase transitions in components leading to the breakup of the crystal lattice (hence to a decrease in the activation energy). Although all of these factors do accelerate diffusion, they seem insufficient for explaining the entire set of experimental data.

To rationalize the data, we suggested the fluid-dynamic model of ultrafast (forced) diffusion caused by a difference in the particle velocity of the components in shock-compressed matter [23, 54]. This model has been confirmed experimentally [55–58]. In terms of this model, complete intermixing of particles is achieved owing to penetration of rapidly moving particles into slowly moving particles. The relative velocity of their motion can be determined as follows.

Let us consider rapid compression of some elementary volume of matter from p_0 to some p ($p_0 \ll p$). Given that the compression time $t \ll d/c$ (d is the characteristic diameter of the elementary volume, c is the sound velocity), the work of pressure forces will be roughly equally distributed between the internal and the kinetic energy. The specific kinetic energy will attain an approximate value of

$$\bar{u}^2/2 = -\left(\Delta p/\Delta v\right)/2, \tag{5.1}$$

where $\bar{u}$ is the rms particle velocity.

Expression (5.1) coincides with that determining the particle velocity of a shock [1]; hence, the value of $\bar{u}$ for each reagent of a powder mixture can be estimated from its shock-compression curve. Given that both reagents have identical shock compression curves, their $\bar{u}$ values are identical, so the forced diffusion is nearly absent. Conversely, in the case of different shock adiabats, the forced diffusion does take place. In this case, $\Delta\bar{u}$ can be regarded as the velocity of forced diffusion.

Accordingly, the time of diffusion τ (intermixing), and hence the reaction time, can be estimated from the expression [59]

$$\tau = 2d/\Delta\bar{u}. \tag{5.2}$$

At $\Delta\bar{u} \approx 1\,\mathrm{km\,s^{-1}}$ and $d \approx 0.1$ mm, we obtain $\tau \approx 10^{-7}$ s. Given that the shock adiabats for Sn, S, and Te are known, we obtain that for the Sn–S mixture

($d = 100\,\mu$m) a value of $\tau = 0.3\,\mu$s can be attained at $p = 10$ GPa, while for the Sn–Te mixture ($d = 100\,\mu$m), a value of $\tau = 0.8\,\mu$s can be attained at $p = 50$ GPa. These estimates agree reasonably with experiment.

It follows that, under conditions of shock compression, the velocity of forced diffusion can be sufficiently large. This means that the SSD in charges with reasonable dimensions can be expected to occur, provided the thermodynamic criterion suggested in [4] is satisfied.

5.3 Shock-Induced Solid–Solid Detonation in Zinc–Sulfur Powders

As can be inferred from Sect. 5.1, SSD at some certain conditions can be expected to occur when the criterion of detonation ability [4] is satisfied, although the SSD in a given system may proceed exceedingly slowly. But this means that in the systems that do not fit the thermodynamic criterion [4], SSD will never occur in principle.

The criterion [4] can be written in the form

$$Q_{\mathrm{p,v}} > 0 \,\mathrm{or}\, \Delta v_{\mathrm{p,h}} > 0. \tag{5.3}$$

Unfortunately, the reference data necessary for calculating (5.3) are often lacking in the literature. For the Mn–S and Al–S systems, we could perform only rough estimations: the Mn–S system was found to be unsuitable, but the Al–S system was suitable for observation of SSD. We managed to strictly apply criterion (5.3) only to the systems Zn–Se, Zn–Te, Cu–S, Ti–C–Al–paraffin, and Zn–S.

For control experiments, we chose the reaction Zn+S $\rightarrow$ ZnS [60]. We have calculated the parameters of ideal detonation in the compact Zn–S system ($\rho_0 = 3.87\,\mathrm{g\,cm^{-3}}$): $D_0 = 2{,}500\,\mathrm{m\,s^{-1}}$, $p_{\mathrm{CJ}} = 3$ GPa [61]. These values are close to those typical of detonation in condensed HEs. The possibility of SSD in the Zn–S system was also supported by the experimental data [62].

We assume that the most convincing evidence for the occurrence of SSD is the reacceleration and intensification of the shock wave in a given medium [4].

5.3.1 Initiation of Detonation

Our preliminary results [61] confirmed the possibility of SSD in Zn–S charges ($\rho_0 = 1.33\,\mathrm{g\,cm^{-3}}$, $20 \times 20 \times 60\,\mathrm{mm^3}$ in size) placed into a thick-walled (5-mm) steel ampoule tightly closed with a 2-mm cover plate. Explosion was initiated with an electrically exploding wire from the bottom. In this geometry, we observed the following two phenomena.

One was a loud clap and the other was strong plastic deformation of the cover plate (Fig. 5.5). Our rich experience in the field suggests that such a

Fig. 5.5. A deformed cover

deformation can be caused by a pressure of several hundred atmospheres. The resultant temperature of the ampoule was below 50° C.

In other runs, the process proceeded smoothly, without noticeable deformation of the cover plate. The resultant temperature was above 100℃, but below 800℃. The extent of conversion into ZnS attained a value above 80%. According to [61], the former process can be regarded as detonation and the latter one as slow combustion of the Zn–S powder mixture.

In further experiments, we measured wave velocities for both processes. A low velocity can be due to the fact that the detonation develops in two stages: first, only a small amount of the Zn–S mixture reacts, and this wave of incomplete combustion turns off our gauges; then the detonation wave begins to propagate over a preheated mixture.

We have evaluated the lower limit for the pressure p that caused the deformation of the cover (Fig. 5.5): $p = 360$ atm. In the case of a slow reaction, the p value evaluated did not exceed 80 atm.

At present, similar experiments with longer ampoules (where the probability of detonation onset is higher) are in progress.

5.3.2 Direct Measurement of Detonation Velocity

In our experiments [63], we used an equimolar powder mixture of Zn and S. The highly exothermic reaction $Zn + S \rightarrow ZnS$ proceeds without gas evolution and, *owing to thermal expansion of the product*, satisfies condition (5.3). The particle size of Zn and S powders was 3–5 μm, while the sample density (ρ) was 0.6–0.7 of the theoretical value.

The experimental setup is shown in Fig. 5.6. A mixture was pelleted into cylinders 16.5 mm in diameter and 40–200 mm long. Samples were placed in a tubular container made of a porous composite (with a low velocity of sound) to exclude the effect of elastic waves in the container walls on the results

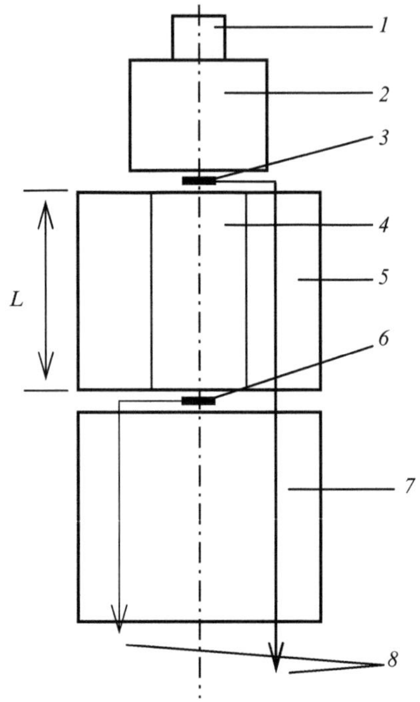

Fig. 5.6. Experimental setup. *1* detonator, *2* high explosive, *3*, *6* contact sensors, *4* Zn–S sample, *5* container (made of porous composite), *7* stand, *8* wire leads to oscilloscopes, L gauge length

of the measurements. Shocks were generated by detonation of the charge (TNT/RDX, 40 mm in diameter, 35 mm in height, $D_0 = 7.85\,\mathrm{km\,s^{-1}}$). In experiments, we measured the time interval τ between the arrival of the shock at sensors (contact gauges). To avoid the sample discontinuity, the gauges were only placed on the sample top and bottom. The average velocity of shock propagation through a sample was determined from the expression $\bar{D} = L/\tau$. For a sensor thickness of 200 μm, the contact gap was 100 μm. The estimated measurement error (δ) was 0.6% at $L = 40$ mm and 0.1% at $L = 200$ mm. Signals from the sensors were recorded with two oscilloscopes (Tektronix TDS 1012). The experiments were carried out at 14°C. The data obtained are presented in Table 5.1.

In the absence of a chemical reaction, the shock wave generated in the sample could be expected to decay at a distance of 25–30 mm from the top. So, a minimal sample length (40 mm) was chosen so that in the absence of chemical replenishment the sensor numbered 6 in Fig. 5.6 would give no signal at all. Experimental data for the trials when both sensors produced signals are presented in Table 5.1. These data suggest that the process of shock propagation was supported by the energy released in the chemical reaction taking place in the zone of high dynamic pressure.

Table 5.1. Experimental data for direct measurement of detonation velocity

Experimental run	L (mm)	$\bar{D}$ ($km\,s^{-1}$)	ρ (%)	δ (%)
1	40	2.27	68.1	0.6
2	60	1.30	63.0	0.2
3	75	1.64	71.6	0.2
4	90	1.39	60.7	0.2
5	100	2.55	62.4	0.2
6	150	2.195	63.5	0.1
7	150	1.915	62.4	0.1
8	200	2.169	59.4	0.1

As follows from Table 5.1, the shock velocity $\bar{D}$ initially drops sharply from a starting value of $7.85\,km\,s^{-1}$ at the bottom of the charge to $1.30\,km\,s^{-1}$ (at $L = 60\,mm$) and then increases to above $2\,km\,s^{-1}$ (for $L = 100$–$200\,mm$). Some scattering can be attributed (a) to a random character of detonation initiation at some points behind the shock wave front, (b) possible occurrence of several detonation modes and transition processes between them [64], and (c) some variation in the induction period for the chemical reaction. Theoretical estimation for an ideal detonation in the monolith matter under study gives a value of $D_0 = 2.486\,km\,s^{-1}$ [61]. If we take into consideration that in our experiments $\rho = 60$–70%, the measured $\bar{D}$ values agree well with the theoretical prediction for the detonation process.

The X-ray diffraction data for the products taken at the bottom of the container (numbered 5 in Fig. 5.6) are indicative of virtually complete conversion of the starting mixture into ZnS (only trace amounts of Zn were detected), which confirms the occurrence of a chemical reaction within the shock. The effects of high temperature were also noticed on the surface of the sensor numbered 6 in Fig. 5.6, which also supports the occurrence of a highly exothermic reaction.

The observed acceleration of the shock can be regarded as experimental evidence for the occurrence of SSD in the system under consideration. When we record a steady propagation of shock, there always exists a probability that the observed process is weakly decaying and hence is not self-sustaining, but the experimental accuracy is not sufficient to observe this on a limited gauge length. In contrast, the observation of acceleration (as in our experiments) leaves no grounds for doubting the occurrence of detonation in the material under study.

SSD can be regarded as a new type of transport phenomenon in reactive media. The phenomenon may find an application (e.g., in mining) where the shattering action of explosives is being used while the presence of gaseous products is not desirable. Just like solid-state synthesis by combustion, detonation-mediated synthesis in the solid state may also prove useful for preparation of various compounds and materials.

5.4 Thermodynamic Fundamentals of Solid–Solid Detonation

5.4.1 Basic Assumptions

The general theory of shock wave processes [2] does not differentiate between gases and condensed media, including powder systems. In numerous monographs on condensed HEs (e.g., [64,65]), little or no attention has been given to the thermochemical aspects of shock wave processes. In this section, we will try to fill this gap.

The notion of the thermal effect of a reaction is basic in the physics of explosions [64,65]. Meanwhile, this notion has not been strictly defined yet in relation to shock wave processes. To fill this gap, let us apply the first law of thermodynamics to an element or microscopic particle of a reactive medium. For irreversible processes starting in the metastable state, it can be written in the form

$$\Delta e = \Delta q - \Delta A. \tag{5.4}$$

Expression (5.4) follows from the classical definition of internal energy [2,66]: the specific internal energy e is a specific measure of the entire internal motion in the element under consideration, including the energy of chemical bonds.

Let us consider the consequences of (5.4) under the following assumptions adopted in thermochemistry:

(a) The mechanical work A in (5.4) is defined by the expression

$$\mathrm{d}A = -p\,\mathrm{d}v. \tag{5.5}$$

(b) In our case, local thermodynamic states of matter can be assumed to be in equilibrium or quasi-equilibrium (metastable). For each of these states, the following thermodynamic functions (variables) can be defined: specific internal energy e, pressure p, specific volume v or density $\rho = 1/v$, absolute temperature T, and specific entropy s. Accordingly, the following relationships hold: $p \geq 0$, $v > 0$, $s \geq 0$, $T \geq 0$ and $s = 0$ at $T = 0$.

(c) Each local thermodynamic state is completely defined by a finite number of parameters. First, these are any pair of the above thermodynamic variables which fully define only equilibrium states of medium. Second, this is some set ($n \geq 1$) of independent scalar inner characteristics of matter (chemical/phase composition, porosity, grain size, etc.). The chemical/phase composition can be characterized by a set of parameters written in the form $\langle\eta\rangle = \langle\eta_1,\ \eta_2,\ \eta_3, \ldots,\ \eta_n\rangle$.
For nonequilibrium states, all the functions of a local thermodynamic state, by definition, are the function of the $n + 2$ parameters above. For instance,

$$e = e(s, v, \langle\eta\rangle) = e(p, v, \langle\eta\rangle) = e(p, T, \langle\eta\rangle) = e(v, T, \langle\eta\rangle), \tag{5.6}$$

where the variables s, v, p, and T refer to the same state of matter.

(d) The choice of the n independent characteristics above is ambiguous. Among all possible sets $\langle\eta\rangle$, there exists at least one for which, at $\langle\eta\rangle$ = const, the thermodynamic parameters satisfy the same expressions that are valid for equilibrium states. Let us term such a set the "inner variables." Hereinafter, such a set $\langle\eta\rangle$ will be regarded as a unique characteristic of a physicochemical transformation in the system under consideration.
Therefore, the following relationships will be assumed to hold:

$$T = \left(\frac{\partial e}{\partial s}\right)_{v,\langle\eta\rangle}, \; p = -\left(\frac{\partial e}{\partial v}\right)_{s,\langle\eta\rangle}, \tag{5.7}$$

$$c_p = T\left(\frac{\partial s}{\partial T}\right)_{p,\langle\eta\rangle} \geq c_v = T\left(\frac{\partial s}{\partial T}\right)_{v,\langle\eta\rangle} \geq 0, \tag{5.8}$$

$$\left(\frac{\partial p}{\partial v}\right)_{s,\langle\eta\rangle} \leqslant \left(\frac{\partial p}{\partial v}\right)_{T,\langle\eta\rangle} \leqslant 0. \tag{5.9}$$

For equilibrium $\langle\eta\rangle$, there is no need for a partial derivative at $\langle\eta\rangle$.

(e) We will consider only conventional media that obey relations (5.10)–(5.12):

$$\beta = \frac{1}{v}\left(\frac{\partial v}{\partial T}\right)_{p,\langle\eta\rangle} > 0, \quad \left(\frac{\partial p}{\partial T}\right)_{v,\langle\eta\rangle} > 0, \quad \left(\frac{\partial e}{\partial p}\right)_{v,\langle\eta\rangle} > 0, \tag{5.10}$$

where β is the volumetric coefficient of thermal expansion,

$$\left(\frac{\partial e}{\partial v}\right)_{p,\langle\eta\rangle} > 0, \tag{5.11}$$

$$\left(\frac{\partial^2 p}{\partial v^2}\right)_{s,\langle\eta\rangle} > 0. \tag{5.12}$$

To date, assumptions a–d seem reasonable only in cases of homogeneous media for which the concentration of components can be regarded as an internal thermodynamic variable. Meanwhile, the physical meaning of internal variables is of no significance for our analysis. This implies that the results obtained are applicable to any system satisfying assumptions a–e.

5.4.2 Thermal Effects of Physicochemical Transformation

For a given transformation $\langle\eta_0\rangle \to \langle\eta\rangle$, we can obtain different thermal characteristics of the process from (5.4) and (5.5).

Physicochemical transformations are normally characterized by the thermal effects $Q_v \equiv Q_{v,T}$ (at v, T = const) or $\Delta H_{\mathrm{r}} \equiv Q_p \equiv Q_{p,T}$ (at p, T = const). According to (5.4) and (5.5), these terms can be defined as

$$Q_{v,T} = e_0 - e(v_0, T_0, \langle\eta\rangle), \tag{5.13}$$

$$Q_{p,T} = e_0 - e(p_0, T_0, \langle\eta\rangle) - p_0 \Delta v_{p,T}, \tag{5.14}$$

where $e_0 = e(V_0, T_0, \langle\eta_0\rangle) = e(p_0, T_0, \langle\eta_0\rangle) = e(p_0, v_0, \langle\eta_0\rangle$ and $\Delta v_{p,T}$ is the increment in the specific volume during the transformation $\langle\eta_0\rangle \to \langle\eta\rangle$ at given p_0 and T_0. Transformations (media) satisfying the conditions $Q_{v,T} > 0$, $Q_{p,T} > 0$ are termed "exothermic."

It was suggested [67] characterizing chemical reactions in shock wave processes by the value of $Q_{p,v}$:

$$Q_{p,v} = e_0 - e\left(p_0, v_0, \langle\eta\rangle\right). \tag{5.15}$$

According to [67], the sign of $Q_{p,v}$ defines the type of shock (see later).

Still another heat parameter of the chemical transformation $\langle\eta_0\rangle \to \langle\eta\rangle$, the value of Q_{cal} (calorimetric), can be defined as

$$Q_{\text{cal}} = e_0 - e\left(p_0, T_0, \langle\eta\rangle\right). \tag{5.16}$$

It is the value of Q_{cal} that is measured in experiments with calorimetric bombs. From (5.14) and (5.16), we obtain

$$Q_{\text{cal}} - Q_{p,T} = p_0 \Delta v_{p,T}. \tag{5.17}$$

This expression is used to determine $Q_{p,T}$ from a measured value of Q_{cal}.

The thermal effects $Q_{v,T}$, $Q_{p,T}$, $Q_{p,v}$, and Q_{cal} may turn out to be identical in their magnitude. According to (5.13)–(5.16), this may happen only as a rare case when $\Delta v_{p,T} = 0$. But when $\Delta v_{p,T} \neq 0$, these thermal effects have different magnitudes. This can be demonstrated by using the diagram presented in Fig. 5.7.

A situation here corresponds to the inequality $\Delta v_{p,T} > 0$ (e.g., numerous exothermic SSRs, decomposition of HE, or combustion of some gaseous mixtures). The $\langle\eta_0\rangle \to \langle\eta\rangle$ processes taking place at v, $T = \text{const}$ or p, $T = \text{const}$ correspond to the solid arrows between the point $(p_0,\ v_0)$ and the end points $(p_1,\ v_0)$ and $(p_0,\ v_2)$, respectively. According to (5.13) and (5.14), the end points belong to the isotherm line T_0 of the product. For $\Delta v_{p,T} > 0$, we have $v_2 > v_0$. In virtue of inequality (5.9), $(\partial p/\partial v)_{T,}\langle\eta\rangle \leqslant 0$, it follows that $p_1 > p_0$. The $\langle\eta_0\rangle \to \langle\eta\rangle$ process taking place at p, $v = \text{const}$ is presented by the point $(p_0,\ v_0)$.

The $\langle\eta_0\rangle \to \langle\eta\rangle$ process in a calorimetric bomb corresponds to the dashed arrow between the point (p_0,v_0) and the end point (p_3,v_3) on the isotherm line T_0. In such experiments, the magnitude of q is normally determined as $q = e_0 - e_3$. It is well known that the weight and volume of a sample are taken to be much smaller than the weight and inner volume of a bomb. Hence, in the case of a SSR, $p_3 = p_0$, $v_3 = v_2$, and $e_3 = e(p_0, v_2,\ \langle\eta\rangle) \equiv e(p_0, T_0,\ \langle\eta\rangle)$. In the case of HE decomposition, the product can be regarded as an ideal gas between points 2 and 3 in Fig. 5.7. Then according to the thermodynamic identity

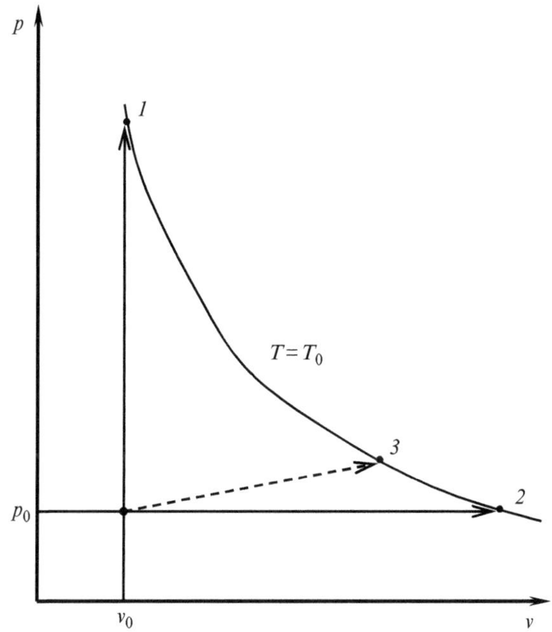

Fig. 5.7. States during transformation of matter for $\Delta v_{p,T} > 0$. *1* point $(p_1,\ v_0)$, *2* point $(p_0,\ v_2)$, *3* point $(p_3,\ v_3)$

$$\left(\frac{\partial e}{\partial v}\right)_{T,\langle\eta\rangle} = T\left(\frac{\partial p}{\partial T}\right)_{v,\langle\eta\rangle} - p \tag{5.18}$$

and the Clapeyron equation, $(\partial e/\partial v)_{T,}\langle\eta\rangle = 0$. Therefore, $e_3 = e(p_3, v_3,\ \langle\eta\rangle) = e(p_0, v_2,\ \langle\eta\rangle) \equiv e(p_0, T_0,\ \langle\eta\rangle)$. According to (5.16), in both cases we obtain $Q_{\mathrm{cal}} = e_0 - e_3 = q$. In other words, experiments with calorimetric bombs indeed yield the values of Q_{cal}.

In publications on the physics of explosions, one often comes across the expression (5.17) in which $Q_v \equiv Q_{v,T}$ is used instead of Q_{cal}, without any indication of the specificity of the equation of state for products and the initial state $(p_0,\ v_0,\ \langle\eta_0\rangle)$. Actually, it is assumed the equalities $Q_{\mathrm{cal}} = Q_{v,T}$ and $e_3 = e_1$ [see (5.13), (5.16)] always hold true. In reality, this is possible only when $(\partial e/\partial v)_{T,}\langle\eta\rangle = 0$ over the entire segment of the isotherm T_0 between points 1 and 2 in Fig. 5.7. In the case of dense gases and condensed media, this condition is certainly not fulfilled. This can be demonstrated by transforming the right-hand side of (5.18) by using the van der Waals equation for dense gases or the inequality $(\partial^2 p/\partial T^2)_{v,\langle\eta\rangle} > 0$, valid for most condensed media at low temperatures [66]. Therefore, for shock wave processes in condensed media, $Q_{\mathrm{cal}} \neq Q_{v,T}$.

In Fig. 5.7, the state of the reaction product $(p_0,\ v_0,\ \langle\eta\rangle)$ satisfies the inequalities $p_0 < p_1$ and $v_0 < v_2$; therefore, it follows from (5.10) and (5.11) that $e(p_0, v_0, \langle\eta\rangle) < e_1$ and $e(p_0, v_0, \langle\eta\rangle) < e_2$. Accordingly, from (5.13), (5.15), and (5.16), we obtain $Q_{p,v} > Q_{v,T}$ and $Q_{p,v} > Q_{\text{cal}}$.

Having plotted a similar diagram for $\Delta v_{p,T} < 0$, one can be convinced that in this case the thermal effects $Q_{v,T}$, $Q_{p,T}$, $Q_{p,v}$, and Q_{cal} (5.13)–(5.16) also have different values.

The values of $Q_{v,T}$, $Q_{p,T}$, and $Q_{p,v}$ normally have the same sign, although there are some exclusions. An example is combustion of the thermite system $3Fe_3O_4 + 8Al$. At $p = 1\,\text{atm}$, the combustion temperature rises. To ensure $T = \text{const}$, one has to expel some heat. Then according to (5.4), (5.5), and (5.14), $Q_{p,T} > 0$. But this is accompanied by a volume decrease which has to be replenished upon additional supply of heat. Hence, from (5.4), (5.5), and (5.15), it follows that $Q_{p,v} < 0$ [67].

Therefore, the transformation $\langle\eta_0\rangle \to \langle\eta\rangle$ can be characterized by at least four thermal effects (5.13)–(5.16) having different physical meaning and magnitude. Let us analyze their role in shock wave processes taking place in reactive media. We will begin with a relation between the mechanical and thermodynamic parameters of shocked matter.

5.4.3 Shock Equations

In the reference system attached to the shock, let us consider an elementary cylindrical volume of matter as shown in Fig. 5.8. Applying the conservation of mass and momentum to this element [2], we come to the well-known relationships for the shock:

$$\rho_0 D = \rho\,(D - u)\,, \tag{5.19}$$

$$p - p_0 = \rho_0 D u. \tag{5.20}$$

Applying the principle of energy conservation (upon neglect of heat exchange between elementary volumes, as adopted for very fast processes), we obtain

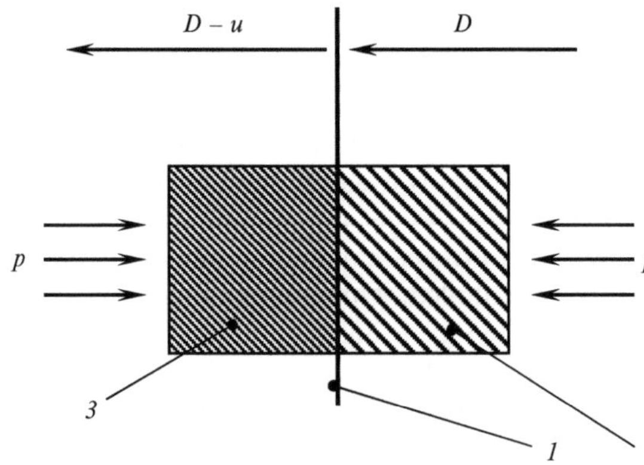

Fig. 5.8. Deriving the shock equation. *1* shock wave, *2* undisturbed material, *3* compressed material

$$\rho_0 D\,(\Delta e + \Delta K) = p_0 D - p\,(D-u)\,,$$

where K is the specific kinetic energy. Introducing $\Delta e = e - e_0$, $\Delta K = (D-u)^2/2 - D^2/2$, and excluding D and u [through the use of (5.19), (5.20)], we obtain

$$e\,(p, v, \langle\eta\rangle) - e_0 = (p - p_0)\,(v_0 - v)\,/2. \qquad (5.21)$$

For the classical definition of e [2, 66], this expression is applicable to shocks with or without transformations [2].

At constant $\langle\eta\rangle$, expression (5.21) describes a family of curves (termed "dynamic adiabats") in the coordinates p–v. According to (5.9)–(5.12), the dynamic adiabats have the form shown in Fig. 5.9 [2].

The character of shock motion in reactive media is known [2] to depend on the position of the adiabat relative to the point (p_0, v_0). This position can be predicted by applying simple criteria (5.3). These criteria can be derived as follows. In view of (5.15), expression (5.21) can be written in the form

$$e(p, v, \langle\eta\rangle) - e(p_0, v_0, \langle\eta\rangle) = (p - p_0)\,(v_0 - v)\,/2 + Q_{p,v}. \qquad (5.22)$$

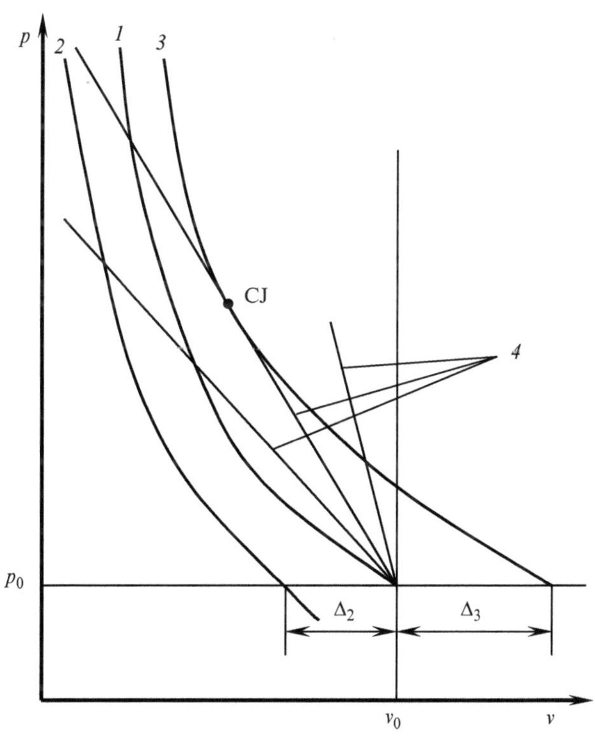

Fig. 5.9. The dynamic adiabats. *1* shock adiabat ($Q_{p,v} = 0$, $\Delta v_{p,h} = 0$), *2* shock adiabat ($Q_{p,v} < 0$, $\Delta v_{p,h} = -\Delta_2 < 0$), *3* detonation adiabat ($Q_{p,v} > 0$, $\Delta v_{p,h} = \Delta_3 > 0$), *4* Rayleigh lines

In (5.22), $Q_{p,v}$ is constant (since it depends on p_0, v_0, $\langle\eta_0\rangle$, and $\langle\eta\rangle$, which are assumed to be constant). This parameter defines a position of the dynamic adiabat relative to the point (p_0, v_0). For $Q_{p,v} < 0$, the adiabat intersects the isobar line p_0 for $v < v_0$ (Fig. 5.9); at $Q_{p,v} = 0$, it passes through the point (p_0, v_0); while for $Q_{p,v} > 0$, it crosses the isobar line p_0 for $v > v_0$ and the isochore line v_0 for $p > p_0$.

Introducing $Q_{v,T}$ (5.13), $Q_{p,T}$ (5.14), or Q_{cal} (5.16) into (5.21), we arrive at a complicated expression that is inconvenient for deriving a simple criterion for the position of the dynamic adiabat relative to the point (p_0, v_0). This implies that *it is the thermal effect* $Q_{p,v}$ (5.15) *that defines the motion of a shock wave in reactive systems.*

Now let us derive another criterion for position of the dynamic adiabat relative to (p_0, v_0). Introducing $p = p_0$ into (5.21) and performing some transformations, we obtain

$$h\left(p_0, v, \langle\eta\rangle\right) = h\left(p_0, v_0, \langle\eta_0\rangle\right), \tag{5.23}$$

where $h(p, v,\ \langle\eta\rangle) = e(p, v,\ \langle\eta\rangle) + pv$ is the specific enthalpy for the matter under consideration. Equation (5.23) defines the point v at which the dynamic adiabat (5.21) intersects the isobar line p_0 (Fig. 5.9). We may also introduce the volume effect of physicochemical transformation:

$$\Delta v_{p,h} = v - v_0 \text{ at } p = \text{const},\ h = \text{const}. \tag{5.24}$$

This is the increment in the specific volume of a medium in the real or imaginary process $\langle\eta_0\rangle \rightarrow \langle\eta\rangle$ taking place at $p_0 = \text{const}$ under adiabatic conditions. In any case, $\Delta v_{p,h}$ is not identical to $\Delta v_{p,T}$. These magnitudes may even have different signs. For example, for a stoichiometric mixture of H_2 with O_2, $\Delta v_{p,h} > 0$, while $\Delta v_{p,T} < 0$.

The value of $\Delta v_{p,h}$ defines the position of the dynamic adiabat relative to the point (p_0, v_0). For $\Delta v_{p,h} < 0$ (Fig. 5.9), the dynamic adiabat intersects the isobar line p_0 for $v < v_0$; at $\Delta v_{p,h} = 0$, it passes through the point (p_0, v_0); and for $\Delta v_{p,h} > 0$, it intersects the isobar line p_0 for $v > v_0$ and the isochore line v_0 for $p > p_0$ (Fig. 5.9). Therefore, the signs of $Q_{p,v}$ and $\Delta v_{p,h}$ are identical. For this reason, both criteria in (5.3) are equivalent.

5.4.4 The Role of Thermal Effects in Laminar Motion of Reacting Matter

Now let us consider the role of the aforementioned four thermal effects in the motion of reactive particles behind the shock. According to (5.4) and (5.5), for an element of a medium, we can write

$$\frac{\mathrm{d}e}{\mathrm{d}t} = \frac{\mathrm{d}q}{\mathrm{d}t} - p\frac{\mathrm{d}v}{\mathrm{d}t}. \tag{5.25}$$

On the other hand, since e is a function of p, v, and $\langle\eta\rangle$, we obtain

$$\frac{\mathrm{d}e}{\mathrm{d}t} = \left(\frac{\partial e}{\partial p}\right)_{v,\langle\eta\rangle} \frac{\mathrm{d}p}{\mathrm{d}t} + \left(\frac{\partial e}{\partial v}\right)_{p,\langle\eta\rangle} \frac{\mathrm{d}v}{\mathrm{d}t} + \left(\frac{\partial e}{\partial\langle\eta\rangle}\right)_{p,v} \frac{\mathrm{d}\langle\eta\rangle}{\mathrm{d}t}. \tag{5.26}$$

Here and hereinafter we will admit that

$$\left(\frac{\partial Z}{\partial\langle\eta\rangle}\right)_{X,Y} \frac{\mathrm{d}\langle\eta\rangle}{\mathrm{d}t} = \sum_{i=1}^{n} \left(\frac{\partial Z}{\partial\eta_i}\right)_{X,Y} \frac{\mathrm{d}\eta_i}{\mathrm{d}t},$$

where Z stands for e or v while X and Y stand for p, v, or T.

The last term in (5.26) (taken with the opposite sign) can be regarded as the rate of heat evolution at p, v = const [cf. (5.15)]:

$$\frac{\mathrm{d}Q_{p,v}}{\mathrm{d}t} = -\left(\frac{\partial e}{\partial<\eta>}\right)_{p,v} \frac{\mathrm{d}<\eta>}{\mathrm{d}t}.$$

Using the thermodynamic identity

$$\left(\frac{\partial p}{\partial v}\right)_{s,\langle\eta\rangle} = -\left(\left(\frac{\partial e}{\partial v}\right)_{p,\langle\eta\rangle} + p\right)\left(\frac{\partial p}{\partial e}\right)_{v,\langle\eta\rangle},$$

and the expression for the Grüneisen coefficient

$$\Gamma = v\left(\frac{\partial p}{\partial e}\right)_{v,\langle\eta\rangle},$$

we obtain

$$\frac{\mathrm{d}p}{\mathrm{d}t} = \left(\frac{\partial p}{\partial v}\right)_{s,\langle\eta\rangle} \frac{\mathrm{d}v}{\mathrm{d}t} + \frac{\Gamma}{v}\left(\frac{\mathrm{d}Q_{p,v}}{\mathrm{d}t} + \frac{\mathrm{d}q}{\mathrm{d}t}\right). \tag{5.27}$$

Therefore, the relation between $\mathrm{d}p$ and $\mathrm{d}v$ is defined by the derivative of $(Q_{p,v} + q)$. In other words, the thermal effect $Q_{p,v}$ (5.15) plays a part similar to that of externally supplied heat q.

For e as a function of v, T, and $\langle\eta\rangle$, we obtain

$$\frac{\mathrm{d}e}{\mathrm{d}t} = \left(\frac{\partial e}{\partial v}\right)_{T,\langle\eta\rangle} \frac{\mathrm{d}v}{\mathrm{d}t} + \left(\frac{\partial e}{\partial T}\right)_{v,\langle\eta\rangle} \frac{\mathrm{d}T}{\mathrm{d}t} + \left(\frac{\partial e}{\partial\langle\eta\rangle}\right)_{v,T} \frac{\mathrm{d}\langle\eta\rangle}{\mathrm{d}t}. \tag{5.28}$$

Combining (5.25) with (5.28), taking into account that $c_v = (\partial e/\partial T)_{v,\langle\nu\rangle}$ and [cf. (5.13)]

$$\frac{\mathrm{d}Q_{v,T}}{\mathrm{d}t} = -\left(\frac{\partial e}{\partial\langle\eta\rangle}\right)_{v,T} \frac{\mathrm{d}\langle\eta\rangle}{\mathrm{d}t},$$

from the thermodynamic identity

$$\left(\frac{\partial T}{\partial v}\right)_{s,\langle\eta\rangle} = -\left(\left(\frac{\partial e}{\partial v}\right)_{T,\langle\eta\rangle} + p\right)\frac{1}{c_v},$$

we obtain

$$\frac{\mathrm{d}T}{\mathrm{d}t} = \left(\frac{\partial T}{\partial v}\right)_{s,\langle\eta\rangle} \frac{\mathrm{d}v}{\mathrm{d}t} + \frac{1}{c_v}\left(\frac{\mathrm{d}Q_{v,T}}{\mathrm{d}t} + \frac{\mathrm{d}q}{\mathrm{d}t}\right). \tag{5.29}$$

Therefore, the relation between $\mathrm{d}T$ and $\mathrm{d}v$ is equally defined by the values of q and $Q_{v,T}$ (5.13).

Since v and e are functions of p, T, and $\langle\eta\rangle$, we obtain

$$\frac{\mathrm{d}v}{\mathrm{d}t} = \left(\frac{\partial v}{\partial p}\right)_{T,\langle\eta\rangle} \frac{\mathrm{d}p}{\mathrm{d}t} + \left(\frac{\partial v}{\partial T}\right)_{p,\langle\eta\rangle} \frac{\mathrm{d}T}{\mathrm{d}t} + \left(\frac{\partial v}{\partial\langle\eta\rangle}\right)_{p,T} \frac{\mathrm{d}\langle\eta\rangle}{\mathrm{d}t}, \tag{5.30}$$

$$\frac{\mathrm{d}e}{\mathrm{d}t} = \left(\frac{\partial e}{\partial p}\right)_{T,\langle\eta\rangle} \frac{\mathrm{d}p}{\mathrm{d}t} + \left(\frac{\partial e}{\partial T}\right)_{p,\langle\eta\rangle} \frac{\mathrm{d}T}{\mathrm{d}t} + \left(\frac{\partial e}{\partial\langle\eta\rangle}\right)_{p,T} \frac{\mathrm{d}\langle\eta\rangle}{\mathrm{d}t}. \tag{5.31}$$

Combining (5.25), (5.30), (5.31), and

$$\frac{\mathrm{d}Q_{p,T}}{\mathrm{d}t} = -\left(\frac{\partial e}{\partial\langle\eta\rangle}\right)_{p,v} \frac{\mathrm{d}\langle\eta\rangle}{\mathrm{d}t} - p\left(\frac{\partial v}{\partial\langle\eta\rangle}\right)_{p,v} \frac{\mathrm{d}\langle\eta\rangle}{\mathrm{d}t}$$

[cf. (5.14)], from

$$c_p = \left(\frac{\partial e}{\partial T}\right)_{p,\langle\nu\rangle} + p\left(\frac{\partial v}{\partial T}\right)_{p,\langle\nu\rangle}$$

and the thermodynamic identity

$$\left(\frac{\partial T}{\partial p}\right)_{s,\langle\eta\rangle} = -\left(\left(\frac{\partial e}{\partial p}\right)_{T,\langle\eta\rangle} + p\left(\frac{\partial v}{\partial p}\right)_{T,\langle\eta\rangle}\right)\frac{1}{c_p},$$

we obtain

$$\frac{\mathrm{d}T}{\mathrm{d}t} = \left(\frac{\partial T}{\partial p}\right)_{s,\langle\eta\rangle} \frac{\mathrm{d}p}{\mathrm{d}t} + \frac{1}{c_p}\left(\frac{\mathrm{d}Q_{p,T}}{\mathrm{d}t} + \frac{\mathrm{d}q}{\mathrm{d}t}\right). \tag{5.32}$$

Here a similar role is played by q and $Q_{p,T}$.

Note that in deriving (5.19)–(5.32) the state of aggregation of matter was never specified; therefore, the conclusions drawn can be equally applied to solid, liquid, and gaseous media. According to (5.22), (5.27), (5.29), and (5.32), any analysis of shock wave processes must be carried out with the highest consideration for a difference between the thermal effects $Q_{v,T}$, $Q_{p,T}$, and $Q_{p,v}$.

5.4.5 Thermal Criterion for Shock or Detonation

Analysis of expressions (5.13)–(5.15) in comparison with (5.22), (5.27), (5.29), and (5.32) shows that none of the thermal effects $Q_{v,T}$, $Q_{p,T}$, and $Q_{p,v}$ can be regarded as the major thermal characteristic of physicochemical transformation. The variable Q_{cal}, defined in (5.16), is absent in all these expressions. It should be noted that thermal effects (5.13)–(5.16) characterize different aspects of the same transformation $\langle\eta_0\rangle \to \langle\eta\rangle$.

Nevertheless, in shock wave processes the thermal effect $Q_{p,v}$ is of key importance. Let us try to distinguish between the shock and detonation waves by the sign of $Q_{p,v}$ (5.15), which defines the type of physicochemical transformation in a shock.

As shown in [2,64,65,67], the parameters of a detonation wave in condensed and gas-phase explosives are defined by the dynamic adiabat (also termed the "detonation adiabat") that intersects the isochore line v_0 for $p > p_0$. In other words, this dynamic adiabat lies above the point (p_0, v_0) (Fig. 5.9). As a result, the detonation adiabat has a point of contact (Chapman–Jouguet point) with one of the Rayleigh lines, as shown in Fig. 5.9. From [2,64,65,67], the detonation wave corresponding to the Chapman–Jouguet point must obey the condition $D - u = c$ (Fig. 5.8) so that the tail rarefaction waves cannot overtake the shock. Such a shock may be expected to become self-sustaining.

Therefore, the detonation adiabats (and respective detonation waves) in HEs and reactive gaseous mixtures must obey the equivalent inequalities given in (5.3):

Note that inequalities (5.3) characterize a given transformation only at p_0. Nevertheless, these can also be used to predict the features of shock-induced transformation for $p \gg p_0$.

We propose regarding inequalities (5.3) as an intrinsic property of the detonation adiabat and/or the detonation wave in any media, including HEs and reactive gaseous mixtures. For example, inequalities (5.3) are applicable to the so-called condensational shocks in a mixture of air with oversaturated water vapor [2]. Therefore, condensation shocks can also be regarded as detonation waves.

We suggest considering inequalities (5.33) as an intrinsic property of the shock wave:

$$Q_{p,v} \leqslant 0, \ \Delta v_{p,h} \leqslant 0. \tag{5.33}$$

For instance, the dynamic adiabats passing through the point (p_0, v_0) (see Fig. 5.9) and respective shocks without transformations can be classified, according to (5.33), as shock adiabats and shock waves, respectively. In this case, criteria (5.33) are consistent with the accepted opinion.

Applying criteria (5.3) and (5.33), one has to keep in mind that, during physicochemical transformation, the values of $Q_{p,v}$ and $\Delta v_{p,h}$ may change their sign. In the case when criterion (5.3) is not applicable to the final products but is applicable to some intermediate products, one can also expect the onset of detonation in this system.

Criteria (5.3) and (5.33) are not always consistent with the definitions of detonation and shock waves adopted by some workers in the physics of explosions: the detonation wave is regarded as any shock accompanied by exothermic reaction ($Q_{v,T} > 0$, $Q_{p,T} > 0$). But not every reaction that satisfies the inequality $Q_{p,T} > 0$ can satisfy the condition $Q_{p,v} > 0$ (5.3). According to (5.33), some shocks that are accompanied by exothermic reactions can nevertheless be classified only as shock waves.

The Chapman–Jouguet detonation theory [2, 64, 65, 67] acknowledges the existence of a detonation adiabat as a prerequisite for the occurrence of detonation in a given medium. This statement is equivalent to criterion (5.3). This condition is a prerequisite for self-sustaining shock propagation in a given medium.

Therefore, the above analysis of dynamic adiabats suggests that either of the inequalities (5.3) is indeed *a sufficient and necessary thermodynamic condition for the occurrence of detonation* in a given medium.

Note in conclusion that criterion (5.3) is strictly thermodynamic in nature. In other words, the criterion does not define specific conditions for realization of a detonation. These conditions have to be determined by experiment. One such condition is a sufficiently high rate of reaction [3, 64, 67]. Some methods for acceleration of shock-induced reactions were suggested in [27, 68, 69], but none of these methods may help in the initiation of detonation if condition (5.3) is not satisfied, at least for intermediate products. When inequalities (5.3) are not fulfilled (at least for one of the intermediate products), all the dynamic adiabats for a given medium will intersect the isobar $p = p_0$ for $v < v_0$. In this case, the Chapman–Jouguet point does not exist, and hence a self-sustaining shock wave process is impossible.

References

1. Lee, J.H.S., Goroshin, S., Yoshinaka, A., Romano, M., Jiang, J., Hooton, I., Zhang, F.: Attempts to initiate detonations in metal–sulfur mixtures. In: Furnish, M.D., Chhabildas, L.C., Hixton, R.S. (eds.) Shock Compression of Condensed Matter-1999, pp. 775–778. American Institute of Physics, Melville (2000)
2. Landau, L.D., Lifshitz, E.M.: Fluid Mechanics, chaps. 9, 14. Butterworth-Heinemann, Oxford (1987)
3. Kuznetsov, N.M.: Detonation and gas-dynamic jumps during phase transformations in metastable compounds. Zh. Eksp. Teor. Fiz. **49**, 1526–1531 (1965)
4. Merzhanov, A.G., Gordopolov, Yu.A., Trofimov, V.S.: On the possibility of gas-free detonation in condensed systems. Shock Waves **6**, 157–159 (1996)
5. Ryabinin, Yu.N.: Sublimation of a crystal lattice under the action of a strong shock wave. Sov. Phys. Dokl. **1**, 424–426 (1956)
6. Trofimov, V.S.: Dynamic study on relaxation processes. Fiz. Goreniya Vzryva **17**(5), 93–101 (1981)
7. Batsanov, S.S., Doronin, G.S., Koshevoi, V.P., Stupnikov, V.P.: Measurement of the residual temperature of substances after shock compression. Combust. Explos. Shock Waves **4**, 64–66 (1968)
8. Batsanov, S.S., Shestakova, N.A., Stupnikov, V.P., Litvak, G.S., Nigmatullina, V.M.: Shock-induced synthesis of chromium chalcogenides. Proc. Acad. Sci. USSR Dokl. Chem. **185**, 174–175 (1969)
9. Gur'ev, D.L., Batsanov, S.S.: Experimental technique for measurement of post-shock temperature. Bull. Am. Phys. Soc. **30**, 1320–1324 (1985)

10. Gur'ev, D.L., Batsanov, S.S.: Measurement of residual temperatures in cylindrical shock recovery capsules. Combust. Explos. Shock Waves **22**, 490–492 (1986)
11. Batsanov, S.S., Gur'ev, D.L.: Interaction of sulfur with tin in shock waves. Combust. Explos. Shock Waves **23**, 236–237 (1987)
12. Hornig, H., Kury, J., Simpson, R., Helm, F., von Holle, W.: Shock ignition by pyrotechnic heat powder. In: Proceedings XI International Pyrotechnics Seminar, pp. 699–720, Vail (1986)
13. Boslough, M.B.: Shock-induced chemical reactions in nickel–aluminum powder mixtures: radiation pyrometer measurements. Chem. Phys. Lett. **160**, 618–622 (1989)
14. Boslough, M.B.: A thermochemical model for shock-induced reactions (heat detonation) in solids. J. Chem. Phys. **92**, 1839–1848 (1990)
15. Batsanov, S.S., Gogulya, M.F., Brazhnikov, M.A., Lazareva, E.V., Doronin, G.S., Klochkov, S.V., Banshchikova, M.B., Fedorov, A.V., Simakov, G.V.: Shock compression of reactive substances in the system tin–chalcogene. Sov. J. Chem. Phys. **10**, 2635–2638 (1993)
16. Batsanov, S.S., Gogulya, M.F., Brazhnikov, M.A., Simakov, G.V., Maksimov, I.I.: Behavior of the reacting system Sn + S in shock waves. Combust. Explos. Shock Waves **30**, 361–365 (1994)
17. Batsanov, S.S.: Solid-phase reactions in shock waves: kinetic studies and mechanism. Combust. Explos. Shock Waves **32**, 102–113 (1996)
18. Gogulya, M.F., Voskoboinikov, I.M., Dolgoborodov, A.Y., Dorokhov, N.S., Brazhnikov, M.A.: Interaction of sulfur with metals under shock loading. Sov. J. Chem. Phys. **11**, 224–228 (1992)
19. Gogulya, M.F., Brazhnikov, M.A.: On the characteristic times of chemical reactions in heterogeneous systems under dynamic load. Chem. Phys. Rep. **13**, 1887–1890 (1995)
20. Tyte, D. Interaction of metallic powders with a shock wave through an oxidizing atmosphere. J. Appl. Phys. **37**, 802–806 (1966)
21. Batsanov, S.S., Doronin, G.S., Nerchenko, A.A., Roman'kov, V.V., Strelyaev, A.E., Stupnikov, V.P.: Breaking of high-speed flows in different media. J. Eng. Phys. **24**, 354–355 (1973)
22. Kovalenko, A.N., Ivanov, G.V.: Physical and chemical transformations of lead nitrate in mixtures with aluminum under shock waves. Combust. Explos. Shock Waves **17**, 82–85 (1981)
23. Batsanov, S.S., Doronin, G.S., Klochkov, S.V., Teut, A.I.: Synthesis reaction behind shock fronts. Combust. Explos. Shock Waves **22**, 765–769 (1986)
24. Gordopolov, A.Yu., Gordopolov, Y.A., Fedorov, V.M., Shikhverdiev, R.M.: Shock-induced chemical transformations in the Ti–C mixture. In: Proceedings XII Symposium on Combustion and Explosion, part II, pp. 190–192, Chernogolovka (2000)
25. Jiang, J., Goroshin, S., Lee, J.H.S.: Shock wave induced chemical reaction in Mn + S mixture. In: Schmidt, S.C., Dankbar, D.P., Forbes, J.W. (eds.) Shock Compression of Matter–1997, pp. 655–658. Elsevier, New York (1998)
26. Gur'ev, D.L., Gordopolov, Yu.A., Batsanov, S.S.: Solid-state synthesis of ZnTe in shock waves. Combust. Explos. Shock Waves **42**, 116–123 (2006)
27. Xu, X., Thadhani, N.N.: Investigation of shock-induced reaction behavior of as-blended and ball-milled Ni + Ti powder mixtures using time-resolved stress measurements. J. Appl. Phys. **96**, 2000–2009 (2004)

28. Graham, R.A., Anderson, M.U., Horie, Y., Yu, S.-K., Holman, G.K.: Pressure measurements in chemically reacting powder mixtures with the Bauer piezoelectric polymer gauge. Shock Waves **3**, 79–82 (1993)
29. Thadhani, N.N., Graham, R.A., Royal, T., Dunbar, E., Anderson, M.U., Holman, G.K.: Shock-induced chemical reactions in titanium–silicon powder mixtures of different morphologies: time-resolved pressure measurements and materials analysis. J. Appl. Phys. **82**, 1113–1128 (1997)
30. Vandersall, K.V., Thadhani, N.N.: Time-resolved measurements of the shock-compression response of Mo + Si elemental powder mixtures. J. Appl. Phys. **94**, 1575–1583 (2003)
31. Batsanov, S.S., Andrianova, E.E., Lazareva, E.V.: Mechanical consequences of chemical conversions in shock-recovery capsules. Sov. J. Chem. Phys. **9**, 2371–2373 (1991)
32. Batsanov, S.S., Gavrilkin, S.M., Gordopolov, A.Y., Gordopolov, Yu.A.: Spalling phenomena in shock-recovery capsules during shock compression of inert and reactive mixtures. Combust. Explos. Shock Waves **40**, 605–611 (2005)
33. Batsanov, S.S., Lazareva, E.V., Kopaneva, L.I.: Shock wave interaction of metals with water. High Energy Chem. **16**, 150–150 (1982)
34. Batsanov, S.S., Lazareva, E.V., Kopaneva, L.I.: Interaction of metals and water under dynamic compression. Sov. J. Chem. Phys. **3**, 1403–1403 (1985)
35. Batsanov, S.S., Gur'ev, D.L., Kopaneva, L.I.: Shock synthesis of the cubic ZrO_2. Combust. Explos. Shock Waves **24**, 505–505 (1988)
36. Sekine, T.: Diamond from shocked magnesite. Naturwissenschaften **75**, 462–463 (1988)
37. Batsanov, S.S., Blokhina, G.E., Deribas, A.A.: The effect of explosion on materials: structural changes in boron nitride. J. Struct. Chem. **6**, 209–213 (1965)
38. Akashi, T., Sawaoka, A., Saito, S., Araki, M.: Structural changes of boron nitride caused by multiple shock-compression. Jpn. J. Appl. Phys. **15**, 891–892 (1976)
39. Sokolowski, M.: Deposition of wurtzite-type boron nitride layers by reactive plus plasma crystallization. J. Cryst. Growth **46**, 136–138 (1979)
40. Fedoseev, D.V., Varshavskaya, I.G., Lavrent'ev, A.V., Deryagin, B.V.: Phase transformations of small-size solid particles during laser heating. Dokl. Phys. Chem. **270**, 416–418 (1983)
41. Sokolowska, A., Wronikowski, M.: The phase diagram (P, T, E) of boron nitride. J. Cryst. Growth **76**, 511–513 (1986)
42. Akashi, T., Pak, H.-R., Sawaoka, A.: Structural changes of wurtzite-type and zinc blend-type boron nitride by shock treatment. J. Mater. Sci. **21**, 4060–4066 (1986)
43. Nameki, H., Sekine, T., Kobayashi, T., Fat'yanov, O.V., Sato, T., Tashiro, S.: Rapid quench formation of E-BN from shocked turbostratic BN precursors. J. Mater. Sci. Lett. **15**, 1492–1494 (1996)
44. Gagnier, M., Szwarc, H., Ronez, A., Low-energy ball-milling: Transformations of boron nitride powders. Crystallographic and chemical characterizations. J. Mater. Sci. **35**, 3003–3009 (2000)
45. Batsanov, S.S.: The E-phase of boron nitride as a fullerene. Combust. Explos. Shock Waves **34**, 106–108 (1998)
46. Batsanov, S.S., Kopaneva, L.I., Lazareva, E.V., Kulikova, I.M., Barinsky, R.L.: On the nature of boron nitride E-phase. Propellants Explos. Pyrotech. **18**, 352–355 (1993)

47. Batsanov, S.S., Deribas, A.A.: The effect of explosion on materials: structural changes in neodymium oxide. Combust. Explos. Shock Waves **1**, 77–79 (1965)
48. Batsanov, S.S., Kopaneva, L.I., Lazareva, E.V., Gavryushin, V.N., Isaev, V.N., Porokhov, P.V.: Formation of the $Zr_{1-x}Fe_xS_2$ solid solution as a result of microspalling at shock compression in recovery ampoules. Combust. Explos. Shock Waves **26**, 614–615 (1990)
49. Kikuchi, M., Syono, Y., Fukuoka, K., Hiraga, K.: Redox reaction between tantalum and silica induced by shock loading. J. Mater. Sci. Lett. **6**, 97–99 (1987)
50. Batsanov, S.S., Shevtsova, N.N., Temnitskii, I.N., Bokarev, V.P.: Thermal and shock-induced synthesis of hybride germanium chalcogenides. Russ. J. Inorg. Chem. **31**, 925–926 (1986)
51. Shevtsova, N.N., Temnitskii, I.N., Batsanov, S.S.: The impact synthesis of the α- and β-forms of GeSeTe. Russ. J. Inorg. Chem. **32**, 1512–1513 (1987)
52. Batsanov, S.S., Bokarev, V.P., Maksimov, I.I., Tumanov, V.A.: Formation of foams upon dynamic/static compression of powdered dioxides of silicium, germanium and tin. Combust. Explos. Shock Waves **29**, 652–652 (1993)
53. Batsanov, S.S.: Specific features of solid-phase reactions induced by shock waves. Combust. Explos. Shock Waves **42**, 237–241 (2006)
54. Batsanov, S.S.: Inorganic chemistry of high dynamic pressures. Russ. Chem. Rev. **55**, 297–315 (1986)
55. Kostyukov, N.A.: Structure of the flow of binary mixtures of solids at two-dimensional shock wave loading. J. Appl. Mech. Tech. Phys. **29**, 362–365 (1988)
56. Batsanov, S.S., Maksimov, I.I.: Forced diffusion in condensed matter at shock compression. Sov. J. Chem. Phys. **8**, 2156–2157 (1991)
57. Dolgoborodov, A.Yu., Voskoboinikov, I.M., Tolstov, I.K., Sudarikov, A.V.: Characteristic properties of shock wave propagation in mixtures. Combust. Explos. Shock Waves **28**, 308–312 (1992)
58. Horie, Y., Mass mixing and nucleation and growth of chemical reactions in shock compression of powder mixtures. In: Murr, L.E., Staudhammer, K.P., Meyers, M.A. (eds.) Metallurgical and Materials Applications of Shock Wave and High-Strain-Rate Phenomena, pp. 603–614. Elsevier, New York (1995)
59. Batsanov, S.S.: Synthesis and modification of materials by shock waves: Real-time measurements and mechanism of reaction. Mater. Sci. Eng. A **210**, 57–63 (1996)
60. Coustal, R., Prevet, F.: Sur un noveau procédé de preparation du sulfure de zinc phosphorescent. C. R. Acad. Sci. **188**, 703–706 (1929)
61. Torunov, S.I., Trofimov, V.S., Calculation of an equilibrium temperature in ideal detonation wave in the zinc–sulfur SHS system. Int. J. SHS **10**, 13–21 (2001)
62. Torunov, S.I., Trofimov, V.S.: Two possible structures of detonation wave in the Zn–S mixture. In: Abstracts XII Symposium on Combustion and Explosion, part II, pp. 165–166, Chernogolovka (2000)
63. Gur'ev, D.L., Gordopolov, Yu.A., Batsanov, S.S., Fortov, V.E.: Solid-state detonation in the zinc–sulfur system. Appl. Phys. Lett. **88**, 024102 (2006)
64. Johansson, C.H., Persson, P.A.: Detonics of High Explosives. Vinterviken, Stockholm (1970)
65. Cook, M.A.: The Science of High Explosives. Reinhold, New York (1958)
66. L.D. Landau, E.M. Lifshits: Statistical Physics, chap. 1. Butterworth-Heinemann, Oxford (1980)

67. Dremin, A.N., Savrov, S.D., Trofimov, V.S., Shvedov, K.K.: Detonatsionnye Volny v Kondensirovannykh Sredakh (Detonation Waves in Condensed Matter), p. 20. Nauka, Moscow (1970)
68. Dolgoborodov, A.Yu., Makhov, M.N., Kolbanev, M.N., Streletskii, A.N., Fortov, V.E.: Detonation in an aluminum-teflon mixture, JETP Lett. **81**, 311–314 (2005)
69. Dolgoborodov, A.Yu., Makhov, M.N., Kolbanev, M.N., Streletskii, A.N., Fortov, V.E.: Detonation in metal-teflon mechanoactivated composites. In: Proc. XIV International Detonation Symposium, Norfolk (2006)

6

Shock Ignition of Particles

S.M. Frolov and A.V. Fedorov

6.1 Introduction

Ignition and combustion of fine solid particles and liquid fuel drops ("particles," for short) are important issues for various branches of science and technology, such as aerospace and chemical technologies, chemical propulsion, ground transportation, and industrial safety. These issues were noticed by numerous researchers both at the end of the foregoing century and presently, and many relevant publications are available. Detailed reviews of the current status of the research can be found elsewhere [1–5].

Particle ignition and combustion are phenomena comprising all the main constituents of the combustion process; namely, fast exothermic chemical reactions complicated by diffusion of reactants and products, thermal energy deposition and spreading of heat in the medium, and convective flows. The classical theory implies that particle combustion is diffusion-limited and therefore chemical kinetic aspects are usually not considered in the analysis [6–15]. In addition, the classical theory considers an isolated particle in an unconfined state. Within these presumptions, notable progress in understanding relevant physical and chemical processes has been achieved recently. However, for problems dealing with transient modes of combustion, such as ignition or extinction, it is necessary to consider the effects of finite-rate chemical kinetics. Moreover, in practice, particle ignition and combustion occur in the presence of neighboring particles or confinement surfaces. The corresponding effects are usually referred to as "spray" (or "collective") and "confinement" effects.

Spray effects manifest themselves in two-phase reactive flows. In existing computational approaches, chemical reaction rates are determined by considering particles as distributed mass, momentum, and energy sources. As a matter of fact, spray ignition and combustion phenomena are a complex combination of chemical kinetics and diffusion-controlled flames around individual particles, their groups, and gas-phase partially premixed flames.

F. Zhang (ed.), *Shock Wave Science and Technology Reference Library: Heterogeneous Detonation*, DOI: 10.1007/978-3-540-88447-7_6,

For modeling ignition and combustion of particles in sprays and clouds, several approaches are used by analyzing the evolution of (1) two or more particles [16, 17], (2) a regular sequence of particles [18, 19], (3) a group of chaotically distributed particles [20, 21], or (4) a particle suspension [22]. Advantages and drawbacks of these approaches are discussed elsewhere [2–7]. Note that only the most detailed description of the flow fields in the space between particles has been obtained in the analysis of two interacting particles and the linear array of particles. "Group" and "suspension" theories do not take into account transient heat and mass transfer processes in the particle vicinity, nor the dependence of the rates of these processes on the distance between particles.

At present, ignition and combustion of particles in combustion chambers are modeled without regard for many factors affecting both local and integral combustor performance. Dynamic and thermal effects of particles on the flow are modeled in terms of the averaged interphase mass, momentum, and energy fluxes [22]. The effect of turbulence on phase interaction is modeled by turbulent dispersion of particles [5]. Mixing of gas-phase species (vapors and reaction products) is modeled by micromixing models which do not account for the difference in species diffusion coefficients. Finite instantaneous depths of penetration of heat and diffusion fluxes from each individual particle are also not taken into account. As a matter of fact, in computational algorithms, the increments of thermal energy and gas-phase species mass arising from particle or drop ignition and combustion are smeared uniformly throughout a corresponding computational cell, which is usually several orders of magnitude larger than the characteristic particle size. As a consequence, gas-phase chemical processes (intrinsically local) are treated as volumetric processes. In view of this, computational cell size becomes an important artificial parameter of a solved problem. Its value determines the dynamics of the local and integral phenomena studied. However, particles are known to be surrounded by finite-dimensional spatial regions with highly nonuniform distributions of temperature and species concentrations. The dimensions of such regions depend on time and the instantaneous mutual position of neighboring particles. In such conditions, preignition processes as well as ignition and combustion cannot be considered as processes in a well-stirred reactor represented by a computational cell.

In this chapter we concentrate on fundamental issues of shock-induced particle and drop ignition and provide an overview of the problems of adequate mathematical modeling of relevant phenomena. Subsequent stages of ignition process development (combustion, extinction, etc.) are beyond the scope of this chapter. As examples of solid particles, magnesium and aluminum are mainly considered. Their oxidation rates either depend on the oxide film thickness (aluminum) or are independent of it (magnesium). As examples of liquid fuel drops, single-component n-alkane drops are considered.

6.2 Ignition of Solid Particles

6.2.1 Experimental and Theoretical Findings

Ignition of solid particles in a high-temperature gas flow behind a shock wave propagating in a gas–particle suspension was studied experimentally elsewhere [23–47]. Figure 6.1 shows a typical schematic of experimental conditions in a shock tube [46, 47]: before firing; after firing, but before the shock wave has reached the end wall; and after shock wave reflection. In the experiments on particle ignition behind the incident shock wave, region 2 is the test region. If particle ignition is studied behind the reflected shock wave, region 5 serves as the test region. Pressure and temperature in regions 2 and 5 are set by adjusting the initial conditions in the driver (region 4) and driven sections (region 1) of the shock tube before firing and by monitoring the shock wave velocity along the tube in the course of the experiments. To estimate the conditions in the experiments, ideal compressible gas relations and Rankine–Hugoniot relationships are usually used.

The shock tube can be either vertical or horizontal. Vertical shock tubes have the advantage of creating dust suspensions, which are homogeneous in tube cross section. However, gravitational dust sedimentation can lead to dust density variation along the tube. To avoid the effect of sedimentation, various dust dispersion techniques are used. For example, in [24, 32, 37, 47], it was reported that dust powders were mounted on a horizontal shelf or knife blade, at a short distance from the end wall. The incident shock wave swept over the particles, entraining them in the postshock flow and thus creating

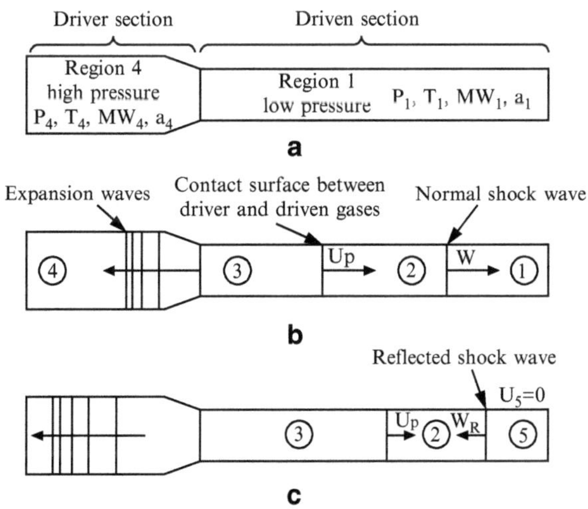

Fig. 6.1. Typical experimental conditions in a shock tube [46, 47]: **a** before firing of the tube; **b** after firing, but before reflection of the incident shock wave; and **c** after reflection of the incident shock wave from the end wall

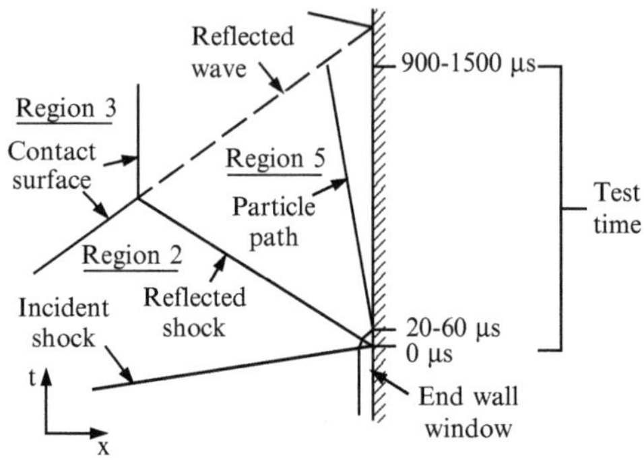

Fig. 6.2. Flow field in time (t)–distance (x) coordinates near the shock-tube end wall showing flow interactions and particle position versus time [46, 47]

a particle suspension. After the shock wave had been reflected off the end wall, the particles in suspension were heated and ignited if the temperature, pressure, and composition conditions behind the reflected shock wave were appropriate for ignition.

Figure 6.2 shows a typical flow field wave diagram for the shock tube near the end wall, with axial position on the horizontal axis and time on the vertical axis [46, 47]. The incident shock wave first approaches and is then reflected from the end wall, positioned on the right side of the x-axis. Then the reflected shock wave moves away from the end wall until it reaches the contact surface, which is moving toward the end wall at the same velocity as the gas in region 2. Through this interaction, a normal shock is transmitted through the contact surface, moving away from the end wall, and a shock wave, Mach wave, or expansion wave, depending on the thermodynamic properties of the gases in regions 2 and 3, is reflected back toward the end wall. This wave then returns to the end wall and is reflected from it, ending the test period. The gas in region 5 is a stagnant, hot, high-pressure mixture in which solid particles are ignited. Thus, the test time is the time between the arrival of the incident shock wave at the end wall and the arrival of the wave reflected from the contact surface at the drifting particles. In experiments [46, 47], the test time ranged from 900 to 1,500 μs.

The experimental data on the ignition delay time t_{ign} as a function of the initiating shock wave Mach number M_0 are available for various metals – volatile (magnesium, zinc, etc.), intermediate (aluminum, iron, copper, lead, titanium, hafnium, etc.), and refractory (tantalum, tungsten, niobium), non-metallic inorganic elements (boron, silicon, carbon, etc.), various organic dusts (cornstarch, wheaten flour, etc.), and monopropellants and high explosives (RDX, TNT, etc.).

For a suspension of magnesium particles of mean initial radius $r_{s0} = 17\,\mu m$, the dependence of t_{ign} on postshock temperature T_0 was obtained, for example, in [27]. At temperatures $T_0 = 1{,}538$–$2{,}500$ K, the ignition delay time of magnesium particles varied from 0.1 to 0.05 ms.

For a suspension of fine aluminum particles ($r_{s0} = 5$–$10\,\mu$m), the t_{ign} versus T_0 dependence was obtained, for example, in [47]. Aluminum particles were ignited in atmospheres consisting of argon and various amounts of H_2O, CO_2, and O_2 at the end wall of a shock tube. At pressures and temperatures of about 0.85 MPa and 2,600 K, the measured ignition delays were in the range from 150 to 900 µs. Interestingly, aluminum particles ignited in a mixture of H_2O/Ar showed ignition delay times 3–6 times greater than those from $CO_2/O_2/Ar$ mixtures.

In the literature, there are still inconsistencies in available experimental data. For example, in some experimental studies the ignition delays were found to reduce when the measurements were made behind the reflected shock wave rather than behind the incident shock wave [36, 37]. In other studies (e.g., [30,34,35,41]) some dusts were more easily ignited behind the incident shock wave than behind the reflected shock wave.

Experimental studies of combustion of various metal particles [48–58] indicate the important role of heterogeneous reactions. It is known that during the oxidation of metal particles in a gaseous oxidizer, an oxide film separates the metal and the oxidizer. The rate of reaction is then governed by the protective properties of this film. In [8,9], this implication was used to model ignition of individual metal particles in a high-temperature oxidizing atmosphere. Particle ignition was found to occur owing to thermal breakdown, when the rate of heat evolution during oxidation becomes larger than the rate of heat removal from the particle surface.

Ignition of volatile metals, owing to their low boiling points (about 1,360 K for Mg), can occur in a vapor phase according to [6]. However, metal oxide layers were always observed on the surface of particles, which indicates the significance of heterogeneous reactions during ignition [55–58]. Ignition of boron particles is inhibited by a thin layer of B_2O_3 on the particle surface, and the high vaporization temperature of boron (about 4,100 K) limits oxidation to slower heterogeneous surface reactions.

There exist a number of theoretical models of particle ignition [59–84]. In most of the models, the metal particle ignition law is represented in the form of an Arrhenius expression with certain values of the preexponential factor K in the oxide film growth rate and the activation energy E of the oxidation process. These kinetic constants are found by fitting experimental and theoretical dependencies of t_{ign} on M_0.

The models differ considerably in their predicting capabilities. For example, the values of t_{ign} predicted in [62] appeared to be an order of magnitude larger than the values measured in dynamic conditions behind a shock wave [27]. The authors of [72–77] found the values of K and E for the dynamic model of particle ignition in the flow behind a propagating shock wave. They took particle motion and low-temperature oxidation into account. The resultant values of K and E appeared to be different from those in the static ignition model, which does not account for the relative motion of solid particles and gas.

A simple estimate for the particle velocity relaxation time, τ_{vel}, in the gas flow using the Stokes law $\tau_{vel} = (2/9)\rho_s r_{s0}^2/\mu$ shows that at $r_{s0} = 1$ and $5\,\mu m$, $\tau_{vel} = 0.02$ and $0.5\,ms$, respectively. Here, ρ_s is the particle material density and μ is the gas viscosity. According to the ignition delay data [63], the effect of velocity relaxation for fine particles may be insignificant at low flow velocities. The velocity of very fine particles rapidly achieves the ambient gas velocity, and the heating conditions quickly become similar to those in a quiescent atmosphere. Besides the differences in K and E values at static and dynamic conditions, there is also a considerable scatter in the values of the determining parameters in available mathematical models of particle ignition. As a result, the constants in the models differ considerably, sometimes by orders of magnitude; therefore, it is worth discussing these problems, first for the example of metal particle ignition in static conditions.

In general, all mathematical models of metal particle ignition in the quiescent atmosphere are based on two approaches [59–71]. The first approach applies the thermal explosion theory developed by Semenov. The second is based on the "reduced film" concept. The critical analysis of these approaches [78] resulted in the following findings:

1. Two unknown constants, K and E, in the models are determined from experimental dependencies at essentially different conditions. In [59, 60], they were found for small values of r_{s0} based on the critical "breakdown" temperature of a liquid magnesium particle. In [65, 66], they were found for large values of r_{s0} for solid magnesium particles. As a result, the values of the constants in these papers are different.
2. Only models based on the second approach contain information on the dependence of the medium critical temperature on oxygen concentration and particle radius as well as the dependence of the ignition delay time on the particle radius.
3. None of the papers cited contain a qualitative study of the solutions of the corresponding Cauchy problem for the models using the first approach or the corresponding boundary-value problem for the models using the second approach. Mathematical issues such as solvability of governing transcendental equations for determining K and E were not addressed. Moreover, no comparison of the data on the temperature growth dynamics by different models has been made [78].

6.2.2 Static Conditions

6.2.2.1 General Mathematical Model

Consider a metal particle placed in a hot flow of oxidizing gas. It is assumed that an exothermic oxidation reaction and an endothermic phase transition (melting, vaporization, etc.) take place on the particle surface. The particle is assumed to occupy region Ω in the space with Cartesian coordinates (x_1, x_2, x_3). The mathematical model governing the evolution of the temperature field in the particle is based on the energy conservation equation [85]

$$c_s\rho_s\frac{\partial T}{\partial t} = \lambda_s\Delta T, \quad t > 0, \quad (x_1, x_2, x_3) \in \Omega \subset R^3 \tag{6.1}$$

supplemented with the initial condition

$$t = 0: \quad T = T_0(x_1, x_2, x_3), \tag{6.2}$$

boundary conditions on the reactive particle surface $\Gamma_1 \subset \partial\Omega$

$$(x_1, x_2, x_3) \in \Gamma_1: \quad \lambda_s\frac{\partial \mathrm{T}}{\partial n} = Q_{\mathrm{chem}} - Q_{\mathrm{conv}} - Q_{\mathrm{phase}}, \tag{6.3}$$

and symmetry condition on the internal surface $\Gamma_2 \subset \partial\Omega$

$$(x_1, x_2, x_3) \in \Gamma_2: \quad \frac{\partial T}{\partial n} = 0. \tag{6.4}$$

To complete the statement of the problem one has to specify source terms in (6.3) as well as the metal oxidation kinetics. The latter is usually expressed in the form of an additional differential equation

$$\frac{\mathrm{d}h}{\mathrm{d}t} = \varphi(h, T) \tag{6.5}$$

with the corresponding initial condition. In (6.1)–(6.5), t is time, $T = T(x_1, x_2, x_3, t)$ is the particle temperature, $h = h(t)$ is the oxide film thickness, $\Delta \equiv \partial^2/\partial x_1^2 + \partial^2/\partial x_2^2 + \partial^2/\partial x_3^2$ is the Laplace operator, $\partial\Omega$ is the boundary of region Ω, ∂n is the element of the normal vector to the particle surface, c is the specific heat, λ is the thermal conductivity, φ is a function, and indices s and 0 relate to particle material and initial conditions, respectively. Equation (6.3) includes source terms corresponding to heat fluxes between the particle and ambient medium owing to convective heat transfer, Q_{conv}, heterogeneous chemical reaction, Q_{chem}, and phase transition, Q_{phase}. The generalized model of (6.1)–(6.5) can be simplified in some particular cases.

6.2.2.2 Point Model of Particle Ignition without Metal Vaporization

Assume that a solid particle is spherical and the temperature field in it is uniform. For the sake of argument, consider a magnesium particle. Then, averaging (6.1)–(6.4) over the particle volume, one obtains the following heat balance equation:

$$\frac{4}{3}\pi r_s^3 c_s\rho_s\frac{\mathrm{d}T}{\mathrm{d}t} = 4\pi r_s^2 q\rho_{\mathrm{ox}}\frac{\mathrm{d}h}{\mathrm{d}t} - 4\pi r_s^2\frac{\lambda Nu}{2r_s}(T - \tilde{T}). \tag{6.6}$$

The kinetic equation (6.5) can be expressed in the form

$$\frac{\mathrm{d}h}{\mathrm{d}t} = k_0 C_{\mathrm{ox}} \psi(T) \exp(-E/RT). \tag{6.7}$$

In (6.6) and (6.7), q is the chemical reaction heat related to metal oxide mass, ρ_{ox} is the metal oxide density, Nu is the Nusselt number determined as $Nu = 2.0 + 0.5\, Re^{0.5}$ with the Reynolds number $Re = 2r_{\mathrm{s}} U \rho/\mu$ based on the relative velocity U of gas and the particle, $\tilde{T}$ is the ambient gas temperature, k_0 and E are the parameters in the empirical kinetic law for the oxide film growth, C_{ox} is the mass concentration of the oxidizer, $\psi(T)$ is a function, and R is the gas constant. To take into account thermal deceleration of oxidation at high temperatures comparable with the magnesium boiling temperature T_{m}, it is usually assumed that $\psi(T) = T_{\mathrm{m}} - T$. At considerably lower temperatures, $\psi(T) = 1$.

6.2.2.3 Point Model of Particle Ignition with Metal Vaporization

In the literature, there are some indications of the importance of metal and its oxide vaporization at the particle surface [77,86]. Mathematical models of this phenomenon are of interest for the general theory of thermal explosion in systems with two chemical reactions exhibiting different activation energies and characteristic times [87]. The model taking into account metal vaporization follows from (6.1)–(6.5) with a nonzero source term describing the phase transition, i.e., $Q_{\mathrm{phase}} \neq 0$. Within this model, particle temperature evolution is governed by the following equation [64]:

$$\frac{1}{3}\frac{c_{\mathrm{s}} r_{\mathrm{s}}}{qk}\frac{\mathrm{d}T}{\mathrm{d}t} = \exp(-E/RT) - c\exp(-L/RT) - \bar{\alpha}(T - \tilde{T}) \equiv g_1(T), \tag{6.8}$$

where $\bar{\alpha} = \lambda Nu/(2qr_{\mathrm{s}}\rho_{\mathrm{s}}k)$, $k = k_0 c_{\mathrm{ox}}$, $c = v/k$, and L and v are the latent heat and reference rate of metal vaporization.

6.2.2.4 Remarks on Point Models

In the models of particle ignition described in the previous sections, a number of simplifying assumptions are adopted which are not always justified.

For metal combustion the assumption of a spherically symmetric ignition process is usually adopted; however, experimental observations show that non-symmetrical combustion modes also exist. The oxide layer possesses spherical symmetry only during the low-temperature induction phase of ignition. At high temperatures, the layer becomes nonuniform, which leads to the appearance of liquid "beads" and "caps" of oxide on the surface of the solid or liquid metal [84,88]. It can therefore become necessary to refrain from making the assumption of spherical symmetry and to include symmetry-breaking conditions when modeling combustion of metal particles. The models which are in current use are lacking in this respect.

To calculate the heat flux to the particle surface, Newton's law of cooling $Q_{\mathrm{conv}} = k_{\mathrm{h}}\left(\tilde{T} - T\right)$, is applied with T treated as the mean particle temperature; however, Newton's law is known to be valid only for steady-state heat transfer and the consequences of its application to intrinsically transient problems of particle heating and ignition are not quite clear. Moreover, the heat flux is determined by the particle surface temperature T_{i} rather than the mean particle temperature. Thus, it is assumed that the thermal conductivity of particle material is infinitely large. However, this assumption can be violated when ignition of a relatively large particle of low-conductivity material is considered or when the chemical reaction rate at a particle surface is sensitive to the surface temperature. Therefore, for better representation of the generalized model of (6.1)–(6.5), the point models have to use some correction factors for Newton's law which would allow one to take into account the transient heat transfer of metal particles with gas and the nonuniform temperature distribution inside the particles.

The application of convective heat transfer correlations of the Ranz–Marshall type [89] for modeling heat fluxes between solid particles and gas under conditions of nonzero velocity slip in two-phase flows is also not fundamentally substantiated. Such correlations were derived from the measurements in steady-state flows and their use in transient two-phase flow conditions is questionable.

Regarding the overall kinetic law of particle ignition, it is usually derived by fitting the measured ignition delays and the results of calculations based on the standard particle ignition model with several unknown Arrhenius parameters (e.g., preexponential factor and activation energy). The Arrhenius parameters of the overall reaction rate constants are then found by applying the least-squares procedure. In view of the above assumptions adopted in the standard model of particle heating, the Arrhenius parameters thus obtained can appear to have little in common with the actual values relevant to the problem under consideration. There are many examples in the literature when the values of preexponential factors and activation energies determined for particular conditions of particle ignition (e.g., large samples of cubic or cylindrical geometry, etc.) were applied for other conditions (fine particles of spherical or flaked shape, thin filaments, etc.). In view of these implications, there is a need for reliable models of solid particle heating to provide the basis for improved modeling of particle ignition and combustion.

One of the promising approaches was reported by Avdeev et al. [90, 91], who derived the correction factors for Newton's law, which allow one to take into account the transient heat transfer of metal particles with gas and the nonuniform temperature distribution inside the particles. It was implied that these correction factors can modify the conditions of particle ignition in the oxidizer gas. The dynamics of metal particle heating in a quiescent gas was calculated using three models: (1) a detailed model based on the conjugate partial derivative equations of thermal conductivity in gas and particles, (2) a standard point model based on the ordinary differential equation for the mean particle temperature T and Newton's law $Q_{\text{conv}} = k_{\text{h}}\left(\tilde{T} - T\right)$, and (3) a new point model, based on the ordinary differential equation for mean particle temperature T and Newton's law in the form $Q_{\text{conv}} = k_{\text{h,eff}}\left(\tilde{T} - T_{\text{i}}\right)$, where $k_{\text{h,eff}}$ is the effective heat transfer coefficient accounting for the transient particle heating. The approximate dependence of $k_{\text{h,eff}}$ on the governing parameters and time was derived from the analytical solution for a particle with constant surface temperature. The dependence of the surface temperature on the mean particle temperature $T_{\text{i}} = T_{\text{i}}(T)$ was determined by generalizing the results of numerical calculations based on the detailed model.

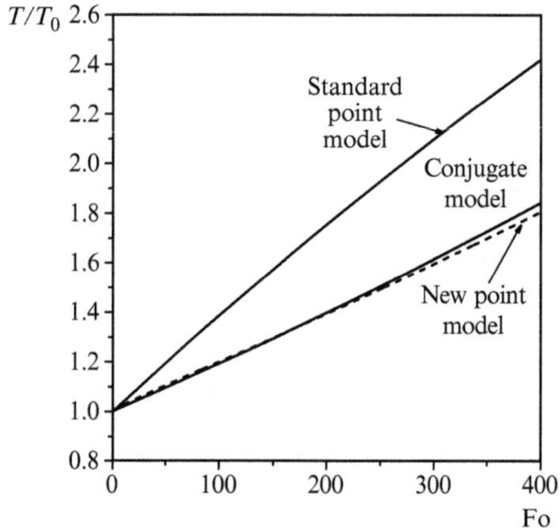

Fig. 6.3. Predicted time histories of aluminum particle heating in air at $\tilde{T}/T_0 = 5.46$ [90]

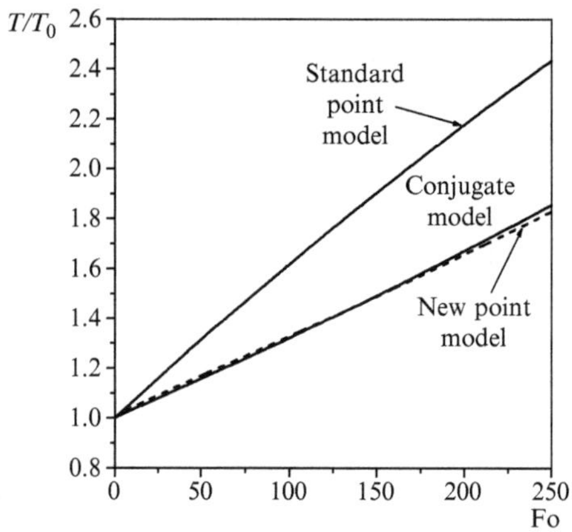

Fig. 6.4. Predicted time history of magnesium particle heating in air at $\tilde{T}/T_0 = 5.46$ [90]

Comparison of the computational results provided by the three models for aluminum, magnesium, boron, and iron particles showed that the new model correlates much better with the detailed model than the standard model. Figures 6.3 and 6.4 show some results of calculations in terms of the mean dimensionless particle temperature T/T_0 versus dimensionless time $Fo = \kappa t/r_{s0}$ for aluminum and magnesium particles predicted by different models.

The maximal deviations of the predicted mean particle temperature from the solution of the conjugate problem were less than 1–2% for the new model and up to 30% for the standard point model. The largest deviations were obtained for higher gas temperatures. The latter finding is particularly important for the problem of metal particle ignition in an oxidizer gas.

The important advantage of the model of Avdeev et al. [90, 91] is that it contains the particle surface temperature $T_{\rm i}$, which may differ considerably from the mean temperature T. When solving a problem on particle ignition, the use of $T_{\rm i}$ instead of T may affect the process evolution in view of the strong dependence of the rate of a heterogeneous reaction on temperature.

6.2.2.5 Two Problems in Particle Ignition Theory

In the Sects. 6.2.2.6 and 6.2.2.7 some important issues are addressed dealing with the solution properties of the point models presented in Sects. 6.2.2.2 and 6.2.2.3. Using the elementary theory of catastrophes, analytical and numerical methods of solving ordinary differential equations, we consider:

1. The structure of catastrophe manifolds (CMs) of the point models and their correspondence to Semenov's critical conditions in the thermal explosion theory.
2. Solvability conditions of transcendental equations for determining parameters K and E. It is implied that there can be nonuniqueness in determining E at a given K.
3. Types of particle temperature histories.
4. Correspondence between particle ignition delay times provided by the models using the first and second approaches.

Consider a small magnesium particle placed into a quiescent gaseous atmosphere at temperature $\tilde{T}$. Within Semenov's model of thermal explosion, the equations of energy balance and oxide film growth kinetics for a particle of initial radius r_{s0} covered with an oxide film of thickness $h \ll r_s$ have the form of (6.6) and (6.7) [64, 68]:

$$mc_{\rm s}\frac{{\rm d}T}{{\rm d}t} = -Sk_{\rm h}(T-\tilde{T}) + Sq\rho_{\rm ox}\frac{{\rm d}h}{{\rm d}t}, \; \frac{{\rm d}h}{{\rm d}t} = K\,{\rm e}^{-E/RT}, \tag{6.9}$$

where $\psi(T) = 1$ is assumed. In (6.9), $k_{\rm h}$ is the heat transfer coefficient, $S = 4\pi r_{\rm s}^2$ is the particle surface area, and $m = (4/3)\pi\rho_{\rm s}r_{\rm s}^3$ is the particle mass. For magnesium particles, $c_s = 1,100\,\mathrm{J\,(kg\ K)^{-1}}$, $\rho_s = 1,740\,\mathrm{kg\,m^{-3}}$, $q = 4.9 \times 10^7\,\mathrm{J\,kg^{-1}}$, and $\rho_{ox} = 3,600\,\mathrm{kg\,m^{-3}}$.

The solution of (6.9) should satisfy the following initial conditions:

$$T(0) = T_0,\ h(0) = h_0. \tag{6.10}$$

Thus, the evolution of particle temperature in the gas with $\tilde{T} > T_0$ is governed by the Cauchy problem of (6.9) and (6.10). The first term of (6.9) is the autonomous ordinary differential equation. It is sufficient to analyze the zero isoclinic line and determine the types of solution of this problem depending on some bifurcation parameters, for example, E, K, and $\tilde{T}$. In view of this, two problems should be solved.

The first problem is a direct problem formulated as follows:

Find the solution of the Cauchy problem (6.9) and (6.10), which is continuously differentiable together with its second derivative in the region $t \geq 0$.

The second problem belongs to the class of inverse problems and is formulated as follows:

Find the solution of the Cauchy problem (6.9) and (6.10) supplemented by some solvability condition with one (E) or two (E and K) unknown parameters such that one or two conditions relating these parameters are satisfied.

6.2.2.6 Direct Problem of Particle Ignition Without Vaporization

The qualitative properties of the direct problem solution can be studied using the methods of elementary catastrophe theory. In the dimensionless form, (6.9) governing particle ignition reads

$$\frac{\mathrm{d}T}{\mathrm{d}t} = \frac{T_{ox}}{t_1}\left(-\alpha(T - \tilde{T}) + \exp(-E/T)\right) \equiv g(T;\alpha,\tilde{T}),$$

where $T_{ox} = 3\xi q/(c_s T_a)$, $\alpha = \lambda\, Nu\, T_a/(2Kr_s\rho_s\xi q)$, and $t_1 = r_s/(Kt_a)$ are some constants; T_a and t_a are the reference values of temperature and time, and $\xi = \rho_{ox}/\rho_s$. Following [78, 92, 93], consider the zero isoclinic line of the equation as a surface in the space of parameters $(T_{ox};\alpha,\, t_1)$. Consider the function $g(T;\alpha,\tilde{T})$ assuming that parameters α and $\tilde{T}$ are the control parameters. This enables one to construct the CM or the equilibrium surface in the $(T_{ox};\alpha,\, t_1)$-space:

$$g(T;\alpha,\tilde{T}) = 0, \tag{6.11}$$

$$\frac{\mathrm{d}g}{\mathrm{d}T} = 0. \tag{6.12}$$

The system of (6.11) and (6.12) then allows determination of double-degenerate critical points (DCPs). If one supplements these equations with the equation

$$\frac{\mathrm{d}^2 g}{\mathrm{d}T^2} = 0 \tag{6.13}$$

and considers the solution of (6.11)–(6.13), then this solution determines triple-degenerate critical points (TCPs).

The solution of (6.11)–(6.13) can be found explicitly:

$$X(T;\alpha,\,\tilde{T}) = \left(\frac{E}{2};\,\frac{4}{E}\mathrm{e}^{-2},\,\frac{E}{4}\right) \equiv (T_*;\alpha_*,\,\tilde{T}_*).$$

The set of DCPs forms the curves of the following folds:

$$\alpha = \alpha_\pm(\tilde{T}) = \frac{\mathrm{e}^{-E/T_\pm}}{T_\pm - \tilde{T}},\quad T_\pm(\tilde{T}) = \frac{E}{2}\left(1 \pm \sqrt{1 - \frac{4\tilde{T}}{E}}\right) \tag{6.14}$$

in the plane of control parameters $(\alpha, \tilde{T})$. Note that the condition $\tilde{T} = E/4$ corresponds to a cusp point, where $\alpha = \alpha_*$ and $T = T_\pm = T_*$. The separatrix on the plane of control parameters $(\alpha, \tilde{T})$ consists of the cusp point $(\alpha_*, \tilde{T}_*)$ and the fold curves (6.14). Note that DCPs of (6.11) and (6.12) on the fold curves (6.14) of the catastrophe surface (6.11) are the solutions of Semenov equations determining the thermal equilibrium breakdown. The curves $\alpha = \alpha_-(\tilde{T}, E)$ and $\alpha = \alpha_+(\tilde{T}, E)$ at the $(\alpha, \tilde{T})$-plane correspond to the ignition and extinction limits, respectively.

To describe ignition of a particle with $\tilde{T} > T_0$ one has to solve the Cauchy problem (6.9) and (6.10). The following analysis of qualitative features of the solution provides the information on the particle ignition criterion [61]. The ignition time is often determined in the literature as the time taken for the rate of particle temperature variation $\dot{T} = \mathrm{d}T/\mathrm{d}t$ or $g(T)$ to attain the maximum value.

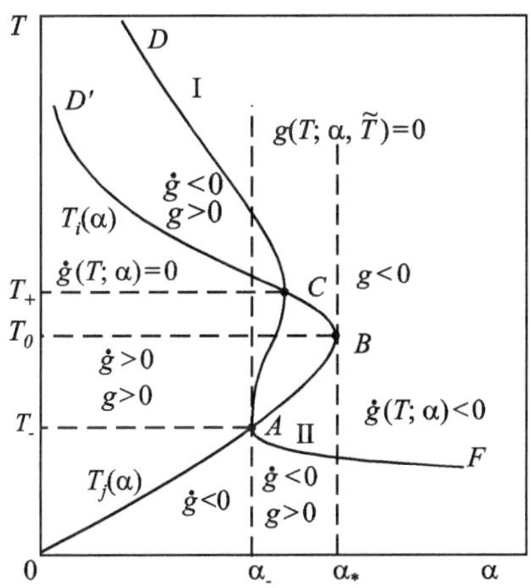

Fig. 6.5. Qualitative shape of ignition manifold in section $\tilde{T} = \mathrm{const}$ $(\tilde{T} < E/4)$

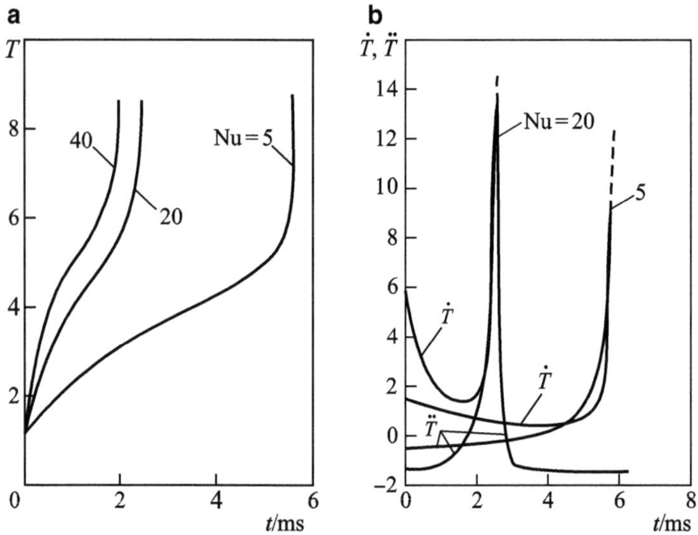

Fig. 6.6. **a** Typical time histories of magnesium particle temperature at $r_{s0} = 17\,\mu$m and $\tilde{T} = 1,538$ K. **b** Time histories of first and second temperature derivatives

Table 6.1. Dependence of t_{ign} on $\tilde{T}$ and Nu at $r_{s0} = 22\,\mu$m for magnesium particles

$\tilde{T}$ (K)	t_{ign} (ms)		
	[63]	$Nu = 2$	$Nu = 5$
1,023	27	67	No ignition
1,083	21	46	41
1,143	17	36	24
1,203	14	30	18
1,263	13	26	14
1,323	11	23	12

A qualitative shape of the manifold M of catastrophes (ignitions) is shown in Fig. 6.5 in the section $\tilde{T} = \text{const}$ at $\tilde{T} < E/4$. The curves $T_i(\alpha)$ and $T_j(\alpha)$ on which the maxima and minima of the function $g(T; \alpha, \tilde{T})$ are reached, and typical time histories of temperature and temperature derivatives of a magnesium particle are also presented in Figs. 6.5 and 6.6. Curve *DC*, denoted as *I*, is a part of the zero isoclinic line of the governing equation, which is responsible for ignition. Curve *AF*, denoted as *II*, corresponds to the lower stationary particle state. Curve *DCAF* is the zero isoclinic line of the first of (6.9), which is denoted as $T^0(\alpha, \tilde{T};\ E)$. Now, one can formulate the following assertion.

Assertion 1: The solution of the Cauchy problem (6.9) and (6.10) at $\alpha < \alpha_-(\tilde{T})$ is stabilized on part *I* of manifold *M*, and at $T_0 < T_j(\alpha)$ has two inflection points: $T_j(\alpha)$ and $T_i(\alpha)$. At $T_0 \in (T_j(\alpha), T_i(\alpha))$, it has one inflection point. At $T_0 > T_i(\alpha)$, there are no inflection points. At $\alpha \geq \alpha_-(\tilde{T})$, the solution of the Cauchy problem is stabilized on part II of manifold M at $T_0 < T^0(\alpha, \tilde{T};\ E)$ (Table 6.1).

Note that in the first three cases, Assertion 1 categorizes the solutions of the Cauchy problem in terms of possible regimes of magnesium particle heating with ignition. The last case corresponds to normal particle heating to some stationary state with $T^0(\alpha, \tilde{T};\ E) < T_-$.

The mathematical model is completely defined when the kinetic parameters in the empirical equation for the oxide film thickness are specified. In this case, it becomes possible to solve the problem numerically and determine the particle temperature history. Here, the ignition time will be treated as an instant at which the second derivative of temperature with respect to time, $\ddot{T}$, vanishes for the second time.

6.2.2.7 Direct Problem of Particle Ignition with Vaporization

In this section, the manifold of catastrophes (ignitions) for a model of thermal explosion of a magnesium particle with metal vaporization is analyzed. The types of particle temperature evolution in the plane of model bifurcation parameters are determined and the results obtained by different models are compared.

Within the frame of the point model, the equation governing particle temperature history has the form of (6.8). To analyze qualitatively the solution of the Cauchy problem of (6.8), the zero isoclinic line of this equation in the domain of variables T, $\bar{\alpha}$, $\tilde{T}$, c, E, and L is considered using elementary catastrophe theory.

Equation (6.8) can be rewritten in the form

$$\frac{\mathrm{d}T}{\mathrm{d}\tau} = Q^+(T) - Q^-(T) = \frac{\partial G_1}{\partial T},$$

where $Q^+(T) = \mathrm{e}^{-2/T} - c\,\mathrm{e}^{-2/\gamma T}$, $Q^-(T) = \alpha(T - \tilde{T})$, $\gamma = E/L$, $\alpha = \bar{\alpha} T_\mathrm{M}$, $T_\mathrm{M} = E/(2R)$ is the reference temperature, and $G_1 = \int g_1(T)\mathrm{d}T$ is the potential function. In a similar way as was done before, consider the conditions enabling the determination of TCPs of the potential function $G_1(T)$:

$$\mathrm{e}^{-2/T} - c\,\mathrm{e}^{-2/\gamma T} = \alpha(T - \tilde{T}),$$

$$\frac{2}{T^2}\mathrm{e}^{-2/T} - \frac{2c}{\gamma T^2}\mathrm{e}^{-2/\gamma T} = \alpha,$$

$$\left(\frac{1}{T} - 1\right)\mathrm{e}^{-2/T} - \frac{c}{\gamma}\left(\frac{1}{\gamma T} - 1\right)\mathrm{e}^{-2/\gamma T} = 0.$$

The solution of this system of equations determining the TCP (T_*, α_*, c_*) location in parameter space (T, α, c) is equivalent to the solution of the equations

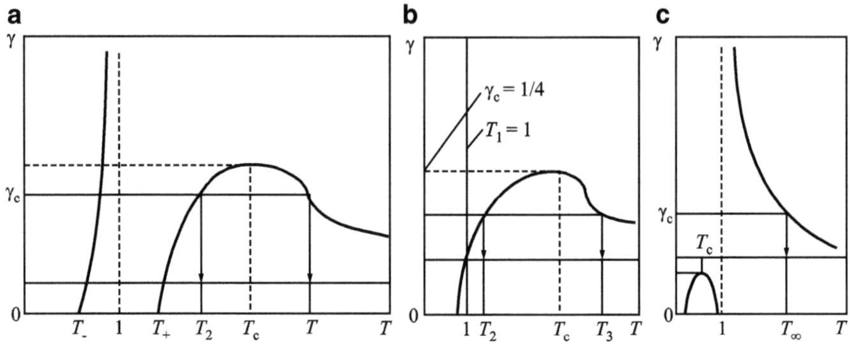

Fig. 6.7. Diagram of zeros for determining the triple-degenerate critical point (TCP) location: **a** $\tilde{T} < 0.5$, **b** $\tilde{T} = 0.5$, and **c** $\tilde{T} > 0.5$

$$\gamma T^2(T-1) = T^2 - 2T + 2\tilde{T}$$

or

$$P(T) \equiv \gamma T^3 - (\gamma+1)T^2 + 2T - 2\tilde{T} = 0. \tag{6.15}$$

It is obvious that the roots of these equations depend parametrically on γ and $\tilde{T}$. At $\tilde{T} < 0.5$, the following representation is valid:

$$\gamma = \frac{(T-T_+)(T-T_-)}{T^2(T-1)}, \tag{6.16}$$

where $T_\pm = 1 \pm \sqrt{1-2\tilde{T}}$. Let us analyze qualitatively the solutions of (6.15) using the diagram of roots $T_* = T_*(\gamma, \tilde{T})$ shown in Fig. 6.7. As a result, the following assertion can be formulated.

Assertion 2: The number and the order of the sequence of the roots of (6.15) are determined depending on γ and $\tilde{T}$ as follows:

1. $\tilde{T} < 0.5$.
   - If $0 < \gamma \le \gamma_c$, there exist three real roots $T_* = T_{1,2,3}(T_- < T_1 < 1 < T_2 < T_c < T_3)$.
   - If $\gamma > \gamma_c$, there exists one real root $T_*(T_- < T_* < 1)$ and two complex conjugate roots.
2. $\tilde{T} = 0.5$.
   - If $0 < \gamma \le \gamma_c = 0.25$, there exist three real roots $T_* = T_{1,2,3}(T_1 = 1 < T_2 < T_c < T_3)$ and $T_2 = T_3$ for $\gamma = \gamma_c$.
   - If $\gamma > \gamma_c$, there exists one real root $T_{*1} = 1$ and two complex conjugate roots.
3. $\tilde{T} > 0.5$. There exists one real root $T_1 > 1$ for all $\gamma > 0$.

The critical parameter here is $\gamma_c = \max\ \gamma(T)|_{T>1} = \gamma_c(\tilde{T})$ at $\tilde{T} \le 0.5$. The function $\gamma_c(\tilde{T})$ is determined parametrically: $\gamma_c = \gamma(T_c)$ as $\tilde{T} = H(T_c)$

owing to the equality $\left.\frac{\mathrm{d}\gamma(T)}{\mathrm{d}T}\right|_{T=T_c} = 0$. It can be readily shown that at realistic $\tilde{T}$ values (which do not exceed 0.2–0.3), the function $\gamma_c(\tilde{T})$ is always less than unity. The proof of Assertion 2 follows from the elementary construction of the function $\gamma(T)$ determined by (6.16) and from its continuity at $T > 1$.

For $\tilde{T} \leq 0.5$, (6.15) has the following approximate solutions:

$$\text{at } \gamma \ll 1,\ T_{1,2} = T_{\mp} + \gamma \frac{T_{\mp}^2}{2} + \gamma^2 \frac{T_{\mp}^3}{2}\left(1 - \frac{3}{2}T_{\mp}\right) + O(\gamma^3),$$

$$T_3 = \frac{\gamma + 1}{\gamma} - (T_{*1} + T_{*2}) + O(\gamma^3) \text{ at } \gamma \gg 1,$$

$$T_* = 1 + \gamma^{-1}(2\tilde{T} - 1) - 2\gamma^{-2}(2\tilde{T} - 1)^2 + O(\gamma^{-3}).$$

In a general case, when γ is finite, the solution of (6.15) can be found numerically. The results are presented in Fig. 6.8 in the form of a diagram of roots $T_* = T_*(\tilde{T}, \gamma)$ for several values of γ.

Note that in cases of practical importance with $\gamma_c < 1 < \gamma$, the CM possesses a unique TCP similar to the CM in the ignition model, which does not take evaporation into account. The existence of a CM with three TCPs is a new feature in the case when the activation energy of metal evaporation is much larger than that of the oxidation process, i.e., $L \gg E$. This implies the possibility of nontrivial scenarios of particle ignition for a given relation between the governing parameters of the system.

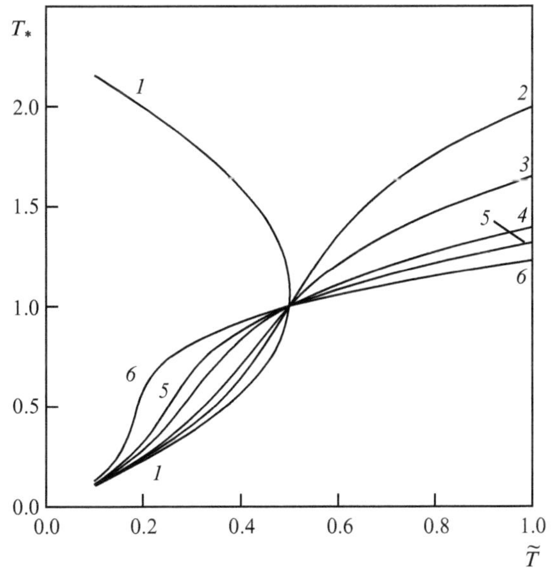

Fig. 6.8. The diagram of roots $T_* = T_*(\tilde{T}, \gamma)$ for determining the TCP location: $\gamma = 0.1$ (*1*), 0.5 (*2*), 0.8 (*3*), 1.5 (*4*), 2.0 (*5*), and 3.0 (*6*) (in the case $\gamma = 0.1$ the branch of the third, largest root is not shown)

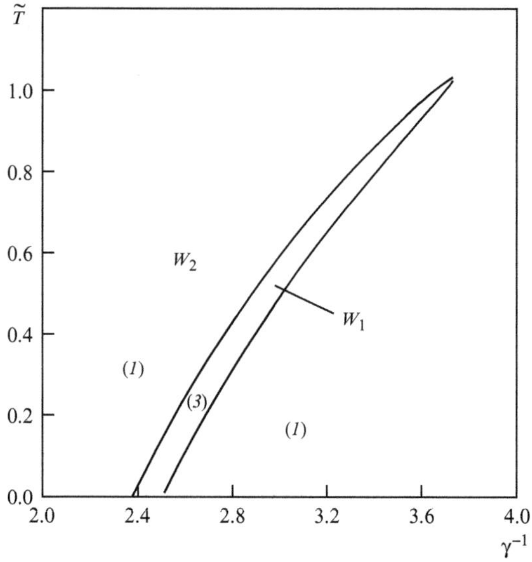

Fig. 6.9. Partition of the parameter plane (γ^{-1}, $\tilde{T}$) into the regions with three (W_1) and one (W_2) TCPs; the *numbers* in *parentheses* show the number of TCPs

Figure 6.9 presents the bifurcation diagram for the roots of (6.15) in the $(\gamma^{-1}, \tilde{T})$-plane. The diagram separates the regions of parameters with different numbers of TCPs and is obtained using curves $\tilde{T} = \tilde{T}_{1,2}(\gamma)$ along which the discriminator of the given cubic equation vanishes. Region W_1 is a set with three TCPs, and region W_2 is a set with one TCP. As a matter of fact, Fig. 6.9 is the Semenov diagram with the curves of total heat supply $Q^+(T)$ and heat removal $Q^-(T)$. The equation $Q^+(T) = 0$ possesses the following roots:

$$T = 0 \text{ and } T = T^0 = \frac{2(\gamma - 1)}{\gamma \ln c^{-1}},$$

where T^0 has the meaning of stationary adiabatic temperature in the system (the temperature up to which the particle is heated in a thermally insulated volume with temperature $\tilde{T}$). It is obvious that depending on the sign of T^0 and the relation between $\tilde{T}$ and T^0, different sorts of tangency of curves $Q^+(T)$ and $Q^-(T)$ can exist (we restrict ourselves to the condition $c < 1$). The analysis of the results can be summarized in the following assertion.

Assertion 3:

1. $T^0 > 0$. Then,
   - If $\tilde{T} < T^0$, there exists one point of tangency (T_K, α_K), and
     - At $\alpha < \alpha_K$, there exist three equilibrium positions: $(0, \tilde{T})$, (T^0, T_K), and (T_K, ∞).
     - At $\alpha > \alpha_K$, there exists one equilibrium position: $(0, \tilde{T})$.

- If $\tilde{T} > T^0$, there exist three points of tangency (T_{K_i}, α_{K_i}), $i = 1, 2, 3$ numbered in the order of increasing α_K, and
  - At $\alpha < \alpha_{K_1}$, there exist three equilibrium positions: $(0, T_{K_1})$, (T_{K_1}, T^0), and (T_{K_3}, ∞).
  - At $\alpha_{K_1} < \alpha < \alpha_{K_2}$, there exists one equilibrium position lying at (T_{K_3}, ∞).
  - At $\alpha_{K_2} < \alpha < \alpha_{K_3}$, there exist three equilibrium positions: $(\tilde{T}, T_{K_2})$, (T_{K_2}, T_{K_3}), and (T_{K_3}, ∞).
  - At $\alpha > \alpha_{K_3}$, there exists one equilibrium position: $(\tilde{T}, T_{K_2})$.

2. $T^0 < 0$. There exist two points of tangency (T_{K_i}, α_{K_i}), $i = 1, 2$ and
   - At $\alpha < \alpha_{K_1}$, there exists a single equilibrium position: (T_{K_2}, ∞).
   - At $\alpha_{K_1} < \alpha < \alpha_{K_2}$, there exist three equilibrium positions: $(\tilde{T}, T_{K_1})$, (T_{K_1}, T_{K_2}), and (T_{K_2}, ∞).
   - At $\alpha > \alpha_{K_2}$, there exists a single equilibrium position: $(\tilde{T}, T_{K_1})$.

The proof of Assertion 3 can be readily seen from curves $Q^{\pm}(T)$ in Fig. 6.10.

Some important properties of the source function $Q^+(T)$ are listed below:

1. $Q^+(T) \to 0$ as $T \to 0$.
2. $Q^+(T) \to (1 - c)$ as $T \to \infty$.
3. If $0 < c < 1$, $\gamma > 1$ or $c > 1$, $\gamma < 1$, then $T^0 > 0$; if $0 < c < 1$, $\gamma < 1$ or $c > 1$, $\gamma > 1$, then $T^0 < 0$.

Turn now to the analysis of particle temperature evolution. For this purpose, consider the typical sections c = const of the CM specified by the equation $Q^+(T) - Q^-(T) = 0$. Upon determining the TCP coordinates, one can construct the images of the fold curves in the (α, c)-plane. They are the projections of the corresponding DCPs onto the equilibrium surfaces (CMs) in the

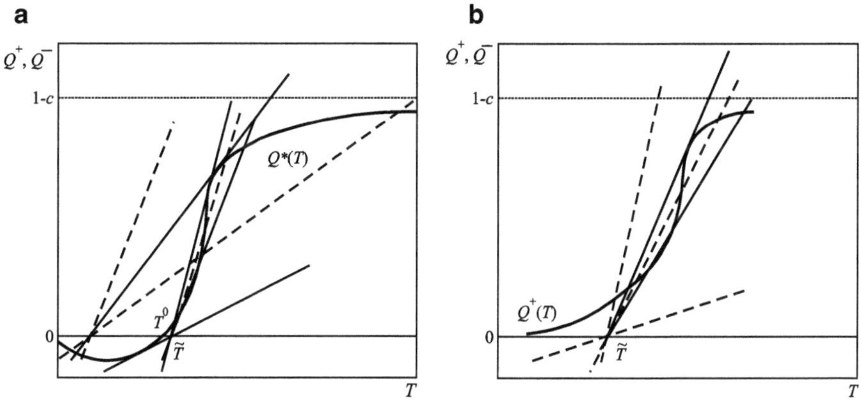

Fig. 6.10. Shapes of Semenov diagrams depending on the stationary adiabatic temperature T^0 (at $c < 1$): **a** $T^0 > 0$, **b** $T^0 < 0$; *dashed curves* show different locations of the heat removal curve $Q^-(T)$, *solid curves* show the limiting (tangent) locations of this curve

(T, α, c)-space. Using the properties of the function $Q^+(T)$ and the estimates for $T_\pm$, one can construct a qualitative shape of the CM in sections $c = \text{const}$ as shown in Fig. 6.11, where typical regions in the phase plane (T, α) are determined. One can observe, for example, the following variants of temperature histories:

- At $(\alpha, T_0) \in D'_1$, a regime with explosion-free particle heating which tends to equilibrium along the lower branch $T'_{\mathrm{I}}(\alpha) < \tilde{T}$
- At $(\alpha, T_0) \in D'_2$, an ignition regime with temperature passage to the upper stable branch $T'_{\mathrm{III}}(\alpha)$ of stationary states
- At $(\alpha, T_0) \in D'_3$, a regime of extinction with temperature stabilization on the branch $T'_{\mathrm{I}}(\alpha)$

The remaining variants can be analyzed in a similar way.

6.2.2.8 Inverse Problem of Particle Ignition

The analysis of the manifold of catastrophes (ignitions) conducted makes it possible to categorize kinetic laws of metal particle oxidation in air with regard to metal vaporization. The equations governing the preignition state of a magnesium particle can be written in the form

$$k\mathrm{e}^{-E'/T} - v\mathrm{e}^{-L'/T} = \alpha_0(T - \tilde{T}),\; k\frac{E'}{T^2}\mathrm{e}^{-E'/T} - v\frac{L'}{T^2}\mathrm{e}^{-L'/T} = \alpha_0, \qquad (6.17)$$

where $\alpha_0 = \lambda\, Nu/c(2r_{\mathrm{s}}\rho_{\mathrm{s}}q)$, E' and L' are the activation energy and evaporation heat normalized by RT_{M}, and $T_{\mathrm{M}} = 300\,\mathrm{K}$ is the reference temperature. Following the data of [27], $L' = 53.333$. Then (6.17) can be written for two arbitrary points $(r_{\mathrm{s}j}, T_j)$, $j = 1, 2$ at the experimental curve describing the dependence of the limiting ignition temperature in air on the particle radius [60]. This allows one to come to a closed system of transcendental equations for determining the unknown quantities E' and k (if they exist) as well as particle temperatures at the ignition limit. In so doing, the following values of the kinetic parameters were derived for $v = 0.15\,\mathrm{m\,s^{-1}}$: $E' = 40.315$ ($E/R = 12{,}094\,\mathrm{K}$), $k = 0.169\,\mathrm{m\,s^{-1}}$ for fine particles (with the radius ranging from 15 to 60 μm), $E' = 96.452$ ($E/R = 28{,}936\mathrm{K}$), and $k = 6.855 \times 10^5\,\mathrm{m\,s^{-1}}$ for large particles (with the radius ranging from 300 to 600 μm).

It is interesting to compare the ignition delay times obtained on the basis of these kinetic parameters with those obtained with the model without metal vaporization taken into account. As is seen from Table 6.2, the differences are insignificant for fine particles. For large particles, the difference does not exceed 11%.

6.2.2.9 Specific Features of Aluminum Particle Ignition

Ignition of aluminum particles differs qualitatively from that of magnesium particles. At normal atmospheric conditions, aluminum particles are known

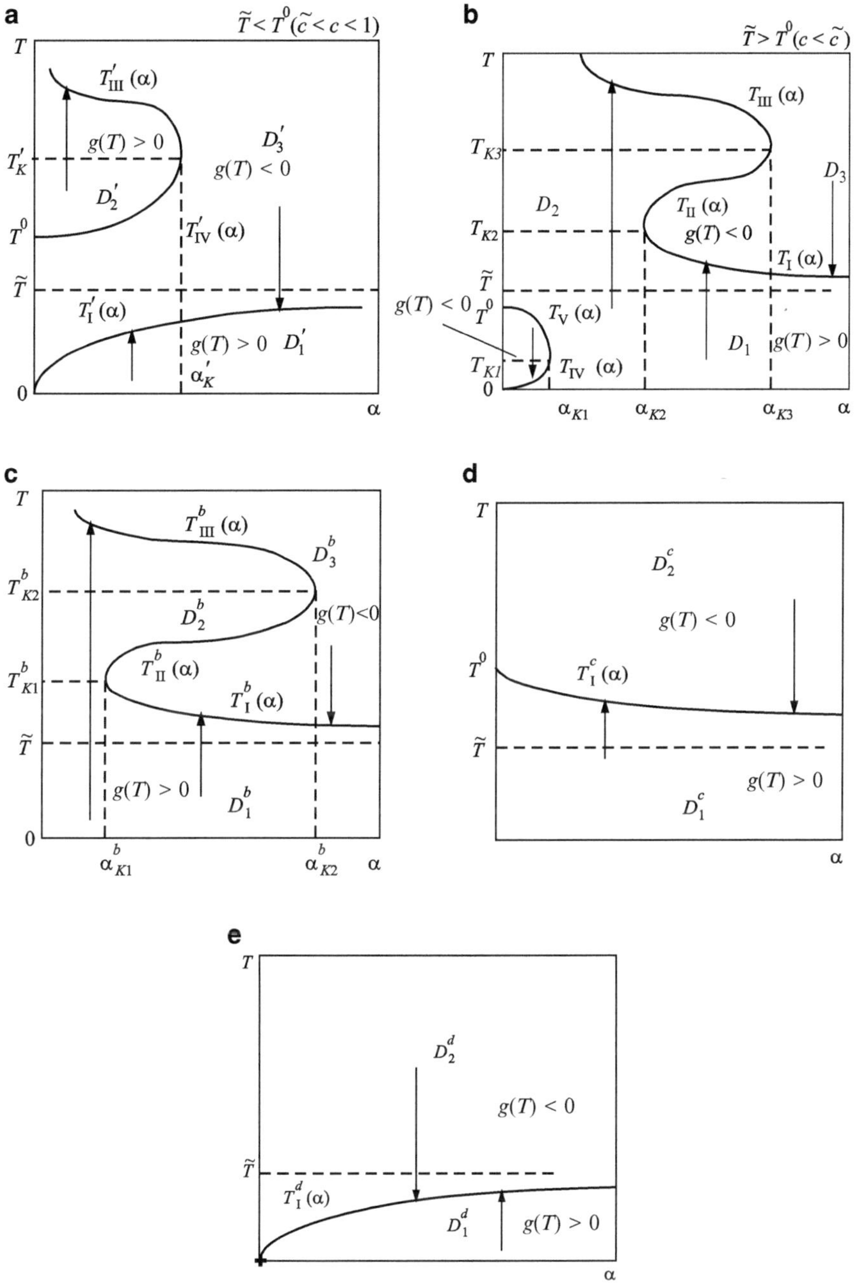

Fig. 6.11. Structure of the ignitions manifold (typical sections c = const): **a**, **b** $0 < c < 1$, $\gamma > 1$; **c** $0 < c < 1$, $\gamma < 1$; **d** $c > 1$, $\gamma < 1$; **e** $c > 1$, $\gamma > 1$; *arrows* indicate possible variants of particle heating/cooling

Table 6.2. Comparison of ignition delay times of a magnesium particle (ms) at $\tilde{T} = 1,538\,\mathrm{K}$ predicted by two models

Particle radius (μm)	15	22	30	60	300	400	500	600
Model (1.2.2)	22.0	39.6	64.0	200.0	2,200	3,800	6,000	8,500
Model (1.2.3)	22.0	39.2	64.1	202.3	1,990	3,460	5,320	7,560

to be covered by an oxide film possessing protective properties which complicate oxygen diffusion to the pure metal. Therefore, the rate of high-temperature oxidation of aluminum particles depends significantly on the oxide film thickness.

Various oxidation laws of aluminum particles have been observed experimentally. In [94, 95], the growth of the oxide film was described by the exponential law

$$\frac{\mathrm{d}h}{\mathrm{d}t} = KC_{\mathrm{ox}}^{n_{\mathrm{ox}}} \exp\left(-\frac{E}{RT}\right) \exp\left(-\frac{h}{h_0}\right),$$

where C_{ox} is the oxidizer concentration near the particle surface and n_{ox} is the reaction order in relation to the oxidizer. Another frequently used equation of oxidation kinetics has the form

$$\frac{\mathrm{d}h}{\mathrm{d}t} = \frac{KC_{\mathrm{ox}}^{n_{\mathrm{ox}}}}{h^n} \exp\left(-\frac{E}{RT}\right).$$

The power exponent n in the last equation determines the dependence of the aluminum oxidation rate on the oxide film thickness. The case with $n = 0$ was considered in Sect. 6.2.2.7 for magnesium particles.

In [96, 97], the oxidation rate of aluminum was assumed to be determined by the kinetics of the heterogeneous reaction and to be independent of the oxide film thickness. It was implied that the heterogeneous reaction occurred only on the portion of the particle surface not covered by oxide crystals. It was assumed in [97] that $n = 1$. This assumption resulted in a parabolic oxidation law, which is valid when the oxidation rate is limited by oxygen diffusion through the oxide film. Thus, the particle ignition model in [97] was based on the heat balance equation supplemented by the equation for the fraction of the particle surface free from the crystalline oxide. The parabolic equation of the oxide-film growth at the aluminum particle surface was also considered in [72], where heat sinks due to particle vaporization and melting as well as radiation heat losses were taken into account in the heat balance equation. Analysis of numerical calculations allowed Medvedev et al. [72] to put forward a particular mechanism of aluminum particle ignition and determine the effect of various initial parameters on the ignition delay.

In the papers mentioned above, various approaches were used to determine the kinetic constants of the empirical ignition law. Among them are the methods of elementary catastrophe theory, the fitting of model predictions with the experimental dependencies of the ignition delay on the shock wave

Mach number [72], and some heuristic conditions relevant to the oxide film melting point [98]. In most papers, the model predictions were compared with the measured ignition delays in terms of a single parameter, namely, the ambient gas temperature. However, it is well known from experiments [99–102] that the ignition delay is affected not only by the ambient gas temperature but also by the particle size and oxidizer concentration in the ambient gas. It is therefore important to have a model capable of predicting aluminum particle ignition delay as a function of different governing parameters of the problem.

6.2.2.10 Point Model of Aluminum Particle Ignition

Consider a spherical aluminum particle of diameter d_s which is suddenly placed in a quiescent gas with temperature $\tilde{T}$ behind a reflected shock wave [103]. Thermal interaction of the particle with hot gas results in a heterogeneous reaction of low-temperature oxidation in a thin (as compared with the particle radius) layer on the particle surface. As a consequence, the mean particle temperature T increases and the particle can be ignited. The heat balance equation for the aluminum particle is expressed in the form

$$mc_{\mathrm{s}}\frac{\mathrm{d}T}{\mathrm{d}t} = -Sk_{\mathrm{h}}(T-\tilde{T}) + Sq\rho_{\mathrm{ox}}\frac{\mathrm{d}h}{\mathrm{d}t}.$$

For the aluminum oxidation kinetics, the following parabolic law is used:

$$\frac{\mathrm{d}h}{\mathrm{d}t} = \frac{KC_{\mathrm{ox}}}{h}\exp\left(-\frac{E}{RT}\right).$$

The initial conditions for these equations are

$$t=0: \quad T=T_0,\ h=h_0,$$

where ρ_{ox} is the aluminum oxide (Al_2O_3) density. Thus, the problem of aluminum particle ignition in a quiescent oxidizing gas is reduced to determining the functions $T(t)$ and $h(t)$ in the region $[0, t_{\mathrm{ign}}]$, where they satisfy the above equations and initial conditions.

There is still a need to determine the unknown parameters K and E. In the literature, different values of these parameters are reported. For example, $E = 17\,\mathrm{kcal\,mol^{-1}}$, $K = 1.9\times10^{-9}\,\mathrm{m^2\,s^{-1}}$, and $n = 1$ are reported in [101], while $E = 20\,\mathrm{kcal\,mol^{-1}}$, $K = 18\times10^{3}\,\mathrm{m^2\,s^{-1}}$, and $n = 0$ are reported in [96]. In general, one cannot expect the values of K and E to be constant over the wide range of governing parameters as the model under consideration oversimplifies the problem. Therefore, to use this simple model to explain the experimental dependencies of the ignition delay t_{ign} and the minimal temperature of the gas required for particle ignition $\tilde{T}_{\mathrm{lim}}$ on the particle size, it is worth considering K as a function of the ambient temperature and particle radius.

The approach suggested herein is as follows. One can determine the value of the preexponential factor K for any two experimental points $(t_{\mathrm{ign1}}, \tilde{T}_1)$ and $(t_{\mathrm{ign2}}, \tilde{T}_2)$ reported, e.g., in [101]. Then $K(\tilde{T})$ can be approximated by

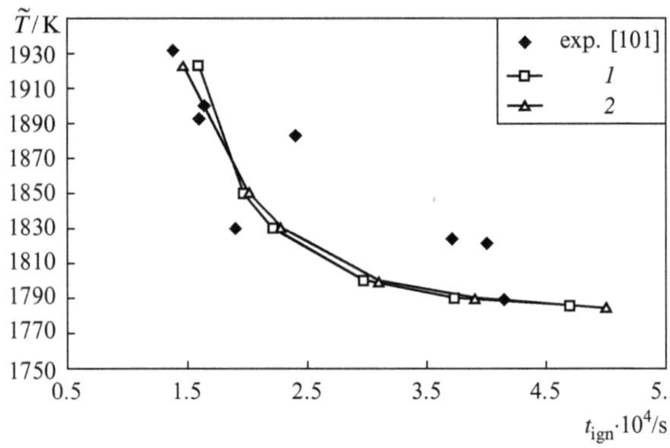

Fig. 6.12. Measured [101] and predicted [103] dependencies of the aluminum particle ($d_s = 6\,\mu$m) ignition delay on the ambient gas (oxygen) temperature. Curves *1* and *2* refer to $K = (a\tilde{T} + b)$ and $K = (a\tilde{T} + b)[\tilde{T} - \tilde{T}_{\rm lim}(d_{\rm s}, C_{\rm ox})]$, respectively. *Solid symbols* correspond to experiments [101]

$$K(\tilde{T}) = \begin{cases} f(\tilde{T}), & \tilde{T} > \tilde{T}_* \\ 0, & \tilde{T} \le \tilde{T}_* \end{cases},$$

where $f(\tilde{T})$ is a linear function, and $\tilde{T}_*$ is some reference temperature. Physically, the reference temperature $\tilde{T}_*$ is the minimum temperature at which particle ignition is still possible. According to the above relationship, at $\tilde{T} < \tilde{T}_*$ the particle temperature rises solely owing to heat transfer from the ambient gas and tends to $\tilde{T}_*$ in the limit.

The results of calculations for aluminum particle ($d_{\rm s} = 6\mu$m) ignition in oxygen are shown in Fig. 6.12 as curve 1 in the plot $\tilde{T}(t_{\rm ign})$. The predictions are compared with the experimental data [101] (solid symbols in Fig. 6.12). A satisfactory agreement of predicted and measured results for intermediate oxygen temperatures exists.

According to experiments [99,100,102], the minimal gas temperature $\tilde{T}_{\rm lim}$ is also not constant and depends on the particle diameter $d_{\rm s}$, oxygen concentration in the ambient gas $C_{\rm ox}$, and ambient gas temperature $\tilde{T}$. Thus, one can represent the reference temperature $\tilde{T}_*$ as $\tilde{T}_* = \tilde{T}_{\rm lim}(d_{\rm s}, C_{\rm ox})$. This dependence can be obtained, e.g., by approximating the experimental data of [102].

Let us study the influence of the particle diameter on the preexponential factor K, using the experimental dependencies of the ignition delay on the particle diameter [99,100]. Note that the results reported in [99,100] are close to each other. As a result of parametric calculations and comparison with the experiments, a certain value of K can be derived, e.g., for $\tilde{T} = 2{,}510$ K. Now, the preexponential factor can be finally represented as

$$K = (a\tilde{T} + b)[\tilde{T} - T_{\rm lim}(d_{\rm s}, C_{\rm ox})].$$

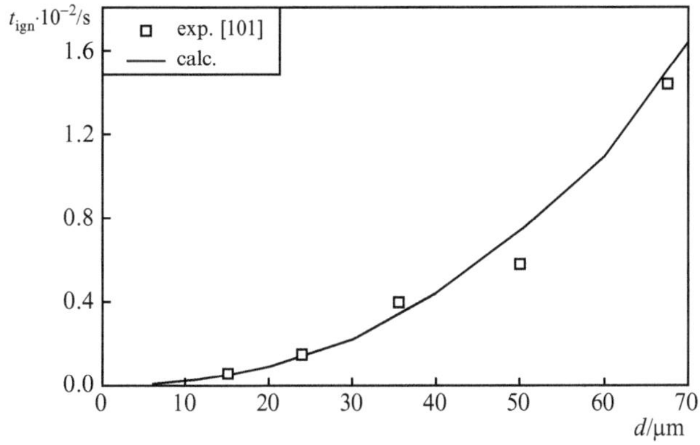

Fig. 6.13. Measured [101] and predicted [103] dependencies of the ignition delay on the aluminum particle diameter

At $\tilde{T} = 2,510\,\mathrm{K}$, the value of K predicted by this empirical formula is the same as that obtained using the experimental data of [100]. The activation energy E entering the governing equations of the model is taken to be $17\,\mathrm{kcal\,mol^{-1}}$ in accordance with [104].

To solve the governing equations, an implicit multistep method [105] was applied. The following values of physical variables were used in the calculations. For aluminum, $\rho = 2,689\ \mathrm{kg\,m^{-3}}$, $c_s = 1,010\ \mathrm{J\,(kg\,K^{-1})}$, and $\rho_{ox} = 3,970\ \mathrm{kg\,m^{-3}}$. For oxygen, $\lambda = 2.4 \times 10^{-2}(\tilde{T}/\tilde{T}_0)^{0.75}\ \mathrm{J\,(m\,s\,K)^{-1}}$, $Nu = 2$, and $q = 35.6 \times 10^6\ \mathrm{J\,kg^{-1}}$.

The results of the calculations are shown in Figs. 6.12 and 6.13. Curves 1 and 2 in Fig. 6.12 are slightly different but still lie within the scatter of experimental data. In general, both figures indicate that the modified ignition model of aluminum particles provides satisfactory agreement with the experimental data on ignition delay as a function of particle size and ignition delay as a function of ambient gas temperature.

6.2.3 Dynamic Conditions

6.2.3.1 Ignition of Metal Particles Behind Reflected Shock Waves

The problem of ignition of fine metal particles in static and dynamic conditions behind planar incident and reflected shock waves was considered theoretically and experimentally in [60, 72, 74–77], and behind detonation and explosion waves in [106, 107]. It was shown in [72, 74] that the consideration of particle motion and low-temperature metal oxidation allows one to reproduce available experimental data on the dependence of the ignition delay time t_{ign} on the shock wave Mach number M_0. To explain the experimental data on ignition of magnesium particles behind reflected shock waves in a shock tube, a hypothesis on cracking of the oxide layer prior to its melting was put forward in [108, 109]. Below we provide another explanation for the effects observed in [108,109] within the framework of a model similar to that in [72, 74], which takes into account particle dynamics.

Consider a gas–particle suspension of magnesium particles filling the half-space bounded by a rigid wall. After a planar shock wave passes through it, the particles start moving and heating up in the shock-induced flow. After the shock wave is reflected from the rigid wall, the gas temperature increases again, while the gas velocity vanishes, i.e., $u_{\mathrm{g}} = 0$. Thus, particles near the wall are subjected to variable dynamic and thermodynamic parameters of the ambient gas. Let the volume fraction of particles be sufficiently small. Then the ignition process can be described by a model similar to that in [72, 74]:

$$mc_{\mathrm{s}}\frac{\mathrm{d}T}{\mathrm{d}t} = 2\pi r_{\mathrm{s}}\lambda Nu(\tilde{T} - T) + Sq\rho_{\mathrm{ox}}\left[k_0 C_{\mathrm{ox}}\exp(-E/RT) - v\exp(-L'/RT)\right], \tag{6.18}$$

$$m\frac{\mathrm{d}u}{\mathrm{d}t} = \frac{1}{2}AC_{\mathrm{D}}\rho U\left|U\right|,$$

where u is the particle velocity, $A = \pi r_{\mathrm{s}}^2$ is the particle cross-sectional area, $U = u_{\mathrm{g}} - u$ is the relative gas–particle velocity, and C_{D} is the particle drag coefficient [74, 110].

The system of (6.18) is supplemented with the initial data

$$t = 0: \quad u = 0, \quad T = T_0, \tag{6.19}$$

which reflect the fact that the velocity and temperature of the particles remain unchanged across the incident shock wave.

The main specific feature of the problem expressed in (6.18) and (6.19) is the presence of multiple stationary states at the variation of a characteristic bifurcation parameter. Let us illustrate this implication of (6.18) and (6.19) for the case when particle vaporization can be neglected and the particle velocity is constant. In this case, there exists a bifurcation parameter $\alpha = c_{\mathrm{s}}\tau_3/(q\tau_2)$, where $\tau_3 = r_{\mathrm{s}}/(3k_0c_{\mathrm{ox}})$ and $\tau_2 = 2c_{\mathrm{s}}\rho_{\mathrm{s}}r_{\mathrm{s}}^2/(3\lambda\, Nu)$. The turning points of the zero curve at the (T, α)-plane were found for the following constants: $T_- = 1{,}158\,\mathrm{K}$, $T_+ = 26{,}960\,\mathrm{K}$, $\alpha_- = 5.946\times10^{-13}$, and $\alpha_+ = 1.363\times10^{-5}$. It turned out that at $\alpha > \alpha_-$, e.g., $\alpha = (1+0.01)\alpha_-$, there is no ignition and the particle temperature tends to its final equilibrium state. At $\alpha < \alpha_-$, e.g., $\alpha = (1 - 0.01)\alpha_-$, particle ignition occurs followed by rapid temperature growth. A comparison with data in [77] for the case of zero particle velocity shows the proximity of predicted and experimental data in terms of the dependence of the ignition delay time on the particle radius.

The results of experimental studies on ignition of gas–particle mixtures of magnesium powders in pure oxygen behind a reflected shock wave were presented in [108] (particles in a shock tube were located initially at a distance of 1 cm from the end wall). For particles with diameter $d_{\mathrm{s0}} = 2r_{\mathrm{s0}} = 90\,\mu\mathrm{m}$, the values of t_{ign} are presented for different initial pressures p_0 and shock wave Mach numbers M_0. In variant I, $t_{\mathrm{ign}} = 1.4 \pm 0.1$ ms at $M_0 = 4.2$ and

Table 6.3. Results of experiments and computations for a monodisperse mixture

Variant	M_0	p_0 (bar)	$\tilde{T}$ (K)	$\tilde{T}_1$ (K)	D (m s^{-1})	D_1 (m s^{-1})	u_g (m s^{-1})	t_{ign} (ms)	
								Experiment	Computation
I	4.2	0.2	1,265	2,470	1,351	484.5	1,071	1.4 ± 0.1	0.65 (1.3)
II	3.9	0.3	1,129	2,162	1,255	457.6	984.6	2 ± 0.2	1 (2.1)
III	3.4	0.3	924.8	1,698	1,094	414.4	839.3	–	—

There is no ignition in variant III. The values in *parentheses* are t_{ign} values at refined parameters behind the shock wave

Table 6.4. Results of experiments and computations for the polydisperse mixture

Run	d_{s0} (μm)	Experiment		Computation run 1		Computation run 2	
		$\tilde{T}_1$ (K)	t_{ign} (ms)	t_{ign} (ms)	d_{s0} (μm)	t_{ign} (ms)	d_{s0} (μm)
1				0.04	20	0.014	4
	1–40	2,173	0.06	0.12	40	0.070	20
						0.210	40
2		1,470	0.11	0.065	20	0.030	4
				0.19	40	0.150	20
						0.480	40
3				0.23	70	0.300	60
		2,325	0.26	0.44	100	0.600	80
	63–100					0.800	100
4				0.47	70	0.900	60
		1,515	0.6	0.90	100	1.500	80
						2.100	100

$p_0 = 0.2$ bar; in variant II, $t_{ign} = 2.0 \pm 0.2$ ms at $M_0 = 3.9$ and $p_0 = 0.3$ bar; in variant III, the particle did not ignite at $M_0 = 3.4$ and $p_0 = 0.3$ bar.

Table 6.3 presents the values of the gas temperature behind the incident shock wave, $\tilde{T}$, and the reflected shock wave, $\tilde{T}_1$, velocities D and D_1 of the waves, and the gas velocity behind the incident shock wave for the above values of M_0 and p_0. The numerical results were obtained using two models (see below) and are presented for each of variants I–III. Without consideration of particle velocity variation, no ignition was observed in the calculations in all three variants and the particle temperature attained its stationary value. The model accounting for particle motion yields the following results: $t_{ign} = 0.65$ and 1.0 ms in variants I and II, respectively, while in variant III there was no ignition.

In [108], the experimental data on ignition delay times for polydisperse magnesium powders were also presented. We have performed numerical calculations of the ignition delay time for each of the variants for some selected values of the particle diameter. The results are presented in Table 6.4 (run 1). The chosen values of a characteristic particle diameter in a polydisperse gas–particle mixture can be seen to yield the upper and lower limits for t_{ign}. The following values of thermophysical parameters were used: $\mu = 2\times10^5\,\text{kg}\,(\text{m}\,\text{s})^{-1}$ and $c_s = 1.2\times10^3\,\text{J}\,(\text{kg}\,\text{K})^{-1}$ (run 1). However, in view of elevated temperatures and pressures behind the incident and the reflected shock wave, there is a need to correct these values. The computations of the ignition delay time with corrected thermophysical data are presented in Table 6.4 (run 2) and in Table 6.3 (numbers in parentheses). The thermophysical data for high-pressure and high-temperature conditions used for the computations reported in Tables 6.3 and 6.4 were taken from [111]. It turns out that taking the dependence of the thermophysical parameters on pressure and temperature into account leads to a better agreement between measured and calculated data.

Thus, the above analyses indicate that the results of the physical experiments reported in [108] on ignition of magnesium particles behind a reflected shock wave in the vicinity of the shock-tube end wall are described satisfactorily by the model accounting for particle motion, low-temperature metal oxidation, and the dependence of the thermophysical properties of the system on temperature and pressure.

The effect of a dense particle cloud on the particle ignition behavior was studied theoretically by Fedorov [112] using the example of a magnesium particle cloud in air. The model included a heterogeneous chemical reaction on the particle surface and was validated against the experimental data on the minimal air temperature required for particle ignition as a function of particle radius. The predicted results are compared with the experimental data in Fig. 6.14.

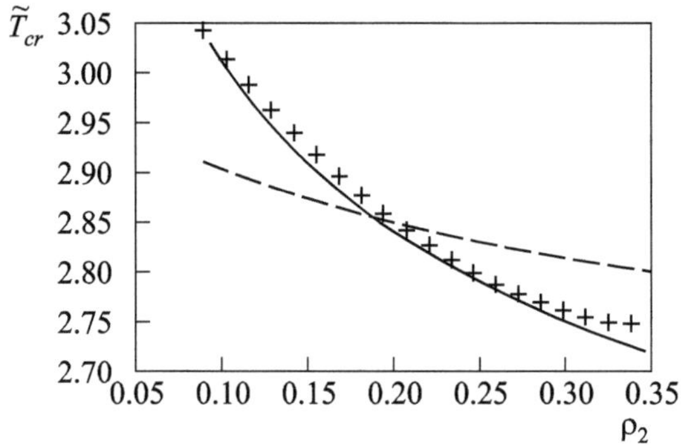

Fig. 6.14. The minimal air temperature $\tilde{T}_{\text{lim}}$ required for ignition of a magnesium particle cloud as a function of cloud density. The particle radius is 5 μm; *crosses* denote experimental data of Matsko et al. [170]; the *dashed curve* corresponds to the calculation with the reaction rate constants derived for oxidation of a single magnesium particle in air; the *solid curve* corresponds to the calculations of magnesium particle cloud ignition by Fedorov [112]

6.2.3.2 Ignition of Metal Particles Behind Incident Shock Waves

Consider now the approaches to mathematical modeling of metal particle ignition in a high-temperature flow behind an incident shock wave. A review of early theoretical and experimental results on particle ignition in steady-state and dynamic conditions is presented elsewhere [87]. Subsequent bibliographic references are presented in [36,54,112–119]. They provide the entire hierarchy of mathematical models of ignition waves in reacting particle suspensions in one- and two-velocity approximations. In the following discussion, two-phase flows with very small concentrations of solid particles are considered. Descriptions of ignition phenomena at finite particle concentrations, when it is necessary to invoke the theory of interpenetrating continua, can be found in the papers cited above.

In the region behind the propagating shock wave, particles accelerate up to the flow velocity and heat up owing to heat exchange with the postshock gas and the oxidation reaction. At a very small particle volume fraction, one can neglect the interaction between particles and the influence of particles on the gas flow. It is implied that the chemical reaction of high-temperature metal oxidation may be activated in the suspension by an incident shock wave. The approximation of “isolated spherical particles” [110] is true and therefore can be adopted. The ignition kinetics (the kinetics of high-temperature oxidation) is assumed to follow the Arrhenius law depending on the oxide film thickness, with the particle radius variation neglected.

The equations governing the flow of such a mixture can be written as

$$\begin{aligned} mc_{\mathrm{s}}\frac{\mathrm{d}T}{\mathrm{d}t} &= Q_{\mathrm{chem}} - Q_{\mathrm{conv}} - Q_{\mathrm{phase}} - Q_{\mathrm{rad}} - Q_{int}, \\ m\frac{\mathrm{d}u}{\mathrm{d}t} &= -F_{\mathrm{S}} - F_{\mathrm{m}} - F_{\mathrm{B}}. \end{aligned} \tag{6.20}$$

The source terms in the first expression in (6.20) correspond to the heat fluxes due to:

1. Heterogeneous chemical reaction:

$$Q_{\mathrm{chem}} = Sq\rho_{\mathrm{ox}}\frac{\mathrm{d}h}{\mathrm{d}t}.$$

2. Convective heat transfer between particles and gas:

$$Q_{\mathrm{conv}} = S\lambda\, Nu(T - \tilde{T}).$$

3. Particle vaporization:

$$Q_{\mathrm{phase}} = \frac{\lambda}{\rho c_p}\frac{S}{Le}\left(\frac{p_*}{p}\right)\exp\left(-\frac{L}{RT}\right).$$

4. Radiation heat loss:

$$Q_{\mathrm{rad}} = S\varepsilon\sigma(T^4 - \tilde{T}^4).$$

5. Particle heating and melting:

$$Q_{\mathrm{int}} = 4\pi r_{\mathrm{s}} \lambda_{\mathrm{s}} (T - T_0) \exp\left(-\frac{t}{t_{\mathrm{h}}}\right).$$

The source terms in the second expression in (6.20) correspond to the forces acting on the particle:

1. Stokes force:

$$F_{\mathrm{s}} = AC_{\mathrm{D}}\rho \frac{(u - u_{\mathrm{g}})\,|u - u_{\mathrm{g}}|}{2}.$$

2. Force of virtual masses:

$$F_{\mathrm{m}} = -\frac{2}{3}\pi r_{\mathrm{s}}^3 \rho \frac{\mathrm{d}u}{\mathrm{d}t}.$$

3. Basset force:

$$F_{\mathrm{B}} = -6r_{\mathrm{s}}^2\sqrt{\pi\rho\mu}\int_0^t (t-\tau)^{-0.5}\frac{\mathrm{d}u}{\mathrm{d}\tau}\mathrm{d}\tau.$$

Le is the Lewis number, c_p is the constant-pressure specific heat of the gas, p_* is the reference pressure, ε is the particle blackness rate, σ is the Stefan–Boltzmann constant, $t_{\mathrm{h}} = r_{\mathrm{s}}^2/3a_{\mathrm{s}}$ is the characteristic time of particle heating, and a_{s} is the thermal diffusivity of the solid.

The following ignition kinetics is specified:

$$\frac{\mathrm{d}h}{\mathrm{d}t} = K_n C_{\mathrm{ox}}^n h^{-n} \exp\left(-\frac{E}{RT}\right), \tag{6.21}$$

where K_n is the preexponential factor and n is the reaction order with respect to the oxidizer: $n = 1$ for aluminum and $n = 0$ for magnesium. The system of (6.20) and (6.21) with supplementary relationships is closed with respect to the functions that are sought.

The parameters characterizing the gas flow can be readily found from standard relationships for the flow behind a shock wave propagating at velocity D. Equations (6.20) and (6.21) written in the frame of reference moving with the shock wave at velocity D must satisfy the following Cauchy conditions:

$$t = t_0: \quad u = D, \quad T = T_0, \quad h = h_0. \tag{6.22}$$

Upon determining the functions $u(x,t)$, $T(x,t)$, and $h(x,t)$, where x is the coordinate, one can find the mean density of the dispersed phase, $\rho_{\mathrm{s}}(x,t)$, from the equation of dispersed-phase mass conservation. Thus, the problem of particle ignition in the shock-induced gas flow can be formulated as follows:

Find functions $u(x,t)$, $T(x,t)$, and $h(x,t)$ $\in C^1(0,t_{\mathrm{ign}})$ satisfying (6.20) and (6.21) in the region $[0,t_{\mathrm{ign}})$ and the Cauchy data (6.22).

The problem formulated above was solved numerically by Gear's method. The calculations were performed for ignition of magnesium and aluminum particles in a shock wave propagating in gaseous oxygen. In the calculations,

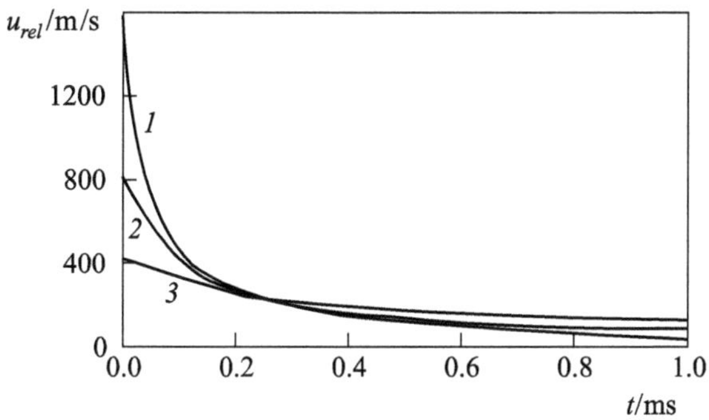

Fig. 6.15. The temporal dependence of the particle slip velocity

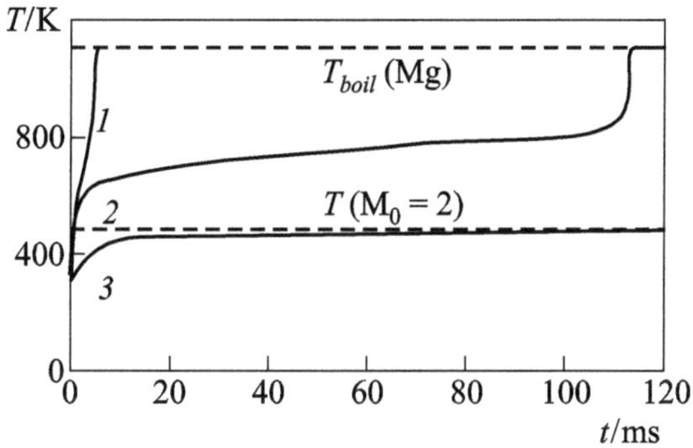

Fig. 6.16. Predicted temperature curves for magnesium particles

the particle size was varied from 1 to 100 μm, the initial pressure was varied from 0.01 to 1 bar, and the shock wave Mach number was varied from 1.1 to 6.0.

The predicted results are shown in Fig. 6.15 as the dependence of the relative particle–gas velocity, $U = u - u_g$, in the postshock flow versus time. Both experiment and calculation were suited for $p_0 = 0.135$ bar and $M_0 = 5.0$. In the experiments, ignition of a cloud of particles with a mean diameter of 17 μm was studied. The calculations show that an increase in the shock wave Mach number leads to a reduction of the velocity relaxation zone length. This can be attributed to an increase in the gas density, and hence the drag force. Note that similar calculations without consideration of the Basset force result in a shorter relaxation zone duration by a factor of about 3. Note also that the effect of the force of virtual masses, F_m, is negligible as $\rho/\rho_s \ll 1$.

Analysis of the calculations enables one to find the conditions and mechanisms relevant to particle ignition. It turned out that magnesium particles ignite at $M_0 \in (2.5, 2.75)$ depending on the particle size. The predicted time histories of particle temperature $T = T(t)$ are presented in Fig. 6.16 to illustrate the effect of M_0 at $d_{s0} = 100$ μm and $p_0 = 1$ bar. In dynamic conditions, the limiting medium temperature $\tilde{T}_{cr}$, at which ignition is still possible, is much lower than in the static conditions. Thus, for magnesium particles with $d_{s0} = 100\,\mu m$, $\tilde{T}_{cr} = 640$–$710\,K$ behind the incident shock wave, while $\tilde{T}_{cr} = 920\,K$ in static conditions. This difference is explained by an increase in the heat flux to the particle owing to higher values of the coefficient for heat transfer between the gas and the particle.

It was found that at $M_0 < 4.5$ magnesium particles ignite according to the thermal explosion mechanism, while at $M_0 > 4.5$ they ignite as a result of fragmentation. The corresponding ignition criteria are the conditions $T \geq T_{mel}$, where T_{mel} is the particle melting temperature, and $We > We_{cr}$, where We is the Weber number (and $We_{cr} \approx 12$ is the critical Weber number at which fragmentation starts). In the calculations, the ignition delay time is taken as the least of the times of fragmentation and ignition by the thermal mechanism. The ignition due to the thermal mechanism was not found under the conditions specified. Note that "thermal explosion" was observed in computations after the particle reached a temperature less than the magnesium melting temperature. Besides comparison of the flow dynamic parameter u, the predicted and measured data of [27] were compared in terms of the dependence of the ignition delay time t_{ign} on the postshock temperature. The latter is shown in Fig. 6.17. With use of the kinetic constants obtained, the computations were carried out, which allowed one to extend the approximation of [27] for the ignition delay time for magnesium particles of different sizes:

$$t_{ign} = A \left(\frac{d_{s0}}{17} \right)^m p_0^n \exp\left(\frac{E_*}{RT} \right),$$

where $A = 1.203 \times 10^4\,\mathrm{ms\,(bar)}^{-n}$, $n = -0.866$, $m = 1.7$, and $E_* = 6.4 \times 10^7\,\mathrm{J\,kmol}^{-1}$. Similar calculations with aluminum particles showed that aluminum particles ignite following the thermal ignition mechanism (Fig. 6.18). The predicted value of t_{ign} for aluminum particles in dynamic conditions

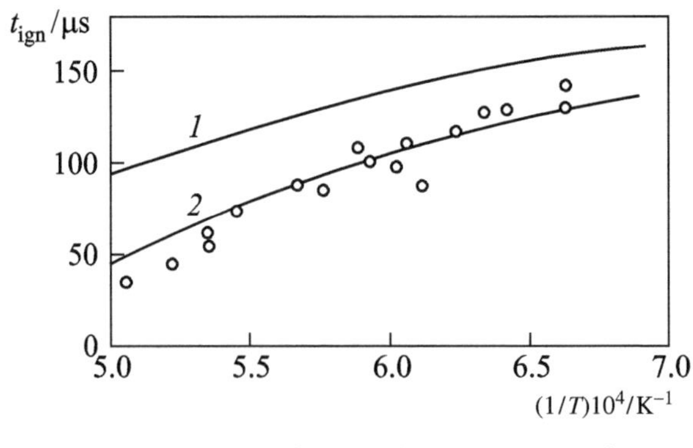

Fig. 6.17. Comparison of measured (*symbols*) and predicted (*solid curves*) data on the dependence of ignition delay time of magnesium particles on the temperature behind the incident shock wave

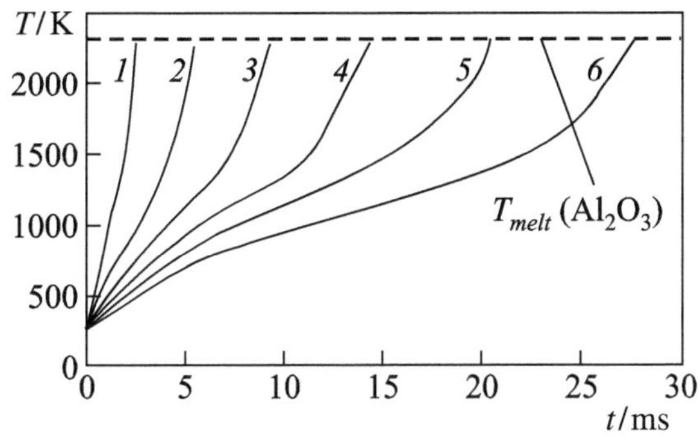

Fig. 6.18. Computed temperature curves for the aluminum particle

proved to be much less than that in static conditions. Aluminum particles with $d_{s0} \leq 100\,\mu m$ ignite in the incident shock wave with a Mach number exceeding 3.2–3.7 (depending on the particle size).

6.3 Ignition of Liquid Drops

Single liquid drop and spray behavior behind an incident shock wave differs considerably from single solid particle and particle suspension behavior. It is commonly accepted that hydrocarbon drop ignition occurs in the vapor phase. Therefore, a considerable increase in the specific surface area of drops caused by their aerodynamic deformation and fragmentation in the shock-induced gas flow is considered as one of the most important phenomena affecting interphase mass, momentum, and energy exchange rates, and finally ignition. Interaction of drops with gaseous flow is governed, on the one hand, by liquid properties, the size and the shape of the drops, and the spacing between drops and, on the other hand, by local properties of gas flow. The situation is significantly complicated by the fact that the phases interact with each other dynamically and thermally. There are hundreds of publications in which these interactions have been studied both experimentally and theoretically.

Here, we analyze only those physical and chemical phenomena which are directly relevant to drop ignition behind shock waves. In particular, we are interested in specific features of the molecular mixing of fuel with air at the mixture formation stage and at autoignition of the two-phase mixture behind the shock wave. Liquid fuels considered are low-viscosity n-alkanes or hydrocarbon fuels such as gasoline and kerosene.

6.3.1 Drop Deformation

Drop deformation in the flow is the process which is capable of influencing the interphase mass, momentum, and energy transfer. In theoretical models, a deformed drop is usually represented as an ellipsoid of revolution with large half-axis b and small half-axis a. The drop deformation degree, Δ, is defined as the ratio of a to radius r_s of a spherical drop of the same volume, $\Delta = a/r_s$. Note that in the course of deformation, a drop can take a shape resembling either an oblate or a prolate ellipsoid of revolution.

One can distinguish at least four mechanisms of drop deformation that can influence heat and mass transfer with the gas flow: (1) variation of the deformed drop cross-sectional area A, (2) variation of the total surface area of the drop S, (3) variation of the aerodynamic drag coefficient C_D of the deformed drop, and (4) variation of the heat transfer coefficient k_h between the gas and the deformed drop.

The maximal variations of cross-sectional area $A = \pi b^2$ and total surface area S are determined by the maximum possible deformation degree at which deformation is still reversible, i.e., there is still no drop fragmentation. According to [120,121], the maximum drop deformation degree is $\Delta = 0.25$–0.44. Hence, the ratio of cross-sectional areas of the deformed and spherical drops can attain values of 2.25–4. The ratio of S to the surface area of the corresponding spherical drop can attain values of 1.3–2.1.

The aerodynamic drag coefficient C_D for a deformed drop differs from that for a spherical drop. At high Reynolds numbers of relative motion of the drop and gas, C_D is approximately equal to 0.44 [122]. At such conditions, C_D for deformed drops can attain values of 1.6–2.2 [121], 1.8–3.0 [123], or 2.3 [124], i.e., values considerably larger than for a spherical drop.

The heat transfer coefficient k_h also depends on the drop shape. For example, if the deformed drop is characterized by the equivalent Sauter mean diameter $d_{eff} = 6V/S$, where V is the ellipsoid volume, then to determine k_h one can use the approximate relationship [125]

$$Nu = \frac{k_h d_{eff}}{\lambda} = 2.0 \exp\left[-2.22\left(\frac{d_{eff}}{2b} - 1\right)\right]. \tag{6.23}$$

This relationship indicates that for a deformed drop with $\Delta = 0.25$–0.44, the heat transfer coefficient can exceed the typical value for the spherical particle by a factor of 3–5.4 even at zero relative velocity of gas and the particle, $U = 0$.

Thus, increasing A and C_D can affect the dynamics of drop motion in the gas flow, leading to faster relaxation of the relative velocity U and therefore a reduction in heat and mass transfer. The reverse, an increase of S and k_h, leads to intensification of heat and mass transfer. The net effect of the deformation of the drop on its heating and vaporization in the gas flow depends on specific conditions.

There exist several models of drop deformation [126–130]. In the existing models of drop ignition behind a shock wave, drop deformation is either not taken into account at all [131–136] or regarded using a simplified model, which does not reflect the features discussed above [137–140]. Note that drop deformation is a complex process including excitation of internal liquid circulations [141, 142] and drop oscillations as a whole. In the course of deformation, a drop can attain asymmetrical shapes which are far from resembling the ellipsoid of revolution [143].

6.3.2 Single Drop Vaporization

In the postshock gas flow, the interfacial strain results in internal liquid circulation in the drop, and the conductive heat transfer is complemented by convective heat transfer [3, 141–143]. Starting from certain values of the liquid Reynolds number, the convective mechanism becomes dominant [3, 142]. Internal liquid circulation in the deformed drop can differ considerably from that in a spherical drop [2, 3]. In [144], a mathematical model of deformed drop heating and vaporization was suggested.

Initially, at time $t = 0$, a spherical drop of radius r_{s0} is placed in the postshock gas flow. Owing to the aerodynamic drag force, the drop accelerates, deforms, heats up, and vaporizes. In addition, internal liquid circulation is induced in the drop. The task is to determine the rate of drop vaporization and the drop lifetime in these conditions. Assume that the drop takes the shape of an ellipsoid of revolution. The properties of the liquid will be denoted by index l and the parameters at the drop surface by index i. In this step, we will restrict ourselves by considering the situations when the drop deformation does not attain the critical stage followed by drop breakup. From Sect. 6.2.3.2, drop breakup occurs at Weber number $We = d_{s0}\rho U_0^2/\sigma_{\mathrm{l}} \geq 12$, where σ_{l} is the surface tension, and $d_{s0} = 2r_{s0}$ is the spherical drop diameter. The critical deformation stage is attained at $\Delta = 0.25$–0.44 or $b/r_{\mathrm{s}} = 1.5$–2.0 [129].

To determine the most important geometrical parameters of the drop – b, a, A, and S – we use the deformation equation [130]:

$$\frac{\mathrm{d}^2 y}{\mathrm{d}t^2} = \frac{C_{\mathrm{F}}}{C_{\mathrm{b}}} \frac{\rho}{\rho_{\mathrm{l}}} \frac{U^2}{r_{\mathrm{s}}^2} - \frac{C_{\mathrm{k}}\sigma_{\mathrm{l}}}{\rho_{\mathrm{l}} r_{\mathrm{s}}^3} y - \frac{C_{\mathrm{d}}\mu_{\mathrm{l}}}{\rho_{\mathrm{l}} r_{\mathrm{s}}^2} \frac{\mathrm{d}y}{\mathrm{d}t}, \tag{6.24}$$

where $y = \delta/(C_{\mathrm{b}} r_{\mathrm{s}})$ is the dimensionless deformation, δ is the displacement of the drop equator from the equilibrium position in the plane normal to the direction of the relative velocity U, and $C_{\mathrm{b}} = 0.5$, $C_{\mathrm{F}} = 0.333$, $C_{\mathrm{k}} = 8$, and $C_{\mathrm{d}} = 5$ are the dimensionless coefficients. The initial conditions for (6.24) are

$$t = 0: \; y = 0; \; \frac{\mathrm{d}y}{\mathrm{d}t} = 0. \tag{6.25}$$

For an ellipsoid of revolution, the large half-axis b is equal to $b = r_{\mathrm{s}} + \delta = r_{\mathrm{s}}(1 + C_{\mathrm{b}} y)$. The small half-axis a can be determined from the condition of constant drop volume during deformation: $V = (4/3)\pi a b^2 = (4/3)\pi r_{\mathrm{s}}^3$. Parameters A and S can then be readily determined.

The equation of heat balance for the deformed drop can be written in the form

$$c_{\mathrm{l}} m \frac{\mathrm{d}T}{\mathrm{d}t} = Q_{\mathrm{i}-} \tag{6.26}$$

where $m = V\rho_{\mathrm{l}}$ is the drop mass, and $Q_{\mathrm{i}-}$ is the total heat flux from the drop surface to drop interior (index $i-$ means that the value of the heat flux is taken at the drop surface from the liquid side). Heat flux $Q_{\mathrm{i}-}$ is a function of temperature and velocity fields inside the drop, i.e., $Q_{\mathrm{i}-} = f(t, T_{\mathrm{i}}, T, u_{\mathrm{i}}, \ldots)$, where u_{i} is the liquid velocity at the drop surface. Heat flux $Q_{\mathrm{i}+}$ from the gas side (index $i+$ means that the value of the heat flux is taken at the drop surface from the gas side) for the evaporating drop is given by the relationship

$$Q_{\mathrm{i}+} = Q + L\frac{\mathrm{d}m}{\mathrm{d}t},$$

where Q is the full heat flux toward the drop from the gas phase and L is the latent heat of vaporization. Owing to continuity of heat flux, $Q_{\mathrm{i}-} = Q_{\mathrm{i}+}$, and (6.26) takes the form

$$c_{\mathrm{l}} m \frac{\mathrm{d}T}{\mathrm{d}t} = Q + L\frac{\mathrm{d}m}{\mathrm{d}t}. \tag{6.27}$$

The initial conditions for (6.27) are formulated as follows:

$$t = 0 : \ T = T_0. \tag{6.28}$$

The heat flux Q in (6.27) is given by the relationship $Q = Sq_{\mathrm{h}}$, where q_{h} is the heat flux per unit surface area of the drop. To determine q_{h} one can apply Newton's law $q_{\mathrm{h}} = k_{\mathrm{h}}(\tilde{T}_\infty - T_i)$ with ∞ denoting gas properties at a large distance from the drop surface. Note that Newton's law is applicable only to the steady-state heat transfer. Nevertheless, as shown in [145], it can be applied to problems of transient heat transfer by introducing an effective thermal conductivity of gas. In the standard drop vaporization model [146], it is assumed that $\lambda = \lambda\left(\overline{T}\right)$, where $\overline{T} = \left(\tilde{T} + T_i\right)/2$ is some characteristic gas temperature. The heat transfer coefficient k_{h} in Newton's law is given by [2,145]

$$k_{\mathrm{h}} = \frac{Nu}{d_{\mathrm{eff}}}\lambda\frac{\ln(1+B)}{B}, \tag{6.29}$$

where B is the mass transfer coefficient. In general, the Nusselt number in (6.29) depends on the drop shape [see (6.23)] and the Reynolds number based on the relative velocity $Re = d_{\mathrm{eff}}\rho\left|U\right|/\mu$. In the absence of proper relationships for deformed drops, it is worthwhile utilizing the corresponding dependence $Nu = Nu(Re)$ for a spherical drop [147] as a first approximation:

$$Nu = 2.0\,\exp\left[-2.22\left(\frac{d_{\mathrm{eff}}}{2b} - 1\right)\right] + 0.6\,Re^{1/2}Pr^{1/3}, \tag{6.30}$$

where $Pr = \mu/\rho a_{\mathrm{T}}$ is the Prandtl number and a_{T} is the gas thermal diffusivity. Relationship (6.30) indicates that, at high Reynolds numbers, the Nusselt number for a spherical drop in air can attain values up to an order of magnitude higher than 2.0.

The Reynolds number can be determined from the solution of the drop motion equations (6.18) with the initial condition

$$t = 0: \quad u = u_0. \tag{6.31}$$

For the evaporating drop, the aerodynamic drag coefficient C_{D} depends not only on the drop shape, but also on the mass transfer coefficient B and on the physical properties of the gas and liquid, which determine the liquid velocity at the drop surface.

The effect of B on C_{D} is usually taken into account as $C_{\mathrm{D}} = C_{\mathrm{D,ne}}/(1+B)$ [2], where $C_{\mathrm{D,ne}}$ is the aerodynamic drag coefficient of a nonevaporating drop. The effect of drop shape on the value of $C_{\mathrm{D,ne}}$ can be taken into account using the relationship [148] $C_{\mathrm{D,ne}} = C_{\mathrm{Ds,ne}}(1+2.632y)$. The aerodynamic drag coefficient of a solid sphere $C_{\mathrm{Ds,ne}}$ is equal to [149]

$$C_{\mathrm{Ds,ne}} = \begin{cases} \dfrac{24}{Re}\left(1 + 0.15\, Re^{0.687}\right) & \text{at } Re < 10^3 \\ 0.44 & \text{at } Re \geq 10^3 \end{cases} .$$

Thus, at high Reynolds numbers and large drop deformations ($y \approx 1$), $C_{\mathrm{D,ne}}$ may attain the value of 1.5 typical for a disk-shaped body. This value is larger by a factor of 3.4 than the value $C_{\mathrm{Ds,ne}} = 0.44$ typical for a spherical body.

The effect of liquid motion on the drop surface on the value of $C_{\mathrm{Ds,ne}}$ can be taken into account by using the results of numerical calculations [143]:

$$\frac{C'_{\mathrm{Ds,ne}}}{C_{\mathrm{Ds,ne}}} = \left(\frac{2 + 3\mu_{\mathrm{l}}/\mu}{3 + 3\mu_{\mathrm{l}}/\mu}\right)\left(1 - 0.03\frac{\mu}{\mu_{\mathrm{l}}} Re^{0.65}\right),$$

where $C'_{\mathrm{Ds,ne}}$ is the aerodynamic drag coefficient of a spherical liquid drop. The latter relationship approximates well the analytical results at low and high Reynolds number, but its application is, in general, limited by the range of viscosity ratios μ_l/μ, studied in [143].

The mass transfer coefficient B is defined as $B = (Y_{\mathrm{v}i} - Y_{\mathrm{v}\infty})/(1 - Y_{\mathrm{v}i})$ [2, 145], where index v relates to the liquid vapor and Y is the mass fraction.

The rate of drop mass variation due to vaporization required in (6.27) is found from the equation

$$\frac{\mathrm{d}m}{\mathrm{d}t} = -Sj, \tag{6.32}$$

where $j = 2\rho D \ln(1+B)/d_{\mathrm{eff}}$ is the vapor mass flow rate per unit drop surface area [2, 145] and D is the binary diffusion coefficient for the gas and vapor.

The initial condition for (6.32) is

$$t = 0: \; m = m_0. \tag{6.33}$$

Many of the relationships discussed above contain, explicitly or implicitly, the drop surface temperature T_i. In particular, T_i plays an important role in determining heat flux Q_{i-}. To determine T_i, the following considerations will be used.

When the mean drop temperature T attains the value of the saturation temperature ("wet-bulb" temperature) $T_i = T_{wb}$, the heat flux to the drop interior vanishes, i.e., $Q_{i-} = Q_{i+} = 0$, and the entire heat flux Q is consumed for liquid vaporization. The initial period in drop evolution, when $Q_{i-} \neq 0$ or $T < T_{wb}$, is referred to as the transient heating period. The period when $Q_{i-} = Q_{i+} = 0$ or $T = T_{wb}$ is referred to as the period of quasi-steady drop vaporization. In the latter case, instead of (6.27), one can write $T = T_i = T_{wb}$ and $Q = \ell dm/dt$. Denote the duration of the transient heating period as t_h, the duration of the quasi-steady vaporization period as Δt, and the total lifetime of the drop as t_l. Then $t_l = t_h + \Delta t$.

In accordance with [144], internal liquid circulation in the drop can result in a considerable decrease of the transient heating period t_h compared with the standard case when only conductive heat transfer is taken into account. At $t \geq t_h$, internal circulation exerts no effect on the drop vaporization dynamics. This means that at $t > t_h$ the drop evaporates independently of internal convective flows. Thus, internal liquid circulation can decrease the drop lifetime by no more than the value of t_h. Note that in the standard model of spherical drop vaporization without internal circulation the duration of the transient heating period for heavy hydrocarbon drops can be rather long (up to $t_h \approx t_l$), in particular at high gas temperatures and pressures [149].

The effect of internal liquid circulation on drop heating during the transient period can be taken into account by introducing the coefficient of internal heat transfer intensification, $\theta = t_h/t_h^o \leq 1$, i.e., $Q_{i-} = \theta^{-1} Q_{i-}^o$, where index o relates to a spherical drop without internal liquid circulation. To determine θ one can use the results of the numerical solution to the problem of spherical drop heating in the gas flow together with the approximation of constant surface temperature $T_i = \text{const}$ [144].

To substantiate the possibility of using such an approximation, let us determine the wet-bulb temperature T_{wb}. Substituting the above relationships into the condition $Q_{i-} = 0$ gives the following algebraic equation for $T_i = T_{wb}$:

$$Nu\,\lambda(\tilde{T}_\infty - T_i) = 2\rho D L \frac{Y_{vi} - Y_{v\infty}}{1 - Y_{vi}}. \tag{6.34}$$

In (6.34), λ, ρ, D, L, and Y_{vi} are the functions of temperature T_i; therefore, its solution is found by iteration. To check the validity of (6.34), detailed numerical calculations of spherical drop heating and vaporization were performed using the model in [149], which does not take into account relative motion of the

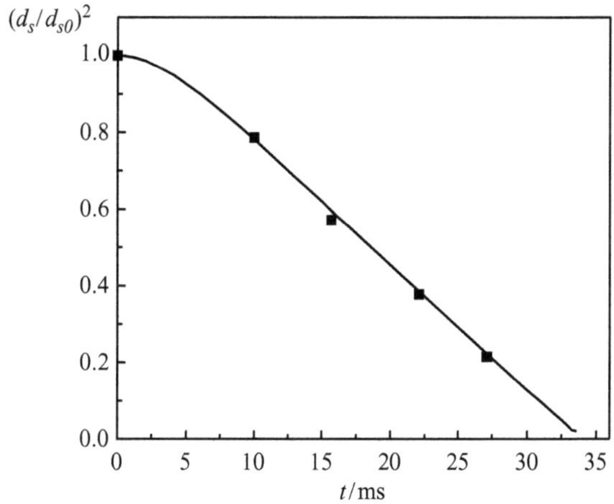

Fig. 6.19. Comparison of predicted (*curve*) and measured (*points*) dynamics of n-heptane drop surface regression at vaporization in air, $d_{s0} = 70\,\mu m$, $T_0 = 293.15\,K$, and $\tilde{T} = 573.15\,K$ [149]

drop and gas. In the computational code [149], partial differential equations for both liquid and gas phases are solved using the approximation of multicomponent diffusion for the gas phase and variable thermophysical properties of phases. To demonstrate the predictive capability of the model [149], Fig. 6.19 shows a comparison of predicted and measured drop surface regression curves for an n-heptane drop. In the simplest standard model of drop vaporization based on the quasi-steady vaporization law, the dependence $d^2(t)$ is known to be linear. In the model of [149] this dependence is more complex: owing to liquid thermal expansion the $d^2(t)$ function exhibits nonlinear behavior.

The solid and dashed curves in Fig. 6.20 show the predicted dynamics of surface temperature $T_{\rm i}$ and mean temperature T for an n-dodecane drop [144]. The horizontal dash-and-dot line corresponds to the solution of (6.34) at $Nu = 2.0$. Termination of the curves in Fig. 6.20 corresponds to complete drop evaporation (lifetime $t_{\rm l}$). Recall that the transient heating period duration t_h is the time taken for the mean drop temperature to attain the value of the wet-bulb temperature.

Analysis of Fig. 6.20 as well as the results of calculations at other values of $\tilde{T}$ and p and for other liquids (n-heptane, n-octane, methanol, etc.) result in the following conclusions:

1. The time taken for the drop surface temperature $T_{\rm i}$ to attain a value close to the wet-bulb temperature $T_{\rm wb}$ is considerably less than the total drop lifetime $t_{\rm l}$, in particular at high gas temperatures $\tilde{T}$.
2. At high gas temperatures, the duration of the transient heating period $t_{\rm h}$ is comparable with the total drop lifetime $t_{\rm l}$.

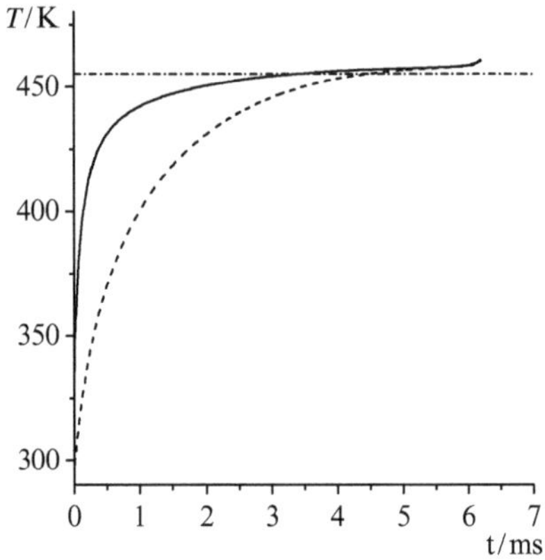

Fig. 6.20. Predicted time histories of surface (*solid curve*) and mean (*dashed curve*) temperatures of a vaporizing n-dodecane drop 50 µm in diameter at $T_0 = 293\,K$, $p = 0.1\,MPa$, and $\tilde{T} = 1,500\,K$ [144]. The *dash-and-dot line* corresponds to the wet-bulb temperature found from (6.34)

3. Equation (6.34) provides a good estimate for the wet-bulb temperature T_{wb} over a wide range of gas temperatures and pressures, as well as for the physical properties of the liquid.

If the drop is placed in the postshock gas flow, the values of the wet-bulb temperature will be somewhat different from the value relevant to drop vaporization in a quiescent atmosphere. This is evident from (6.34). At $Re \neq 0$, the Nusselt number in (6.34) can be estimated on the basis of (6.30) for a spherical drop.

To understand how the Nusselt number affects the wet-bulb temperature, additional calculations have been performed. Variation of the wet-bulb temperature with Nusselt number is most pronounced at high pressures and low gas temperatures. Nevertheless, the wet-bulb temperature for liquid drops in the gas flow can differ by no more than 30–40 K compared with the quiescent conditions with $Nu = 2$. Fig. 6.21 demonstrates this implication for n-dodecane drops.

Thus, the results obtained allow one to adopt an important assumption that the drop surface temperature T_i attains the value of T_{wb} instantaneously, i.e., $T_i = T_{wb} = \text{const}$. This assumption makes it possible to simplify considerably the modeling of drop heating, using (6.34) to determine the wet-bulb temperature at $Nu = 2$. To find the value of θ, one can use the results in [144].

Figure 6.22 shows the predicted dependencies of the mean n-heptane drop temperature on the normalized time t/t_h^o at different liquid Reynolds numbers $Re_l = \rho_l u_i r_s/\mu_l$ [144]. The calculations were performed using the approach in [144]. Clearly, internal circulation leads to a significant reduction of the transient heating period t_h. To generalize the results, detailed calculations of drop heating in constant and variable dynamic conditions were

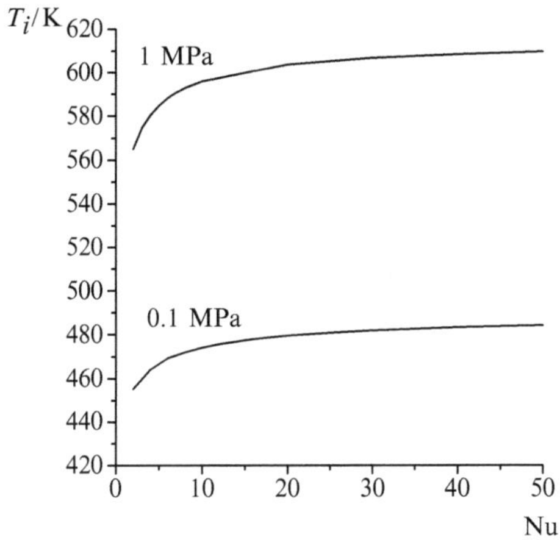

Fig. 6.21. Predicted dependencies of the wet-bulb temperature T_{wb} on the Nusselt number Nu for n-dodecane drops at $\tilde{T} = 1,500\,\mathrm{K}$ and $p = 0.1$, and $1\,\mathrm{MPa}$ [149]

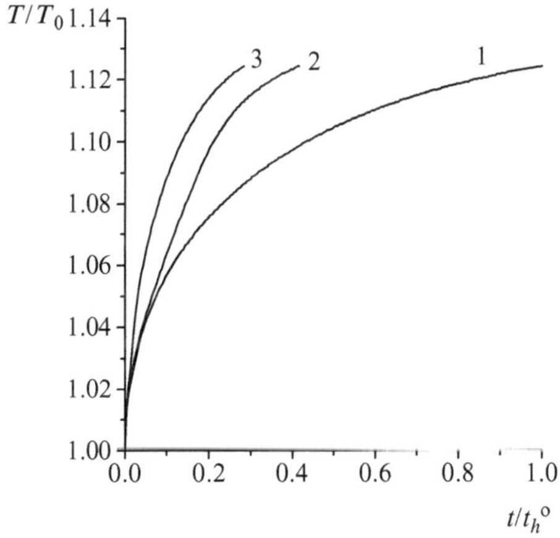

Fig. 6.22. Predicted dependencies of the mean temperature of spherical n-heptane drop on normalized time at $T_0 = 300\,\mathrm{K}$, $\tilde{T} = 750\,\mathrm{K}$, $p = 0.1\,\mathrm{MPa}$, and $d_{s0} = 200\,\mu\mathrm{m}$ [149]. Curve 1 corresponds to $Re_{\mathrm{l}} = 0$, curve 2 to $Re_{\mathrm{l}} = 1.3$, and curve 3 to $Re_{\mathrm{l}} = 13$

performed. In constant dynamic conditions, the relative velocity was kept constant ($U = U_0$), whereas in variable dynamic conditions drop heating was calculated with regard to drop motion according to (6.18) with initial conditions (6.31). The following correlation for the mean coefficient θ was suggested in [144]:

$$\begin{aligned} &\theta = 1 && \text{at} \quad Re_{\mathrm{l}} \le Re_{\mathrm{l}}^{*}, \\ &\theta = C_1 \log Re_{\mathrm{l}} + C_2 && \text{at} \quad Re_{\mathrm{l}}^{*} < Re_{\mathrm{l}} < Re_{\mathrm{l}}^{**}, \\ &\theta = C_3 \log Re_{\mathrm{l}} + C_4 && \text{at} \quad Re_{\mathrm{l}} > Re_{\mathrm{l}}^{**}, \end{aligned} \tag{6.35}$$

Table 6.5. Characteristic Reynolds numbers $Re_\mathrm{l} = \rho_\mathrm{l} u_\mathrm{i} r_\mathrm{s} / \mu_\mathrm{l}$

Liquid	Re_l^*	Re_l^{**}
n-Heptane	0.59	11.60
n-Dodecane	0.19	3.74

Table 6.6. Coefficients in approximation relationships for θ

Liquid	C_1	C_2	C_3	C_4
n-Heptane	−0.410	0.910	−0.182	0.670
n-Dodecane	−0.395	0.700	−0.193	0.600

where Re_l^* and Re_l^{**} are some characteristic values of the Reynolds number. Their values and the values of the constants C_1, C_2, C_3, and C_4 for *n*-heptane and *n*-dodecane drops are presented in Tables 6.5 and 6.6. Thus, (6.35) allows one to account for the effect of internal liquid circulation on the heat and mass transfer of spherical liquid drops with gas flow at $t < t_\mathrm{h}$.

To model deformed drop heating, one can assume, in addition to the assumptions adopted in [144], that drop deformation is quasi-static, i.e., the characteristic time required for the establishment of internal motion is small compared with the characteristic heating time of the drop. The validity of this assumption was checked by 3D numerical simulation of transient heating of deformed liquid drops. The mathematical statement of the problem for the deformed drop was similar to that for a spherical drop [142, 144], with one exception. Since the analytical solution for the internal flow field is available only for a spherical drop, the corresponding flow fields in the ellipsoidal drops were found from the steady-state solution of the Navier–Stokes equations with special boundary conditions on the drop surface.

On the basis of the calculated velocity fields inside the deformed drops, the equation of convective thermal conductivity was solved. As a result, a coefficient of heat transfer intensification for deformed drops θ' was derived. The calculations revealed that $\theta' \approx \theta\theta_\mathrm{f}$, where $\theta_\mathrm{f} = t_\mathrm{h}^\mathrm{e}/t_\mathrm{h}^\mathrm{o}$ (index *e* relates to the deformed drop without internal liquid circulation) is a function of drop shape only and is virtually the same for liquids with different physical properties (*n*-heptane, *n*-octane, *n*-dodecane, water, and methanol) at different intensities of internal liquid circulation. The results of the calculations were approximated by the second-order polynomial

$$\theta_\mathrm{f} = -0.78 + 3.67\Delta - 1.89\Delta^2 \text{ at } \Delta < 1 \tag{6.36}$$

for the oblate ellipsoid of revolution

$$\theta_\mathrm{f} = 0.65 + 0.77\Delta - 0.42\Delta^2 \text{ at } \Delta > 1 \text{ and} \tag{6.37}$$

for the prolate ellipsoid of revolution. The approximation error in (6.36) and (6.37) is less than 6%.

Coefficient θ_{f}^{-1} can be treated as an additional correction factor to the heat flux $\dot{Q}_{\mathrm{i}-}$, i.e.,

$$\dot{Q}_{\mathrm{i}-} = \dot{Q}_{\mathrm{i}-}^{\mathrm{o}}(\theta\theta_{\mathrm{f}})^{-1}, \tag{6.38}$$

where θ and θ_{f} are given by (6.35) and (6.36) or (6.37).

Differential equations (6.24), (6.27), (6.18), and (6.32) with initial conditions (6.25), (6.28), (6.31), and (6.33), as well as the additional relationships presented above allow one to predict the behavior of a liquid drop in the gas flow with regard to drop deformation, motion, transient heat transfer, and vaporization, and therefore to determine the drop lifetime. As compared with the standard vaporization model of a spherical drop, (6.23)–(6.37) include at least eight supplementary factors which can affect the drop lifetime in the gas flow. These are (1) internal liquid circulation, which affects the duration of the transient heating period from the initial temperature T_0 to the wet-bulb temperature T_{wb}, (2) variation of drop surface area S, (3) variation of the heat transfer coefficient k_{h} owing to drop deformation, (4) variation of the heat transfer coefficient k_{h} owing to relative motion of the drop and gas, (5) variation of the deforming drop cross-sectional area A, (6) variation of the aerodynamic drag coefficient C_{D} owing to drop vaporization, (7) variation of C_{D} owing to drop deformation, and (8) variation of C_{D} due to liquid motion at the drop surface.

All these factors can be considered as corrections to a standard model. Since corrections 1–4 relate to the heat balance equation and corrections 5–8 relate to the drop motion equation, they can be conditionally referred to as "thermal" and "dynamic" corrections, respectively. These corrections can be readily introduced into the standard model [146]. Below we show some examples of calculations with and without these corrections. The problem was solved by the Runge–Kutta method of the fourth order for drops of various primary hydrocarbons at different temperatures and pressures of ambient air and at different Weber numbers determining drop deformation. Thermophysical properties of liquids were treated as functions of pressure and temperature.

The results of calculations with corrections 1–8 were compared with the results predicted by the model in [146] at $U \neq 0$ ($We_0 \neq 0$). The calculations revealed that the most pronounced effect on the drop lifetime is produced by "thermal" corrections 1–4, whereas "dynamic" corrections 5–8 play an insignificant role.

Figure 6.23 shows the predicted time histories of "fine" ($d_{\mathrm{s}0} = 25\,\mu\mathrm{m}$) n-dodecane drops at $\tilde{T} = 1{,}000\,\mathrm{K}$, $T_0 = 300\,\mathrm{K}$, and $p = 1\,\mathrm{MPa}$, and different Weber numbers. Solid curves correspond to the model in [146] and dashed curves to the modified model in [146] with "thermal" corrections 1–4 [144].

Analyzing the computational results at different values of d_{s0}, $\tilde{T}$, p, and We, one comes to the following conclusions regarding the effect of "thermal" corrections 1–4 on the drop lifetime:

1. Corrections can result in a considerable reduction of drop lifetime (up to a factor of 2.5).

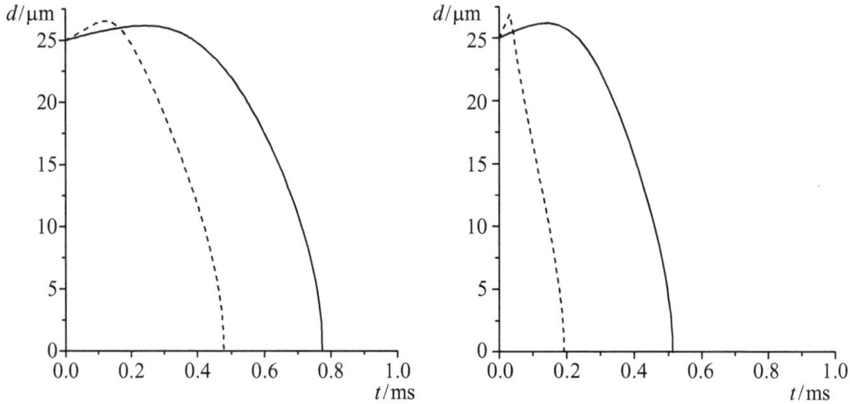

Fig. 6.23. Time histories of *n*-dodecane drop diameter predicted by the standard model (*solid curves*) and the modified standard model with "thermal" corrections 1–4 (*dashed curves*) at $d_{s0} = 25\,\mu m$, $\tilde{T} = 1{,}000\,K$, $T_0 = 300\,K$ and $p = 1\,MPa$ [144]. **a** $We_0 = 0.2$; **b** $We_0 = 12$

2. The effect of corrections on drops of less volatile fuel is more pronounced.
3. The effect of corrections increases with Weber number We.
4. The effect of corrections increases with pressure p.
5. The effect of corrections increases with gas temperature $\tilde{T}$.
6. The variation of d_{s0} is not virtually reflected in the effect of corrections.

The less pronounced effect of "thermal" corrections on the volatile fuel drop lifetime is explained by a shorter transient heating period t_h.

6.3.3 Drop Breakup

From physical reasoning and dimensional analysis, breakup is governed by the following basic dimensionless numbers representing ratios of the forces: aerodynamic force to surface tension, or Weber number $We = \rho U^2 d_s/\sigma_l$, aerodynamic force to viscous force in the gas phase, or Reynolds number $Re = \rho\,|U|\,d_s/\mu$, and viscous force to surface tension in the liquid phase, or Ohnesorge number $Oh = \mu_l/(\rho_l\sigma_l d_s)^{1/2}$. The Ohnesorge number comes into play when the viscosity of a liquid is high; therefore, when breakup of inviscid liquids (such as hydrocarbons) for which $Oh < 0.1$ is considered, its effect on mass transfer can be neglected. Generally, depending on the Weber number, the breakup modes appear as shown in Fig. 6.24 [150].

The figure illustrates schematically the sequence of breakup stages representing only their major features, many details have been omitted. Although the conditions for critical Weber numbers and for each particular breakup mode reported in various publications differ, the discrepancy is not that significant to deserve special discussion. In a strong shock wave spreading in sprays, the main droplet breakup mode is a combination of stripping (Fig. 6.24d)

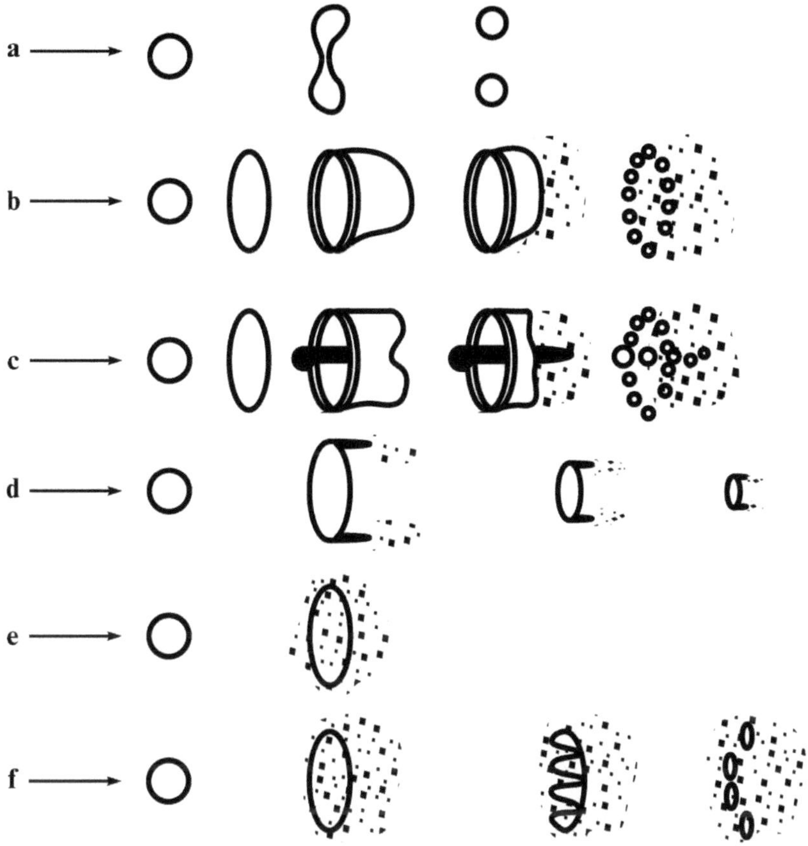

Fig. 6.24. Mechanisms of breakup of low-viscosity liquid drops observed experimentally [150]: **a** vibrational breakup $We < 12$, **b** bag breakup $12 < We < 50$, **c** bag-and-stamen breakup $50 < We < 100$, **d** sheet stripping $100 < We < 350$, **e** wave crest stripping $350 < We < 2,670$, and **f** catastrophic breakup $We > 2,670$. The *arrows* show the flow direction

and Rayleigh–Taylor (Fig. 6.24e) instability modes. These modes produce two sorts of secondary drop: micromist, as a result of disintegration of the stripped sheet, and larger drops arising after perforation of the parent drop with gaseous "fingers" due to Rayleigh–Taylor instability and the disintegration of the perforated disk.

From linear analysis, there is a minimum wavelength of instability waves $\lambda_w = 2\pi\left(\sigma_l/\rho_l a_d\right)^{1/2}$ (where a_d is the drop acceleration) below which their amplitude does not grow [151]; hence, drops of size less than λ_w are not subjected to Rayleigh–Taylor instability. According to [151], behind a shock wave with $M = 3.0$, λ_w ranges between 23 and 63 µm. Unfortunately, available experimental data pertain to larger drops; therefore, breakup patterns of fine drops require additional experimental verification.

Breakup modes inherent in lower Weber numbers are important for weak shock waves. Usually they do not result in such a dramatic increase of the evaporating surface area as do the stripping and Rayleigh–Taylor instability modes. But there is one process where these modes can significantly change the burning process. This is the transition from deflagration of a spray to detonation, which is significantly stimulated by sending weak shock waves into a burning spray.

6.3.4 Cloud of Breakup Fragments

One of the most important questions relevant to the drop ignition phenomenon behind a shock wave is how fast the fuel is mixed with the oxidizer on a molecular level. The answer to this question presumes knowledge of (1) the total breakup times and (2) the state and geometry of the two-phase cloud arising after the breakup. The representative time of drop deformation, and hence breakup, following from dimensional analysis is $t^* = d_s\left(\rho_l/\rho\right)^{1/2}/U$; therefore, for convenience the total breakup time is usually expressed in t^* units. The reduced breakup times τ for low-viscosity liquids range between 3.5 and 6. Pilch and Erdman [150] reported empirical formulas relating τ to Weber number:

$$\begin{aligned}
&\tau = 6\,(We - 12)^{-0.25} \ \text{at}\ 18 \geq We \geq 12,\\
&\tau = 2.45\,(We - 12)^{-0.25} \ \text{at}\ 45 \geq We \geq 18,\\
&\tau = 14.1\,(We - 12)^{-0.25} \ \text{at}\ 351 \geq We \geq 45,\\
&\tau = 0.766\,(We - 12)^{-0.25} \ \text{at}\ 2,670 \geq We \geq 351,\\
&\tau = 5.5 \ \text{at}\ We \geq 2,670.
\end{aligned}$$

Thus, the fluid in the wake of a disintegrated drop is a mixture of the free-stream gas, fuel vapor, and secondary droplets of various sizes. Photographic studies fail to provide information about the state of the material in the wake of the drop because of strong light scattering by dispersed material. Even X-ray diagnostics [152] furnish data on the overall density of the mixture while indicating nothing about the structure and aggregate state of the mixture.

Figure 6.25 shows a photograph of a shattered water drop 0.3 mm in diameter in the shock-induced airflow behind the shock wave of Mach number 2.4. The totally opaque wake of the drop consists of very fine droplets. The amount of air entrained in the wake at $\tau = 3.1$ is estimated at about 450 times the volume of the original drop. If it were a hydrocarbon drop, rather than water, the equivalence ratio Φ averaged over the wake would be about 5. On the basis of a qualitative understanding of the drop breakup mechanism, one can only speculate that the cloud behind the drop has a shape close to conical and the main part of the fragments is concentrated at its periphery.

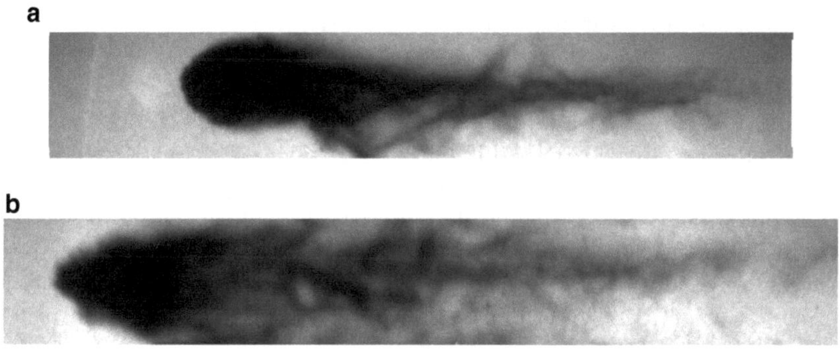

Fig. 6.25. Water drop (0.3 mm in diameter) shattered behind a shock wave with Mach number 2.4 at $\tau = 2.5$ **a** and 3.1 **b**, respectively. The initial air pressure is 1 atm

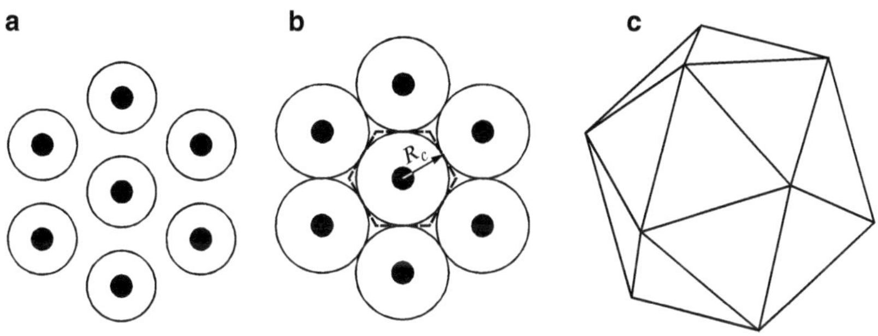

Fig. 6.26. Elementary cell for the uniform monodisperse drop suspension [153,154]. *Black circles* denote drops. *Circumferences around drops* characterize the spread of diffusion fluxes from individual drops. In **a**, spray effects are absent, while in **b** spray effects start to appear. The *dashed line* bounds the elementary cell with zero mass and energy fluxes through its surface. R_c is the characteristic cell size (half-distance between drops). In **c**, the 3D elementary cell is shown in the form of a regular polyhedron with 20 faces

6.3.5 Vaporization of Drops in Clouds

Consider a uniform monodisperse suspension of liquid fuel drops [153, 154]. In such a suspension, all drops have the same diameter and are located at the same distance from each other. The black circles in Fig. 6.26a and b show schematically the drops distributed equidistantly over the plane. The circumferences around the drops characterize the penetration depth of diffusion fluxes, i.e., the conditional boundaries of the regions in which vapor concentration and gas temperature differ from their values in the undisturbed surroundings. The parameters of the surroundings remain stationary until the circumferences do not touch each other (Fig. 6.26a). When the diffusion fluxes from neighboring drops meet each other (Fig. 6.26b), all parameters in the interdrop space start varying in time. Obviously, at the process stage shown in Fig. 6.26a, the spray effects are absent (in the approximation of constant pressure). The spray effects begin to develop at the stage shown in Fig. 6.26b.

Owing to symmetry considerations, an elementary cell in the form of a hexahedron shown by the dashed line in Fig. 6.26b can be constructed around each drop. The mass, momentum, and energy fluxes through the faces of this cell should evidently be zero. The characteristic size of the cell R_c is the half-distance between drops. In 3D space, the elementary cell has the shape of a regular polyhedron with 20 faces in the form of equilateral triangles with face length R_c (Fig. 6.26c). Thus, drop behavior in the suspension can be modeled by solving the governing conservation equations for a single drop with symmetrical boundary conditions at the polyhedron faces. Polyhedron volume and surface area are equal to $V_c = (5\sqrt{2}/3)R_c^3$ and $S_c = 5\sqrt{3}R_c^2$, respectively.

To visualize the flow pattern in such an elementary cell, a transient 3D problem on the flow evolution around a porous sphere modeling an evaporating drop was solved in [153, 154]. The flow field in the computational domain appeared to be very close to the 1D spherically symmetric field. The distortions of the 1D flow field were observed only in the vicinity to the polyhedron vertices where some tangential energy fluxes and insignificant convective flows took place. In view of this, the 3D problem can be reduced to a 1D formulation with zero-flux boundary conditions at the surface of the elementary sphere of radius R_{sc}, volume $V_{sc} = \frac{4}{3}\pi R_{sc}^3$ and surface area $S_{sc} = 4\pi R_{sc}^2$. Using the condition of equal volumes of the elementary sphere and polyhedron $V_{sc} = V_c$, one can readily obtain the radius of the elementary sphere as $R_{sc} = (5\sqrt{2}/4\pi)^{1/3}R_c \approx 0.826R_c$. It appears that the surface areas of the elementary sphere and the polyhedron differ only by 1%, i.e., $S_{sc}/S_c \approx 0.99$. Despite the fact that the adopted approximation does not take into account the tangential mass, momentum, and energy fluxes existing at the periphery of the polyhedron cell, one can anticipate that the approximate solution of the problem will reflect the main features of heat and mass transfer phenomena in drop suspensions.[1]

The statement of the 1D spherically symmetrical problem is the same as that reported in [153, 154]. The model is based on nonstationary differential equations of conservation of mass and energy in liquid and gas phases with variable thermophysical properties. In the statements, a concept of multicomponent diffusion of reactive species is used for the mixture containing fuel vapor, oxygen, nitrogen, and various combustion products. The effect of liquid surface tension on drop evaporation rate is also taken into account. The model is formulated for constant-pressure conditions in the gas–drop system, i.e., $p = p_0 =$ const.

[1] These implications are also valid for a localized region in a suspension with uniform spatial distribution of monodisperse drops. In realistic nonuniform two-phase flows with polydisperse drops, one can also distinguish localized regions with such prerequisites owing to dynamic stratification of drop size fractions.

Initially, at $t = 0$, the radius of an elementary polyhedron cell R_c can be found on the basis of the mass content of liquid in the unit volume of a drop suspension, $\eta \ll \rho_l$, and the initial drop radius r_{s0}:

$$R_c \approx \left(4\pi/5\sqrt{2}\right)^{1/3} r_{s0} \left(\rho_l/\eta\right)^{1/3} \approx 1.211 r_{s0} \left(\rho_l/\eta\right)^{1/3}, \qquad (6.39)$$

or on the basis of the fuel–air ratio $\Phi = \eta/(\phi_{st}\rho)$:

$$R_c \approx \left(4\pi/5\sqrt{2}\right)^{1/3} r_{s0} \left[\rho_l/\left(\rho\Phi\varphi_{st}\right)\right]^{1/3} \approx 1.211 r_{s0} \left[\rho_l/\left(\rho\Phi\varphi_{st}\right)\right]^{1/3}, \qquad (6.40)$$

where ϕ_{st} is the stoichiometric fuel–air ratio.[2] The radius of the spherical elementary cell can be derived from (6.39) and (6.40):

$$R_{sc} \approx r_{s0} \left(\rho_l/\eta\right)^{1/3} \approx r_{s0} \left[\rho_l/\left(\rho\Phi\phi_{st}\right)\right]^{1/3} \qquad (6.41)$$

At normal conditions, for stoichiometric mixtures of hydrocarbon fuels with air $\rho = 1.19\,\mathrm{kg\,m^{-3}}$, $\rho_l = 700$–$800\,\mathrm{kg\,m^{-3}}$, $\phi_{st} \approx 0.06$, and $\Phi = 1$; hence, $\eta = \eta_{st} \approx 0.07$–$0.08\,\mathrm{kg\ m^{-3}}$, $R_c/r_{s0} \approx 25$–27, and $R_{sc}/r_{s0} \approx 21$–22. At elevated pressures, for example, at the end of the compression stroke in a diesel engine ($\rho_g \approx 30\,\mathrm{kg\,m^{-3}}$), $R_c/r_{s0} \approx 9$ and $R_{sc}/r_{s0} \approx 8$.

Since the statement of the problem implies that $p = \mathrm{const}$, R_{sc} (and R_c) is time-dependent, i.e., $R_{sc} = R_{sc}(t)$. The value of $R_{sc}(t)$ should be determined in the course of the solution allowing the boundary of the spherical elementary cell $r = R_{sc}$ to move with gas. Thus, the boundary conditions at $r = R_{sc}$ are written in the form

$$r = R_{sc}(t): \quad \frac{\partial \tilde{T}}{\partial r} = 0, \quad \frac{\partial Y_j}{\partial r} = 0 (j = 1, 2, \ldots, M), \qquad (6.42)$$

where M is the number of gaseous species. The conditions (6.42) differ from the conditions for a single isolated drop [155] as the mass and energy fluxes vanish at a finite distance $r = R_{sc}$ from the drop rather than at $r \to \infty$.

The initial conditions at $t = 0$ are written as

$$\begin{array}{rl} & r_s(0) = r_{s0} \\ r < r_{s0}, & T(r,0) = T_0 \\ r > r_{s0}, & \tilde{T}(r,0) = \tilde{T}_0 \\ r_{s0} < r \le R_{sc}, & Y_j(r,0) = Y_{j0} \quad j = 1, 2, \ldots, M \end{array} . \qquad (6.43)$$

Conditions (6.43) make a provision for nonzero initial vapor content Y_{v0} in the gas phase. For numerical solution of the problem, a finite-difference scheme and iterative procedure are used. The details are reported elsewhere [153–155].

Analysis of drop evaporation in dense suspensions shows [153, 154] that the drop vaporization rate and lifetime in suspension may differ considerably compared with those relevant to a single drop placed in an unconfined

[2] η and Φ can be treated as local parameters in nonuniform drop suspensions.

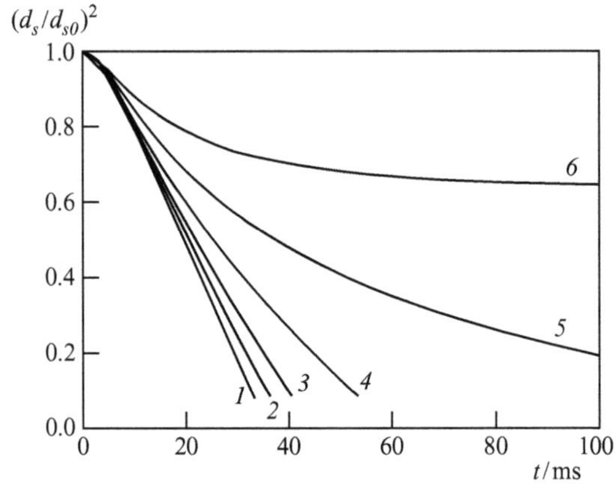

Fig. 6.27. Predicted dynamics of drop surface area for *n*-tetradecane drop suspensions evaporating in air at different fuel–air ratios Φ (distance between drops R_c/r_{s0}) at $d_{s0} = 70\,\mu$m, $T_0 = 293.15$ K, $\tilde{T} = 573.15$ K, and $p = 0.1$ MPa [153,154]. *1* $\Phi = 9.5$ ($R_c/r_{s0} \to \infty$), *2* $\Phi = 0.6$ (40.0), *3* $\Phi = 1.2$ (31.5), *4* $\Phi = 2.4$ (25), *5* $\Phi = 4.75$ (20), and *6* $\Phi = 9.5$ (15.7)

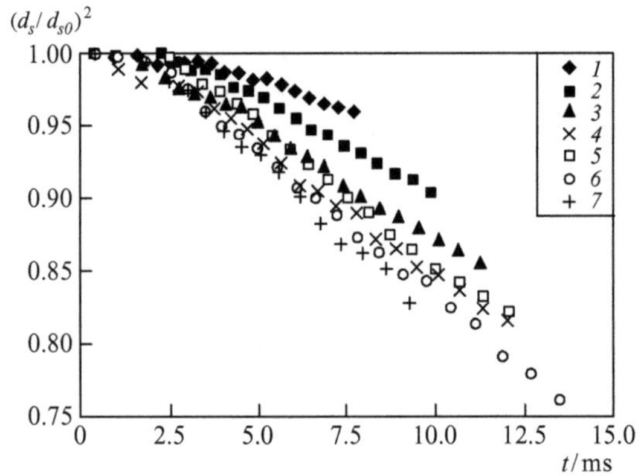

Fig. 6.28. Measured vaporization dynamics of monodisperse ethanol drops in the linear array issued vertically upward along the thermal boundary layer at the vertical heated plate [171]. Points *1*–*7* correspond to different spacings between drops R_c/r_{s0}: *1* 2.6, *2* 5, *3* 7.2, *4* 9.4, *5* 11.5, *6* 13.6, and *7* 15.8

atmosphere. Figure 6.27 shows the predicted drop surface regression curves for *n*-tetradecane drops in suspensions of different densities. Three important findings follow from Fig. 6.27: (1) the inclusion of spray effects results in slower drop vaporization; (2) the quasi-steady drop vaporization law is, in general, not valid; and (3) for each liquid there exists a fuel–air ratio Φ_m (or η_m) such that at $\Phi > \Phi_m$ (or $\eta > \eta_m$), drops evaporate only partly. The first and second findings have been confirmed experimentally. For example, Fig. 6.28 shows the results of measurements [140] of ethanol drop vaporization dynamics in linear arrays with different spacings between drops. It is seen that a decrease in the drop spacing leads to a decrease in the drop vaporization rate just at the beginning of the process and the dependence of d_s^2/d_{s0}^2 on time deviates more greatly from the straight line. Note that the screening effect of neighboring drops in the linear array is considerably smaller than that in the schematics of Fig. 6.26b and c.

It is natural to anticipate that similar effects are inherent in the micromist behind a shattered parent drop. Unfortunately, this stage of the process has not been sufficiently well studied to allow reliable relationships for calculation of its rate to be proposed.

6.3.6 Kinetic Mechanisms of Drop Ignition and Combustion

To study gas-phase autoignition behind a shock wave, detailed reaction mechanisms are widely used. As for the fuel drops and sprays, detailed reaction mechanisms have not been used so far except in several recent publications, e.g., [156]. This is caused by the fact that the kinetic mechanisms of high hydrocarbons are very complex and, in addition to chemical complications, drop combustion is accompanied by complex physical processes. Therefore, for modeling n-alkane drop ignition and combustion, reduced or overall reaction mechanisms are usually applied [154, 157, 158]. A reaction mechanism in [154,157,158] contains ten reactions with ten species (fuel, O_2, N_2, CO, CO_2, H_2, H_2O, NO, soot, and generalized radical). The mechanism was validated on the problems of premixed laminar flame propagation and a laminar counterflow diffusion flame, as well as fuel drop autoignition and combustion [154, 157–161]. It was found to be well applicable to the description of high-temperature oxidation of n-alkanes at temperatures exceeding 1,200–1,300 K. At temperatures below 900 K, the kinetics of n-alkane oxidation changes and so-called multistage behavior accelerating the chemical process start to appear. To model multistage autoignition, more complex kinetic mechanisms are required.

There are several publications on detailed and semiempirical reaction mechanisms for heavy hydrocarbons. For example, in [162] oxidation mechanisms of n-heptane and isooctane were suggested. The mechanism [162] is composed of two reaction blocks: a detailed mechanism of oxidation of C_1–C_2 hydrocarbons and an overall mechanism of low-temperature oxidation and decomposition of n-heptane and isooctane. In [163], a kinetic mechanism for low-temperature n-tetradecane oxidation was developed, and is also composed of two reaction blocks. The first block is the detailed reaction mechanism of oxidation of C_1–C_2 hydrocarbons. It includes 119 reversible elementary reactions with 29 species. This block is the same as that used in [162]. The second block is the overall mechanism of low-temperature oxidation and decomposition of $C_{14}H_{30}$. It contains 15 reactions with seven new species.

6.3.7 High-Temperature Drop Ignition

High-temperature autoignition of n-heptane drops was modeled in [154,157, 158] using the overall kinetic mechanism. For a single drop in microgravity conditions, satisfactory agreement between predicted and measured [164,165] ignition delay times was obtained (Table 6.7). In the calculations, the initial temperature of the drops was 293 K. The ambient air temperature was assumed to be spatially uniform. The size of the computational domain around the drop R_{sc} was sufficiently large compared with the initial drop radius r_{s0}, so the values of all parameters on the external boundary were constant in the course of the calculations.

After a certain time, referred to as the ignition delay time, t_{ign}, at a certain distance from the drop surface, autoignition of fuel vapor was detected in the calculations. The ignition delay time t_{ign} was defined as the time interval for the rate of maximal temperature rise to attain a predefined value of $T'_{max} = 10^6\,Ks^{-1}$. Other reasonable definitions were also used, but they resulted in very similar values of t_{ign}. Figure 6.29 shows predicted dependencies of the

Table 6.7. Comparison of predicted and measured autoignition delay times for isolated n-heptane drops at a pressure of 0.1 MPa under microgravity conditions [154,158]

Initial drop diameter (μm)	Air temperature (K)	t_{ign} (s)	
		Measured	Calculated
700	1,000	0.30 [164]	0.19
1,000	960	0.58 [165]	0.48

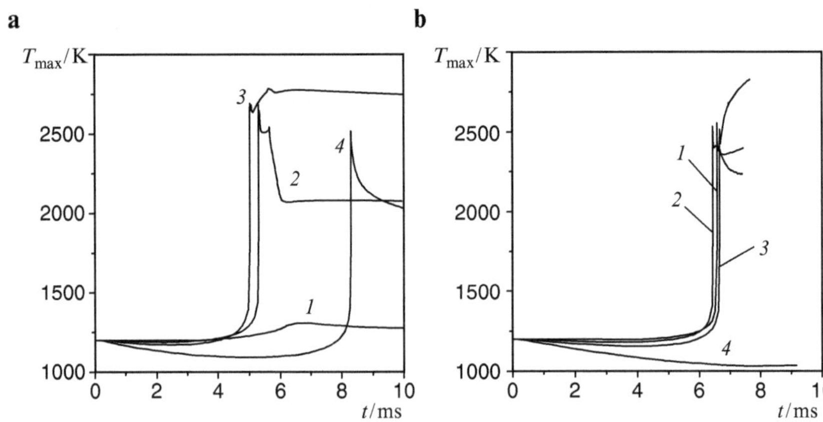

Fig. 6.29. Predicted dependencies of maximal temperature in the vicinity of n-heptane **a** and n-tetradecane **b** drops at autoignition conditions. Initial data: $d_{s0} = 50\,\mu m$, $T_0 = 293.15\,K$, $\tilde{T} = 1,200\,K$, and $p = 0.1\,MPa$. *1* $\Phi = 0$ (single drop), *2* $\Phi = 0.5$, *3* $\Phi = 1.0$, and *4* $\Phi = 2.0$

maximal temperature in the vicinity of an n-heptane (Fig. 6.29a) and an n-tetradecane (Fig. 6.29b) drop at $T_0 = 293.15\,\mathrm{K}$, $\tilde{T} = 1{,}200\,\mathrm{K}$, and $p = 0.1\,\mathrm{MPa}$. A single n-heptane drop fails to ignite at these conditions, while a single n-tetradecane drop ignites (see curves 1). Curves 2–4 correspond to $\Phi = 0.5$, 1.0, and 2.0 and demonstrate the spray effect on autoignition of drop suspensions of these two fuels. For n-heptane drop suspensions, the shortest ignition delay is attained in the stoichiometric mixture (curve 3 in Fig. 6.29a). Fuel-rich drop suspensions of n-heptane ($\Phi = 2.0$) ignite only after complete drop vaporization. For n-tetradecane drop suspensions, the shortest ignition delays are attained in fuel-lean suspensions (curve 2 in Fig. 6.29b, $\Phi = 0.5$). Fuel-rich n-tetradecane suspensions ($\Phi = 2.0$) fail to ignite, at least in 10 ms. Thus, autoignition of drop suspensions is very sensitive to the suspension density and liquid fuel properties.

Following [157, 158], let us introduce the concept of normalized mass content of the jth species, $I_j(t)$, as the ratio of the total mass of this component in the gas phase at time t to the initial drop mass, $m_0 = (4/3)\,\pi r_{s0}^3 \rho_l$, i.e.,

$$I_j(t) = m_0^{-1} \int\limits_{r(t)}^{R_{sc}} 4\pi\xi^2 Y_j(\xi, t)\mathrm{d}\xi.$$

As before, index v will be used for fuel vapor. It is instructive to distinguish between fuel which is initially in the liquid phase and fuel which is initially in the vapor phase. The former and the latter will be marked by indices vd and vg, respectively. Thus,

$$I_v(t) = I_{vd}(t) + I_{vg}(t).$$

The first term on the right-hand side of this equation relates to the part of I_v corresponding to the fuel which is initially contained in the liquid drop. The second term relates to the part of I_v which is initially contained in the vapor phase, e.g., owing to preliminary partial drop vaporization. By definition, the prevaporization degree is equal to $I_{vg}(0)$. The normalized fuel content varies with time because of two processes: drop vaporization and chemical reaction. When a liquid drop evaporates completely but the fuel does not react chemically, the value of I_{vd} tends to unity. If there are gas-phase reactions of fuel oxidation, then fuel vapor is depleted in the course of the reactions and the value of I_v decreases (in the case of a single drop, I_v tends to zero).

Figure 6.30a presents an example of predicted time histories of a squared drop diameter $d^2(t)$ at autoignition of a single n-hexane drop. At completion of the ignition delay time ($t_{ign} = 160\,\mu\mathrm{s}$), the $d^2(t)$ curve exhibits a kink and the dependence becomes linear. The drop lifetime is about $250\,\mu\mathrm{s}$. Figure 6.30b shows the corresponding dependencies $I_j(t)$ for different species: fuel (C_6H_{14}), CO, H_2, CO_2, and H_2O at $I_{vg}(0) = 0$. It is seen that during the ignition delay the fuel vapor accumulates in the gas phase (curve C_6H_{14}). After autoignition, some of fuel vapor burns out rapidly and then the rate of fuel vapor consumption stabilizes and becomes nearly constant. This stage

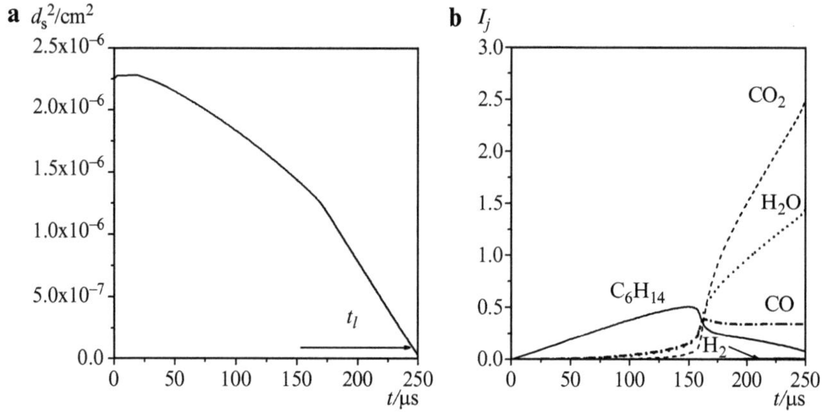

Fig. 6.30. Predicted time histories of a squared n-hexane drop diameter **a** and normalized mass contents of various combustion products in the gas phase **b**; t_l is the drop lifetime; $d_{s0} = 15\,\mu m$, $\tilde{T} = 1,500\,K$, and $p = 3\,MPa$ [157, 158]

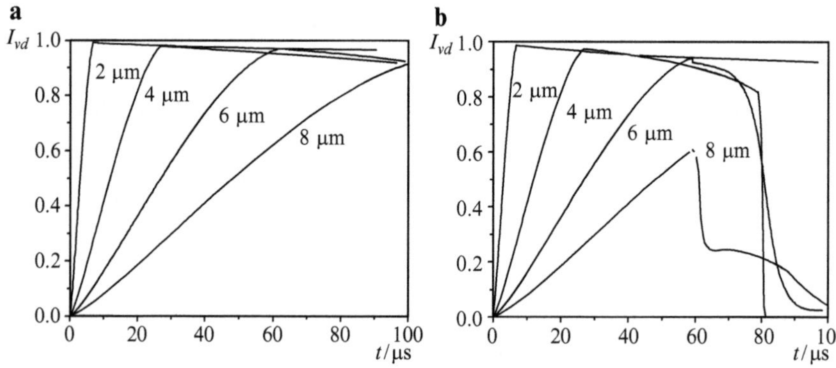

Fig. 6.31. Predicted time histories of normalized mass content of fuel I_{fd} for fine n-heptane drops of initial diameter 2, 4, 6, and 8 μm at $\tilde{T} = 1,500\,K$ and $p = 3\,MPa$: **a**$I_{fg} = 0$ and **b** 0.25 [157, 158]

corresponds to the diffusion-controlled combustion of the drop. CO_2 and H_2O accumulate with time, while the yields of other products (CO and H_2) are insignificant.

Figure 6.31a shows the predicted time history of the normalized mass content of fuel vapor I_{vd} (at $I_{vg} = 0$) in a stoichiometric suspension of fine n-heptane drops of initial diameter 2, 4, 6, and 8 μm in air. Such drops exhibit very short velocity relaxation times behind incident shock waves. In view of this, their ignition can be approximately treated as ignition in a quiescent oxidizing atmosphere. Drops of an initial diameter 2 μm first evaporate very fast ($t_l = 5.8\,\mu s$) and then the normalized fuel vapor content decreases very slowly over 100 μs, i.e., the oxidation reaction is slow. The analysis shows that owing to fast drop evaporation, the temperature in the elementary sphere of radius R_{sc} around the initial drop decreases by about 300 K, i.e., to a value less than 1,225 K. At such a low temperature, the oxidation reaction takes considerably longer than 100 μs. It is seen from Fig. 6.31a that all the specific features found for 2-μm drops are valid for drops 4, 6, and 8 μm in diameter.

The drop lifetime decreases considerably if the gas phase initially contains fuel vapor, e.g., owing to partial drop prevaporization. In this case, $I_{vg} \neq 0$. Figure 6.31b shows predicted time histories of I_{vd} for stoichiometric suspensions of fine n-heptane drops of initial diameter 2, 4, 6, and 8 μm in air at $I_{vg}(0) = 0.25$. Such a value of $I_{vg}(0)$ corresponds to the case when 25% of the total fuel mass is initially in the vapor phase.

When comparing Fig. 6.31a and b, one notices that at $I_{vg} \neq 0$, the $I_{vd}(t)$ curves exhibit a new feature: starting from a certain time instant, the value of the normalized mass content of fuel decreases drastically, i.e., autoignition occurs. For example, at $I_{vg}(0) = 0.25$ (Fig. 6.31b), the drops of diameter 6 μm are ignited at $t = 60$ μs owing to fast oxidation of prevaporized fuel. After the autoignition event, the rate of fuel burnout slows and a transition to diffusion-controlled drop combustion is observed. At $t = 100$ μs, there is only 2.5% unburned fuel. Autoignition of prevaporized fuel in the stoichiometric suspension with drops 8 μm in diameter occurs approximately at the same time: in 60 μs. However, in 100 μs, nearly 4.6% of the fuel remains unburned. Autoignition of drops 4 μm in diameter occurs with a delay time of about 80 μs (the strong influence of initial cooling caused by fast vaporization appears); however, toward 100 μs, the fuel is completely burned. Drops 2 μm in diameter are ignited with an ignition delay exceeding 100 μs. The dynamics of suspension ignition change with increasing prevaporization degree. At $I_{vg}(0) = 0.5$, the drops 2, 4, and 6 μm in diameter burn out completely in 100 μs, while the 8 − μm drops burn out only partly: about 2.5% of the fuel remains unburned. Even at $I_{vg}(0) = 0.75$, 8-μm drops burn out incompletely in 100 μs (about 1.5% of the fuel remain unburned). It is interesting that larger drops (6 and 8 μm) exhibit combustion modes with several autoignition events [157, 158].

6.3.8 Low-Temperature Drop Ignition

In the calculations of low-temperature drop autoignition, the semiempirical kinetic mechanism of n-tetradecane was used [163]. The calculations were performed for the stoichiometric drop suspension, i.e., the radius of the computational domain R_{sc} around a drop was equal to the radius of the stoichiometric elementary sphere.

Figure 6.32 shows examples of predicted time histories of squared drop diameter $(d_s/d_{s0})^2$, maximal gas temperature $\tilde{T}_{max}$, normalized mass content of hydroxyl, and normalized mass contents of alkylhydroperoxide and hydrogen peroxide at autoignition of an n-tetradecane drop with $d_{s0} = 20$ μm at $p = 2.5$ MPa and $\tilde{T} = 650$ K.

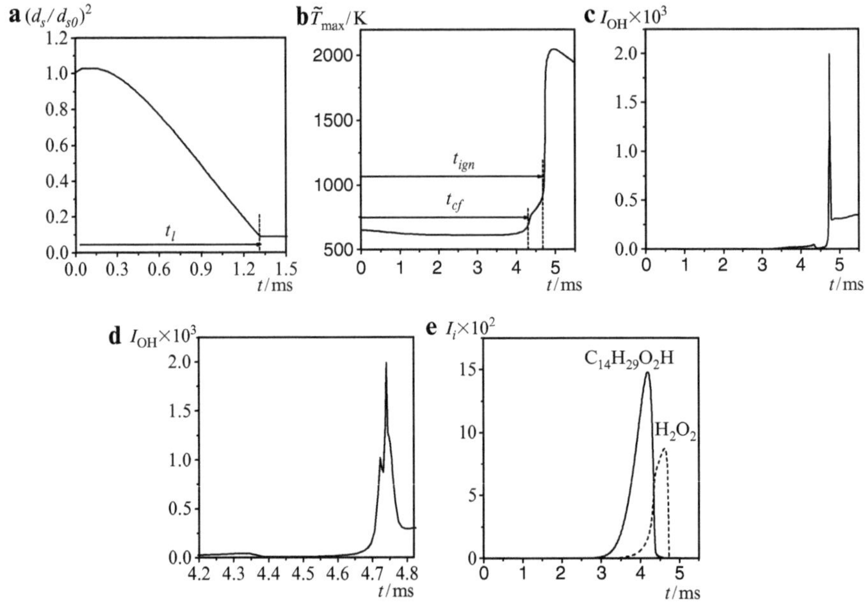

Fig. 6.32. Predicted time histories of the squared drop diameter $(d_s/d_{s0})^2$ **a**, maximum gas temperature $\tilde{T}_{max}$ **b**, normalized mass content of OH **c**, **d**, and normalized mass contents of $C_{14}H_{29}O_2H$ and H_2O_2 **e** at autoignition of an n-tetradecane drop with $d_{s0} = 20\,\mu$m ($p = 2.5$ MPa and $\tilde{T} = 650$ K) [163]

The drop lifetime t_l was determined as the time for the drop mass to decrease by a factor of 30. The arrow in Fig. 6.32a indicates the calculated drop lifetime with $d_{s0} = 20\,\mu$m ($t_l \approx 1.32$ ms). It is seen from Fig. 6.32b that at the beginning of the process the maximal gas temperature around the drop (at the external boundary of the computational domain) decreases owing to drop vaporization and heating. Until a time of 1.7 ms, the value of $\tilde{T}_{max}$ decreases by about 40 K and further temperature decrease stops owing to the drop vanishing and a growing heat release caused by chemical reactions. Autoignition occurs in two stages. At first, a cool flame comes into effect after a delay time of $t_{cf} = 4.33$ ms, which is accompanied by a temperature rise of about 200 K. Then a hot explosion occurs with a total delay time of $t_{ign} = 4.71$ ms, which is accompanied by a temperature rise to approximately 2,200 K.

Comparing Fig. 6.32a and b one can see that $t_{cf} > t_l$ for the drop with $d_{s0} = 20\,\mu$m. This means that fine drops completely evaporate before the development of the cool flame, and the cool-flame oxidation of fuel vapor proceeds in a relatively large volume. Such conditions are close to the conditions of cool-flame oxidation in a homogeneous mixture, where it was discovered and observed repeatedly. Further analysis of Fig. 6.32b and c or d (Fig. 6.32d is a zoomed view of Fig. 6.32c) indicates that the temperature rise in the cool flame coincides with the hydroxyl concentration buildup. On the other hand, when comparing Fig. 6.32c or d with e, one notices that the buildup of hydroxyl concentration coincides with decomposition of alkylhydroperoxide. In homogeneous mixtures, the latter process is responsible for a cool flame appearance. Note that at completion of the total ignition delay in Fig. 6.32d ($t_{ign} = 4.71$ ms), one can clearly see a local maximum in hydroxyl concentration, coinciding with hydrogen peroxide decomposition (owing to the reaction $H_2O_2 = OH+OH$ in the block of the detailed reaction mechanism of oxidation of C_1–C_2 hydrocarbons). In homogeneous mixtures, this process is responsible for a blue flame appearance. The blue flame appearance is not evident on the temperature curve of Fig. 6.32b because it virtually coincides with a hot explosion.

Figure 6.33 presents the predicted dependencies of the total ignition delay t_{ign} and cool flame ignition delay t_{cf} on the initial drop diameter d_{s0} at $p = 2.5$ MPa and $\tilde{T} = 650$ K. In addition to curves $t_{ign}(d_{s0})$ and $t_{cf}(d_{s0})$, Fig. 6.33 shows the predicted dependence of the drop lifetime on the initial drop diameter $t_l(d_{s0})$. As shown, $t_{cf} > t_l$ only for drops with $d_{s0} < 35\,\mu$m. At $d_{s0} < 35\,\mu$m, the total ignition delay is a weak function of drop diameter and the cool-flame ignition delay t_{cf} is a main contributor to t_{ign}. Note that the use of the temperature curve for determining t_{cf} becomes complicated for drops with $d_{s0} > 30\,\mu$m because the temperature curve starts to resemble the curve of single-stage high-temperature oxidation.

Consider now the case when $t_{cf} > t_l$ in the example of autoignition of an n-tetradecane drop $d_{s0} = 40\,\mu$m in diameter at $p = 2.5$ MPa and

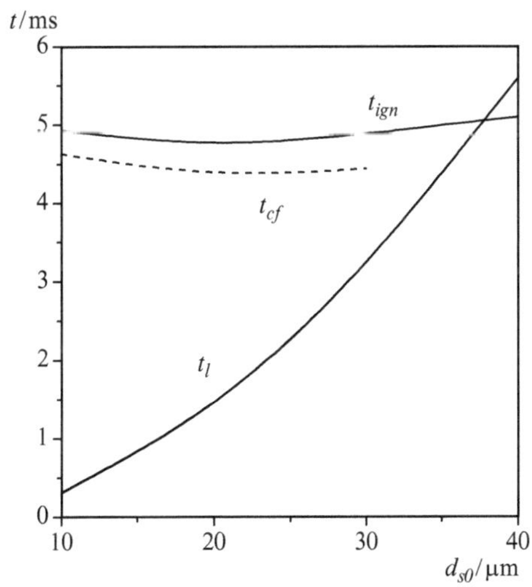

Fig. 6.33. Predicted dependencies of the total ignition delay t_{ign}, ignition delay of the cool flame t_{cf}, and drop lifetime t_l on initial drop diameter d_{s0} at $p_0 = 2.5$ MPa and $T_{g0} = 650$ K [163]

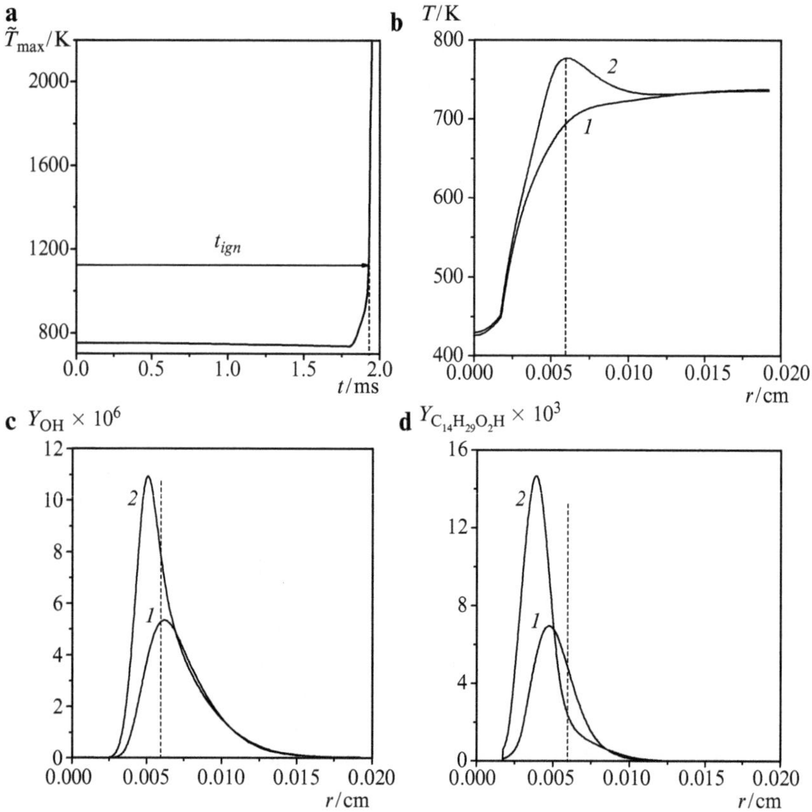

Fig. 6.34. Predicted time histories of maximum gas temperature $\tilde{T}_{max}$ **a**, instantaneous radial distributions of temperature **b**, and mass fractions of hydroxyl Y_{OH} **c** and alkylhydroperoxide $Y_{C_{14}H_{29}O_2H}$ **d** at $t = 1.75$ (curves *1*) and 1.85 ms (curves *2*) at *n*-tetradecane drop autoignition ($d_{s0} = 40\,\mu m$, $p_0 = 2.5$ MPa and $T_{g0} = 750$ K) [163]

$\tilde{T} = 750$ K. It follows from Fig. 6.34a that the total ignition delay in this case is $t_{ign} = 1.94$ ms (shown by an arrow). At the instant of autoignition the surface area of the drop decreased by about 20% and nearly 30% of the drop mass has evaporated. The cool flame is not evident at the temperature curve $\tilde{T}_{max}(t)$. It cannot be distinguished in the time histories of normalized mass contents of hydroxyl and alkylhydroperoxide. This is caused by a significant nonuniformity of the temperature and concentration fields around the drop. Nevertheless, careful analysis of these fields reveals that all features relevant to cool flame development in homogeneous mixtures are exhibited locally.

Figure 6.34b–d shows the instantaneous radial distributions of temperature, and mass fractions of hydroxyl Y_{OH} and alkylhydroperoxide $Y_{C_{14}H_{29}O_2H}$ at time instants $t = 1.75$ (curves 1) and 1.85 ms (curves 2). It follows from

Fig. 6.34b that before ignition, at a time instant between $t = 1.75$ and 1.85 ms, a localized temperature rise occurs at a distance $r = 0.006$ cm from the drop center (denoted by the dashed line). This temperature rise corresponds to a runaway reaction owing to a localized increase in the hydroxyl mass fraction at the same distance from the drop center (Fig. 6.34c), which, in turn, is caused by a localized decomposition of alkylhydroperoxide (Fig. 6.34d). A special computational experiment indicates that, when starting from $t = 1.75$ ms, the active decomposition products of alkylperoxide in the reaction $C_{14}H_{29}O + OH$ are replaced by considerably less reactive initial species $C_{14}H_{30} + O_2$, the reaction slows down sharply instead of being a runaway reaction, and the localized temperature rise is replaced by mixture cooling caused by drop vaporization. Thus, the reaction proceeds in a way similar to that in a homogeneous mixture, but locally. This is the reason why the cool-flame stage cannot be distinguished in the time histories of characteristic parameters such as maximum temperature of normalized mass contents of the various species.

6.3.9 Ignition of Disintegrating Drops

The characteristic breakup time of relatively large liquid drops behind a strong shock wave relevant to spray detonations is usually very small compared with the ignition delay. Whatever the prevailing mechanism of liquid drop atomization behind a shock wave, ignition occurs in the boundary layer surrounding the wake of the drop, which, according to experimental observations, is nearly conical in shape. Thus, both numerical simulations and evidence derived from shock-tube experiments suggest three major stages of the ignition and combustion process: (1) drop breakup, (2) ignition of the mixture at the wake periphery, and (3) diffusion-controlled fuel burning in the wake.

The rate control can be determined from simple considerations. Breakup of drops 100 μm in diameter is completed within less than 10 μs after the shock wave spreading at a velocity of 1,600 m s^{-1} at atmospheric pressure: the total breakup time of smaller drops is even shorter. However, the pressure measured behind detonation waves in sprays of low-volatility fuels shows that the so-called von Neumann spike, where the contribution of the reaction heat to flow characteristics is insignificant, lasts no less than 100 μs [166]. Hence, the heat release rate behind fuel–air detonation waves in sprays with physically attainable drop sizes is controlled by the other two stages.

Mixture autoignition is reasonably assumed to occur at the periphery of the cloud arising in the course of drop breakup because the gas in the wake is colder than the free-stream gas and fuel–oxidizer mixture. Therefore, the capability for autoignition most probably exists at the boundary between the free-stream and stripped droplet material. Rough estimates [167] show that the amount of fuel mixed with air in this boundary layer is small: no more than 16% over the total breakup time for 100-μm particles.

Autoignition in the boundary layer is controlled by chemical kinetics and may occur either during or after breakup subsequent to an induction period,

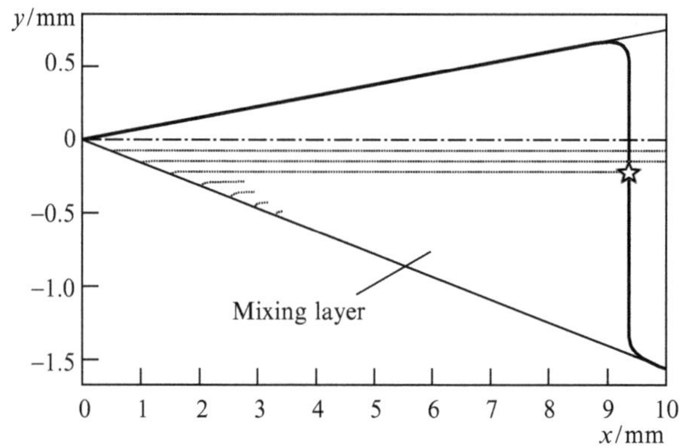

Fig. 6.35. Predicted autoignition location in the planar, isothermal, turbulent mixing layer of a fine fuel spray with air [168, 169]. *Thin solid lines* show the boundaries of the mixing layer. The *thick solid curve* denotes the locus of the limiting normalized preexplosion heating $R\tilde{T}/E = 0.05$. *Dashed curves* show the predicted mean trajectories of notional particles in the mixing layer. The *star* denotes the self-ignition location, where the notional particle residence time t is equal to the ignition delay t_{ign}

which can be calculated approximately using the free-stream temperature and pressure. For example, Fig. 6.35 shows the predicted autoignition location (denoted by a star) in the planar, isothermal, turbulent mixing layer of a fine fuel spray in air [168,169]. In the calculations, the following values of the governing parameters were used: air density $30\,\mathrm{kg\,m^{-3}}$, fuel density $850\,\mathrm{kg\,m^{-3}}$, activation energy $8{,}800\,\mathrm{cal\,mol^{-1}}$, temperature 900 K, pressure 4 MPa, and fuel spray velocity at the nozzle exit $100\,\mathrm{m\,s^{-1}}$. Ignition occurs at the periphery of the mixing layer at a distance of about 9 mm from the layer origin and 1.2 mm from the outer (air) boundary of the layer. Although the study in [168, 169] dealt with diesel spray, the results obtained can be readily applied to the autoignition phenomenon in the conelike mixing layer attached to shattered drops behind a shock wave.

6.4 Concluding Remarks

Current understanding of the physical and chemical processes accompanying solid and liquid particle ignition behind shock waves is based on the simplified treatment of various interactions between particles and the shock-induced flow. Analysis of these interactions indicates that the phenomena encountered are very complex and interrelated. Even simple particle/drop ignition models exhibit numerous scenarios of particle/drop temperature evolution behind incident and reflected shock waves. For liquid drops, the phenomena of drop breakup combined with the formation of a mixing layer of micromist droplets with air, micromist vaporization, ignition, and combustion exhibit many features that have not yet been studied. The essential role is played by various local rather than averaged processes, implying that a multidimensional treatment of the problem is inevitably required.

References

1. Beckstead, M.W.: A summary of aluminum combustion. RTO-EN 023 Paper presented at the RTO/VKI special course on internal aerodynamics in solid rocket propulsion, Rhode-Saint-Genese, 27–31 May 2002
2. Law, C.K.: Recent advances in droplet vaporization and combustion. Prog. Energy Combust. Sci. **8**, 171–201 (1982)
3. Sirignano, W.A.: Fuel droplet vaporization and spray combustion theory. Prog. Energy Combust. Sci. **9**, 291–322 (1983)
4. Bachalo, W.D.: Injection, dispersion, and combustion of liquid fuels. In: Proceedings 25th Symposium (International) on Combustion, pp. 333–344. The Combustion Institute, Pittsburgh (1994)
5. Mashayek, F., Pandya, R.V.R.: Analytical description of particle laden flows. Prog. Energy Combust. Sci. **29**, 329–378 (2003)
6. Brzustowski, T.A., Glassman, I.: Vapor phase. diffusion flames. In: Wolfhard, H.G., Glassman, I., Green, L. (eds.) Heterogeneous Combustion, pp. 75–115. Academic, New York (1964)
7. Klyachko, L.A.: Ignition of particle group with heterogeneous reaction. Teploenergetika **8**, 65 (1966)
8. Khaikin, B.I., Bloshenko, V.N., Merzhanov, A.G.: On the ignition of metal particles. Combust. Explos. Shock Waves **6**, 412–422 (1970)
9. Aldushin, A.P., Bloshenko, V.N., Seplyarskii, B.S.: Ignition of metal particles with a logarithmic oxidation law. Combust. Explos. Shock Waves **9**, 423–428 (1973)
10. Law, C.K.: A simplified theoretical model for the vapor-phase combustion of metal particles. Combust. Sci. Technol. **7**, 197–212 (1973)
11. Varshavskii, G.A.: Drop combustion. Trans. NII-1 **6**, 30 (1945)
12. Godsave, G.A.E.: Studies of the combustion of drops in a fuel spray – the burning of single drops of fuel. In: Proceedings 4th Symposium (International) on Combustion, pp. 818–830. Williams & Wilkins, Baltimore (1953)
13. Spalding, D.B.: The combustion of liquid fuels. In: Proceedings 4th Symposium (International) on Combustion, pp. 847–864. Williams & Wilkins, Baltimore (1953)
14. Goldsmith, M., Penner, S.S.: On the burning of single drops of fuel in an oxidizing atmosphere. Jet Propuls. **24**(4), 245 (1954)
15. Frank-Kamenetskii, D.A.: Diffusion and Heat Transfer in Chemical Kinetics. Nauka, Moscow (1961)
16. Twardus, E.M., Brzustowski, T.A.: The interaction between two burning fuel droplets. Arch. Processow Spal. **8**, 347–358 (1977)
17. Dwyer, H.A., Nirschl, H., Kerschl, P., Denk, V.: Heat, mass, and momentum transfer about arbitrary groups of particles. In: Proceedings 25th Symposium (International) on Combustion, pp. 389–396. The Combustion Institute, Pittsburgh (1994)

18. Marberry, M., Ray, A.K., Leung, K.: Effect of multiple particle interactions on burning droplets. Combust. Flame **57**, 237 (1984)
19. Cuadros, J., Linares, J., Sivasankaran, K., Seetharamu, K.N., Natarajan, R.: Numerical investigation of the interference effects between two burning fuel spheres. Int. J. Heat Mass Transfer **39**, 3949–3957 (1996)
20. Chiu, H.H., Liu, T.M.: Group combustion of liquid droplets. Combust. Sci. Technol. **17**, 127–142 (1977)
21. Correa, S.M., Sichel, M.: The group combustion of a spherical cloud of monodisperse fuel droplets. In: Proceedings 19th Symposium (International) on Combustion, pp. 856–863. The Combustion Institute, Pittsburgh (1983)
22. Nigmatulin, R.I.: Dynamics of Multiphase Media, vol. 1. Nauka, Moscow (1987)
23. Borisov, A.A., Kogarko, S.M., Kozenko, V.P.: Detonation limits of methane–oxygen mixtures diluted with argon or helium. Combust. Explos. Shock Waves **2**, 55 (1967)
24. Nettleton, M.A., Stirling, R.: The ignition of clouds of particles in shock-heated oxygen. Proc. R. Soc. Lond. A **300**, 62–77 (1967)
25. Macek, A., Semple, J.M.: Combustion of boron particles at atmospheric pressure. Combust. Sci. Technol. **1**(3), 181–191 (1969)
26. Ryzhik, A.B., Yurmanov, Yu.A., Limonov, B.S., et al.: Ignition of dispersed magnesium behind shock waves. Phys. Aerodispersed Syst. **9**, 115 (1973)
27. Fox, T.W., Rackett, C.W., Nicholls, J.A.: Shock wave ignition of magnesium powders. In: Proceedings 11th International Shock Tubes and Waves Symposium, pp. 262–268, Seattle (1978)
28. Cohen, A., Decker, L.: Shock tube ignition of nitrocellulose and nitramines. In: Proceedings 10th JANAF Meeting, CPIA Publ., Laurel, MD (1978)
29. Cohen, A., Decker, L.: Shock tube ignition of nitrocellulose. In: Proceedings 12th Symposium (International) Shock Tubes, Jerusalem, 16–19, pp. 514–520 July 1979
30. Kauffman, C.W., Wolanski, P., Ural, E., et al.: Shock wave initiated combustion of grain dust. In: Proceedings Symposium (International) on Grain Dust, pp. 164–190, Manhattan (1979)
31. Breipohl, G., Lester, T.W., Merklin, J.F.: Shock tube studies of the mechanism of grain dust ignition. In: Proceedings Symposium (International) on Grain Dust, pp. 191–211, Manhattan (1979)
32. Seeker, W.R., Lester, T.W., Merklin, J.F.: Shock tube. techniques in the study of pulverized coal ignition and burnout. Rev. Sci. Instrum. **51**(11), 1523–1531 (1980)
33. Ural, E.A., Sichel, M., Kauffman, C.W.: Shock wave ignition of pulverized coal. In: Treanor, C.E., Hall, J.G. (eds.) Shock Tubes and Waves, pp. 809–817. State University of New York Press, Albany (1981)
34. Wolanski, P.: Problems of dust explosions. In: Proceedings 1st Specialists Meeting (International) of the Combustion Institute, pp. 497–502, Bordeaux (1981)
35. Wolanski, P.: Fundamental problems of dust explosions. In: Lee, J.H.S., Guirao C.M. (eds.) Fuel–Air Explosions, pp. 349–373. University of Waterloo Press, Waterloo (1982)
36. Borisov, A.A., Gelfand, B.E., Tsyganov, S.A., et al.: Ignition of dusts behind shock waves. Chem. Phys. Rep. **2**(8), 40 (1983)
37. Borisov, A.A., Gelfand, B.E., Timofeev, E.I., et al.: Ignition of dust suspensions behind shock waves. In: Bowen, J.R., Manson, N., Oppenheim, A.K.,

Soloukhin, R.I. (eds.) Dynamics of Shock Waves, Explosions, and Detonations. Progress in Astronautics and Aeronautics, vol. 94, pp. 332–339. American Institute of Aeronautics and Astronautics, New York (1984)
38. Sichel, M., Baek, S.W., Kauffman, C.W., et al.: The shock wave ignition of dusts. AIAA J. **23**(9), 1374–1380 (1985)
39. Boiko, V.M., Lotov, V.V., Papyrin, S.V.: Ignition of metal powders in reflected shock waves. Arch. Combust. **8**(2), 101–114 (1988)
40. Li, S.C., Williams, F.A., Takahashi, F.: In: Proceedings 22nd Symposium (International) on Combustion, pp. 1951–1960. The Combustion Institute, Pittsburgh (1988)
41. Wolanski, P.: Deflagration and detonation combustion of dust mixtures. In: Kuhl, A. L., Leyer, J.-C., Borisov, A.A., Sirignano, W.A. (eds.) Dynamics of Deflagrations and Reactive Systems: Heterogeneous Combustion. Progress in Astronautics and Aeronautics, vol. 132, pp. 3–31. American Institute of Aeronautics and Astronautics, New York (1991)
42. Boiko, V.M., Papyrin, S.V., Poplavskii, S.V.: Mechanism of dust ignition in incident shock waves. Combust. Explos. Shock Waves **29**(3), 389–394 (1993)
43. Boiko, V.M., Papyrin, S.V., Poplavskii, S.V.: Ignition mechanism of coal suspension in shock waves. In: Kuhl, A.L., Leyer, J.-C., Borisov, A.A., Sirignano, W.A. (eds.) Dynamic Aspects of Explosion Phenomena. Progress in Astronautics and Aeronautics, vol. 154, pp. 278–290. American Institute of Aeronautics and Astronautics, New York (1993)
44. Roberts, T.A., Burton, R.L., Krier, H.: Ignition and combustion of aluminum/magnesium alloy particles in O_2 at high pressures. Combust. Flame **92**(1–2), 125–143 (1993)
45. Geng, J.H., van de Ven, A., Zhang, F., et al.: A new setup to measure ignition delay of a dust suspension behind an incident shock wave. In: Xufan, D., Wolanski, P. (eds.) Proceedings 6th Colloquium (International) on Dust Explosions, pp. 309–314. Northeastern University Press, Shenyang (1994)
46. Spalding, M. J., Krier, H., Burton, R. L.: Boron suboxides measured during ignition and combustion of boron in shocked $Ar/F/O_2$ and $Ar/N_2/O_2$ mixtures. Combust. Flame **120**, 200–210 (2000)
47. Servaites, J., Krier, H., Melcher, J. C., et al.: Ignition and combustion of aluminum particles in shocked $H_2O/O_2/Ar$ and $CO_2/O_2/Ar$ mixtures. Combust. Flame **125**, 1040–1054 (2001)
48. Dreizin, E.L., Trunov, M.A., Suslov, A.V.: General trends in metal particles heterogeneous combustion. Combust. Sci. Technol. **90**, 79–99 (1993)
49. Dreizin, E.L., Trunov, M.A.: Surface phenomena in aluminum combustion. Combust. Flame **101**, 378–382 (1995)
50. Temos, J., Pratt, H.R.C., Stevens, G.W., Dreizin, E.L.: Experimental study of stages in aluminum particle combustion in air. Combust. Flame **105**(4), 541–556 (1996)
51. Molodetsky, I.E., Vicenzi, E.P., Dreizin, E.L., Law, C.K.: Phases of titanium combustion in air. Combust. Flame **112**(4), 522–532 (1998)
52. Dreizin, E.L.: Experimental study of aluminum particle flame evolution in normal and micro-gravity – progress in astronautics and aeronautics. Combust. Flame **116**(3), 323–333 (1999)
53. Dreizin, E.L., Berman, C.H., Vicenzi, E.P.: Condensed-phase modifications in magnesium particle combustion in air. Combust. Flame **122**(1), 30–42 (2000)

54. Trunov, A., Schoenitz, M., Zhu, X., et al.: Effect of polymorphic phase transformations in Al_2O_3 film on oxidation kinetics of aluminium powders. Combust. Flame **140**, 310–318 (2005)
55. Takeno, T., Yuasa, S.: Ignition of magnesium and magnesium-aluminum alloy by impinging hot-air stream. Combust. Sci. Technol. **21**, 109–121 (1980)
56. Fassel, W.M., Papp, C.A., Hildenbrand, D.L., Sernka, R.P.: The experimental nature of the combustion of metal powders. In: Summerfield, M. (ed.) Solid Propellant Rocket Research, pp. 259–270. Academic, New York (1960)
57. Gusachenko, E.I., Stesik, L.N., Fursov, V.P., et al.: Combust. Explosion Shock Waves **10**, 548 (1974)
58. Shevtsov, V.I., Fursov, V.P., Stesik, L.N.: Combust. Explos. Shock Waves **12**, 859 (1976)
59. Cassel, H.V., Liebman, I.: The cooperative mechanism in the ignition of dust dispersions. Combust. Flame **3**(4), 467–472 (1959)
60. Cassel, H.V., Liebman, I.: Combustion of magnesium particles, II. Ignition temperatures and thermal conductivities of ambient atmospheres. Combust. Flame **7**(1), 79–81 (1963)
61. Lermant, J.C., Yip, S.: A generalized Semenov model for thermal ignition in nonuniform temperature systems. Combust. Flame **7**(1), 21–30 (1963)
62. Gurevich, M.A., Stepanov, A.M.: Metal particle ignition. Combust. Explos. Shock Waves **3**, 334–342 (1968)
63. Gurevich, M.A., Stepanov, A.M.: The limiting conditions of metal particle ignition. Combust. Explos. Shock Waves **4**, 189–185 (1970)
64. Khaikin, B.I., Bloshenko, V.I., Merzhanov, A.G.: On the ignition of metal particles. Combust. Explos. Shock Waves **4**, 474–488 (1970)
65. Ezhovsky, G.K., Mochalova, A.S., Ozerov, E.S., et al.: Ignition and combustion of magnesium particle. In: Combustion and Explosion, pp. 234–240. Nedra, Moscow (1972)
66. Ezhovsky, G.K., Ozerov, E.S.: Ignition of powder-like magnesium. Combust. Explos. Shock Waves **6**, 845–851 (1974)
67. Gremyachkin, V.M., Istratov, O.I.: Combust. Explos. Shock Waves **11**, 366 (1974)
68. Bloshenko, V.I., Merzhanov, A.G., Khaikin, B.I.: On the question of determining the kinetic parameters of high-temperature oxidation of magnesium. Combust. Explos. Shock Waves **5**, 682–688 (1976)
69. Derevyaga, M.V., Stesik, L.N., Fedorin, N.A.: Experimental investigation of critical conditions for magnesium ignition. Combust. Explos. Shock Waves **6**, 44–49 (1978)
70. King, M.K.: Modeling of single-particle aluminum combustion in O_2–N_2 atmospheres. In: Proceedings 17th Symposium (International) on Combustion, pp. 1317–1328. The Combustion Institute, Pittsburgh (1979)
71. Fursov, V.I., Shevtsov, V.I., Gusachenko, E.I., et al.: The role of the process of evaporation of high-volatile metals in the mechanism of their high-temperature oxidation and ignition. Combust. Explos. Shock Waves **3**, 3–12 (1980)
72. Medvedev, A.E., Fedorov, A.V., Fomin, V.M.: Mathematical modeling of metal particle ignition in the high-temperature flow behind a shock. Combust. Explos. Shock Waves **18**(3), 261–265 (1982)
73. Medvedev, A.E., Fedorov, A.V., Fomin, V.M., et al.: Mathematical modeling of ignition process of aerodispersed systems. In: Proceedings 3rd International School of Industrial Dust Explosions, pp. 65–79, Turava (1982)

74. Boiko, V.M., Fedorov, A.V., Fomin, V.M., et al.: Ignition of small solid particles behind shock waves. In: Bowen, J.R., Manson, N., Oppenheim, A.K., Soloukhin, R.I. (eds.) Shock Waves, Explosions, and Detonations. Progress in Astronautics and Aeronautics, vol. 87, pp. 71–87. American Institute of Aeronautics and Astronautics, New York (1983)
75. Kazakov, Yu.V., Medvedev, A.E., Fedorov, A.V., et al.: Mathematical modelling of ignition in dusty gases. Arch. Combust. **7**(1–2), 7–17 (1987)
76. Fedorov, A.V., Tetenov, E.V., Veyssiere, B.: Ignition of a suspension of metal particles with an actual explosion. I. Statement of the problem and solution in a self-modeling approximation. Combust. Explos. Shock Waves **27**(5), 527–532 (1991)
77. Petukhova, E.V., Fedorov, A.V.: Ignition of magnesium particles near the end of a shock tube. Combust. Explos. Shock Waves **27**(6), 778–780 (1991)
78. Fedorov, A.V.: Physical and Mathematical Modeling of Ignition of Fine Magnesium Particles. Preprint no. 12–94, pp. 1–30. Institute of Theoretical and Applied Mechanics of the Siberian Branch of the Russian Academy of Sciences, Novosibirsk (1994)
79. Makino, A.: An approximation explicit expression for the combustion rate of a small carbon particle. Combust. Flame **90**, 143–154 (1992)
80. Zhou, W., Yetter, R.A., et al.: A comprehensive physical and numerical model of boron particle ignition. In: Proceedings 26th Symposium (International) on Combustion, pp. 1909–1918. The Combustion Institute, Pittsburgh (1996)
81. Hayhurst, A.N.: The mass transfer coefficient for oxygen reacting with a carbon particle in a fluidized or packed bed. Combust. Flame **121**(4), 679–688 (2000)
82. Foertsch, D., Schnell, U., Hein, K.R.G.: The mass transfer coefficient for the combustion of pulverized carbon particles. Combust. Flame **126**, 1662–1668 (2001)
83. Tao, H.: Shock wave ignition of aluminum particles. J. Phys. IV **12**(7), 105–112 (2002)
84. Meinkoh, D.: Liquid oxide surface layer in metal combustion. Combust. Theory Modelling **8**, 315–338 (2004)
85. Vilyunov, V.N.: Theory of Ignition of Condensed Substances. Nauka, Novosibirsk (1984)
86. Gosteev, Yu.A., Fedorov, A.V.: Magnesium-particle ignition (distributed model). Combust. Explos. Shock Waves **32**(4), 363–369 (1996)
87. Borovikov, M.B., Goldschleger, U.I.: Critical phenomena in a system with two parallel exo-and endothermal reactions. Dokl. USSR Acad. Sci. **261**(2), 392 (1981)
88. Meinkoehn, D.: Metal particle ignition and oxide layer instability. In: Borisov, A., Frolov, S., Kuhl, A. (eds.) Progress in Combustion and Detonation, pp. 180–181. Torus, Moscow (2004)
89. Ranz, W.E., Marshall, W.R., Jr.: Evaporation from drops, part I. Chem. Eng. Prog. **48**(3), 141–146 (1952)
90. Avdeev, K.A., Frolov, F.S., Frolov, S.M.: Effect of transient heat transfer on metal particle ignition. In: Roy, G., Frolov, S., Sinibaldi, J. (eds.) Pulsed and Continuous Detonations, pp. 72–83. Torus, Moscow (2006)
91. Avdeev, K.A., Frolov, F.S., Frolov, S.M., Basara, B.: Effect of transient heat transfer on metal particle ignition. In: Hanjalic, K., Nagano, Y., Jakirlic, G. (eds.) Turbulence, Heat and Mass Transfer, vol. 5, pp. 581–584. Begell House, New York-Wallingford (UK) (2006)

92. Gilmore, R.: Catastrophe Theory for Scientists and Engineers. Wiley-Interscience, New York (1981)
93. Fedorov, A.V.: Numerical and analytical study of magnesium particle ignition. Combust. Explos. Shock Waves **32**(1), 64–72 (1996)
94. Merzhanov, A.G.: Thermal theory of metal particle ignition. AIAA J. **13**(2), 209–214 (1975)
95. Grigor'ev, Yu.M., Vakina, Z.G.: Critical ignition conditions for metals with a logarithmic oxidation law. Combust. Explos. Shock Waves **15**(1), 51–53 (1979)
96. Alekseeva, T.I., Gurevich, M.A., Ozerov, E.S.: Ignition of an aluminum particle. Tr. Leningr. Politekh. Inst. **280**, 98–106 (1967)
97. Gurevich, M.A., Ozerov, E.S., Yurinov, A.A.: Effect of an oxide film on the inflammation characteristics of aluminum. Combust. Explos. Shock Waves **14**(4), 448–451 (1978)
98. Gurevich, M.A., Ozerova, G.E., Stepanov, A.M.: Heterogeneous ignition of an aluminum particle in oxygen and water vapor. Fiz. Goreniya Vzryva **6**(3), 326–335 (1970)
99. Friedman, R., Macek, A.: Ignition and combustion of aluminum particles in hot ambient gases. Combust. Flame **6**, 9–19 (1962)
100. Pokhil, P.F., Belyaev, A.F., Frolov, Yu.V., et al.: Combustion of Powdered Metals in Active Media. Nauka, Moscow (1972)
101. Boiko, V.M., Lotov, V.V., Papyrin, A.N.: Ignition of gas suspensions of metallic powders in reflected shock waves. Combust. Explos. Shock Waves **25**(2), 193–199 (1989)
102. Gurevich, M.A., Lapkina, K.I., Ozerov, E.S.: Limiting ignition conditions of an aluminum particle. Fiz. Goreniya Vzryva **6**(2), 172–176 (1970)
103. Fedorov, A.V., Kharlamova, Yu.V.: Ignition of an aluminum particle. Combust. Explos. Shock Waves **39**(5), 544–547 (2003)
104. Grigor'ev, Yu.M., Gal'chenko, Yu.A., Merzhanov, A.G.: Investigation of the rate of the high-temperature reaction between aluminum and oxygen using the ignition method. Combust. Explos. Shock Waves **9**(2), 162–167 (1973)
105. Gear, C.W.: The automatic integration of ordinary differential equations. Comput. Struct. **20**(6), 915–920 (1985)
106. Afanasieva, K.A., Levin, V.A.: Ignition of aluminum particles behind detonation and explosion waves. Chem. Phys. Rep. **3**(9), 1328–1332 (1984)
107. Fedorov, A.V., Tetenov, E.V., Veyssiere, B.: Dynamics and ignition of metal particles dispersed in atmosphere at a real explosion. Preprint no. 6–90. Institute of Theoretical and Applied Mechanics of the Siberian Branch of the Russian Academy of Sciences, Novosibirsk (1990)
108. Boiko, V.M., Papyrin, A.N.: On the ignition of magnesium particles behind reflected shock waves. In: Proceedings VIII All-Union Symposium Combustion and Explosion, pp. 65–68, Chernogolovka, 13–17 Oct 1986
109. Rosenband, V.: Thermo-mechanical aspects of the heterogeneous ignition of metals. Combust. Flame **137**(3), 366–375 (2004)
110. Yanenko, N.N., Soloukhin, R.I., Fomin, V.M., et al.: Supersonic Two-Phase Flows in Conditions of the Particles Velocity Non-Equilibrium. Nauka, Novosibirsk (1980)
111. Zubarev, V.N., Kozlov, A.D., Kuznetsov, V.M., et al. Thermophysical properties of technically important gases. Reference Book, p. 232. Energoatomizdat, Moscow (1989)

112. Fedorov, A.V.: Ignition of gaseous suspensions in an interacting continuum regime. Combust. Explos. Shock Waves **34**(4), 418–425 (1998)
113. Medvedev, A.E., Fedorov, A.V., Fomin, V.M.: Ignition of metal particles in high-temperature flow behind the shock wave. Preprint no. 33. Institute of Theoretical and Applied Mechanics of the Siberian Branch of the Russian Academy of Sciences Siberian Branch, Novosibirsk (1981)
114. Fedorov, A.V., Fomin, V.M.: Modeling of detonation flows in reacting gas-particle mixtures. In: Borisov, A., Frolov, S., Kuhl, A. (eds.) Progress in Combustion and Detonation, pp. 329–330. Torus, Moscow (2004)
115. Fedorov, A.V., Fomin, V.M.: Numerical study of flows of reacting composite mixtures. J. Appl. Mech. Tech. Phys. **40**(2), 300–307 (1999)
116. Fedorov, A.V., Gosteev, Yu.A.: A numerical study of ignition of magnesium samples in spherical and cylindrical symmetry. In: Proceedings 16th International Colloquium on the Dynamics of Explosions and Reactive Systems, pp. 299–302. AGH, Cracow (1997)
117. Gosteev, Yu.A., Fedorov, A.V.: Ignition of a cloud of metal particles in the continuum regime I. Adiabatic flow. Combust. Explos. Shock Waves **35**(5), 493–500 (1999)
118. Gosteev, Yu.A., Fedorov, A.V.: Ignition of a cloud of metal particles in the continuum regime. II. Nonadiabatic flow. Combust. Explos. Shock Waves **35**(6), 684–689 (1999)
119. Gosteev, Y.A., Fedorov, A.V.: Mathematical study of thermal explosion of a magnesium particle with allowance for metal evaporation. Combust. Explos. Shock Waves **34**(2), 151–158 (1998)
120. Wierzba, A.: Deformation and breakup of liquid drops in a gas stream at nearly critical Weber numbers. Exp. Fluids **9**(1–2), 59–64 (1990)
121. Gelfand, B.E., Gubin, S.A., Kogarko, S.M.: Specific features of the breakup of viscous liquid drop. Eng. Phys. J. **25**(3), 467 (1973)
122. Olim, M., Igra, O., Mond, M., Ben-Dor, G.: A general attenuation law of planar shock waves propagating into dusty gases. In: Groenig, H. (ed.) Shock Tubes and Waves, pp. 217–225. HI, Aachen (1987)
123. Simpkins, P.G., Bales, E.L.: Water-drop response to sudden accelerations. J. Fluid Mech. **55**(4), 629–639 (1972)
124. Boiko, V.M., Papyrin, A.N., Poplavskii, S.V.: Dynamics of droplet breakup in shock waves. J. Appl. Mech. Tech. Phys. **28**, 263–269 (1987)
125. Wadewitz, A., Specht, E.: Limit value of Nusselt number for particle of different shape. Int. J. Heat Mass Transfer **44**, 967–975 (2001)
126. Gordon, G.D.: Mechanism and speed of breakup of drops. J. Appl. Phys. **30**, 1759 (1959)
127. Volynskii, M.S., Lipatov, A.S.: Deformation and breakup of drops in gas flow. Eng. Phys. J. **18**(5), 838 (1970)
128. Davidson, V.E.: On drop deformation in gas flow. In: Belyaev, N.M., et al. (eds.) Jets and Flow in Tubes, pp. 3–35, Dnipropetrovsk(1974)
129. Detkovskii, D.A., Frolov, S.M.: Model of the deformation of a liquid droplet in a gas flow. J. Appl. Mech. Tech. Phys. **35**(6), 911–919 (1994)
130. O'Rourke, P.J., Amsden, A.A.: The TAB method for numerical calculation of spray droplet breakup. SAE Pap. 872089 (1987)
131. Borisov, A.A., Gelfand, B.E., Gubin, S.A., et al.: Detonation reaction zone in two-phase mixtures. Combust. Explos. Shock Waves **6**, 327–336 (1970)
132. Zhdan, S.A.: Calculation of a spherical heterogeneous detonation. Combust. Explos. Shock Waves **12**(4), 586–594 (1976)

133. Eidelman, S., Burkat, A.: Evolution of a detonation wave in a cloud of fuel droplets. I – influence of igniting explosion. AIAA J. **18**(9), 1103–1109 (1980)
134. Burkat, A., Eidelman, S.: Evolution of a detonation wave in a cloud of fuel droplets. II – influence of fuel droplets. AIAA J. **18**(10), 1233–1236 (1980)
135. Sichel, M.: In: Oran, E.S., Boris, J.P. (eds.) Numerical Approaches to Combustion Modeling. Numerical modeling of heterogeneous detonations. Progress in Astronautics and Aeronautics, vol. 135, pp. 447–458. American Institute of Aeronautics and Astronautics, New York (1991)
136. Gubin, S.A., Sichel, M.: Calculations of the detonation velocity of a mixture of liquid fuel droplets and a gaseous oxidizer. Combust. Sci. Technol. **17**(3–4), 109–111 (1977)
137. Borisov, A.A., Gelfand, B.E., Gubin, S.A, et al.: On drop deformation in the reaction zone of two-phase detonation. J. Appl. Mech. Tech. Phys. **5**, 39 (1970)
138. Borisov, A.A., Gelfand, B.E., Gubin, S.A., et al.: The reaction zone of two-phase detonations. Astronaut. Acta **15**, 411 (1970)
139. Zhdan, S.A.: Calculation of heterogeneous detonation taking into account deformation and breakdown of fuel droplets. Combust. Explos. Shock Waves **13**(2), 217–221 (1977)
140. Voronin, D.V., Zhdan, S.A.: Calculation of heterogeneous detonation initiation for a hydrogen-oxygen mixture in an explosion tube. Combust. Explos. Shock Waves **20**(4), 461–465 (1984)
141. Prakash, S., Sirignano, W.A.: Liquid fuel droplet heating with internal circulation. J. Heat Mass Transfer **21**, 885–895 (1978)
142. Smetanyuk, V.A., Frolov, S.M.: Hydrocarbon fuel drop vaporization and combustion. III: drop heating in gas flow with regard for internal liquid circulation. Khim. Fiz. **23**(7), 40 (2004)
143. Helenbrook, B.T., Edwards, C.F.: Quasi-steady deformation and drag of uncontaminated liquid drops. Int. J. Multiphase Flow **28**(10), 1631–1657 (2002)
144. Frolov, S.M., Smetanyuk, V.A.: Heat and mass transfer of a drop with gas flow. Khim. Fiz. **25**(4), 42 (2006)
145. Frolov, S.M., Frolov, F.S., Basara, B.: Mathematical model for transient droplet vaporization. In: Roy, G., Frolov, S., Starik, A. (eds.) Nonequilibrium Processes. Combustion and Detonation, vol. 1, pp. 179–193. Torus, Moscow (2005)
146. Dukowicz, J.K.: Quasi-steady droplet phase change in the presence of convection. Informal report LA7997-MS. Los Alamos Science Laboratory, Los Alamos (1979)
147. Liu, A.B., Mather, D., Reitz, R.D.: Modeling the effects of drop drag and breakup on fuel sprays. SAE Pap. 930072 (1993)
148. Clift, R., Grace, J.R., Weber, M.E.: Bubbles, Drops and Particles. Academic, New York (1978)
149. Frolov, S.M., Posvianskii, V.S., Basevich, V.Ya., et al.: Hydrocarbon fuel drop vaporization and combustion. II: nonempirical model of drop vaporization with regard for multicomponent diffusion. Khim. Fiz. **23**(4), 75 (2004)
150. Pilch, M., Erdman, C.: Use of breakup time data and velocity history data to predict the maximum size of stable fragments for acceleration-induced breakup of a liquid drop. Int. J. Multiphase Flow **13**, 741–757 (1987)
151. Joseph, D.D., Belanger, J., Beavers, G.S.: Vaporization of a liquid drop suddenly exposed to a high-speed airstream, July 15, 2003, http://www.aem.umn.edu/research/Aerodynamic_Breakup/

152. Reinecke, W.G., Waldman, G.D.: A study of drop breakup behind strong shocks with applications to flight. Technical report SAMSO-TR-70142. Avco Systems Division, Wilmington, MA (1970)
153. Frolov, S.M., Basevich, V.Ya., Posvianskii, V.S., et al.: Hydrocarbon fuel drop vaporization and combustion. IV: drop vaporization with regard for spray effects. Khim. Fiz. **23**(7), 49 (2004)
154. Frolov, S.M., Basevich, V.Ya., Belyaev, A.A., et al.: Modeling of drop vaporization and combustion with regard for spray effects. In: Roy, G.D., Frolov, S.M., Starik, A.M. (eds.) Combustion and Pollution: Environmental Effect, pp. 117–132. Torus, Moscow (2005)
155. Basevich, V.Y., Belyaev, A.A., Evlampiev, A.V., et al.: Hydrocarbon fuel drop vaporization and combustion. I: nonempirical model of single-component drop vaporization. Khim. Fiz. **21**(3), 58 (2002)
156. Yang, J.-R., Wong, S.C.: On the suppression of negative temperature coefficient (NTC) in autoignition of n-heptane droplets. Combust. Flame **132**(3), 475–491 (2003)
157. Frolov, S.M., Basevich, V.Ya., Posvianskii, V.S.: Limiting drop size and prevaporization degree required for spray detonation. In: Roy, G.D., Frolov, S.M., Shepherd, J. (eds.) Application of Detonation to Propulsion, pp. 110–119. Torus, Moscow (2004)
158. Basevich, V.Ya., Frolov, S.M., Posvianskii, V.S.: Existance conditions for stationary heterogeneous detonations. Khim. Fiz. **24**(7), 60 (2005)
159. Basevich, V.Ya., Belyaev, A.A., Frolov, S.M.: Overall kinetic mechanisms for modeling turbulent reactive flows. Part I: main chemical process of heat release. Chem. Phys. Rep. **17**(9), 112 (1998)
160. Basevich, V.Ya., Belyaev, A.A., Frolov, S.M.: Overall kinetic mechanisms for modeling turbulent reactive flows. Part II: nitrogen oxide formation. Chem. Phys. Rep. **17**(10), 71 (1998)
161. Evlampiev, A.V., Frolov, S.M., Basevich, V.Ya., et al.: Overall kinetic mechanisms for modeling turbulent reactive flows. Part IV: diffusion combustion. Chem. Phys. Rep. **20**(11), 21 (2001)
162. Basevich, V.Ya., Belyaev, A.A., Brandstaetter, W., et al.: Kinetic scheme for modeling autoignition of iso-octane and n-heptane blends in air. Combust. Explos. Shock Waves **30**(6), 15 (1994)
163. Basevich, V.Ya., Frolov, S.M., Posvianskii, V.S., et al.: Low-temperature drop autoignition. Khim. Fiz. **24**(5), 71 (2005)
164. Takei, M., Kobayashi, H., Niioka, T.: Ignition experiment of a blended-fuel droplet in a microgravity field. Int. J. Microgravity Res. Appl. Microgravity Sci. Technol. **VI**(3), 184–187 (1993)
165. Niioka, T., Kobayashi, H., Mito, D.: Ignition experiments on droplet array in normal and microgravity environments. In: Proceedings IUTAM Symposium Mechanics and Combustion of Droplets and Sprays Proceedings, pp. 367–377, Tainan (1994)
166. Roy, G.D., Frolov, S.M., Borisov, A.A., Netzer, D.W.: Pulse detonation propulsion: Challenges, current status, and future perspective. Prog. Energy Combust. Sci. **30**, 545–672 (2004)
167. Borisov, A.A., Komissarov, P.V., Kulikov, A.V., et al.: Kinetics of heat release in heterogeneous mixtures. Chem. Phys. Rep. **17**(3), 49 (1998)

168. Frolov, S.M., Scripnik, A.A., Kavtaradze, R.Z.: Modeling of diesel spray ignition. In: Roy, G.D., Frolov, S.M., Starik, A.M.: Combustion and Atmospheric Pollution, pp. 220–227. Torus, Moscow (2003)
169. Scripnik, A.A., Frolov, S.M., Kavtaradze, R.Z., et al.: Modeling of autoignition in liquid fuel spray. Khim. Fiz. **23**(1), 54 (2004)
170. Matsko, A.M., Kopeika, K.M., Polishchuk, D.I., et al.: Effect of external heat transfer on the critical conditions for ignition of gaseous suspension of magnesium particles. Phys. Aerodispersed Syst. **20**, 53–56 (1980)
171. Atthasit, A., Doue, N., Biscos, Y., et al.: Influence of drop concentration on the dynamics and evaporation of a monodisperse stream of drops in evaporation regime. In: Roy, G.D., Frolov, S.M., Starik, A.M. (eds.) Combustion and Atmospheric Pollution, pp. 214–219. Torus, Moscow (2003)

Index

Shock Wave Science and Technology Reference Library, Volume 4
Heterogeneous Detonation

About the Authors

Chapter 1

Stephen B. Murray

Defence R&D Canada – Suffield, PO Box 4000
Medicine Hat, Alberta
T1A 8K6 Canada
Stephen.Murray@drdc-rddc.gc.ca

Dr. Murray served in the position of Head of the Threat Assessment Group at Defence R&D Canada – Suffield for 11 years. He initiated a R&D program in fuel-air explosives (FAE) at Suffield in 1979 and has been conducting research in shock waves and combustion since that time. The FAE program focused on explosive minefield breaching during the 1980s, weapon threat assessment during the 1990s, and currently on public safety and force protection. Dr. Murray has represented Canada on numerous NATO, TTCP and OECD panels and has served as the Deputy Thrust Leader for the Army's R&D program on Munitions and Firepower. He was awarded the 'Golden Hands' R.I. Soloukhin Medal by the Institute for the Dynamics of Explosions and Reactive Systems in 2005 for his career accomplishments in experimental combustion.

Paul A. Thibault

TimeScales Scientific Ltd.
554 Aberdeen St. SE
Medicine Hat, Alberta
T1A 0R7 Canada
paulth@timescales-sci.com

Dr. Thibault received his Ph.D. from McGill University in 1978. He founded Combustion Dynamics Ltd. in 1984 and TimeScales Scientific Ltd. in 2001. He has been active in the areas of combustion, explosions, shock waves, propulsion and Computational Fluid Dynamics, with particular emphasis on numerical modeling and the development of scientific simulation software.

Chapter 2

Fan Zhang

Defence R&D Canada – Suffield, PO Box 4000
Medicine Hat
Alberta, T1A 8K6
Canada
fan.zhang@drdc-rddc.gc.ca

Dr. Fan Zhang is a Senior Scientist in the Department of National Defence at Defence Research and Development Canada – Suffield and an adjunct Professor at the University of Waterloo in the Department of Mechanical Engineering. He specializes in shock waves, detonations and explosions, more specifically in multiphase reactive flow, heterogeneous explosives and high energy density systems. He earned his doctoral degree in science in 1989 from the University of Technology Aachen (RWTH), Germany, and received a Borchers Medal, a Friedrich-Wilhelm Prize and several best paper awards. He published more than 150 refereed journal and proceedings papers, book chapters and journal special issues. He has served for a number of international defense panels and academic committees.

Chapter 3

David L. Frost

McGill University
Department of
Mechanical Engineering
817 Sherbrooke St. W.
Montreal, Quebec, H3A
2K6 Canada
david.frost@mcgill.ca

Dr. David Frost is an Associate Professor of Mechanical Engineering in the Shock Wave Physics Group at McGill University in Montreal. He received his PhD from Caltech in 1985 in the area of explosive vaporization and then spent a year as a postdoctoral fellow at GALCIT. At McGill, his research interests have ranged from experimental work in molten metal/water interactions to high-speed combustion processes in multiphase media and shock/material interactions. His research on heterogeneous explosives was initiated during sabbatical visits to DRDC-Suffield and Los Alamos National Laboratory. He has published more than 70 refereed journal and proceedings papers and book chapters.

Fan Zhang

Defence R&D Canada –
Suffield, PO Box 4000
Medicine Hat
Alberta, T1A 8K6
Canada
fan.zhang@drdc-rddc.gc.ca

Dr. Fan Zhang is a Senior Scientist in the Department of National Defence at Defence Research and Development Canada – Suffield and an adjunct Professor at the University of Waterloo in the Department of Mechanical Engineering. He specializes in shock waves, detonations and explosions, more specifically in multiphase reactive flow, heterogeneous explosives and high energy density systems. He earned his doctoral degree in science in 1989 from the University of Technology Aachen (RWTH), Germany, and received a Borchers Medal, a Friedrich-Wilhelm Prize and several best paper awards. He published more than 150 refereed journal and proceedings papers, book chapters and journal special issues. He has served for a number of international defense panels and academic committees.

Chapter 4

Michael F. Gogulya

Semenov Institute of
Chemical Physics
Russian Academy of
Sciences
Moscow, 119991 Russia
gogul@polymer.chph.ras.ru

Dr. Michael F. Gogulya is the Co-Head of the Department of Burning and Explosion at Semenov Institute of Chemical Physics of Russian Academy of Sciences, and a member of the Scientific Counsel at the Institute. He has been an Associate Professor of Moscow Engineering Physical Institute since 1988. His research interests include all aspects of shock and detonation wave in condensed matter: shock-induced polymorphic transformations, shock-induced reactions, and application of optical pyrometry to study of shock and detonation waves. He is the author of a textbook and more than 150 publications on the aforementioned problems.

Chapter 4, cont.

Michael A. Brazhnikov

Semenov Institute
of Chemical Physics
Russian Academy
of Sciences
Moscow, 119991 Russia
bragario@inbox.ru

Michael Brazhnikov is a senior researcher at Semenov Institute of Chemical Physics of Russian Academy of Sciences. He graduated from Moscow Engineering Physical Institute. His research interests include all aspects of shock and detonation wave in condensed matter and shock-induced reactions. He is the author of more than 40 publications on the aforementioned problems.

Chapter 5

Yury A. Gordopolov

Institute of Structural
Macrokinetics and
Materials Science
Russian Academy
of Sciences
Chernogolovka, Moscow
142432 Russia
gordop@ism.ac.ru

Professor Gordopolov's main area of research is the chemical physics of shock waves. His research interests include shock-assisted chemical/structural transformations in reactive and chemically inert materials. He obtained his Cand. Sc. degree from the Moscow Institute of Physics and Technology in 1977 and D. Sc. degree from the Institute of Chemical Physics, Russian Academy of Sciences in 1991, both in Chemical Physics, including Combustion and Explosion Physics. He is Director of the Institute of Structural Macrokinetics and Materials Science, Russian Academy of Sciences.

Stepan S. Batsanov

Center for High Dynamic
Pressures
Mendeleevo, Moscow,
141570 Russia
batsanov@gol.ru

Professor Batsanov is an expert in the crystal chemistry and physics of shock waves. He obtained his D.Sc. degree in solid state physics (1960) and a position of Professor of Physical Chemistry at the Novosibirsk University (1965). He is a founder (1970) of the Center for High Dynamic Pressures at Mendeleevo, Moscow. His most important publications include the following monographs: Chemical Structure and Refractometry (Moscow, 1959, 1976) translated in USA (1961, 1966) and China (1962), Electronegativity and Chemical Bond (Novosibirsk, 1962, 1972) translated in China and Czechoslovakia, Overlap Integrals (Novosibirsk, 1966 and 1969), Experimental Foundations of Structural Chemistry (Moscow, 1986, 2000), and Effects of Explosions on Materials (New York, 1994).

Chapter 5, cont.

Vladimir S. Trofimov

Institute of Structural Macrokinetics and Materials Science
Russian Academy of Sciences
Chernogolovka, Moscow
142432 Russia
www.ism.ac.ru

Professor Trofimov's area of research is the physics of shock wave processes in reactive condensed media. He obtained his Cand. Sc. (1965) and D. Sc. (1983) degrees from the Institute of Chemical Physics, Russian Academy of Sciences. Most of his papers are devoted to the detonation of high explosives. He extended the Zel'dovich-Neumann-Dring theory of detonation to the case of turbulent motion of reactive media. For many years, he has been a lecturer on shock wave processes at the Moscow Institute of Physics and Technology.

Chapter 6

Sergey M. Frolov

Semenov Institute of Chemical Physics
Russian Academy of Sciences
Moscow, 119991 Russia
smfrol@chph.ras.ru

Dr. Sergey Frolov is a Director of Laboratory of Explosion Processes in Gaseous and Heterogeneous Media at Semenov Institute of Chemical Physics and Professor at Moscow Physical Engineering Institute (State University). He received his Ph.D. and D.Sc. degrees in Chemical Physics in 1987 and 1992, respectively. His major research and teaching interests include shock waves and detonations in gases and multiphase media; combustion of gases, dusts, and liquid fuel sprays; chemical thermodynamics and kinetics; and computational fluid dynamics. He has over 300 publications in journals, books, and conference proceedings, 20 edited books, and about 200 research seminars and presentations.

Alexander V. Fedorov

Institute of Theoretical and Applied Mechanics
Russian Academy of Sciences, Siberian Branch, Novosibirsk
630090 Russia
fedorov@itam.nsc.ru

Professor Alexander Fedorov is the Head of the Laboratory of Wave Processes in Fine-Dispersed Media in the Institute of Theoretical and Applied Mechanics of Siberian Branch of Russian Academy of Sciences. He has a Doctor degree in physical-mathematical sciences. He is an expert in mathematical modeling of reactive heterogeneous media problems and the author of more than 200 scientific papers and one monograph. His theoretical scientific contribution includes heat explosion of fine metal and organic particles in shock waves and the Chapman-Jouguet ideal and non-ideal detonation in particles (inert and reactive) and reactive gas mixtures.

MIX
Papier aus verantwortungsvollen Quellen
Paper from responsible sources
FSC® C105338

If you have any concerns about our products,
you can contact us on
ProductSafety@springernature.com

In case Publisher is established outside the EU,
the EU authorized representative is:
Springer Nature Customer Service Center GmbH
Europaplatz 3, 69115 Heidelberg, Germany

Printed by Libri Plureos GmbH
in Hamburg, Germany